**POTENTIAL INDUSTRIAL
CARCINOGENS AND MUTAGENS**

Studies in Environmental Science

Volume 1   **Atmospheric Pollution 1978**
            Proceedings of the 13th International Colloquium, held in Paris,
            April 25–28, 1978
            edited by M.M. Benarie

Volume 2   **Air Pollution Reference Measurement Methods and Systems**
            Proceedings of the International Workshop, held in Bilthoven,
            December 12–16, 1977
            edited by T. Schneider, H.W. de Koning and L.J. Brasser

Volume 3   **Biogeochemical Cycling of Mineral-Forming Elements**
            edited by P.A. Trudinger and D.J. Swaine

Volume 4   **Potential Industrial Carcinogens and Mutagens**
            by L. Fishbein

Studies in Environmental Science 4

POTENTIAL INDUSTRIAL CARCINOGENS AND MUTAGENS

Lawrence Fishbein

*Assistant to Director for Environmental Surveillance
National Center for Toxicological Research,
Jefferson, Arkansas 72079, U.S.A.
Adjunct Professor of Pharmacology
University of Arkansas Medical Center,
Little Rock, Arkansas, U.S.A.*

ELSEVIER SCIENTIFIC PUBLISHING COMPANY
Amsterdam — Oxford — New York   1979

ELSEVIER SCIENTIFIC PUBLISHING COMPANY
335 Jan van Galenstraat
P.O. Box 211, 1000 AE Amsterdam, The Netherlands

Distributors for the United States and Canada:

ELSEVIER/NORTH-HOLLAND INC.
52, Vanderbilt Avenue
New York, N.Y. 10017.

Library of Congress Cataloging in Publication Data

Fishbein, Lawrence.
  Potential industrial carcinogens and mutagens.

  (Studies in environmental science ; 4)
  Includes index.
  1. Carcinogens. 2. Chemical mutagenesis.
3. Industrial toxicology. I. Title. II. Series.
[DNLM: 1. Carcinogens, Environmental. 2. Industrial medicine. 3. Mutagens. QZ202.3 F532p]
RC268.6.F57   615.9'02   78-27560
ISBN 0-444-41777-X

ISBN 0-444-41777-X (Vol. 4)
ISBN 0-444-41696-X (Series)

© Elsevier Scientific Publishing Company, 1979
All rights reserved. No part of this publication may be reproduced, stored in a retrieval system or transmitted in any form or by any means, electronic, mechanical, photocopying, recording or otherwise, without the prior written permission of the publisher, Elsevier Scientific Publishing Company, P.O. Box 330, 1000 AH Amsterdam, The Netherlands

Printed in The Netherlands

TABLE OF CONTENTS

Preface IX

Chapters

1. Introduction 1
   References 21
2. Combination effects in chemical carcinogenesis 31
   References 38
3. Aspects of epidemiology, risk-assessment and "threshold dose" . . . 42
   References 73
4. Tabular summaries of potential industrial carcinogens and mutagens . 79
5. Alkylating agents — epoxides and lactones 93
   A. Epoxides 93
      1. Ethylene oxide 93
      2. Propylene oxide 96
      3. 1,2-Butylene oxide 97
      4. Epichlorohydrin 97
      5. Glycidol 99
      6. Glycidaldehyde 99
      7. Styrene oxide 99
      8. Diglycidyl resorcinol ether 102
      9. Miscellaneous epoxides 103
   B. Lactones 104
      1. β-Propiolactone 104
      2. β-Butyrolactone 107
      3. γ-Butyrolactone 107
   References 109
6. Aziridines, aliphatic sulfuric acid esters, sultones, diazoalkanes and arylalkyltriazenes 118
   A. Aziridines 118
      1. Aziridine 118
      2. 2-Methylaziridine 119
      3. 2-(1-Aziridinyl)ethanol 120
   B. Aliphatic sulfuric acid esters 120
      1. Dimethyl sulfate 120
      2. Diethyl sulfate 121
      3. Cyclic aliphatic sulfuric acid esters 122
   C. Sultones 123
      1. 1,3-Propane sultone 123
   D. Diazoalkanes 124
      1. Diazomethane 125
   E. Aryldialkyltriazenes 126
   References 129
7. Phosphoric acid esters 135
      1. Trimethylphosphate 135
      2. Triethylphosphate 135
      3. Tris(2,3-dibromopropyl)phosphate 135
   References 139
8. Aldehydes 142
      1. Formaldehyde 142
      2. Acrolein 147
      3. Acetaldehyde 149
   References 151

9. Acylating agents 155
   1. Dimethylcarbamoylchloride 155
   2. Diethylcarbamoyl chloride 155
   3. Benzoyl chloride 155
   4. Ketene 156
   References 157
10. Peroxides 158
    1. Di-tert.-butyl peroxide 158
    2. tert.-Butyl hydroperoxide 158
    3. Cumene hydroperoxide 158
    4. Succinic acid peroxide 159
    5. Peracetic acid 159
    6. Hydrogen peroxide 159
    7. Benzoyl peroxide 160
    References 162
11. Halogenated unsaturated hydrocarbons 165
    1. Vinyl chloride 165
    2. Vinylidene chloride 178
    3. Trichloroethylene 182
    4. Tetrachloroethylene 187
    5. Chloroprene 189
    6. *Trans*-1,4-dichlorobutene 193
    7. Hexachlorobutadiene 193
    8. Allyl chloride 194
    References 197
12. Halogenated saturated hydrocarbons 211
    1. Methyl chloride 211
    2. Methylene chloride 212
    3. Chloroform 213
    4. Carbon tetrachloride 217
    5. Methyl chloroform 220
    6. 1,1,2-Trichloroethane 222
    7. 1,1,2,2-Tetrachloroethane 223
    8. Hexachloroethane 224
    9. Miscellaneous chloro, bromo, iodo derivatives 224
    10. Fluorocarbons 227
    References 232
13. Alkane halides, halogenated alkanols and halogenated ethers .. 240
    A. Alkane halides 240
       1. Ethylene dichloride 240
       2. Ethylene dibromide 241
       3. Propylene dichloride 245
       4. Dibromochloropropane 245
    B. Halogenated alkanols 248
       1. 2-Chloroethanol 248
       2. 1-Chloro-2-propanol 249
       3. 2,3-Dibromo-1-propanol 250
    C. Haloethers 251
       1. Bis(chloromethyl)ether 251
       2. Miscellaneous haloethers 256
    References 258
14. Halogenated aryl derivatives 266
    A. Chlorinated benzenes 266
       1. Chlorobenzene 266
       2. *ortho*-Dichlorobenzene 266
       3. *para*-Dichlorobenzene 268

	4. 1,2,4-Trichlorobenzene	269
	5. Hexachlorobenzene	270
	6. Brominated benzenes	277
	7. Benzyl halides	278
	References	280
15.	Halogenated polyaromatic derivatives	286
	1. Polychlorinated biphenyls	286
	2. Polybrominated biphenyls	294
	3. Polychlorinated terphenyls	297
	4. Chlorinated naphthalenes	298
	References	300
16.	Hydrazines, hydroxylamines, carbamates, acetamides, thioacetamides and thioureas	307
	A. Hydrazine and derivatives	307
	1. Hydrazine	308
	2. 1,1-Dimethylhydrazine	308
	3. 1,2-Dimethylhydrazine	308
	B. Hydroxylamines	310
	C. Carbamates	315
	1. Urethan	315
	D. Acetamides and thioacetamides	316
	1. Acetamide	316
	2. Thioacetamide	317
	E. Thioureas	317
	1. Thiourea	317
	2. Ethylenethiourea	318
	References	320
17.	Nitrosamines	331
	References	352
18.	Aromatic amines and azo dyes	356
	A. Aromatic amines	356
	1. Benzidine	356
	2. 3,3'-Dichlorobenzidine	361
	3. 3,3'-Dimethylbenzidine	366
	4. 3,3'-Dimethoxybenzidine	367
	5. 1-Naphthylamine	368
	6. 2-Naphthylamine	370
	7. N-Phenyl-2-naphthylamine	373
	8. 4,4'-Methylenedianiline	374
	9. 4,4'-Methylene bis(2-chloroaniline)	375
	10. 4,4'-Methylene bis(2-methylaniline)	377
	11. 4-Aminobiphenyl	377
	12. 2,4-Diaminotoluene	378
	13. 2,4-Diaminoanisole	380
	14. *para*-Phenylenediamine	382
	15. *meta*-Phenylenediamine	383
	16. 2-Nitro-*para*-phenylenediamine	384
	17. 4-Nitro-*ortho*-phenylenediamine	385
	18. *ortho*-Toluidine	386
	19. Miscellaneous aromatic amines	388
	B. Azo dyes	388
	1. Azobenzene	389
	2. *para*-Aminoazobenzene	389
	3. *ortho*-Aminoazotoluene	390
	4. *para*-Dimethylaminoazobenzene	391

	5. Chrysoidine	394
	6. Sudan II	394
	6a. Ponceau MX	395
	7. Trypan blue	396
	8. Amaranth	396
	9. Miscellaneous azo dyes	397
	References	399
19.	Heterocyclic amines and nitrofurans	417
	A. Amines	417
	1. Quinoline	417
	2. 8-Hydroxyquinoline	417
	B. Nitrofurans	419
	References	427
20.	Nitroaromatics and nitroalkanes	432
	A. Nitroaromatics	432
	1. 2,4-Dinitrotoluene	432
	2. 2,4,5-Trinitrotoluene	432
	B. Nitroalkanes	433
	1. 2-Nitropropane	433
	References	435
21.	Unsaturated nitriles and azides	436
	A. Unsaturated nitriles	436
	B. Azides	440
	References	442
22.	Aromatic hydrocarbons	445
	1. Benzene	445
	2. Toluene	454
	References	456
23.	Anthraquinones, quinones, and polyhydric phenols	460
	A. 9,10-Anthraquinones	460
	B. Quinones	467
	1. *para*-Quinone	467
	C. Polyhydric phenols	468
	1. Catechol	468
	2. Hydroquinone	469
	References	470
24.	Phthalic and adipic acid esters	473
	A. Phthalic acid esters	473
	B. Adipic acid esters	477
	References	479
25.	Cyclic ethers	481
	1. 1,4-Dioxane	481
	References	485
26.	Phosphoramides and phosphonium salts	487
	A. Hexamethyl phosphoramide	487
	B. Phosphonium salts	488
	1. Tetrakis(hydroxymethyl)phosphonium salts	488
	References	491
27.	Miscellaneous potential carcinogens and mutagens	493
	1. Vinylic derivatives	493
	2. Allylic derivatives	496
	3. Diisocyanates	497
	4. Chlorophenols	500
	References	506
Summary		512
Subject Index		518

Preface

The spectrum of chemicals and the general benefits that have accrued from their use are both generally acknowledged to be very broad indeed. However, we are becoming increasingly aware that living in an environment into which more and more chemicals have been and are continuously introduced can, in some measure, be attendant with risks because of the possible hazards to health and safety of a number of these same chemicals.

We are daily exposed to literally hundreds of chemical agents, in perhaps staggering numbers of combinations via the air we breathe, the water and food we consume, by skin contact, medication, and in industrial environments. The types of compounds which may be encountered cover the expanse of inorganic and organic, simple and complex structures of natural or synthetic origin.

Health problems which can thus arise from exposure to toxicants may include neurologial and cardiovascular diseases, tissue and organ failures, birth defects, or mutations and cancers.

We are particularly concerned with carcinogens and mutagens because of the widely held belief that 60-90% of all human cancers have their etiology in environmental factors. Many believe that the cancer patterns of today are dependent on agents which entered the environment over 20 years ago.

It would appear prudent that steps be taken as rapidly as possible to identify deleterious agents in existence as well as those that may be introduced in order to minimize the burden and risk of potentially carcinogenic and mutagenic agents in the environment, lest we bequeath to subsequent generations the risk of catastrophic exposures for which redress would be either difficult or impossible.

We should especially require greater focus on potent carcinogens compared to weak carcinogens with requisite greater concern given to a carcinogen to which comparatively large segments of the population are potentially exposed to rather than one to which a relatively small number may come into contact.

The central theme and overriding purpose of this book was to bring together relevant facts about primarily carcinogenic and/or mutagenic organic chemicals and related compounds of industrial utility, arrange them by structural categories, wherever feasible listing the reported mutagenic test system and carcinogenic responses for more facile identification and recognition of the current risk and possible prioritization for additional short-term and chronic bioassay where deemed necessary. It is also the obvious wish that this structural arrangement could be of potential predictive utility for advance recognition of toxic compounds (of all use categories) considered for introduction into the environment. Additionally germane aspects of the synthesis of these agents, primarily in terms of the nature of the possible trace impurities, chemical and biological reactivity, production volumes and use patterns, environmental occurrence, suggested or estimated populations at risk and national permissible worker exposure levels (TLV's and MAC's) have been included for cohesiveness of treatment and hopefully for enhanced utility for a broader spectrum of scientists and public health officials. Directly bearing on the scope and interpretation of the potential hazards of industrial carcinogens and mutagens are the areas of combination effects in chemical carcinogenesis, epidemiology and risk-assessment. These considerations were thus included in the book as well.

Although industrial chemical carcinogens and mutagens were considered as discrete entities, it is well recognized that man can be exposed to a broad galaxy of these agents and thus considerations relative to possible synergistic, potentiating, co-carcinogen, co-mutagenic and/or antagonistic interactions of these agents as well as with non-carcinogenic and non-mutagenic agents are of vital importance. It is imperative that we endeavor to improve as rapidly as possible the understanding of the underlying molecular mechanisms by which these effects are mediated, as well as to rationally delineate those chemical agents of principal carcinogenic and mutagenic risk and the scope of the potential risk.

A special acknowledgement of gratitude is extended to Mrs. Dabra Huneycutt for her patience and skill in the typing, assembly and proofing of the manuscript. However, the author should bear the burden and responsibility for all mistakes and errors of omission and commission.

Lawrence Fishbein

CHAPTER 1

INTRODUCTION

There is a continuing need to assess the status of existing potentially hazardous chemicals and strategies as well as those that may be introduced in the future that may impact on man and the environment, primarily from a predictive view of avoiding or ameliorating catastrophic episodes similar to those which have occurred in the past involving agents such as methyl mercury, cadmium, PCBs, PBBs, vinyl chloride, chlorinated dibenzodioxins (e.g., TCDD) and a number of chlorinated pesticides (e.g., hexachlorobenzene (HCB), dibromochloropropane (DBCP), Mirex and Kepone).

It is generally acknowledged that the past few decades have witnessed an unparalleled expansion of chemical industry with the concomitant development of many new organic chemical products as well as enhanced produce application. Hence, the number and amounts of chemicals and end products that man is potentially exposed to is staggering.

There are approximately 3,500,000 known chemicals with about 25,000 - 30,000 chemicals in significant production in the United States alone. It is estimated that about 700[1,2] to 3,000[3] new industrial chemicals are introduced per year. In the past 10 years, the production of synthetic organic chemicals has expanded by 255 percent[4]. In 1975, the world production of plastics, rubber products and fiber products was 40, 10 and 30 million tons, respectively[5].

Some 2,500 individual chemicals or mixtures are reported to be in current use by the plastics industry alone as antioxidants, antistatics, blowing agents, catalysts, colorants, flame retardants, lubricants, plasticizers, stabilizers and U.V. absorbers[6]. The number of possible combinations of these agents can be staggering[5].

It is estimated that of all the chemicals on the market, a relatively small proportion[7], approximately 6,000 have been tested to determine their cancer-causing potential[8,9], of these approximately 1,000 compounds have thus far been found to be tumorigenic in test animals[9].

We do not know with precision what percentage of existing chemicals as well as those which enter the environment annually may be hazardous, primarily in terms of their potential carcinogenicity, mutagenicity, and teratogenicity. For example, although the etiology of human neoplasia, with rare exceptions, is unknown, it has been stated repeatedly that a large number of cancers can be directly or indirectly attributed to environmental factors (using the term in its widest connotation) in proportions that can vary from 75%[10], to 80%[11,12] to 90%[13]. Although these percentages can be disputed, the fact remains that a certain proportion of human cancers can be attributed to environmental factors[14-16] (e.g., exposure to toxic chemicals, including: benzidine, 2-naphthylamine, 4-aminobiphenyl, 4-nitrobiphenyl, bis(chloromethyl)-ether, vinyl chloride, auramine, chromium and inorganic arsenic).

If estimates are correct that 60 percent or more of all human cancers are due to environmental agents, then about 500,000 cases per year may be involved[17]. Exposure to chemical agents is known to cause a range of occupational cancers, mainly of the skin, bladder, lungs and nasal sinuses[14-16].

Geographic analysis of U.S. cancer mortality, 1950-1969, has revealed excess rates for bladder, lung, liver, and certain other cancers among males in 139 counties where the chemical industry is most highly concentrated[17-20]. The correlation was limited to counties with specific categories of the chemical industry. Of particular note are the elevated rates of (a) bladder cancer in counties manufacturing cosmetics, industrial gases, soaps, and detergents; (b) lung cancer in counties producing pharmaceutical preparations, soaps and detergents, paints, inorganic pigments, and synthetic rubber; and (c) liver cancer in counties manufacturing cosmetics, soaps and detergents and printing inks[19].

Cancer epidemiology has demonstrated striking differences in cancer incidence for certain organs in different geographic regions (nationally and internationally). Migrant studies have demonstrated that migrating populations acquire the predominant cancer pattern of the host countries and lose that of their home country. This can best be explained by a dominant influence of the respective environment.

Modes of exposure to chemicals include diet, personal habits such as smoking, and external environmental air and water pollution. At present, the most influential single carcinogenic exposure appears still to be cigarette smoking.

The risk of cancer in the human population from the large and ever increasing multitude of chemicals entering the environment poses a serious problem. It is recognized that any direct assessment of human risk requires epidemiologic data, which unfortunately are only available for only a relatively small number of industrial chemials or drugs[21]. It should also be noted that because of the typical latency period of 15 to 40 years for cancer, it can be assumed that much of the cancer from recent industrial development is not yet observable[22,23]. However, it has also been suggested that as this "lag period" for chemical carcinogenesis is almost over, a steep increase in the human cancer rate from suspect chemicals may soon occur[24].

A NIOSH survey conducted in 1972-1974 showed that more than 7 million workers in the U.S. are exposed to 20,000 trade name products containing toxic substances regulated by the Occupational Safety and Health Administration (OSHA). Another 300,000 workers may be exposed to similar products containing one of the 16 cancer-causing agents regulated by OSHA[25,26].

Concomitant with the potential cancer risk of environmental agents, is the growing concern over the possibility that future generations may suffer from genetic damage by mutation-inducing chemical substances to which large segments of the population may unwittingly be exposed[27-31].

Few can dispute the desirability and sense of urgency in controlling the number of carcinogenic and/or mutagenic agents that are already in the environment or that which may be introduced in the future.

The major objectives of this book are to: (1) consider and collate a number of industrially significant compounds encompassing a spectrum of structural categories that have been reported to be carcinogenic and/or mutagenic in order to better assess the nature of the <u>present</u> potential risk; and (2) to determine whether there are structural and biological similarities amongst these agents which would better permit a measure of predictability and prioritization in both the screening of new or untested compounds and the determination of which of the existing potential chemical carcinogens to investigate in long-term bioassays and (3) to highlight the possible modifying factors of carcinogenesis and mutagenesis, assess the various aspects of risk assessment in order to bring into better perspective the utility of potential chemical carcinogens and/or mutagens versus the possible societal risk in their continued employment.

The cost of examining for carcinogenicity (by existing long-term testing procedures) large numbers of suspect chemicals already in the environment as well as those that will be introduced, is expensive and time consuming. For example, animal tests of a substance can take as much as 2 to 3 years and cost upward of 100,000[32,33] to 200,000 dollars[34]. Even if all the screening facilities in the country were mobilized, by one estimate, it would still be possible to screen only about 700 chemicals per year[35]. Bartsch[35] estimated that the world capacity for testing carcinogenicity (in long-term tests) is only about 500 compounds per year[35].

Currently, there are about 450 chemicals now being screened for carcinogenicity at 28 different U.S. laboratories under the sponsorship of the National Cancer Institutes Carcinogenesis Program[33].

Chemical carcinogens and mutagens represent a spectrum of agents varying in quantitative requirements by a factor of at least 10^7, (Table 1)[34] with strikingly different biological activities, ranging from highly reactive molecules that can alkylate macromolecules and cause mutations in many organisms to compounds that are hormonally active and have neither of these actions[34-38].

Approximately 100 chemicals have been shown to be definitely carcinogenic in experimental animals[14,15]. In many carcinogenesis studies, the type of cancer observed is the same as that found in human studies (e.g., bladder cancer is produced in man, monkey, dog and hamster by 2-naphthylamine) while in other instances, species variations can exist resulting in the induction of different types of neoplasms at different locations by the same carcinogen (e.g., benzidine causes liver cell carcinoma in the rat and bladder carcinoma in dog and man)[14,15].

A program aimed at identifying and eliminating exposure to potential carcinogens and/or mutagens undoubtedly requires the development of rapid, inexpensive screening methods to augment long-term animal tests (for potential carcinogens) in order to focus on the hazardous chemicals among the many thousands to which humans are exposed.

Mutagenicity screening is now apparently both feasible and necessary for chemicals now in, and those which will enter the environment. The mutagenic activity of certain reactive chemicals can be detected in prokaryotic and eukaryotic cells. Short-term microbial tests (in <u>Salmonella typhimurium</u>, <u>Escherichia coli</u> and <u>Bacillus subtilus</u> in combination with <u>in vitro</u> metabolic activation) for mutation-induction include assays for both forward and reverse mutation at specific loci, as well as tests for inhibition of DNA repair[27-29,35,37,39,44].

TABLE 1
RANGE OF POTENCY OF CARCINOGENS IN ANIMALS[34]

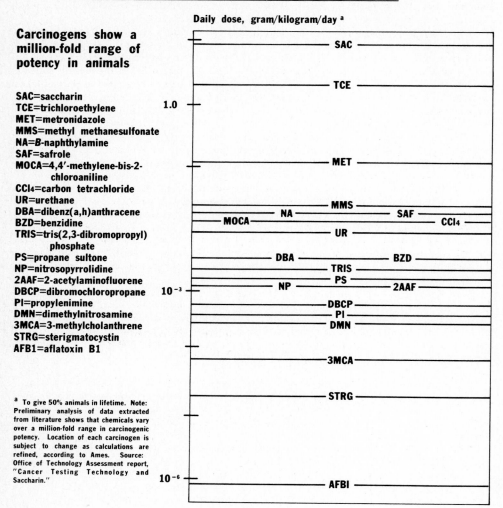

Carcinogens show a million-fold range of potency in animals

SAC=saccharin
TCE=trichloroethylene
MET=metronidazole
MMS=methyl methanesulfonate
NA=*B*-naphthylamine
SAF=safrole
MOCA=4,4'-methylene-bis-2-chloroaniline
CCl₄=carbon tetrachloride
UR=urethane
DBA=dibenz(a,h)anthracene
BZD=benzidine
TRIS=tris(2,3-dibromopropyl) phosphate
PS=propane sultone
NP=nitrosopyrrolidine
2AAF=2-acetylaminofluorene
DBCP=dibromochloropropane
PI=propylenimine
DMN=dimethylnitrosamine
3MCA=3-methylcholanthrene
STRG=sterigmatocystin
AFB1=aflatoxin B1

[a] To give 50% animals in lifetime. Note: Preliminary analysis of data extracted from literature shows that chemicals vary over a million-fold range in carcinogenic potency. Location of each carcinogen is subject to change as calculations are refined, according to Ames. Source: Office of Technology Assessment report, "Cancer Testing Technology and Saccharin."

The mutagenic activity of some chemicals have also been detected in Saccharomyces[27,28,45,46], Neurospora[27,28,47] and Drosophila[27,28,48].

Chemically-induced stable phenotypic changes have been induced in mammalian cell culture systems that include Chinese hamster cells[28,49-53], L5178Y mouse lymphoma cells[28,44,54-57], human skin fibroblasts[28,58,59] and a human lymphoblastoid cell line[60].

Unscheduled DNA synthesis (a measure of excision repair) in human fibroblasts has been used as a prescreen for chemical carcinogens and mutagens, both with and without metabolic alteration[61-65].

Although at present there are many test systems available that involve different genetic indicators and metabolic activation systems for detecting mutagenic activity, all appear to possess individual advantages and limitations[27,28,35,43].

Hence, the belief is generally held that a battery of test systems is needed to detect the genetic hazards caused by chemicals[27-29,34,35,42-44,65-68]. The utilization of a battery of tests should provide confirmation of positive test data as well as reduce the possibility of false negative tests.

Tier systems (hierarchical) approaches to mutagenicity testing and potential regulatory control of environmental chemicals that have been proposed by Dean[70], Bartsch[35], Bridges[67,68] and Bora[69] are illustrated in Tables 2 and 3 and Figures 1 and 2 respectively. A modified scheme for three- tier testing of Claxton and Barry[71] is shown in Figure 3 and Table 4 summarizes further information regarding costs and time requirements for each tier. Flamm[66] stressed that "the genetic effects of concern to man would include the entire myriad of mutational events known to occur in man such as base-pair substitutions, base additions or deletions, which comprise the category referred to as point mutations, as well as other category of mutations that are chromosomal in nature and are represented by chromosome deletions, re-arrangements or non-disfunctions".

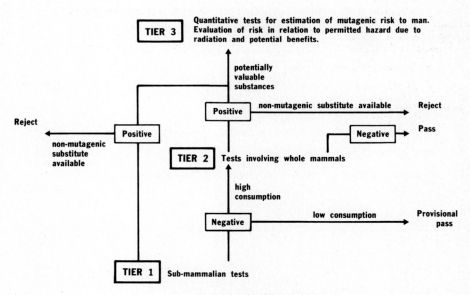

Figure 1. Three-tier framework for mutagenicity screening. [67] [68]

TABLE 2
A PREDICTIVE TESTING SCHEME FOR CARCINOGENICITY OR MUTAGENICITY
OF INDUSTRIAL CHEMICALS [70]

Phase 1: Initial Screen

(a) Screening test with sensitive micro-organisms
    (i) Salmonella tyhpimurium TA 1538 (frame shift)
    (ii) Escherichia coli WP2 (base-pair substitution)
    (iii) Saccharomyces cerevisiae (mitotic gene conversion)

(b) Microsomal assay using rat liver homogenate with the above four microorganisms

(c) Cytotoxicity study with HeLa cells and cultured rat liver (RL_1) cells

(d) Chromosome study in cultured rat liver cells

(e) Short-term exposure of rats by a relevant route to the highest tolerated dose followed by histological examination and analysis of chromosome damage

Phase 2:

(a) Microsomal assay using liver homogenates from mice and other species

(b) Dominant lethal assay in male mice

(c) Assay of gene mutation in cultured mammalian cells

(d) Assay of malignant transformation in cultured cells or by a host-mediated approach

Phase 3:

(a) An <u>in vivo</u> assay of gene mutation

(b) Dominant lethal assay in male rats

(c) Dominant lethal assay in female rats

(d) In vivo chromosome study in Chinese hamsters or mice or both

(e) Long-term carcinogenicity studies in one or two species

(f) Pharmacokinetic studies and biochemical studies at the sub-cellular level

Aspects of the evaluation of environmental mutagens have been described in regard to the estimation of human risk[27,28,35,66-68]. Of all the test systems currently employed, the Ames test using a rat-liver microsome activation has been evaluated in the greatest detail[24,37,39,40,43,44,65,79-83]. The Ames assay determines the mutagenic activity of chemicals by measuring gene mutations manifested as revertant colonies growing or a selective azar medium. In a test of a non-mutagenic chemical, a predictable number of revertant colonies are formed spontaneously in a culture plate which contrasts dramatically with the high population density obtained in a plate

TABLE 3
FRAMEWORK OF CARCINOGENICITY TEST PROCEDURES [35]

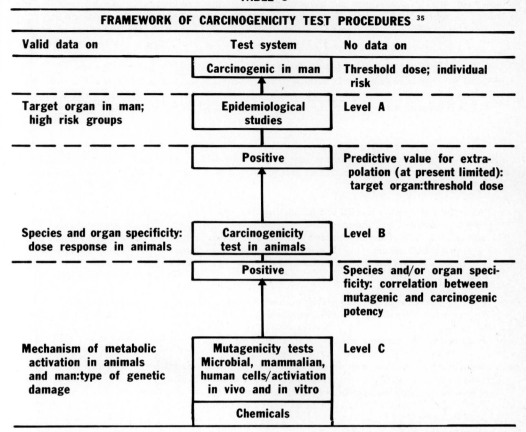

from a test with a mutagenic chemical in which several thousand revertants have back-mutated in response to the compound. While chemicals can be rapidly detected if they induce mutations at levels at least five times greater than spontaneous levels, chemicals that have weak mutagenic activity (e.g., inducing only twofold or threefold increases in mutation rates) may not be detected because the assay results fall within reliable confidence limits of the control tests. However, in some cases modifications of the tests can be apparently constructed in a manner that weak mutagens can be confidently detected[83].

Results accumulated up to the present time using a rat-liver microsome test in vitro with Salmonella typhimurium strains developed by Ames[34,37,39,40,43,44,79-82] have shown that about 80-90% of the carcinogens tested were also mutagens, while the number of false positives and false negatives was much lower, ranging from 10 to 15%[37,39-41,65,79-82]. For example, the assay of 300 chemicals utilizing Salmonella/microsome test in Ames' laboratory[37,39,40,43,79-82] included almost all of the known

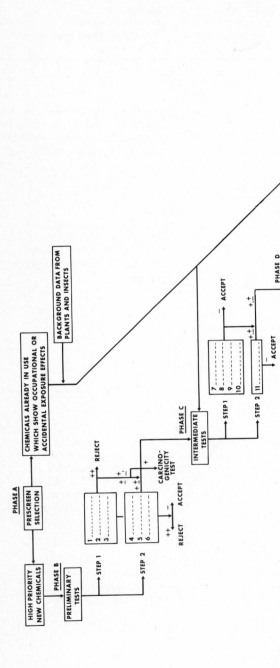

Scheme 2. Phase B. 1. In vitro bacterial/microsomal/urine assay for detection of gene mutations. 2. Host-mediated assay (HMA) detection of gene mutations in bacteria. 3 In vitro (direct and microsomal) and in vivo (HMA) detection of gene mutation, gene conversion and mitotic recombination in eukaryotes. 4. In vitro detection of DNA damage and repair in human and mammalian cells. 5. In vitro detection of chromosome aberrations in human and mammalian cell lines. 6. In vitro detection of gene mutations in cultured human and mammalian cell lines. Phase C. 7. In vivo (bone marrow and peripheral blood) detection of chromosome aberrations and/or micronuclei in experimental mammals. 8. In vivo (HMA) detection of chromosome aberrations in human cells. 9. Detection of dominant lethals in experimental mammals. 10 In vivo detection of heritable translocations in mammals germ cells. 11. In vivo detection of chromosome aberrations in exposed human populations. Phase D. 12. In vivo detection of gene mutations in mammals (specific locus test). 13. In vivo detection of genetic defects in exposed human populations. Phase E. (a) Establishment of dose response relationships. (b) Risk benefit considerations and evaluation of genetic risk to human population. (c) Estimation of acceptable safe dose levels. Phase F. (d) Recommendations. (e) Regulatory actions. (++) Positive-substitute available-reject; (+) positive-new chemical-has considerable industrial potentials-chemical in use-considerable social and economic benefit; (+) results inconclusive or negative but other considerations e.g. structural similarity makes it a suspect chemical-further test; (−) negative in all tests no reasonable doubt-accept.

TABLE 4
Description of the Tier System Concept[71]

	Types of Screening Systems Used	Average Costs	Time Needed	Use
Tier Three	Complex mammalian systems	> $50,000 per compound	> 6 mo.	Testing of needed or beneficial chemicals under use-like conditions for good benefit-risk estimate
Tier Two	Simpler mammalian systems, some plant, animal, and tissue culture systems including cytogenetic screening	$5000 to $50,000 per compound	> 2 mo.	Definitive testing of compounds
Tier One	Simple microbial, plant, and animal systems	< $5000 per compound	< 2 mo.	Pre-screen for other tier levels; definitive testing of some compounds

human carcinogens (e.g., 4-aminibiphenyl, β-naphthylamine, benzidine, bischloromethylether and vinyl chloride) and hence demonstrated a definite correlation between carcinogenicity and mutagenicity in the testing. A tabluation of these data[39,65,80] as a function of chemical class (Table 5) showed a high level of ascertainment in classes such as aromatic amines (A), alkyl halides (B), polycyclic aromatic (C), nitroaromatics and heterocyclics (E), nitrosamines (G), fungal toxins and antibiotics (H), mixtures (I), and azo dyes and diazo dyes (L) and lower positive response for esters, epoxides and carbamates (D, 76%), miscellaneous heterocyclics (J, 25%) and miscellaneous nitrogen compounds (K, 78%). It should be stressed that for all the classes tested, the number of compound within each class is small, and does not permit, at present, the distinction that the level of ascertainment varies markedly as a function of chemical class[65]. Similarly, the data on 63 non-carcinogenic chemicals tested which show that 22.5% (14/62) are mutagens do not indicate compelling data that the positive test data occur in any particular chemical class.

As a consequence of the rat-liver microsome test in vitro with S. typhimurium strains, Ames has supported the theory that most carcinogens cause cancer through somatic mutation[37,39,40]. Genetic susceptibility to cancer is also suggested by the fact that rates of leukemia are associated with many genetic disorders including Down's syndrome, Tusomy D, Kleinfelter's syndrome, Falconi's syndrome and Bloom's syndrome[71,84]. Several tumors and tumor syndromes such as retinoblastoma, pheochromcytoma, and polyposis of the colon are inherited as though caused by dominant gene mutations[84]. Hence increases in mutation rates could increase the incidence of these conditions[71].

The recent evaluation of 6 short-term tests for detecting organic chemical carcinogens by Purchase et al[82] using 120 organic chemicals, 58 of which are known human or animal carcinogens, disclosed a 93% of accurate predictions employing the Ames test[79]

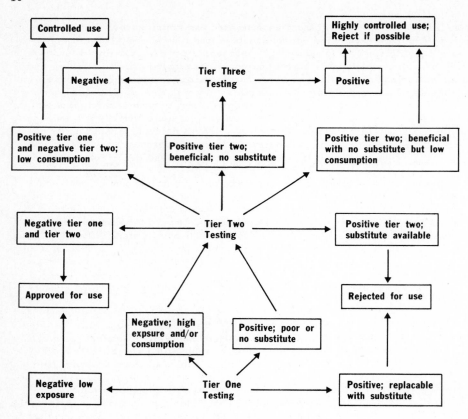

FIGURE 3 - Modified Scheme for Tier Testing[71]

TABLE 5

MUTAGENICITY OF CHEMICAL CARCINOGENS[a] [65]

Class	Type	Carcinogens			Non-carcinogens		
		Total N. tested	Positive number	Response percentage	Total No. tested	Positive number	Response percentage
A.	Aromatic amines	25	23	92	11	2	18
B.	Alkyl halides	20	18	90	3	2	67
C.	Polycyclic aromatics	26	26	100	8	2	12.5
D.	Esters, epoxides and carbamates	17	13	76	8	2	25
E.	Nitro aromatics and heterocycles	28	28	100	4	3	75
F.	Miscellaneous aliphatics and aromatics	5	1	20	12	0	0
G.	Nitrosamines	21	20	95	0	0	0
H.	Fungal toxins and antibiotics	8	8	100	2	0	0
I.	Mixtures	1	1	100	—	—	—
J.	Miscellaneous heterocycles	4	1	25	7	0	0
K.	Miscellaneous nitrogen compounds	9	7	78	4	2	50
L.	Azo dyes and diazo dyes	11	11	100	3	1	33
		175	157	89.7	62	14	22.5

[a] Adapted from McCann et. al.[39] McCann and Ames[80]

(S. typhimurium strains TA 1535, 1538, 98 and 100 with rat liver microsomal preparation S-9 fraction: cofactor 1:3). In the study of Purchase et al[76], a cell transformation assay with neonatal Syrian hamster kidney fibroblasts (BHK 21/C 13) and either human diploid lung fibroblasts (W1-38) or human liver cells (Chang) were treated with the above test compounds in liquid tissue culture medium (without serum) and the S-9 mix of the Ames test[79] to aid in the metabolism, yielded a 94% of accurate predictions. When the responses of the Ames test and cell transformation assay were compared, it was found that they agree with each other in correctly predicting the activity of 106 of the 120 compounds (88%), while in contrast, they both disagree in only 2 cases, viz., those of diethylstilbestrol and vinyl chloride[82].

The prospect is widely held that short term tests will offer a method of rapidly searching a group of compounds for potential carcinogens in order that priorities may be set for conventional long term studies[15,27-29,34,35,43,66-82]. It has been stressed that in tests used for preliminary screening, a small proportion of false positive and false negative results may be acceptable, but for a final test, no false negative results can be accepted[35]. However, it is also conceded that despite the extent to which short-term test systems might be improved, there will always remain a finite level of false positives and negatives. Hence, a number of limitations of mutagenicity test systems are acknowledged, e.g., some of the factors that determine the processes of cancer development in vivo cannot be duplicated by mutagenicity systems in vitro. Other determining factors are: biological absorption and distribution; the concentration of ultimate reactive metabolites available for reaction in organs and animal species with cellular macromolecules; the biological half-life of metabolites; DNA repair mechanisms between the test system and the whole animal (excision, strand break, post-replication and photoreactivation); immuno-surveillance; and organ-specific release of proximate or ultimate carcinogens by enzymic deconjugation[35,42].

Criticisms of submammallian testing have been reported and range from the current limitations of microbial assays to assess the carcinogenic potential of metals, organometallics, hormones and particulates[85] and the lack of correlation (in potency or activity) between microbial mutagenicity and rodent carcinogenicity in a group of direct-acting and metabolically activated agents (polycyclic hydrocarbons)[86]. Additionally, although it is conceded that many chemical carcinogens may exhibit mutagenic activity in certain assay procedures, there are exceptions such as the nucleic acid base analogs and the acridines which are excellent mutagens but are not known to be carcinogenic in vivo. The view is also held by some that the mutational origin of cancer remains an unproven hypothesis with evidence in support of other mechanisms[86-90].

It is useful to briefly summarize several key considerations of reactivity of chemical carcinogens that are particularly germane to the predictive value of mutagenicity tests in chemical carcinogenesis. Many chemical carcinogens are reactive electrophiles per se, e.g., alkylating agents, acylating agents, and other electrophiles[35,43,91-98].

Despite the diversity in chemical structures of known carcinogens and mutagens, such as alkylating agents, N-nitrosamines and N-nitrosamides, nitro aryl- and furan derivatives, aromatic and heterocyclic amines and azo dyes, carbamates, polycyclic aromatic hydrocarbons, chlorinated hydrocarbons and naturally occurring compounds (e.g., pyrolizidine alkaloids, aflatoxin), recognition of a common element in chemical carcinogens and mutagens has rapidly progressed since it was understood that the majority of carcinogens (procarcinogens) and many mutagens need metabolic activation in the host for transformation to their so-called ultimate reactive forms[35,91-99]. Some procarcinogens are often chemically or spontaneously converted to ultimate carcinogens by hydrolytic reactions and often exhibit a broad spectrum of activity in many species and target organs[91,100-103]. Other procarcinogens which require host-controlled biochemical activation (dependent on specific enzyme systems)[102-106] may exhibit more specific and/or restricted carcinogenic activity[95,107]. It should also be noted that the procarcinogen (and its derivatives) are subject to deactivation reactions which can lead to compounds possessing either no carcinogenic activity or less carcinogenic potential than the parent compound[93].

The common denominator of these ultimate reactive metabolites of carcinogens is their electrophilicity (electron-deficient reactants). They are compounds which react with electron-rich sites in cellular nucleic acids and proteins causing mutagenic effects frequently paralled by the onset of DNA repair processes[35,91-93].

As summarized by Brusick[43], "if chemical carcinogens or their electrophilic metabolites induce genetic changes which directly or in association with other cellular dysfunctions result in the malignant transformation of normal cells to potential tumor cells, than by the detection of mutagenic activity, potential carcinogens can be identified".

Figures 4 to 9 illustrate generalized schemes of the metabolic activation of chemical carcinogens, possible mechanisms of action of these agents, and steps in the carcinogenic process respectively[93,95]. Typical activation reactions (procarcinogens → proximate carcinogen → ultimate carcinogen) for a variety of agents are shown in Figure 10 (A-D)[97]. An illustration of some of the factors influencing the formation of reactive metabolites and their interaction with biological functions in liver cells has been provided by Arrhenius[98] for the case of aromatic amines (Figure 11). Figure 12 illustrates the site of interaction of a number of chemical carcinogens with DNA <u>in vivo</u> and <u>in vitro</u>[96], although all four bases of DNA and in some instances the phosphodiester backbone are targets for one or more carcinogens under some circumstances, by far the most reactive groups are the purine nitrogens. The N-7 of guanine appears to be the most reactive site, followed by the N-3 and N-7 positions of adenine[96].

It is recognized that although many bioassays and safety assessments have considered single agents (principally purified materials), the role of trace contaminants, continuous exposure to low levels of multiple agents, co-carcinogens and other factors are of importance in the evaluation of the sequence of chemical carcinogenesis and the etiology of human cancer[93,95,107,108].

Despite the converging tendency of chemicals to be both carcinogenic and mutagenic, it cannot be known at present whether all carcinogens will be found to be mutagens and all mutagens, carcinogens, e.g., for classes of compounds such as base analogs which do not act via electrophilic intermediates and steroidal sex hormones which are carcinogenic in animals and have not yet been shown to be mutagens, differeent cancer-inducing mechanisms may be implied.

The industrial chemicals considered in this report were limited to organic compounds and selected on factors including: their reported carcinogenicity and/or mutagenicity, their chemical structures and relationships to known chemical carcinogens and mutagens, their volume or use characteristics and suggested or estimated potential populations at risk.

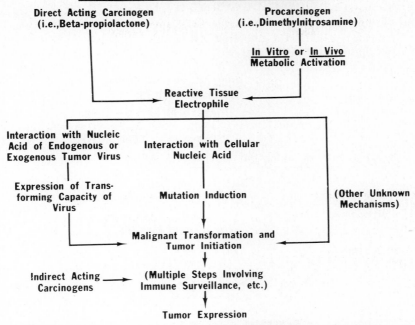

MUTAGENICITY ASSAYS ARE DESIGNED TO IDENTIFY THE REACTIVE ELECTROPHILE, WHICH IS ALSO HYPOTHESIZED TO BE INVOLVED IN TUMOR INITIATION.

It should also be noted that while the numbers of individuals directly involved in the preparation of these chemicals and their byproducts (e.g., plastics, polymers, etc.) are relatively small in number compared to many industrial segments and processes, the degree of exposure to potentially hazardous (carcinogenic and mutagenic) substances can be very substantial indeed. Substantially greater numbers of individuals may be indirectly exposed to these potential carcinogens and mutagens via (1) use applications which may contain entrained materials, (2) inhalation, ingestion or absorption of these agent via air, water and food sources resulting from escape into the atmosphere, leaching into water and food, etc.

In terms of worker exposure, the predominant rats of exposure are via inhalation and dermal absorption and secondarily from ingestion of food and water.

It should also be stressed that there are instances where the volume produced of a potentially hazardous chemical is not the overriding consideration. This would pertain to materials that also have broad utility as laboratory and analytical reagents

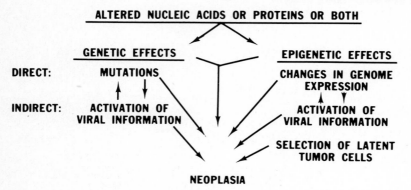

METABOLIC ACTIVATION OF CHEMICAL CARCINOGENS AND POSSIBLE MECHANISMS OF ACTION OF THESE AGENTS

(e.g., sodium azide, semicarbazide, hydroxylamine, hydrazine, etc.) and hence the number of individuals in contact with these substances may be relatively larger than for a high-volume hazardous monomer handled in a closed system.

This volume is arranged in four major categories for what is desired to be an enhancement of cohesiveness and subject treatment. These are: (1) an initial broad perspective of the interrelationships between chemical carcinogens and mutagens, metabolic activation of carcinogens and short-term bioassays for mutagenesis (Chapter 1); (2) a consideration of combination effects in chemical carcinogenesis (Chapter 2); (3) an overview of current considerations in aspects of epidemiology and risk assessment (Chapter 3) and (4) consideration of individual classes of chemical carcinogens and mutagens based on aspects of their structure and activity as: a) alkylating agents, acylating agents and peroxides (Chapters 5-10); b) halogenated derivatives (Chapters 11-15); c) nitrogen derivatives (Chapters 16-21); d) oxygenated derivatives and e) phosphorous derivatives and miscellaneous carcinogens and mutagens.

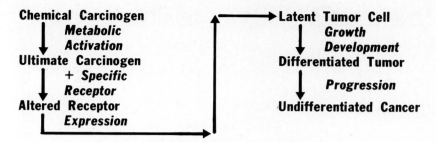

Figure 6. Sequence of complex events during chemical carcinogenesis

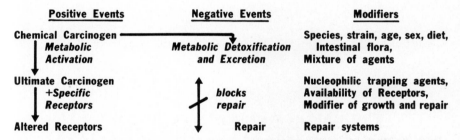

Figure 7. First steps of chemical carcinogenesis, involving activation of procarcinogen and reaction of resulting ultimate or primary carcinogen with specific cellular receptors, including DNA. These reactions are controlled and modified by numerous factors, some of which are noted.

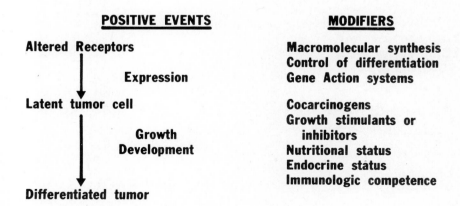

Figure 8. Later steps in carcinogenic process involving elements affecting the development and growth of carcinogen-modified cells and constituents.

POSITIVE EVENTS	MODIFIERS
Differentiated Tumor	Immunologic Status
↓ Progression	Gene and chromatin stability
Undifferentiated Cancer	Surgery Radiation Chemotherapy Immunotherapy

Figure 9. Last steps in carcinogenic process leading to malignancy, including spread by metastasis. Much more fundamental information is required to understand fully these steps. It is these terminal steps which often are responsible for the fatal outcome if not controlled by the modifiers listed.

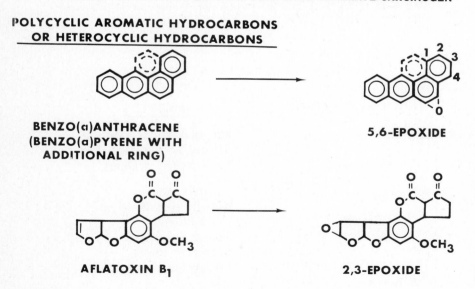

TYPICAL ACTIVATION REACTIONS
PROCARCINOGEN ► PROXIMATE CARCINOGEN ► ULTIMATE CARCINOGEN

POLYCYCLIC AROMATIC HYDROCARBONS OR HETEROCYCLIC HYDROCARBONS

BENZO(a)ANTHRACENE
(BENZO(a)PYRENE WITH ADDITIONAL RING)

5,6-EPOXIDE

AFLATOXIN B_1

2,3-EPOXIDE

TYPICAL ACTIVATION REACTIONS
PROCARCINOGEN▸PROXIMATE CARCINOGEN▸ULTIMATE CARCINOGEN

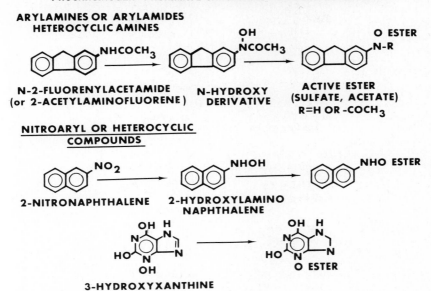

TYPICAL ACTIVATION REACTIONS
PROCARCINOGEN▸PROXIMATE CARCINOGEN▸ULTIMATE CARCINOGEN

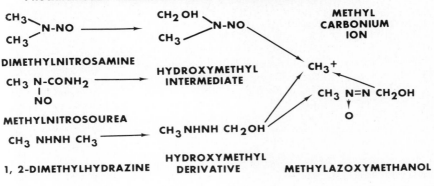

TYPICAL ACTIVATION REACTIONS
PROCARCINOGEN ► PROXIMATE CARCINOGEN ► ULTIMATE CARCINOGEN

SAFROLE → 1'-HYDROXY DERIVATIVE → O ESTER

ETHYL CARBAMATE ($H_2N\,COOC_2H_5$) → N-HYDROXY DERIVATIVE (HN(OH)$COOC_2H_5$) → CARBONIUM RESIDUE (HN(OH)CO^+)

PYRROLIZIDINE ALKALOID → PYRROLIC DERIVATIVE → CARBONIUM RESIDUE

No attempt has been made to present an <u>exhaustive</u> account of the theories of carcinogenesis and mutagenesis, environmental and genetic determinants of cancer, protocols for carcinogenicity and mutagenicity testing or the status of statistical tools in application to safety evaluation.

A number of books and reviews have been published in the above regard that are particularly germane. Foremost are the IARC Monographs on the Evaluation of Carcinogenic Risk of Chemicals to Man in which a total of 368 chemicals were evaluated in the first 16 volumes starting in 1971[109-124] and a review of the Monograph Program from 1971-1977[125,126]. Other recommended literature includes treatises on chemical carcinogens and carcinogenesis and the genetics, etiology and epidemiology of cancer[126-147]; chemical mutagens and mutagenesis[148-154]; the testing of chemicals for carcinogenicity and mutagenicity[153-175]; and aspects of biometry and risk-assessment[176-187]; and aspects of selection, evaluation, classification and regulation of potentially hazardous chemicals in the environment[188-197].

FIG. 11 FACTORS INFLUENCING THE FORMATION OF REACTIVE METABOLITES AND THEIR INTERACTION WITH BIOLOGICAL FUNCTIONS IN LIVER CELLS [98]

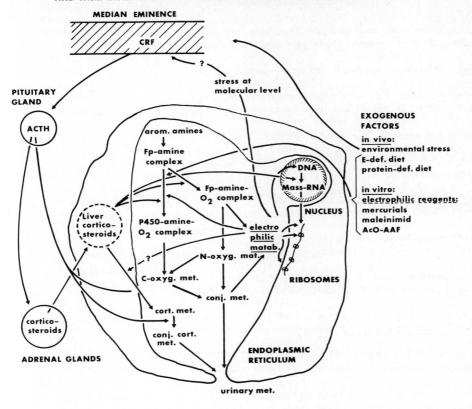

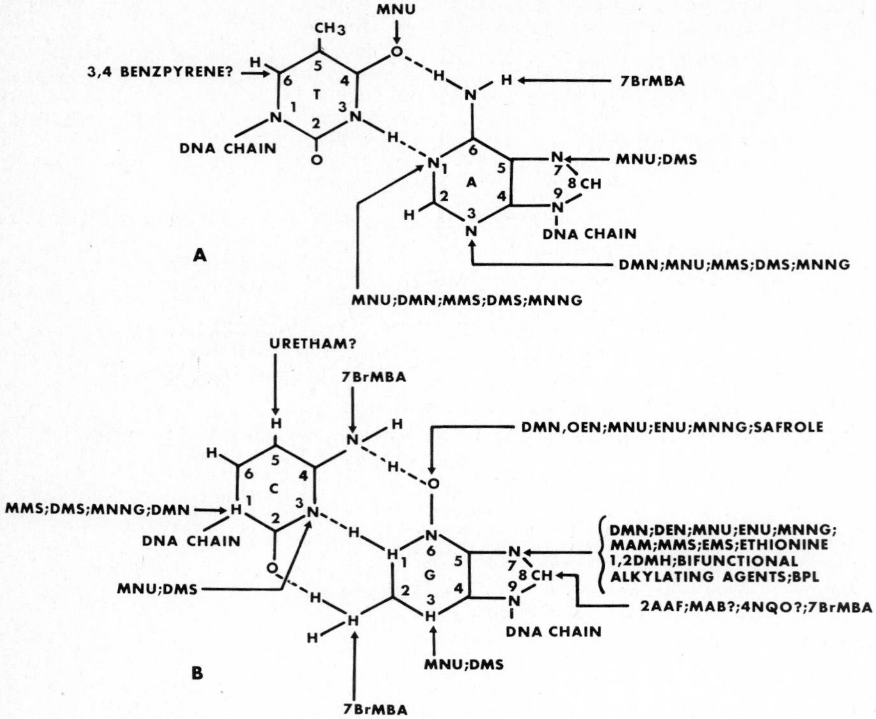

FIGURE 12 (A,B) Sites of interaction of chemical carcinogens with DNA in vivo and in vitro.

References

1. National Cancer Institute, 4th Annual Collaborative Conference, Carcinogenesis Program, Orlando, Fla., Feb. 22-26 (1976)
2. Clayson, D., Human epidemiology and animal laboratory correlations in chemical carcinogenesis meeting, Mescalero, New Mexico, June 1-4 (1977)
3. Anon, Kennedy adds chemical amendments to clean air act; hears EPA's current efforts, Toxic Materials News, 4 (20), (1977) 127
4. U.S. Tariff Commission, "Synthetic Organic Chemicals", U.S. Production and Sales, Annual, Government Printing Office, Washington, DC 1964 and 1974
5. Vouk, V. B., WHO-NIEHS Symposium on Potential Health Hazards from Technological Developments in Plastic and Synthetic Rubber Industries, Research Triangle Park, NC, March 1-3 (1976)
6. Withey, J. R., Mutagenic, Carcinogenic and Teratogenic Hazards Arising from Human Exposure to Plastics Additives, Origins of Human Cancer Meeting, Cold Spring Harbor, NY, Sept. 7-14 (1976)
7. Council on Environmental Quality, 6th Annual Rept., Washington, DC December (1975) p. 23
8. Anon, The spector of cancer, Env. Sci. Technol., 9 (1975) 116
9. Toth, B., Synthetic and naturally occurring hydrazines as possible causative agents, Cancer Res., 35 (1975) 3693
10. World Health Organization, Prevention of Cancer, Tech. Rept. Ser. No. 276, Geneva, Switzerland (1964)
11. Haddow, A., Proceedings of the 9th Int. Cancer Congr. Tokyo, Japan, 1966, R. J. C. Harris, ed., UICC Monograph Series Vol. 9: 111-115, Springer-Verlag New York, NY (1967)
12. Boyland, E., The correlation of experimental carcinogenesis and cancer in man, Progr. Exp. Tumor Res., 11 (1969) 222-234
13. Higginson, J., Present Trends in Cancer Epidemiology, Proc. Eighth Canadian Cancer Conf. Honey Harbour, 1968: 40-75, Pergamon Press, NY (1968)
14. Tomatis, L., the IARC Program on the Evaluation of the Carcinogenic Risk of Chemicals to Man, Ann. NY Acad. Sci., 271 (1976) 296
15. WHO, "Health Hazard of the Human Environment", World Health Organization, Geneva (1972) 213-323
16. Hueper, W. C., Occupational and Environmental Cancer of the Urinary System, Yale University Press, New Haven (1969)
17. Train, R. E., Environmental cancer, Science, 195 (1977) 443
18. Hoover, R., Mason, T. J., McKay, F. W., and Fraumeni, J. F., Jr., Geographical patterns of cancer mortality in the United States, In "Persons at High Risk of Cancer: An Approach to Cancer Etiology and Control" Academic Press, New York, San Francisco, London (1975) pp. 343-360
19. Hoover, R., and Fraumeni, J. F., Jr., Cancer mortality in U.S. counties with chemical industries, Environ. Res., 9 (1975) 196-207
20. Mason, T., Statistics point to high-cancer localities, Occup. Hlth. Safety, May/June (1977) 44-48
21. Weinhouse, S., Problems in the Assessment of Human Risk of Carcinogenesis by Chemicals, Origins of Human Cancer Meeting, Cold Spring Harbor, NY Sept. 7-14 (1976)
22. Hueper, W. C., Medicolegal considerations of occupational and nonoccupational environmental cancers, In "Lawyers Medical Encyclopedia", (in press) cited in reference 7

23. Selikoff, I. J., and Hammond, E. C., Environmental cancer in the year 2000, Presented at the Seventh National Cancer Conference, Los Angeles, Sept. 29 (1972)
24. Blum, A., and Ames, B. N., Flame-retardant additives as possible cancer hazards, Science, 195 (1977) 17-22
25. Anon, NIOSH says workers must know hazards, Chem. Ecology, June (1977) p. 5
26. Anon, Chem. Eng. News, May 2 (1977) 10
27. Committee 17, Environmental mutagenic hazards, Science, 187 (1975) 503-514
28. DHEW, Approaches to Determining the Mutagenic Properties of Chemicals: Risk to Future Generations, DHEW Subcommittee on Environmental Mutagenesis, Washington, DC, April (1977)
29. DeSerres, F. J., Prospects for a revolution in the methods of toxicological evaluation, Mutation Res., 38 (1976) 165-176
30. Crow, J. F., Chemical risk to future generations, Scientist and Citizen, 10 (1968) 113-117
31. Auerbach, C., The chemical production of mutations, Science, 158 (1967) 1141
32. Anon, Application for Ames test expanded, Chem. Eng. News, 54 (1976) 18
33. Maugh, T. H., Can potential carcinogens be detected more quickly? Science, 183 (1974) 94
34. Fox, J. L., Ames Test Success Paves Way for Short-Term Cancer Testing, Chem. Eng. News, Dec. 12 (1977) 34-36, 42-46
35. Bartsch, H., Predictive value for mutagenicity tests for chemical carcinogenesis, Mutation Res., 38 (1976) 177-190
36. Shubik, P., and Clayson, D. B., Environmental cancer and chemical agents, Ann. Intern. Med., 85 (1976) 120-121
37. McCann, J., Ames, B. N., The Salmonella/Microsome Mutagenicity Test: Predictive Value for Animal Carcinogenicity, In Proceedings of Conference on Origins of Human Cancer, Cold Spring Harbor Laboratory, New York, Sept. 7-14 (1976)
38. Russell, K., and Meselson, M., Quantitative Measures of Carcinogenic and Mutagenic Potency, In Proceedings of Conference on Origins of Human Cancer, Cold Spring Harbor Laboratory, New York, Sept. 7-14 (1976)
39. McCann, J., Choi, E., Yamasaki, E., and Ames, B. N., Detection of Carcinogens as mutagens in the Salmonella/microsome test: Assay of 300 chemicals, Proc. Natl. Acad. Sci., 72 (1975) 5135-5139
40. Ames, B. N., and McCann, J., Carcinogens are mutagens: A simple test system, in R. Montesano, H. Bartsch and L. Tomatis (eds) Screening Tests in Chemical Carcinogenesis, Lyon (IARC Scientific Publications No. 12) 1976
41. Sigimura, T., Yahagi, T., Nagao, M., Takeuchi, M., Kawachi, T., Hara, K., Yamasaki, E., Mtsushima, T., Hashimoto, Y., and Okada, M., Validity of Mutagenicity Tests Using Microbes as a Rapid Screening Method for Environmental Chemicals, In: R. Montesano, H. Bartsch, and L. Tomatis (eds) Screening Tests in Chemical Carcinogenesis, Lyon, IARC Scientific Publications No. 12 1976
42. Stoltz, D. R., Poirier, L. A., Iving, C. C., Strich, H. F., Weisburger, J. H., and Grice, H. C., Evaluation of short-term tests for carcinogenicity, Toxicol. Appl. Pharmacol., 29 (1974) 157-180
43. Brusick, D. J., In vitro mutagenesis assays as predictors of chemical carcinogenesis in mammals, Clin. Toxicol., 10 (1977) 79-109

44. Poirier, L. A., and Simmon, V. F., Mutagenic-carcinogenic relationships and the role of mutagenic screening tests for carcinogenicity, Clin. Toxicol., 9 (1976) 761-777
45. Mortimer, R. K. and Manny, T. R., Mutation induction in yeast, in A. Hollaender (ed) Chemical Mutagens: Principles and Methods for their Detection, Plenum Press, New York, (1976) pp. 289-310
46. Brusick, D., and Andrews, H., Comparison of the genetic activity of dimethylnitrosamines, ethyl methanesulfonate, 2-acetylaminofluorene and ICR-170 in Saccharomyces cerevisiae strains D3, D4 and D5 using in vitro assays with and without metabolic activation, Mutation Res., 26 (1974) 491-500
47. Ong, T., and De Serres, F. J., Mutagenicity of chemical carcinogens in Neurospora crassa, Cancer Res., 32 (1972) 1890-1893
48. Sobels, F. H., and Vogel, E., Assaying potential carcinogens with Drosophila, Env. Hlth. Persp., 15 (1976) 141-146
49. Duncan, M. E., and Brookes, P., The induction of azaguanine-resistant mutants in cultured Chinese hamster cells by reactive derivatives of carcinogenic hydrocarbons, Mutation Res., 21 (1973) 107-118
50. Huberman, E., Aspiras, L., Heidelberger, C., Grover, P. L., and Sims, P., Mutagenicity to mammalian cells of epoxides and other derivatives of polycyclic hydrocarbons, Proc. Natl. Acad. Sci. (US) 68 (1971) 3195-3199
51. Huberman, E., Donovan, P. J., and DiPaolo, J. A., Mutation and transformation of cultured mammalian cells by N-acetoxy-N-2-fluorenylacetamide, J. Natl. Cancer Inst., 48 (1972) 837-840
52. Kao, F. T., and Puck, T. T., Genetics of somatic mammalian cells, VII. Induction and isolation of nutritional mutants in Chinese hamster cells, Proc. Natl. Acad. Sci. (US) 60 (1968) 1275-1281
53. Kao, F. T., Cell mutagenesis studies in vitro using auxotrophic markers, in P.O.P. Ts'O and J. A. DiPaolo (eds) Chemical Carcinogenesis, Part B, Marcel Dekker, Inc., New York (1974) pp. 565-573
54. Clive, D., Flamm, W. G., Machesko, M. R., and Bernheim, N. J., A mutational assay system using the thymidine kinase locus in mouse lymphoma cells, Mutation Res., 16 (1972) 77-87
55. Capizzi, R. L., Papirmeister, B., Mullins, J. M., and Cheng, E., The detection of chemical mutagens using the L5178Y/Asn- murine leukemia in vitro and in a host-mediated assay, Cancer Res., 34 (1974) 3073-3082
56. Fischer, G. A., Predictive tests in culture of drug-resistant mutants selected in vivo, Natl. Cancer Inst. Monograph, 34 (1971) 131-134
57. Summers, W. P., Chemically induced mutation of L5178Y mouse leukemia cells from asparagine-dependence to asparagine-independence, Mutation Res., 20 (1973) 377-385
58. Rappaport, H., and DeMars, R., Diaminopurine-resistant mutants of cultured, diploid human fibroblasts, Genetics, 75 (1973) 335-345
59. Maher, V. M., and Wessel, J. E., Mutations to azaguanine resistance induced in cultured diploid human fibroblasts by the carcinogen, N-acetoxy-2-acetylaminofluorene, Mutation Res., 28 (1975) 277-284
60. Sato, K., Slesinski, R. S., and Littlefield, J. W., Chemical mutagenesis at the phosphoribosyltransferase locus in cultured human lymphoblasts, Proc. Natl. Acad. Sci., (US) 69 (1972) 1244-1248
61. San, R. H., and Stich, H. F., DNA repair synthesis of cultured human cells as a rapid bioassay for chemical carcinogens, Int. J. Cancer, 16 (1975) 284-291

62. Stich, H. F., San, R.H., and Kawazoe, Y., DNA repair synthesis in mammalian cells exposed to a series of oncogenic and non-oncogenic derivatives of 4-nitroquinoline-1-oxide, Nature, 229 (1971) 416-419
63. Stich, H. F., and San, R. H., DNA repair and chromatin anomalies in mammalian cells exposed to 4-nitroquinoline-1-oxide, Mutation Res., 10 (1970) 389
64. Ahmed, F. E., Hart, R. W., and Lewis, N. J., Pesticide induced DNA damage and its repair in cultured human cells, Mutation Res., 42 (1977) 161-174
65. DeSerres, F. J., Mutagenicity of chemical carcinogens, Mutation Res., (1976) 41 43-50
66. Flamm, W. G., A tier system approach to mutagen testing, Mutation Res., 26 (1974) 329-333
67. Bridges, B. A., The three-tier approach to mutagenicity screening and the concept of radiation-equivalent dose, Mutation Res., 26 (1974) 335-340
68. Bridges, B. A., Use of a three-tier protocol for evaluation of long-term toxic hazards particularly mutagenicity and carcinogenicity, In Screening Tests in Chemical Carcinogenesis, WHO/IARC Publ. No. 12, eds Montesano, R., Bartsch, H., and Tomatis, L., Lyon (1976) pp. 549-568
69. Bora, K. C., A hierarchical approach to mutagenicity testing and regulatory control of environmental control chemicals, Mutation Res., 41 (1976) 73-82
70. Dean, B. J., A predictive testing scheme for carcinogenicity and mutagenicity of industrial chemicals, Mutation Res., 41 (1976) 83-88
71. Claxton, L. D., and Barry, P. Z., Chemical mutagenesis: An emerging issue for public health, Am. J. Publ. Hlth., 67 (1977) 1037-1042
72. Bochkov, N. P., Sram, R. J., Kuleshov, N. P. and Zhurkov, V. S., System for the evaluation of the risk from chemical mutagens for man: Basic principles and practical recommendations, Mutation Res., 38 (1976) 191-202
73. Ehling, U. H., Mutagenicity testing and risk estimation with mammals, Mutation Res., 41 (1976) 113-122
74. Drake, J. W., Environmental mutagenesis: Evolving strategies in the USA, Mutation Res., 33 (1975) 65-72
75. Brewen, J. G., Practical evaluation of mutagenicity data in mammals for estimating human risk, Mutation Res., 41 (1976) 15-24
76. Purchase, I. F. H., Longstaff, E., Asby, J., Styles, J. A., Anderson, D., LeFevre, P. A., and Westwood, Evaluation of six short term tests for detecting organic chemical carcinogens and recommendations for their use, Nature, 264 (1976) 624-627
77. Sobels, F. H., Some thoughts on the evaluation of environmental mutagens, Mutation Res., 38 (1976) 361-366
78. Sram, R. J., Genetic risk from chemicals: Mutagenicity studies and evaluation, Rev. Czeck. Med., 21 (1975) 186-193
79. Ames, B. N., McCann, J., and Yamasaki, E., Methods for detecting carcinogens and mutagens with the Salmonella/mammalian microsome mutagenicity test, Mutation Res., 31 (1975) 347-364
80. McCann, J., and Ames, B. N., Detection of carcinogens as mutagens in the Salmonella/microsome test: Assay of 300 chemicals, Part II, Proc. Natl. Acad. Sci., 73 (1976) 950-954
81. McCann, J., Spingarn, N. E., Kobori, J., and Ames, B. N., Detection of carcinogens as mutagens: Bacterial tester strains with R factor plasmids, Proc. Nat. Acad. Sci., 72 (1975) 979-983

82. Purchase, I. F. H., Longstaff, E., Ashby, J., Styles, J. A., Anderson, D., Lefevre, P. A., and Westwood, F. R., Evaluation of six short-term tests for detecting organic chemical carcinogens and recommendations for their use, Nature, 264 (1976) 624-627
83. Brusick, D. J., Automated plate-counting assays in the Ames Salmonella/microsome mutagenicity test, Pharm. Tech., 2 (1978) 37-40, 65
84. Knudson, A., Mutation and human cancer, Adv. In Cancer Res., 17 (1973) 317-352
85. Legator, M. S., The inadequacies of sub-mammalian tests for mutagenicity, In "Molecular and Environmental Aspects of Mutagenesis", Thomas, Springfield (1974)
86. Sivak, A., The Ames Assay, Science, 193 (1976) 271
87. Pitot, H. C., Neoplasia: A somatic mutation or a heritable change in cytoplasmic membranes? J. Natl. Cancer Inst., 53 (1974) 905-911
88. Pierce, G. B., Neoplasms, differentiation and mutations, Am. J. Pathol., 77 (1974) 103-118
89. Mintz, B., and Illmensee, K., Normal genetically mosaic mice produced from malignant teratocarcinoma cells, Proc. Natl. Acad. Sci., 72 (1975) 3585
90. Nery, R., Carcinogenic mechanisms: A critical review and a suggestion that oncogenesis may be adaptive oncogenesis, Chem. Biol. Interactions, 12 (1976) 145-169
91. Miller, J. A., and Miller, E. C., Metabolic and reactivity of chemical carcinogens, Mutation Res., 33 (1975) 25
92. Miller, J. A., Carcinogenesis by chemicals: An overview, G.H.A. Cowes Memorial Lecture, Cancer Res., 30 (1970) 559-576
93. Miller, E. C., and Miller, J. A., The metabolism of chemical carcinogens to reactive electrophiles and their possible mechanisms of action in carcinogenesis, In "Chemical Carcinogens" ed. Searle, C. E., ACS Monograph No. 173, American Chemical Society, Washington, D.C. (1976) 737-762
94. Lawley, P. D., Carcinogenesis by alkylating agents, In "Chemical Carcinogenesis" ed. Searle, C. E., ACS Monograph No. 173, American Chemical Society, Washington, D.C. (1976) pp. 83-244
95. Weisburger, J. H., Bioassays and tests for chemical carcinogens, In "Chemical Carcinogens" ed. Searle, C. E., ACS Monograph No. 173, American Chemical Society, Washington, D.C. (1976) pp. 1-23
96. Sarma, D. S. R., Rajalakshmi, S., and Farber, E., Chemical carcinogenesis: Interactions of carcinogens with nucleic acids, In "Cancer" ed. Becker (1975) pp. 235-287
97. Weisburger, J. H., and Williams, G. M., Metabolism of chemical carcinogens, In "Cancer" ed. Becker (1975) pp. 185-234
98. Arrhenius, E., Comparative metabolism of aromatic amines, In "Chemical Carcinogenesis Assays", eds. Montesano, R., and Tomatis, L., IARC Publ. No. 10, International Agency for Research on Cancer, Lyon (1974) pp. 25-37
99. Selkirk, J. K., Huberman, E., and Heidelberger, C., Epoxide is an intermediate in the microsomal metabolism of the chemical carcinogen dibenz(a,h)-anthracene Biochem. Biophys. Res. Commun., 43 (1971) 1010
100. Kriek, E., Carcinogenesis by aromatic amines, Biochim. Biophys. Acta., 355 (1974) 177

101. Boutwell, R. K., Colburn, N. J., Muckerman, C. S., In vivo reactions of β-propiolactone, Ann. NY Acad. Sci., 163 (1969) 751
102. Clayson, D. B., and Garner, R. C., Carcinogenic aromatic amines and related compounds, In "Chemical Carcinogens", ed. Searle, C. E., ACS Monograph No. 173, American Chemical Society, Washington, DC (1976) 366-461
103. Sugimura, T., Kawachi, T., Experimental stomach cancer, Methods Cancer Res., 7 (1973) 245-308
104. Oesch, F., Mammalian epoxide hydrases: Inducible enzymes catalyzing the inaction of carcinogenic and cytotoxic metabolites derived from aromatic and olefinic compounds, Xenobiotica, 3 (1972) 305
105. Weisburger, J. H., Weisburger, E. K., Biochemical formation and pharmacological, toxicological and pathological properties of hydroxylamines and hydroxamic acids, Pharm. Rev., 25 (1973) 1
106. Grover, P. L., Hewer, A., Sims, P., Metabolism of polycyclic hydrocarbons by rat-lung preparations, Biochem. Pharmacol., 23 (1974) 323
107. Falk, H. L., Possible mechanisms of combination effects in chemical carcinogenesis, Oncology, 33 (1976) 77-85
108. Nashed, N., Evaluation of a possible role for antimutagens, antiteratogens, and anticarcinogens in reducing environmental health hazards, Env. Hlth. Persp. 14 (1976) 193-200
109. IARC, Monographs on the Evaluation of the Carcinogenic Risk of Chemicals to Man, Vol. 1, Some Inorganic Substances, Chlorinated Hydrocarbons, Aromatic Amines, N-Nitroso Compounds and Natural Products, International Agency for Research on Cancer, Lyon, France (1972)
110. IARC, Monographs on the Evaluation of the Carcinogenic Risk of Chemicals to Man, Vol 2., Some Inorganic and Organometallic Compounds, International Agency for Research on Cancer, Lyon, France (1973)
111. IARC, Monographs on the Evaluation of Carcinogenic Risk of Chemicals to Man, Vol. 3, Certain Polycyclic Aromatic Hydrocarbons and Heterocyclic Compounds, International Agency for Research on Cancer, Lyon France (1973)
112. IARC, Monographs on the Evaluation of the Carcinogenic Risk of Chemicals to Man, Vol. 4, Some Aromatic Amines, Hydrazine and Related Substances, N-Nitroso Compounds and Miscellaneous Alkylating Agents, International Agency for Research on Cancer, Lyon, France, (1974)
113. IARC, Monographs on the Evaluation of the Carcinogenic Risk of Chemicals to Man, Vol. 5, Some Organochlorine Pesticides, International Agency for Research on Cancer, Lyon, France (1974)
114. IARC, Monographs on the Evaluation of the Carcinogenic Risk of Chemicals to Man, Vol. 6, Sex Hormones, International Agency for Research on Cancer, Lyon, France, (1974)
115. IARC, Monographs on the Evaluation of the Carcinogenic Risk of Chemicals to Man, Vol. 7, Some Anti-thyroid and Related Substances, Nitrofurans and Industrial Chemicals, International Agency for Research on Cancer, Lyon, France (1974)
116. IARC, Monographs on the Evaluation of the Carcinogenic Risk of Chemicals to Man, Vol. 8, Some Aromatic Azo Compounds, International Agency for Research on Cancer, Lyon, France (1975)
117. IARC, Monographs on the Evaluation of the Carcinogenic Risk of Chemicals to Man, Vol. 9, Some Aziridines, N-, S- and O-Mustards and Selenium, International Agency for Research on Cancer, Lyon, France, (1975)

118. IARC, Monographs on the Evaluation of the Carcinogenic Risk of Chemicals to Man, Vol. 10, Some Naturally Occurring Substances, International Agency for Research on Cancer, Lyon, France (1976)
119. IARC, Monographs on the Evaluation of the Carcinogenic Risk of Chemicals to Man, Vol. 11, Cadmium, Nickel, Some Epoxides, Miscellaneous Industrial Chemicals and General Considerations on Volatile Anaesthetics, International Agency for Research on Cancer, Lyon, France (1976)
120. IARC, Monographs on the Evaluation of the Carcinogenic Risk of Chemicals to Man, Vol. 12, Some Carbamates, Thiocarbamates and Carbazides, International Agency for Research on Cancer, Lyon, France (1976)
121. IARC, Monographs on the Evaluation of the Carcinogenic Risk of Chemicals to Man, Vol. 13, Some Miscellaneous Pharmaceutical Substances, International Agency for Research on Cancer, Lyon, France, (1977)
122. IARC, Monographs on the Evaluation of the Carcinogenic Risk of Chemicals to Man, Vol. 14, Asbestos, International Agency for Research on Cancer, Lyon, France (1977)
123. IARC, Monographs on the Evaluation of the Carcinogenic Risk of Chemicals to Man, Vol. 15, Some Fumigants, the Herbicides 2,4-D and 2,4,5-T, Chlorinated Dibenzo-dioxins and Miscellaneous Industrial Chemicals, International Agency for Research on Cancer, Lyon, France (1977)
124. IARC, Monographs on the Evaluation of the Carcinogenic Risk of Chemicals to Man, Vol. 16, Some Aromatic Amines and Related Nitro Compounds: Hair Dyes, Colouring Agents and Miscellaneous Industrial Chemicals, International Agency for Reserach on Cancer, Lyon, France (1978)
125. Tomatis, L., Agthe, C., Bartsch, H., Huff, J., Montesano, R., Saracci, R., Walker, E., and Wilbourn, Evaluation of the carcinogenicity of chemicals: A review of the Monograph Program of the International Agency for Research on Cancer (1971-1977), Cancer Res., $\underline{38}$ (1978) 877-885
126. Searle, C. E. (ed) "Chemical Carcinogens", ACS Monograph No. 173, American Chemical Society, Washington, DC (1976)
127. Arcos, J. C., and Argus, M. F., Chemical Induction of Cancer, Vols. IIA, IIB, Academic Press, New York (1974)
128. Becker, F. F., (ed) "Cancer-A Comprehensive Treatise", Vol. 1, Etiology: Chemical and Physical Carcinogenesis, Plenum Press, New York and London (1975)
129. Ts'O, P. O. P., and DiPaolo, J. A., "Chemical Carcinogenesis", Parts A and B, Marcel Dekker, New York (1974)
130. Knudson, A. G., Jr., Genetics and Etiology of Human Cancer, In: Advances in Human Genetics, $\underline{8}$ (1977) pp. 1-66
131. Agthe, C., and Tomatis, L., Evaluation of the Carcinogenicity of Environmental Chemicals and the Possibility of Cancer Prevention. In: "The Physiopathology of Cancer" (ed) Homburger, F., Vol. 2, S. Karger, A. G. Basel (1976) 345-358
132. Tomatis, L., The IARC Program on the Evaluation of the Carcinogenic Risk of Chemicals to Man, Ann. NY Acad. Sci., $\underline{27}$ (1976) 396-409
133. Boyland, E., The correlation of experimental carcinogenesis and cancer in man, Progr. Exptrl. Tumor Res., $\underline{11}$ (1969) 222-234
134. Clayson, D. B., "Chemical Carcinogenesis", J. & A. Churchill, London (1962)
135. Stewart, B. W., and Sarfaty, G. A., Environmental chemical carcinogenesis, Med. J. Aust., $\underline{1}$ (1978) 92-95
136. Mulvihill, J. J., Miller, R. W., and Fraumeni, J. F., Jr., (eds), Genetics of the Human Cancer, Papers from a Conference, Raven Press, New York (1977)
137. Muir, C. S., Predictive value of cancer statistics, J. Environ. Pathol. Toxicol., $\underline{1}$ (1977) 3-10

138. Wynder, E. L., Cultural and behavioral aspects of risk factors: Society's obligation, J. Environ. Pathol. Toxicol., 1 (1977) 11-18
139. Fraumeni, J. F., Jr., Environmental and genetic determinants of cancer, 1 (1977) 19-30
140. Fraumeni, J. F., Jr. (ed) Persons at High Risk of Cancer: An Approach to Cancer Etiology and Control, Academic Press, New York (1975)
141. Levin, D. L., Devesa, S. S., Godwin, J. D., and Silverman, D. T., Cancer rates and risks, 2nd ed., U.S. Dept. of Health, Education and Welfare, Washington, DC (1974)
142. UICC, "Cancer Incidence in Five Continents" (eds) Doll, R., Muir, C., and Waterhouse, S., Vol. II, International Union of Cancer, Springer-Verlag, New York, Berlin-Heidelberg (1970)
143. Doll, R., "Prevention of Cancer, Pointers from Epidemiology", Nuffield Provincial Hospital Trust, Nuffield, England (1967)
144. Doll, R., Strategy for detection of cancer hazards to man, Nature, 265 (1977) 589-596
145. Higginson, J., The role of geographical pathology in environmental carcinogenesis, In: "Environments and Cancer", A Collection of Papers Presented at the 24th Annual Symposium on Fundamental Cancer Research, 1971; Houston, Texas, The Williams and Wilkins Co., Baltimore (1972)
146. Higginson, J., Present trends in cancer epidemiology, Can. Cancer Conf., 8 (1968) 40-75
147. WHO, Report of WHO Scientific Group on the Assessment of Carcinogenicity and Mutagenicity of Chemicals, WHO Tech. Reptr. Ser. No. 546, World Health Organization, Geneva, August 13-17 (1973)
148. Fishbein, L., Flamm, W. G., and Falk, H. L., "Chemical Mutagens", Academic Press, New York and London (1970)
149. Hollaender, A., (ed) "Chemical Mutagens: Principles and Methods for their Detection", Vols. I-VI, Plenum Press, New York (1971-1978)
150. Vogel, F., and Rohrborn, G. (ed) "Chemical Mutagenesis in Mammals and Man', Springer-Verlag, New York-Berlin-Heidelberg (1970)
151. Banks, G. R., Mutagenesis: A review of some molecular aspects, Sci. Progr. Oxf., 59 (1971) 475-503
152. Drake, J. W., The molecular basis of mutation, Holden-Day, San Francisco, CA (1970)
153. DHEW Working Group of Subcommittee on Environmental Mutagenesis, Approaches to Determining the Mutagenic Properties of Chemicals: Risk to Future Generations, J. Environ. Pathol. Toxicol., 1 (1977) 301-352
154. deSerres, F. J., and Sheridan, W., (eds), Proceedings of the Workshop on the Evaluation of Chemical Mutagenicity Data in Relation to Population Risk, Research Triangle Park, NC, April 26-28 (1973), Environ. Hlth. Persp., 6 (1973)
155. Health and Welfare of Canada, Testing of Chemicals for Carcinogenicity, Mutagenicity, and Teratogenicity, Ottawa, Canada, Sept. (1973)
156. Page, N. P., Concepts of a bioassay program in environmental carcinogenesis, In: Advances in Modern Toxicology, Vol. 3, Environmental Cancer (eds) Kraybill, H. F., and Mehlman, M. A., Hemisphere Publ., Washington, DC (1977) pp. 87-171
157. Page, N. P., Chronic toxicity and carcinogenicity guidelines, J. Environ. Pathol. Toxicol., 1 (1977) 161-182
158. Munro, I. C., Considerations in chronic toxicity testing: The chemical, the dose, the design, J. Environ. Pathol. Toxicol., 1 (1977) 183-197

159. Clayson, D. B., Relationships between laboratory and human studies, J. Environ. Pathol. Toxicol., 1 (1977) 31-40
160. Clive, D., Flamm, W. G., and Patterson, J. B., Specific locus mutational assay system for mouse lymphoma cells, In: Chemical Mutagens, Principles and Methods for the Detection (ed) Hollaender, A., Plenum Press, New York (1973) 79-104
161. DeMars, R., Mutation studies with human fibroblasts, Environ. Hlth. Persp., 6 (1973) 127-136
162. Jacobs, L., and DeMars, R., Chemical mutagenesis with diploid human fibroblasts, In: Handbook of Mutagen Testing (ed) Kilbey, B., et al., Elsevier Press, Amsterdam (1977)
163. Sutton, E. H., Prospects of monitoring environmental mutagenesis through somatic mutations,
164. Bridges, B. A., Some principles of mutagenicity screening and a possible framework for testing procedures, Environ. Hlth. Persp., 6 (1973) 221-227
165. Abrahamson, S., and Lewis, E. B., The detection of mutations in Drosophila melanogaster, In: Chemical Mutagens: Principles and Methods for their Detection, (ed) Hollaender, A., Vol. 2, Plenum Press, New York (1971) 461-487
166. Mohn, C., Ellenberger, J., and McGregor, D., Development of mutagenicity tests using Escherichia coli K-12 as indicator organism, Mutation Research, 25 (1974) 187-196
167. Marquardt, H., Mutation and recombination experiments with yeast as prescreening tests for carcinogenic effects, Z. Krebsforsch., 81 (1974) 333-346
168. Ehrenberg, L., Higher plants, In: Hollaender, A. (ed) Chemical Mutagens: Principles and Methods for their Detection, Plenum Press, New York, Vol. 2, (1971) pp. 365-386
169. Schmid, W., The micronucleus test, Mutation Res., 31 (1975) 9-15
170. Allen, J. W., and Latt, S. A., Analysis of sister chromatid exchange formation in vivo in mouse spermatogonia as a new test system for environmental mutagens, Nature, 260 (1976) 449-451
171. Dunkel, V. C., Wolff, J. S., and Pienta, R. J., In vitro transformation as a presumptive test for detecting chemical carcinogens, Cancer Bulletin, 29 (1977) 167-174
172. Bateman, A. J., and Epstein, S. S., Dominant lethal mutations in mammals, In: "Chemical Mutagens: Principles and Methods for their Detection, Vol. 2, (ed) Hollaender, A., Plenum Press, New York (1971) 541-568
173. Brusick, D. J., and Mayer, V. W., New developments in mutagenicity screening techniques with yeast, Environ. Hlth. Persp., 6 (1973) 83-96
174. Fahrig, R., Development of host-mediated mutagenicity test-yeast systems II. Recovery of yeast cells out of testes, liver, lung and peritoneum of rats, Mutation Res., 31 (1975) 381-394
175. Legator, M. S., and Malling, H. V., The host-mediated assay, a practical procedure for evaluating potential mutagenic agents in mammals, In: Chemical Mutagens: Principles and Methods for their Detection, Vol. 2, (ed) Hollaender, A., Plenum Press, New York (1973) 21-56
176. Hoel, D. G., Gaylor, D. W., Kirschstein, R. L., Saffiotti, U., and Schneiderman, M. A., Estimation of risks of irreversible, delayed toxicity, J. Toxicol. Environ. Hlth., 1 (1975) 133-151
177. Crump, K. S., Hoel, D. G., Langley, C. H., and Peto, R., Fundamental carcinogenic processes and their implications for low dose risk assessment, Cancer Res., 36 (1976) 2973-2979

178. Hartley, H. O., and Sielken, R. L., Jr., Estimation of "Safe Doses" in carcinogenic experiments, J. Environ. Pathol. Toxicol., 1 (1977) 241-278
179. Mantel, N., Bohidar, N. R., Brown, C. C., Ciminera, J. C., and Tukey, J. W., An improved Mantel-Bryan procedure for "safety" testing of carcinogens, Cancer Res., 35 (1975) 865-872
180. Guess, H. A., and Hoel, D. G., The effect of dose on cancer latency period, J. Environ. Pathol. Toxicol., 1 (1977) 279-286
181. Armitage, P., and Doll, R., Stochastic models for carcinogenesis, In: Proceedings of the Fourth Berkeley Symposium on Mathematical Statistics and Probability, Vol. 4, University of California Press, Berkeley and Los Angeles, (1961) 19-38
182. Chand, N., and Hoel, D., A comparison of models for determining safe levels of environmental agents, In: "Reliability and Biometry", (eds) Proschan, F., and Serfling, R. J., Siam, Philadelphia (1974) pp. 681-700
183. Druckrey, H., Quantitative aspects of chemical carcinogenesis, In: "Potential Carcinogenic Hazards from Drugs (Evaluation of Risks)" (ed) Truhaut, R., UICC Monograph Series, Vol. 7, Springer-Verlag (1967) pp. 60-78
184. Mantel, N., and Bryan, W. R., "Safety" testing of carcinogenic agents, J. Natl. Cancer Inst., 27 (1961) 455
185. Gross, M. A., Fitzhugh, O.G., and Mantel, N., Evaluation of safety for food "additives", Biometrics, 26 (1970) 181
186. Truhaut, R., Can permissible levels of carcinogenic compounds in the environment be envisaged?, Ecotoxicol. Environ. Safety, 1 (1977) 31-37
187. Falk, H. L., Considerations of risks versus benefits, Environ. Hlth. Persp., 11 (1975) 1-5
188. World Health Organization, Health Hazards of the Human Environment, Geneva (1972)
189. Stephenson, M. E., An approach to the identification of organic compounds hazardous to the environment and human health, Ecotoxicol. Environ. Safety, 1 (1977) 39-48
190. Selikoff, I. J., Perspectives in the investigation of health hazards in the chemical industry, Ecotoxicol. Environ. Safety, 1 (1977) 387-397
191. National Science Foundation, Final Report of NSF Workshop Panel to Select Organic Compounds Hazardous to the Environment, Washington, DC, October (1975)
192. National Academy of Sciences, Decision Making for Regulating Chemicals in the Environment, Washington, DC (1975)
193. Watson, J. D., Hiatt, H., (eds) Abstracts of Papers of Meeting on Origins of Human Cancer, Cold Spring Harbor Laboratory, Cold Spring Harbor, New York Sept. 7-14 (1976)
194. National Academy of Sciences, Principles for Evaluating Chemicals in the Environment, Washington, DC (1975)
195. U.S. Congress, "Toxic Substances Control Act", Public Law 94-469, 94th Congress, Washington, DC, October 11 (1976)
196. U.S. Dept. of Labor, Identification, Classification and Regulation of Toxic Substances, Posing A Potential Occupational Carcinogenic Risk, Federal Register, 42 (192) 54148-54247; October 4 (1977)
197. Montesano, R., and Tomatis, L., Legislation concerning chemical carcinogens in several industrialized countries, Cancer Research, 37 (1977) 310-316

CHAPTER 2

COMBINATION EFFECTS IN CHEMICAL CARCINOGENESIS

The term "carcinogen" as generally used, refers to an agent, exposure to which results in the eventual appearance of cancer which would not have occurred in the absence of such exposure[1]. However, this allows no implication concerning its mode of action in particular whether chemical carcinogens are direct acting, or are capable of initiating malignant change only by some indirect mechanism (e.g., via activation, activation of latent oncogenic viruses, depression of genes or interference with normal immunologic defences against malignancy)[1].

Clayson in 1962 has defined a carcinogen as an agent or process which significantly increases the yield of neoplasms in a population[2]. An agent may be regarded as carcinogenic even though its presence in a host has been shown only to induce benign growths (tumors that do not show invasiveness or metastasis)[3]. The National Institute for Occupational Safety and Health (NIOSH)[4] has published a list of suspected carcinogens, and the 1500 chemical substances listed include agents causing either benign or malignant tumors consistant with Claysons[2] definition and the views of Miller and Miller[3].

Although there is much controversy as to the _precise_ scientific definition of a carcinogen, the term must be defined for regulatory purposes. OSHA considers a chemical to be a carcinogen if properly designed studies show that exposure to the chemical causes cancer in: humans; two different animal (mammal) species; one animal species if the results are duplicated in separate studies; and one animal species if the results are supported by multi-test evidence of mutagenicity[5].

Carcinogens are characterized as viral, chemical, or physical agents and may occur as relatively simple, well defined forms or in more complex undefined forms[3]. Viral carcinogens are considered to be capable of using processes in living cells to reproduce, and quantitative limits for viral exposure are not practical.

Ambrose and Roe in 1966[6] described cancer as a group of many diseases with a variety of etiologic factors. Cancer cells may be distinguished by microscopic and histologic characteristics which show differences from more normal counterparts. Cell growths are regarded as hyperplastic, metaplastic, or neoplastic[3]. Hyperplasia describes a state produced by an increased rate of cellular proliferation, stimulated for a variety of reasons, including repair processes, hormones, and disturbance of biologic regulation of cellular reactions. Removing the stimulus can cause a hyperplastic proliferation to subside.

The origin of cancer, either from one single cell or from a number of cells which transform independently is compatible with its development following a similar or same mechanism. According to Tomatis[7], if we accept the hypothesis that cancers of clonal (single) cell origin are mainly spontaneous and that cancers originating from multiple cells are in many cases induced, then we should accept that spontaneous and induced cancers share the same or similar mechanistic steps. This implies that exposure to environmental carcinogens can add to and speed up a process already spontaneously initiated, as well as initiate it de novo.

The interaction of multiple factors in carcinogenesis was recently reviewed by Falk[8] and Bigham et al[9]. Exposure of an individual to a carcinogenic agent can be modified by genetic determinants, such as enzyme levels, immune defense capacity, and hormonal imbalance; cultural influences, such as smoking, diet, and other environmental factors; and health status. As examples of such modification, smoking synergistically increases lung cancer incidence in asbestos workers and uranium miners.

Species and strain differences illustrate the major role of genetic background in determining susceptibility to carcinogenesis. One major source of variation is the known difference in activity of the drug-metabolizing enzymes among different species which can account for differences of carcinogenic potency, if the proximate carcinogen is metabolized to an active product, or if an active carcinogen has to reach a remote site of action before it is transformed to an inert metabolite.

There may in addition to species differences in the metabolism of compounds to active carcinogens, intrinsic differences in sensitivity of tissues to the carcinogenic process.

The influence of genetic constitution is revealed in marked strain differences within a single species. The pattern of genetic variation in susceptibility to cancer in general, susceptiblity to cancer of particular tissues, and susceptibility to individual chemical carcinogens, may well also apply in the human.

While classical epidemiologic studies have shown that a number of occupations increase the risk of cancer in man, most human cancers are related to man's life style, especially in terms of what he smokes, eats and drinks[10].

While excessive tobacco usage has been shown to be a major contributor to cancer, particularly among men and increasingly among women, nutrition has only recently been viewed in this light[10,11]. Nutrition could effect the incidence of cancer in a number of ways including specific carcinogens, either naturally occurring in food or present as additives. For example, mycotoxins and cycasin have been implicated in liver cancer, nitrate, nitrite and nitrosamines. With the possible exception of aflatoxin and related mycotoxins which have been shown to play a significant role in the incidence of liver cancer in some parts of the world[12], these carcinogens are believed to make a relatively minor contribution to the total incidence of human cancer[10].

Deficiences of specific nutritional substances might also contribute to the transformation of normal cells into neoplastic cells; e.g., alcoholism and riboflavin deficiency with cancer of the upper gastrointestinal tract; vitamin A deficiency and low-fat intake with cancer of the cervix and stomach and pyridoxine deficiency with liver cancer[10]. Wynder[10] believes that the effect of alcohol on cancer of the esophagus, oral cavity, and larnyx relates to the nutritional deficiencies commonly associated with alcoholism, a condition which appears to make the epithelia cells more susceptible to tobacco carcinogens.

Wynder[10] has also stressed the role of overnutrition in the etiology of cancer. (Overnutrition as referred to by Wynder does not pertain to obesity, but instead refers to the excessive intake of specific components such as dietary fat and cholesterol).

The major differences in incidence rates between the Japanese and American populations for cancer of the colon, breast, and prostrate can be best explained by significant differences in the consumption of fat and fat-related variables[13]. It is also important to note that increased consumption in Japan of fat and fat-related products is paralleled by an increased incidence of cancer of the colon, breast and prostrate[14]. The possible environmental influence of nutrition has also been suggested[10] to be a factor to account for the marked difference which exist in the incidence of breast cancer between American and Japanese populations[15,16]. The largest differences exist in the area of dietary fat and cholesterol intake, principally from meat and dairy products. These broad-based leads from epidemiology have been apparently supported by experimental studies. For example, both spontaneous breast tumors in mice and chemically induced tumors in rats have been accelerated by a high-fat diet[17].

Modan[18] views the current evidence for the involvement of diet in cancer etiology to be based on indirect relationships between the consumption of selected food constituents and incidence, dietary studies, and laboratory data. The indirect evidence most often referred to is the suggested correlation between the complex of fats-meat-egg-animal protein and the risk for cancer of the colon. Such observations are, however, hampered by the fact that human diet does not consist of isolated food components. Case control studies implicate a higher intake of starchy foods in gastric cancer, a lower intake of fiber in colon cancer, and possibly coffee in renal cancer. Carcinogenic agents identified include food additives, plant toxicants, aflatoxins, polycyclic hydrocarbons, nitrosamines, and certain normal major food constituents. The experimental evidence is augmented by studies indicating an inter-relationship between dietary constituents, intestinal flora, and bile acid metabolism. A synergistic action of ingested or metabolized carcinogens and a co-carcinogenic function of certain dietary components are suggested[18].

One of the most significant advances in understanding the mechanism of carcinogenesis was the discovery[19] that the chemical induction of cancer involved two distinguishable processes, designated initiation and promotion. Initiation is the production of an irreversible cellular change, which is a necessary but not a sufficient condition for the development of cancer. Promotion is the process whereby a tumor is caused to develop in tissue in which initiation has alrady occurred. Promotion is thought to be "reversible" while initiation probably is not. Promoters work only at sustained and comparatively high doses, while initiators may need only "one hit" at a relatively low dose to produce cancer.

There have been a number of recent reviews on tumor promoters, including those of Van Duuren[20-22], Hecker[23] and Berenblum[24]. The term "tumor promoter" is used at times interchangeably with co-carcinogen, and accelerating agent. However, there is a clearcut distinction in the method of biological testing for tumor promoters and for co-carcinogens[20] which may apply to their mode of action.

There have been many studies concerning dose-response of initiator with constant dose of croton oil and vice-versa, the high frequency of tumor regression with croton oil as promoter, the role of mouse-skin inflammatory responses in tumor promotion and the induction of tumors by croton oil alone[20-24].

The various active compounds from croton oil are diesters of phorbol and were found to be potent tumor-promoting agents[20,25,26]. Phorbol-phorbol myristate acetate (PMA) (I) is the most active agent known in two-stage carcinogenesis experiments on mouse skin[20-22].

At present, it is not possible to define initiating and promoting agents which are involved in two-stage carcinogenesis in terms of their mechanism of action. Two-stage carcinogenesis on mouse skin, for example, involves an initiating agent and promoting agent. An initiator or initiating agent is one which results in the induction of benign and malignant tumors when applied on mouse skin in a single dose followed by repeated applications of a promoting agent[20].

Many initiating agents used in two-stage carcinogenesis are carcinogenic (e.g., the aromatic hydrocarbons, benzo(a)pyrene and 7,12-dimethylbenz(a)anthracene) although a number of non-carcinogenic mouse skin initiating agents are also known (e.g., dibenz(a,c)anthracene, chrysene, urethane). Not all chemical carcinogens are initiating agents. There are usually carcinogens which have to be metabolized in vivo to activated carcinogenic intermediates. A promoting agent is one which is applied repeatedly after a single dose of an initiating agent and results in benign and malignant tumors[20]. All known promoting agents also have weak tumorigenic activity[21]. The two terms initiator and tumor promoter are mechanistically and temporally inter-related and cannot be separated[20]. When the sequence of applications is reversed, few if any tumors are noted[20]. In co-carcinogenesis experiments, two agents are administered simultaneously and usually repeatedly and result in significantly higher tumor incidences than either agent alone. Hence, there is a operational difference and probably a real difference in mode of action since not all co-carcinogens are tumor promoters and vice-versa[20,27].

Co-carcinogenic response is not limited to one tissue or one species. Promotion by croton oil is not limited to polycyclic hydrocarbons as initiators. For example, urethane application to the skin only produced skin tumors on croton oil treatment[28]; a single intraperitoneal injection of urethane resulted in the development of skin tumors, provided the skin was painted repeatedly with croton oil[29]. Ritchie and Saffiotti[30] reported that oral administration of acetylaminofluorene for 50 days caused skin tumor development in mice, following 41 weeks of painting with croton oil.

Aliphatic hydrocarbons are ubiquitous in the environment and may possess good promoting activity[8,31-33] for polycyclic hydrocarbons. Simple phenols and chlorinated phenols are also promoters for polycyclic hydrocarbon carcinogenesis on mouse skin[34].

Cigarette smoke inhalation has been considered as a potent promoter for man[8,35,36]. Selikoff found that cigarette smoking workers in the asbestos industry had 100 times the risk of bronchogenic carcinoma, compared to their non-smoking fellow-workers[35]. Similar, but somewhat less convincing observations were made by Evans[36] regarding the increased lung cancer risk in smoking uranium miners.

The promoting effect by carbon tetrachloride on hepatoma formation[37] and oral tumors in mice[38] by benzo(a)pyrene; the increase in hepatomas in female mice by 2-acetylaminofluorene (2-AAF)[39] and dimethylaminobenzene (DAB)[40] have all been reported.

The polychlorinated biphenyl (PCB) Arochlor 1254 has been considered to be a promoter of carcinogenesis. The administration of the PCB to rats which previously had been treated with a known carcinogen, enhanced the carcinogenic response[41].

Enzyme induction is a very important factor influencing carcinogenesis by many precarcinogens. Depending upon the nature of the inducer, the nature of the precarcinogen, the responsiveness of the target tissue, and the nutritional adequacy of the diet, a number of variations in carcinogenic response can be possibly explained[8].

Organochlorine pesticides (e.g., DDT, DDE, DDD, chlordane, methoxychlor, endrin, aldrin, dieldrin, hexachlorocyclohexane, heptachlor and its epoxide) have been found to be very effective inducers[42,43].

Of the polycyclic hydrocarbons, methyl cholanthrene and benz(a)anthracene were the most effective inducers while benzo(a)pyrene (BaP) and chrysene were decidedly less; anthracene, fluoranthene, pyrene, perylene, phenanthrene, fluorene and naphthalene required very large doses for a weak response or were entirely inactive[44,45].

Many drugs are potent enzyme inducers, e.g., analgesics, hypnotics, hypoglycemics, antihistamines, anticovulsants, and anti-inflammatory agents[46]. For example, phenobarbital is frequently employed as an inducer and has been demonstrated to be a potent epoxide hydrase and glutathione-S-transferase inducer[47].

Cigarette smoking increased benzpyrene hydroxylase and 3'-methyl-4-monomethyl-aminoazobenzene N-demethylase activity of human placenta, while in non-smokers no such activity was detected[44]. Aromatic polycyclic hydrocarbons were suggested as responsible for the induction[44], but nicotine may well be the probable inducer[8,48] resulting from cigarette smoking chemicals belonging to other classes have also been shown to be enzyme inducers. These include steroid hormones[40], flavones[50] and among the food additives, butylated hydroxytoluene (BHT)[51].

In addition to the liver, other organs such as the lung, adrenal and mammary gland of rats[52], rat and the rabbit kidney[53], skin of mice[54], are capable of enzyme induction. Aryl hydroxylase activity was demonstrated for the duodenal mucosa of laboratory animals and man[54].

The list of inducible enzymes also includes the following: some N-, O-, and S-oxidative demethylases; o- and p-aromatic ring hydroxylases, S-gluathione transferase and glucuronide transferase, a number of which are very substrate specific[8]. Conditions for maximal enzyme induction may not also be encountered[8]; e.g., in absence of inducers and in protein-free diets[55,56]. The dietary influence on hepatic microsomal enzyme activity in relation to the mutagenicity of N-nitroso compounds for S. typhimurium in the host-mediated assay has been elaborated by Zeiger[57].

A number of studies have revealed alterations in tumor incidence resulting from interactions for which enzyme induction with resulting enhanced detoxification was the most likely explanation[8,48]. For example, simultaneous feeding of an azo dye and methyl cholanthrene produced a far lower incidence of hepatomas and bile duct carcinomas in rats than feeding the azo dye alone[58].

Antagonism has been observed with polycyclic hydrocarbons when a weak and a strong carcinogen, benz(a)anthracene and dibenz(a,b)anthracene were injected subcutaneously into mice. The tumor yield was that expected of the weak carcinogen only[59].

There is a paucity of data concerning the significance of repair inhibitors on human susceptibility to carcinogenicity and mutagenicity, except where repair is inhibited on a genetic basis[8].

While most of the changes in DNA brought about many mutagens can apparently be remedied by the work of several repair enzymes, it has become apparent that physical or chemical factors can be encountered which can interfere with these enzyme activities[60-62].

The inhibition of tumor induction has been reviewed recently[63,64]. Agents that have been found to inhibit two-stage carcinogenesis include: protease inhibitors (e.g., toxyl lysine chloromethyl ketone); anti-inflammatory steroid hormones (e.g., dexamethasone); and sulfur mustard (bis(β-chloroethyl)sulfide). The mechanism of action of these agents remains largely unknown.

Numerous studies have utilized natural metabolites and synthetic analogs of vitamin A ("retinoids") for prevention and therapy of benign and malignant tumors[65]. Retinoids are capable of retarding the appearance, reducing the incidence, or inhibiting the in vivo growth of chemically induced tumors of the skin[66-68], lungs[69,70] intestines[71], mammary glands[72,73] and urinary bladder[74] and virally induced[75,76] tumors. Also, the growth of transplantable chondrosarcomas in rats fed retinoic acid[77-79] and of mammary adenocarcinomas in mice fed retinyl palmitate was reduced[80], and the development of tumors by S91 melanoma cells injected into mice pretreated with retinyl palmitate was inhibited[81]. However, in other studies tumor development was not inhibited by orally or ip administered retinoic acid[82].

Newberne[83,84] reported that a defiency of dietary vitamin A can increase susceptibility to dimethylhydrazine (DMH) induced colon tumors in rats, resulting in colon tumors as well as liver tumors in rats exposed to aflatoxin B_1 and that the retinoid, 13-cis-retinoic acid can inhibit colon tumor induction by DMH. These findings further substantiate the importance of adequate vitamin A and the potential for therapeutic use of B-cis-retinoic acid.

The inhibition of chemical carcinogenesis by additional classes of agents such as phenolic antioxidants (e.g., butylated hydroxyanisole (BHA), butylated hydroxytoluene (BHT) and ethoxyquin), disulfiram and related compounds, organic isothiocyanates (e.g., benzyl-, phenethyl-, 1-, and 2-naphthylisothiocyanates) and thiocyanates, lactones (coumarin), selenium salts, and inducers of microsomal mixed-function oxidases (e.g., polycyclic hydrocarbons, PCBs, phenobarbital phenothiazines) has recently been reviewed by Wattenberg[85].

BHA is of particular interest because of its extensive use as an additive in food for human consumption. In the U.S., the human consumption of BHA is estimated to be of the order of several milligrams a day[85]. It was shown that at a concentration of 5 mg BHA/g diet, the carcinogenic effect of 1 mg benzo(a)pyrene/g diet on the forestomach of the mouse is inhibited[86]. If one assumes that the results of animal experiments hold for man, then this amount of antioxidant could be of importance in the inhibition of the effects of chronic exposure to low doses of carcinogens, the type of exposure that is most likely to occur in human populations.

An important aspect of the role of inhibitors is a consideration of toxicity. Inhibitors currently identified have, to a greater or lesser degree, other biological activities. A number have toxic properties, and some are even carcinogens or co-carcinogens. Also of critical importance is the need for a full elucidation of mechanisms of inhibition. Such an understanding could thus provide information that could aid in the elaboration of compounds in the environment that are inhibitors. Additionally, mechanisms that involve measurable biochemical parameters could provide a basis for assessment of the susceptibility of particular population groups or individuals to neoplasia from chemical carcinogens.

References

1. Searle, C. E. (ed) "Chemical Carcinogens", ACS Monograph No. 173, American Chemical Society, Washington, DC (1976) ix-xxvii
2. Clayson, D. B., Chemical Carcinogenesis, J. and A., Churchill, Ltd., London (1962)
3. Miller, E. C. and Miller, J. A., A survey of molecular aspects of chemical carcinogenesis, Carcinogenesis, 15 (1) (1966) 217-235
4. Christensen, H. E., Luginbyhl, T. T., (eds), Suspected Carcinogens: A Subfile of the NIOSH Toxic Substances List, U. S. DHEW, PHS, Center for Disease Control, National Institute for Occupational Safety and Health, Rockville, MD, June (1975)
5. NIOSH, Carcinogens-Regulation and Control, National Institute for Occupational Safety and Health, Rockville, Maryland, (1977)
6. Ambrose, E. J., and Roe, F. J. C. (eds), The Biology of Cancer, D. Van Nostrand Co., Ltd., London (1966)
7. Tomatis, L., Extrapolation of test results to man, Inserm Symposia, Series No. 52 (1976) 261-262, IARC Publ. No 13, International Agency for Research on Cancer, Lyon France (1976)
8. Falk, H. L., Possible mechanisms of combination effects in chemical carcinogenesis, Oncology, 33 (1976) 77-89
9. Bingham, E., Niemeier, R. W., and Reid, J. B., Multiple factors in carcinogenesis, Ann. NY Acad. Sci., 271 (1976) 14-21
10. Wynder, E. L., The dietary environment and cancer, J. Am. Dietetic Assoc., 71 (1977) 385-391
11. Symposium: Nutrition in the causation of cancer, Cancer Res., 35 (11) Pt. 2 (Nov.) (1975)
12. Wogan, G. N., Dietary factors and special epidemiological situations of liver cancer in Thailand and Africa, Cancer Res., 35 (1977) 3499
13. Oiso, T., Incidence of stomach cancer and its relation to dietary habits and nutrition in Japan between 1900 and 1975, Cancer Res., 35 (1975) 3254
14. Hirayama, T., Prospective studies on cancer epidemiology based on census populations in Japan, Excerpta Med., 3 (1975) 26
15. MacMahon, B., Cole, P., and Brown, J., Etiology of human breast cancer: A review, J. Natl. Cancer Inst., 50 (1973) 21
16. Wynder, E. L., Bross, I. J., and Hirayama, T., A study of the epidemiology of cancer of the breast, Cancer, 13 (1960) 559
17. Chan, P. C., and Cohen, L. A., Effect of dietary fat antiestrogen, and antiprolactin on the development of mammary tumors in rats, J. Natl. Cancer Inst., 52 (1974) 25
18. Modan, B., Role of diet in cancer etiology, Cancer, 40 (1977) 1887-1891
19. Berenblum, I., and Shubik, P., A new quantitative approach to the study of the stages of chemical carcinogenesis in the mouses' skin, Brit. J. Cancer, 1 (1948) 383
20. Van Duuren, B. L., Tumor-promoting and co-carcinogenic agents in chemical carcinogenesis, In "Chemical Carcinogenesis" ed. Searle, C. E., ACS Monograph No. 173, American Chemical Society, Washington, DC (1976) pp. 24-51
21. Van Duuren, B. L., Tumor-promoting agents in two-stage carcinogenesis, Prog. Exp. Tumor Res., 11 (1969) 31-68
22. Van Duuren, B. L., and Sivak, A., Tumor-promoting agents from Croton tiglium L and their mode of action, Cancer Res., 28 (1968) 2349=2356
23. Hecker, E., Cocarcinogens from Euphorbiaceae and Thymelaeceae, IN: Symposium on Pharmacognosy and Phytochemistry, Springer Verlag, Berlin/New Yor, (1971) pp. 147-165
24. Berenblum, I., A re-evaluation of the concept of cocarcinogenesis, Prog. Exp. Tumor Res., 11 (1969) 21-30

25. Van Duuren, B. L., and Orris, L., The tumor-enhancing principles of Croton Tiglium L, Cancer Res., 25 (1965) 1871-1875
26. Hecker, E., Kubinyi, H., and Bresch, H., A new group of co-carcinogens from croton oil, Angewchem. Int. Ed., 3 (1964) 747-748
27. Van Duuren, B. L., Blazesj, T., Goldschmidt, B. M., Katz, C., Melchionne, S., and Sivak, A., Cocarcinogenesis studies in mouse skin and inhibition tumor induction, J. Natl. Cancer Inst., 46 (1971) 1039-1044
28. Berenblum, I., and Haran, N., The initiating action of ethyl carbamate (urethane) on mouse skin, Brit. J. Cancer, 9 (1955) 453-456
29. Ritchie, A. C., Epidermal carcinogenesis in the mouse by intraperitoneally administered urethane followed by repeated application of croton oil, Brit. J. Cancer, 11 (1957) 206-211
30. Ritchie, A. C., and Saffiotti, U., Orally administered 2-acetylaminofluorene as an initiator and as a promoter in epidermal carcinogenesis in the mouse, Cancer Res., 15 (1955) 84-88
31. Horton, A. W., Carcinogenesis of the skin II. The accelerating properties of aliphatic and related hydrocarbons, Cancer Res., 17 (1957) 758-766
32. Bingham, E. and Falk, H.L., Environmental carcinogens: The modifying effect of cocarcinogens on the threshold response, Arch. Environ. Hlth., 19 (1969) 779-783
33. Sice, J., Tumor-promoting activity of n-alkanes and 1-alkanols, Toxicol. Appl. Pharmacol., 9 (1966) 70-74
34. Boutwell, R. K., and Bosch, D. K., The tumor-promoting action of phenol and related compounds for mouse skin, Cancer Res., 19 (1959) 413-424
35. Selikoff, I. J., Hammond, C., Churg, J., Asbestos exposure, smoking and neoplasia, J. Amer. Med. Assoc., 204 (1968) 106-112
36. Evans, R. D., Congressional record-radiation exposure of uranium workers, 19th Congr. Part I (1967) 650-661
37. Kotin, P., et al., Effect of carbon tetrachloride intoxication on metabolism of benzo(a)pyrene in rats and mice, J. Natl. Cancer Inst., 28 (1962) 725-742
38. Proetzel, M., Giardina, A. C., and Albano, E. H., The effect of liver inbalance on the development of oral tumors in mice following the application of benzpyrene or tobacco tar, Oral Surg., 18 (1964) 622-635
39. Weisburger, J. H., Hadidan, Z., Fredrickson, T. N., and Weisburger, E. K., Carcinogenicity simultaneous action of several agents, Toxicol. Appl. Pharm., 7 1965, 502
40. Ueda, U., The relation between experimental liver carcinoma and liver cirrhosis induced by simultaneous administration of p-dimethylaminoazobenzene and carbon tetrachloride, Tohoku J. Exp. Med., 92 (1967) 83-107
41. Anon, PCB is a cancer "promoter" according to NCI Clearinghouse Use Subgroup. Pesticide & Toxic Chem. News, 6 (1977) 3-4
42. Fouts, J. R., Factors influencing the metabolism of drugs in liver microsomes, Ann. NY Acad. Sci., 104 (1963) 875
43. Hart, L. G., Fouts, J. R., Further studies on the stimulation of hepatic microsomal drug metabolizing enzymes by DDT and its analogs, Arch. Exp. Path. Pharmakol., 249 (1965) 486
44. Welch, R. M., Harrison, Y. E., Gommi, B. W., Poppers, P. J., Finster, M., Conney, A. H., Stimulatory effect of cigarette smoking on the hydroxylation of 3,4-benzpyrene and N-demethylation of 3'-methyl-4-monomethyl aminoazobenzene by enzymes in human placenta, Clin. Pharmacol. Therap., 10 (1969) 100
45. Miller, E. C., et al., On the protective action of certain polycyclic aromatic hydrocarbons against carcinogenesis by amino azo dyes and 2-acetylaminofluorene, Cancer Res., 18 (1958) 469

46. Conney, A. H., Implications of drug induced changes in microsomal enzymes to toxicity of drugs; in Drugs and Enzymes, Vol. 4., Macmillian New York (1965) pp. 277-296
47. Breshnik, K., Personal communication (1976) in Falk, H. L., Possible mechanisms of combination effects in chemical carcinogenesis, Oncology, 33 (1976) 77-85
48. Wenzel, D. G., and Broadie, L. L., Stimulatory effect of nicotine on the metabolism of meprobamate, Toxicol. Appl. Pharmacol., 8 (1966) 455
49. Adlercreutz, H., and Tenhunen, R., Some aspects of the interaction between natural and synthetic female sex hormones and the liver, Am. J. Med., 49 (1970) 630
50. Wattenberg, L. W., Page, M. A., Leong, J. L., Induction of increased benzpyrene hydroxylase activity by flavones and related compounds, Cancer Res., 28 (1968) 934
51. Gilbert, D., and Golberg, L., Liver response tests, III. Liver enlargement and stimulation of microsomal processing enzyme activity, Food Cosmet. Toxicol., 3 (1965) 417
52. Huggins, C., and Fukunishi, R., Induced protection of adrenal cortex against 7,12-dimethylbenz(a)anthracene, J. Exp. Med., 119 (1964) 923
53. Uehleke, H., Extrahepatic microsomal drug metabolism in sensitization to drugs, Proc. Europ. Society Study Drug Toxicity, 10 (1969) 94
54. Wattenberg, L. W., Leon, J. L., Strand, P. J., Benzpyrene hydroxylase activity in the gastrointestinal tract, Cancer Res., 22 (1962) 1120
55. Judah, J. D., McLean, A. E., and McLean, E. K., Biochemical mechanisms of liver injury, Am. J. Med., 49 (1970) 609
56. McLean, A. E. M., The effect of protein deficiency and microsomal enzyme induction by DDT and phenobarbitone on the acute toxicity of chloroform and a pyrrolizidine alkaloid, retrosine, Brit. J. Exp. Pathol., 51 (1970) 317-321
57. Zeiger, E., Dietary effects on the mutagenicity of N-nitroso compounds in the host-mediated assay, 1st Int. Conf. Environmental Mutagens, Pacific Grove, CA (1973)
58. Richardson, H. L., Stier, A. R., and Borsosnachtnebel, E., Liver tumor inhibition and adrenal histological responses in rats to which 3'-methyl-4-dimethylaminoazobenzene and 20-methylcholanthrene were simultaneously administered, Cancer Res., 12 (1952) 356
59. Steiner P. E., and Falk, H. L., Summation and inhibition effects of weak and strong carcinogenic hydrocarbons: 1,2-benzanthracene, chrysens, 1,2,5,6-dibenzanthracene and 20-methylcholanthrene, Cancer Res., 11 (1951) 56-63
60. Kilbey, B. J., Allele-specific responses to factors that modify U.V. mutagenesis, In: "Mutational Cellular Process" eds. Wolsteinholme and O'Conner, Churchill, London (1969) pp. 50-62
61. Witking, E. M., Modification of mutagenesis initiated by ultraviolet light through post-treatment of bacteria with basic dyes, J. Cell. Comp. Physiol., 58 (1961) 135-144
62. Pazmino, N. H., and Yuhas, J. M., Chloroquine: Nonselective inhibition of recovery from radiation injury in tumors and normal tissues, Radiat. Res., 60 (1974) 54-61
63. Falk, H. L., Anticarcinogenesis-An alternative, Progr. Exp. Tumor Res., 14 (1971) 105-137
64. Van Duuren, B. L., and Melchionne, S., Inhibition of tumorigenesis, Progr. Exp. Tumor Res., 12 (1969) 55-101
65. Sporn, M. B., Dunlop, N. M., Newton, D. L., et al., Prevention of chemical carcinogenesis by vitamin A and its synthetic analogs (retinoids), Fed. Proc., 35 (1976) 1332-1338
66. Davies, R. E., Effect of vitamin A on 7,12-dimethylbenz(a)anthracene-induced papillomas in Rhino mouse skin, Cancer Res., 27 (1967) 237-241

67. Bollag, W., Prophylasis of chemically induced benign and malignant epithelial tumors by vitamin A acid (retinoic acid), Eur. J. Cancer, 8 (1972) 689-693
68. Bollag, W., Therapeutic effects of an aromatic acid analog on chemical induced skin papillomas and carcinomas of mice. Eur. J. Cancer, 10 (1974) 731-737
69. Saffiotti, U., Montesano, R., Sellakumar, R., et al., Experimental cancer of the lung. Inhibition of vitamin A of the induction of tracheobronchial squamous metaplasia and squamous cell tumors, Cancer, 20 (1967) 857-864
70. Nettesheim, P., Cone, M. V., Snyder, C., The influence of retinyl acetate on the postinitiation phase of preneoplastic lung nodules in rats, Cancer Res., 36 (1976) 996-1002
71. Chu, E. W., Malmgren, R. A., An inhibitory effect of vitamin A on the induction of tumors of forestomach and cervix in the Syrian hamster by carcinogenic polycyclic hydrocarbons, Cancer Res., 25 (1965) 884-915
72. Moon, R. C., Grubbs, C. J., Sporn, M. B., Inhibition of 7,12-dimethylbenz(a)-anthracene-induced mammary carcinogenesis by retinyl acetate, Cancer Res., (1976) 2626-2630
73. Grubbs, C. J., Noon, R. C., Sporn, M. B., et al., Inhibition of mammary cancer by retinyl methyl ether, Cancer Res., 37 (1977) 599-602
74. Sporn, M. B., Squire, R. A., Brown, C. C., et al., 13-cis-Retinoic acid inhibition of bladder carcinogenesis in the rat, Science, 195 (1976) 487-489
75. McMichael, H., Inhibition of growth of Shope rabbit papilloma by hypervitaminosis A, Cancer Res., 25 (1965) 947-952
76. Seifter, E., Rettura, G., Padawar, J., et al., Antipyretic and antiviral action of vitamin A in Moloney sarcoma virus- and poxvirus-inoculated mice. J. Natl. Cancer Inst., 57 (1976) 355-359
77. Hellman, C., Swarm, R. L., Effects of thirteen-cis-vitamin A acid on chondrosarcoma, Fed. Proc., 34 (1975) 822
78. Shapiro, S. S., Bishop, M., Poon, J. P., et al., Effect of aromatic retinoids on rat chondrosarcoma glycosaminoglycan bisynthesis, Cancer Res., 36 (1976) 3702-3706
79. Trown, P. W., Buck, M. J., Hansen, R., Inhibition of growth and progression of a transplantable rat chondrosarcoma by three retinoids, Cancer Treat Rep., 60 (1976) 1647-1653
80. Rettura, G., Schitter, A., Hardy, M., et al., Antitumor action of vitamin A in mice inoculated with adenocarcinoma cells, J. Natl. Cancer Inst., 54 (1975) 1489-1491
81. Felix, E. L., Foyd, B., Cohen, M. H., Inhibition of the growth and development of a transplantable murine melanoma by vitamin A, Science, 189 (1975) 886-888
82. Bollag, W., Effects of vitamin A acid on transplantable and chemically induced tumors, Cancer Chemother Reo., 55 (1971) 53-55
83. Newberne, P. M., and Suphakarn, V., Preventive role of vitamin A in colon carcinogenesis in rats, Cancer, 40 (1977) 2553-2556
84. Newberne, P. M., and Rogers, A. E., Nutritional modulation of carcinogenesis In Fundamentals in Cancer Prevention (eds) Magee, P. N., Takayama, S., Sugamura, T., and Matsushima, T., University of Tokyo Press, Tokyo (1976) pp. 15-40
85. Wattenberg, L. W., Inhibition of chemical carcinogenesis, J. Natl. Cancer Inst., 60 (1978) 11-17
86. Wattenberg, L. W., Inhibition of carcinogenic and toxic effects of polycyclic hydrocarbons by phenolic antioxidants and ethoxyquin, J. Natl. Cancer Inst. 48 (1972) 1425-1430

CHAPTER 3

ASPECTS OF EPIDEMIOLOGY, RISK-ASSESSMENT AND "THRESHOLD DOSE"

Carcinogenesis and mutagenesis etiology involves the interplay of genetic and environmental factors. Wide variations in cancer patterns have been correlated with geographical differences on the national and international level[1]. Changes in incidence of cancers at one site were in general unrelated to variations at other sites suggesting that separate etiological factors were involved[2]. These observations have led epidemiologists to infer that probably the major etiologic component is environmental[1-3]. Under this aegis are included both chemical and irradiation exposures.

Higginson[3] classified the environment into two components, viz., (a) macro- which applies to the individual's total general environmental (e.g., air and water pollution, general food supplies, etc.) which the individual cannot usually significantly modify and (b) the micro-environment which refer to the personal environment created by the individual and includes cultural habits (e.g., cigarette smoking and individual drinking and eating habits). It can also include occupation where the degree of individual control varies according to local political and socio-economic conditions.

In addition to site, cancer may be classified according to etiology[4]. Cancers can be due to factors including: (a) the cultural environment-this accounts for the most important cancers of known etiology. For example, cancer of the lung due to cigarette smoking accounts for approximately 40% of cancer deaths in males in the United Kingdom or 11% of all deaths. Excessive ingestion of alcoholic beverages has been shown to lead to cancer of the mouth, esophagus and liver; (b) occupational environment- although a relatively small proportion of all cancers (e.g., 1-3%) has been attributed to occupational environment, this group has been suggested by Higginson[3] to be important since control and prevention may be possible, e.g., bladder cancer in the rubber industry[5]. It was stressed that what may be initially an occupational hazard with a high potential high risk may eventually involve large population groups at lower levels[3]. Hence, the existence of high risk groups has been of importance in identifying the actual risk in man. Other occupational risks identified include: lung cancer in workers exposed to bis(chloromethylether) or chromium; leukemia in benzene workers; angiosarcoma of the liver in PVC production workers; and nasal cancer in nickel workers or woodworkers; (c) iatrogenic factors-until recently, cancers of this type were limited to certain specific agents, e.g., nitrogen mustards, chlornaphazine and radiation. However, increasing long-term use of certain drugs (e.g., tranquillizers and contraceptives) in healthy persons has given rise to concern regarding potential hazards[3]; (d) idiopathic cancers-at present, the etiology of approximately only 40% (excluding skin) of human cancers has been identified[3]. Of the remainder, between half and two-thirds and possibly more are believed[2,6,7] to be related directly or indirectly to environmental factors. Higginson[3] stressed the fact that the cancer patterns of today are dependent on agents entering the environment over 20 years ago, and that those new agents at present giving rise to the greatest disquiet are unlikely to be involved.

Epidemiologic studies seek to identify the extent to which particular agents contribute to causation of particular types of cancer. Methods necessarily involve indirect observations. These include both prospective and retrospective analyses of potential cause-effect relationships through cohort and case control studies, in addition to analyses correlating geographic and temperal trends with selected environmental features. Modes of exposure to chemicals include diet, personal habits such as smoking, and external environmental air and water pollution. While epidemiologic studies will continue to be an essential means for monitoring potential human risks, it is recognized that the long latent periods involved in human carcinogenesis severely limit the usefulness of such approaches.

As Shubik and Clayson[8] have stated the epidemiological discovery of a carcinogen in man often does not follow until the substance has been in the environment for as long as 20-40 years, during which time hundreds, thousands or even millions of people may have been exposed and are at potential risk from its effects.

While it was postulated that a number of investigators believe that the overwhelming majority of human cancers are due to environmental factors, it then follows that environmental carcinogens must be common or even ubiquitous. Hammond[9], as well as others[10] have attempted to resolve the dichotomy of why some people get cancer while others do not, if carcinogens are indeed common. It is as important to know who gets cancer and why, as it is to know what environmental agents induce neoplasia in man[11]. If as much as 85% of human cancer can be caused by environmental factors, as Hammond has suggested[9], why don't monitoring systems detect more time/space clustering and why can't we identify the underlying carcinogen?[10]

Hammond[9] pointed out that man has evolved in an environment which presents a high level and great variety of carcinogenic potential. Cancer incidence is uniformly low until one reaches the end of the reproductive period, at which time cancer incidence begins an almost geometric increase during the remaining years of life[9]. An explanation for these age-related rates is the existence of extended latency periods between initial exposure to environmental carcinogens and the onset of disease. Another explanation offered by Hammond[9] lies in the development of genetically transmitted defense mechanisms which have evolved through natural selection.

Given the apparently complex situation in which the environment may be heavily laden with both natural and synthetic carcinogens, what are the epidemiological strategies for identifying environmental carcinogens? Janerich and Lawrence[11] stress that epidemiology has not yet devised a fool-proof systematic substitute for the enlightened clinical observation which is usually the first way in which our attention has focused on that fact that a specific environmental agent can cause some type of cancer. They further state that while epidemiology is the essential tool for evaluation of these associations, only rarely does the first recognition of the associations come from a systematic epidemiological study. A case in point is the newly recognized association between vinyl chloride and angiosarcoma of the liver based on the original observation of Creech and Johnson[12] in 1974 of this rare form of cancer from one plant manufacturing polyvinylchloride. Selikoff[13] later provided an initial report on the study of a cohort of 257 PVC workers in which 3 deaths from angiosarcoma of the liver were noted. The PVC issue is still at a comparatively early stage and the essential epidemiological studies are barely under way. The limits of risk latency period and many other important parameters remain to be unambigously defined.

IARC (International Agency for Research on Cancer) recently summarized the evidence of carcinogenicity in humans to be derived from three types of study, the first two of which usually provide only suggestive evidence: (1) reports concerning particular cancer patients (case reports) with a history of exposure to the supposed carcinogenic agent; (2) descriptive epidemiological studies in which the incidence of cancer in human populations is found to vary (spatiatly or temporally) with exposure to the agent; and (3) analytical epidemiological studies (e.g., case-control or cohort studies) in which individual exposure to the agent is found to be associated with an increased risk of cancer[14].

Experience with human cancers of known etiology suggests that the period from first exposure to a chemical carcinogen to development of clinically observed cancer is usually measured in decades and may be in excess of 30 years[14].

IARC[14] contends that while no adequate criteria are presently available to interpret experimental carcinogenicity data directly in terms of the carcinogenic potential for humans, positive extrapolations to human risk can be reasonably approximated utilizing data collected from appropriate animal tests.

As a result of epidemiological studies, between 20 and 50 chemicals, or mixtures, of substances are known to induce cancer in man. According to Shubik and Clayson[8] the precise number depends on the level of evidence deemed necessary to incriminate a chemical of carcinogenicity to man. This information (Table 1) has been assembled as a result of the consequences of occupational exposure to specific substances, the use of drugs, or natural food contaminants, and personal habits.

Information compiled from the first 16 IARC monographs[15,16] published during the period 1972-1977 reveals that of about 25 chemicals or manufacturing processes now generally accepted as causing cancer in humans, all but possibly two (arsenic and benzene) of those which have been tested appropriately have been shown to be carcinogenic in at least one animal species[14]. It should be noted that for several compounds (e.g., 4-aminobiphenyl, mustard gas, vinyl chloride, diethylstilbestrol, and aflatoxins) evidence of carcinogenicity in experimental animals preceded evidence obtained from epidemiological studies or case reports. (See Table 2)

Table 1. Examples of Carcinogens in Man[8]

Occupational Exposure	Tumor Site
Soot, tar, pitch and oil	Skin, lung
Benzidine, 2-naphthylamine, 4-aminobiphenyl	Bladder
Asbestos	Lung, pleura
Vinyl chloride	Angiosarcoma
Drugs	
Chlornaphazine	Bladder
Diethylstilbestrol	Vagina (offspring)
Analgesics (abuse)	Renal pelvis
Natural Food Contaminant	
Aflatoxin	Liver
Habits	
Cigarette smoking	Lung, bladder, pancreas, etc.

Table 2

Chemicals or industrial processes associated with cancer induction in humans: comparison of target organs and main routes of exposure in animals and humans

Chemical or industrial process	Humans			Animals		
	Main type of exposure	Target organ	Main route of exposure	Animal	Target organ	Route of exposure
1. Aflatoxins	Environmental, occupational	Liver	p.o., inhalation[c]	Rat	Liver, stomach, colon, kidney	p.o.
				Fish, duck, marmoset, tree shrew, monkey	Liver	p.o.
				Rat	Liver, trachea	i.t.
					Liver	i.p.
				Mouse, rat	Local	s.c. injection
				Mouse	Lung	i.p.
2. 4-Aminobiphenyl	Occupational	Bladder	inhalation, skin, p.o.	Mouse, rabbit, dog	Bladder	p.o.
				Newborn mouse	Liver	s.c. injection
				Rat	Mammary gland, intestine	s.c. injection
3. Arsenic compounds	Occupational, medicinal, and environmental	Skin, lung, liver[c]	Inhalation, p.o., skin	Mouse, rat, dog	Inadequate, negative	p.o.
				Mouse	Inadequate, negative	Topical, i.v.
4. Asbestos	Occupational	Lung, pleural cavity, gastrointestinal tract	Inhalation p.o.	Mouse, rat, hamster, rabbit	Lung, pleura	Inhalation or i.t.
				Rat, hamster	Local	Intrapleural
				Rat	Local	i.p., s.c. injection
					Various sitet	p.o.
5. Auramine (munufacture of)	Occupational	Bladder	Inhalation, skin, p.o.	Mouse, rat	Liver	p.o.
				Rabbit, dog	Negative	p.o.
				Rat	Local, liver, intestine	s.c. injection
6. Benzene	Occupational	Hemopoietic system	Inhalation, skin	Mouse	Inadequate	Topical, s.c. injection
7. Benzidine	Occupational	Bladder	Inhalation, skin, p.o.	Mouse	Liver	s.c. injection
				Rat	Liver	p.o.
					Zymbal gland, liver, colon	s.c. injection
				Hamster	Liver	p.o.
				Dog	Bladder	p.o.
8. Bis(chloromethyl) ether	Occupational	Lung	Inhalation	Mouse, rat	Lung, nasal cavity	Inhalation
				Mouse	Skin	Topical
					Local, lung	s.c. injection
				Rat	Local	s.c. injection
9. Cadmium-using industreis (possibly cadmium oxide)	Occupational	Prostate, lung	Inhalation, p.o.	Rat	Local, testis	s.c. or i.m. injection
10. Chloramphenicol	Medicinal	Hemopoietic system	p.o., injection		No adequate	
11. Chloromethyl emthyl ether (possibly associated with bis(chloroemthyl) ether	Occupational	Lung	Inhalation	Mouse	Initiator	Skin
					Lung	Inhalation
					Local, lung	s.c. injection
				Rat	Local	s.c. injection
12. Chromium (chromate-producing industries)	Occupational	Lung, nasal cavities[c]	Inhalation	Mouse, rat	Local	s.c.,i.m. injection
				Rat	Lung	Intrabronchial implantation

Table 2 - Continued

	Humans			Animals		
Chemical or industrial process	Main type of exposure[a]	Target organ	Main route of exposure[b]	Animal	Target organ	Route of exposure
13. Cyclophosphamide	Medicinal	Bladder	p.o., injection	Mouse	Hemopoietic system, lung	i.p., s.c. injection
					Various sites	p.o.
				Rat	Bladder[c]	i.p.
					Mammary gland	i.p.
					Various sites	i.v.
14. Diethylstilbestrol	Medicinal	Uterus, vagina	p.o.	Mouse	Mammary	p.o.
				Mouse	Mammary, lymphoreticular, testis vagina	s.c. injection, s.c. implantation Local
				Rat	Mammary, hypophysis bladder	s.c. implantation
				Hamster	Kidney	s.c. injection, s.c. implantation
				Squirrel monkey	Uterine serosa	s.c. implantation
15. Hematite mining (? radon)	Occupational	Lung	Inhalation	Mouse, hamster, guinea pig	Negative	Inhalation, i.t.
				Rat	Negative	s.c. injection
16. Isopropyl oils	Occupational	Nasal cavity, larynx	Inhalation		No adequate tests	
17. Melphalan	Medicinal	Hemopoietic system	p.o., injection	Mouse	Initiator Ling, lymphosarcomas	Skin i.p.
				Rat	Local	i.p.
18. Mustard gas	Occupational	Lung, larynx	Inhalation	Mouse	Lung Local, mammary	inhalation, i.v. s.c. injection
19. 2-Naphthylamine	Occupational	Bladder	Inhalation, skin, p.o.	Hamster, dog, monkey	Bladder	p.o.
				Mouse	Liver, lung	s.c. injection
				Rat, rabbit	Inadequate	p.o.
20. Nickel (nickel refining)	Occupational	Nasal cavity, lung	Inhalation	Rat	Lung	Inhalation
				Mouse, rat, hamster	Local	s.c., i.m. injection
				Mouse, rat	Local	i.m. implantation
21. N,N-Bis(2-chloroethyl)-2-naphthylamine	Medicinal	Bladder	p.o.	Mouse	Lung	i.p.
				Rat	Local	s.c. injection
22. Oxymetholone	Medicinal	Liver	p.o.		No adequate tests	
23. Phenacetin	Medicinal	Kidney	p.o.		No adequate tests[d]	
24. Phenytoin	Medicinal	Lymphoreticular tissues	p.o., injection	Mouse	Lymphoreticular tissues	p.o., i.p.
25. Soot, tars, and oils	Occupational, environmental	Lung, skin (scrotum)	Inhalation, skin	Mouse, rabbit	Skin	Topical
25. Vinyl chloride	Occupational	Liver, brain[c] lung	Inhalation, skin	Mouse, rat	Lung, liver, blood vessels, mammary, Zymbal gland, kidney	Inhalation

[a] The main types of exposure mentioned are those by which the association has been demonstrated; exposures other than those mentioned may also occur
[b] The main routes of exposure given may not be the only ones by which such effects could occur.
[c] Indicative evidence
[d] The induction of tumors of the nasal cavities in rats given phenacetin has been reported recently (S. Odashima, personal communication, 1977).

It is generally acknowledged that animal experiments to test for the possible carcinogenic activity of chemicals provide the best and only deeply researched method for the detection of environmental carcinogens for man[8,14-18].

The fact that most known human carcinogens give tumors in animals encourages the belief that these tests have validity. However, there are significant differences in the numbers of the exposed populations of men and animals, in the part of the lifespan during which each is exposed, in the metabolic activation of the carcinogens, and in the longevity of men and experimental rodents. For regulatory purposes, Shubik and Clayson[8] suggest that we must assume that the results of bioassays in rodents will closely parallel tumor induction in man, although we cannot be sure of this. In some cases, such as rodent bladder tumors associated with bladder stone, or subcutaneous sarcomas arising locally to massive injection of food dyes, there may be reason to reject an association[8].

IARC[14] suggests that chemicals which have been shown to produce neoplasms in experimental animals can be assigned to two groups: (1) chemicals that have been shown unequivocally to produce malignant neoplasms ("strong evidence" of carcinogenicity); and (2) chemicals for which the evidence is based solely on the appearance of such neoplastic lesions as lung adenomas or hepatomas in mice (often considered as "weak" evidence of carcinogenicity). For many of the "strong evidence" chemicals (reviewed in the first 326 individual chemical monographs) which are not among those known to cause cancer in humans, conclusions for or against human carcinogenicity could not be reached because the epidemiological data were insufficient or nonexistent. In the presence of appropriate carcinogenicity animal data, and in the absence of adequate human data, IARC suggests that it is reasonable to regard such chemicals as if they were carcinogenic to humans[14]. However, exceptions should be noted. For example, to date evidence of a carcinogenic effect in man has not been demonstrated for certain recognized animal carcinogens, e.g., isoniazide and sodium penicillin[3].

Weinhouse[19] observed that while a large number of chemicals are suspected to be carcinogenic on the basis of animal assays, their role in human cancer remains undetermined and possibly undeterminable. To obtain statistically significant data with a reasonable number of mice or rats, cancer incidences of 5 to 10% must be obtained necessitating the administering of large doses[8,14]. (Unless there are very high occupational exposures of limited numbers of men, there are few instances of carcinogen exposure which affect such a high proportion of the human population)[8]. Thus assessment of risk in man requires extrapolation of these data to a human population approximately 10^6 times larger, with exposure levels likely to be approximately 10^4 times lower, and with a lifespan 30 times longer[19].

Whereas the animal population under test is genetically and environmentally homogeneous, human populations differ in age, diet, geography, life-style, occupation, health and particularly in genetic makeup which may put certain subpopulations at especially high risk.

It is useful at this point to briefly describe some of the common design faults in carcinogenicity testing as delineated by Roe and Tucker[20]. These include: (1) inadequate randomization; (2) unintended variation (e.g., position on racks, room differences, operator differences, and observer differences); (3) high loss of animals without post-mortem examination; (4) poor records of necropsy findings-position and size of lesions not recorded; (5) non-standard post-mortem techniques; (6) failure to match microscopic with macroscopic findings and (7) failure to take survival differences into account in expressing results. Some discrepancies between the logical basis of routine carcinogenicity testing and current knowledge of mechanisms of carcinogenesis as summarized by Roe and Tucker[20] include: (1) cocarcinogens and immune suppressants make increase risk of cancer development; (2) under laboratory conditions, rats of many strains have incidence of mammary, pituitary, adrenal and other tumors, suggesting a highly abnormal hormonal status; (3) "spontaneous" tumor incidence may be greatly influenced by dietary intake and (4) the presence of C-type viruses may greatly influence the risk of malignant transformation of cells on exposure in vitro to chemical agents.

How is one to extrapolate the result to man of a substance proven to be carcinogenic at some particular dosage and route of administration in laboratory animals?

We do not even know for certain if there is a threshold of dosage below which no cancers would occur, regardless of the size of the exposed population.

There exists considerable disagreement as to the acceptance and interpretation of animal data, and in their extrapolation to man. For example, IARC[14] states that with the present state of knowledge, it would be difficult to define a predictable relationship between the dose (mg/kg bw/day) of a particular chemical required to produce cancer in test animals and the dose which may produce a similar incidence of cancer in humans. The limited data available suggest, however, that such a relationship may exist[17], at least within certain classes of carcinogenic chemicals. Data that provide "strong evidence" of carcinogenicity in test animals may therefore be used in an approximate quantitative evaluation of the human risk at some given exposure level, provided that account is taken of the nature of the chemical concerned and of the physiological, pharmacological and toxicological differences between the test animals and humans[14]. However, no acceptable methods are currently available for quantifying the possible errors in such a procedure, either for a generalization between species or for the extrapolation from high to low doses. The methodology for such quantitative extrapolation to humans needs further consideration[14].

A view is held by many that no rational methods yet exist for calculating a non-effect dose in man for animal data[3,14,15,21]. According to Higginson[3], if "socially acceptable" levels of exposure for man are described, such levels are to a large extent dependent either on epidemiological data, e.g., aflatoxin[22] or ionizing radiation[23] or "guestimates" in which the dose carcinogenic for animals is treated according to certain mathematical formulae[24]. Nonetheless, the latter, no matter how sophisticated the mathematical treatment, remains dependent for confirmation on epidemiological investigations[3].

Some investigators have assumed that the dose-response curve can be extrapolated downwards to zero from the 5-10% level, and using statistical devices such as those proposed by Mantel et al[25], have tried to calculate an acceptable risk.

The justification for the concept that no dose of a carcinogen is safe appears to rest on the work of Druckrey[26] showing that the effect of carcinogens persist in an irreversible fashion, and the support for this concept gained from the two-stage initiation-promotion model of mouse skin carcinogenesis[27,28]. It must be remembered, however, that these observations were made with large experimental doses of carcinogens and do little to answer the question of what happens at really low exposures, for example, ppb or ppt levels of carcinogens that possess weak carcinogenicity in animals[8]. The theoretical concept of zero tolerance to carcinogens was initiated before DNA repair, the ability of the cell to repair lesions induced in its own genetic material was discovered. According to Shubik and Clayson[8] we should be looking into the practicability of determining whether pre-replication DNA repair mechanisms are capable only of removing a certain proportion of induced lesions in the genome no matter how few are present, or whether DNA repair becomes progressively more efficient as the number of induced lesions diminishes.

The determination of minimum effective doses and the whole concept of "threshold dose" have been the subject of considerable discussion (often heated) in recent years in the context of dose-response relationships in chemical carcinogenesis. The question according to D'Aguanno[29] is whether or not there exists a "safe" subthreshold dose, or rate of intake, of carcinogenic chemicals[29]. Based on the persistence of effect and the additive effect of repeated doses for a sufficiently long time, some investigators conclude that no "safe" dose for carcinogenic substances may be set. The problem of existence of biological threshold has been discussed by Dinman[30]. Using the suggestion that 10^4 drug molecules per cell is the limiting concentration for biological activity, Friedman[31] calculated 8.6×10^{15} molecules per kilogram of cells as the minimum number to evoke a biological effect.

Aspects of extrapolation and risk estimation for carcinogenesis have been recently reviewed by Gaylor and Shapiro[32], Brown[33], Albert et al[34], Cornfield[35], Newill[36], Albert and Altchuler[37], Hoel[38], Crump et al[39], Gehring and Blau[40], and Hoel et al[41].

Hoel et al[41] reviewed the current statistical methods used to establish dose-response relationships for irreversible self-replicating toxic effects (i.e., carcinogenesis) in laboratory animals and extrapolation of these data to man. The method or procedure for estimating human risk based on animal studies is basically carried out in 4 sequential steps: (1) Design and/or assessment of laboratory experiments as to quality and biological appropriateness; (2) statistical extrapolation of the experimental results to low dose levels; (3) extrapolation of the estimated results in animals at the low level to man; and (4) assessment of the risk to man based upon experimental data.

Use of data from animal experiments to estimate the human risk from long-term exposure to very low doses of environmental carcinogens poses a number of biological, pharmacological, and statistical problems. One of the statistical problems is to extrapolate the animal dose-response relations from the high dose range where animal test data are available to low doses which humans might encounter. Different techniques which would seem to fit the test data equally well can lead to low dose risk estimates that differ by several orders of magnitude[42]. Carcinogenesis data is typically of two types with regard to modeling for extrapolation purposes. The usual form presents life-time incidences at various dose levels, used for estimation of a dose-incidence curve. From this estimated curve and its associated error probabilities, either an estimated dose corresponding to a given level of incidence or an estimated incidence at a given dose level are statistically obtained. For purposes of regulatory standards, these estimates are extrapolated values, since the fitted data is most often at doses corresponding to high response levels whereas the need is for estimated dose levels corresponding to very low incidence or risk levels. Thus we are predicting responses well outside the usual experimental range[38].

The second type of dose-response modeling that is currently receiving attention deals with the distribution of the "time-to-occurrence" and its relationships to dose. This "time-to-occurrence" would typically be defined as the time of death due to the type of carcinogenesis of interest, or it may be defined as the time of first appearance of detection of a particular tumour type. Two distributions of time-to-occurrence of cancer have been receiving particular attention recently. Blum[43] and Druckrey[44] have considered a lognormal distribution with median time-to-cancer depending upon dose. Albert and Altshuler[45] have continued theoretical work on this model and have applied it to several sets of carcinogenesis data. They relate the dose level in the model by assuming that the logarithm of the median single-risk time of appearance of cancer is linear in log dose, while the standard deviation is independent of dose.

A number of other investigators[46-49] have studied the Weibull distribution as a time-to-tumor distribution. One method of including dose in the Weibull model is to assume that the logarithm of the Weibull's scale parameter is linear in log dose, whereas the other parameters are assumed independent of dose.

For purposes of risk assessment, the more sophisticated time-to-tumor models are more appropriate than the simplistic dose-incidence approach. Questions relevant to risk assessment, such as tumor incidence in a particular age group, amount of life-shortening by age group, and so on, can be answered at least theoretically by time-to-occurrence models[38]. With these advantages there are still serious practical difficulties with these models. First, more reserach and experience with the models is required and should be encouraged before one would have much confidence in their use as an extrapolation method. The second difficulty is that in being a more complicated model than the dose-incidence approach, they seem to require considerably more data to obtain proper fits. Unfortunately, however, extrapolations and risk estimates are often required when only a small amount of laboratory data is available.

A number of investigators[50-52] have considered using probabilistic models to describe the carcinogenic process. These models are generally of a multistage nature, with each stage having an incidence rate approximated by a linear function of dose. Except for a single-stage or one-hit model and possibly a two-stage model, there is a large number of unknown parameters associated with these multistage models, so many parameters, in fact, that it is unrealistic to attempt to estimate them from experimental data other than in exceptional cases[38].

Currently, mathematical investigations of stochastic models for carcinogenesis are being continued. In one area attempts are being made to modify the hit and multistage models in order to permit cell repair by some stochastic mechanism. These models are then studied to see if repair does have a significant qualitative effect upon the dose-response curve. A second area of consideration has to do with the implication of variance among individuals in sensitivities to a carcinogen insult upon low dose risk estimation[38].

An approach was presented by Albert and Altshuler[37] to the assessment of carcinogen risks in which the dominant effect of carcinogen exposure is life shortening and the impact falls both on those individuals who would have gotten cancer without the carcinogen exposure as well as the new cancer cases. This analysis was based on the interaction of age-specific tumor incidence rates and population survival in terms of age-specific mortality rates without the induced risk from carcinogen exposure. The analysis yielded estimates for lifetime probability of developing cancer, average lifespan loss of cancer cases. The approach utilized the animal response data to assign, to the existing human cancer occurrence, and equivalent dose of the same carcinogen which is under consideration in terms of risk evaluation. The approach has the advantages of keying the estimates of carcinogen risks to those which already exist in the environment, avoiding large extrapolations from animal data, and encompassing the variability in susceptibility and carcinogen exposure in humans.

Many quantitative theories of carcinogenesis have been proposed which relate the frequency of occurrence of detectable tumors to the level of the carcinogen and the duration of exposure to the carcinogen. The purposes of these theories are twofold, to elucidate the mode of action of the carcinogen and the nature of the neoplastic change and to estimate, from animal experimentation, the risk to human populations exposed to environmental concentrations of the carcinogen[32,33].

Gaylor and Shapiro summarized a number of different mathematical models which have been proposed to describe the dose-response for carcinogenic purposes[32].

One-hit (linear) model. The one-hit model is based upon the concept that a tumor can be induced after a single receptor has been exposed to a single quantum of a chemical. Thus, the probability of a tumor induced by exposure to a chemical at a dose of d is: $P(d) = 1-\exp(-\lambda d)$ where, $d \geq 0$, $0 \leq P(d) \leq 1$, and λ is an unknown constant which represents the rate of change (slope) of the dose-response curve at $d = 0$. Dose is used here in a very general sense. It may be the total accululated dose or the dosage rate in terms of body weight, surface area approximations or concentration in the diet. Hence this model implies that the expected number of tumors within a lifetime will depend only upon the total dose and not upon the pattern of exposure[50,53].

The model is referred to as the linear model, since the probability of a tumor is directly proportional to dose with a slope of λ.

Multi-hit (k-hit) model. Where k hits of a receptor are required to induce a cancer, the probability of a tumor as a function of exposure to a dose (d) is given by
$$P(d) = 1-\exp(-\lambda d^k).$$
For small values of λd^k, the k-hit model may be approximated by
$$P(d) = \lambda d^k$$
or
$$\log P(d) = \log \lambda + k \log d,$$
thus k represents the slope of log P(d) versus log d.

Multi-stage model. A generalization of the multi-hit model is given by the multi-stage model
$$P(d) = 1 - \exp[-(\alpha_1 + \beta_1 d)(\alpha_2 + \beta_2 d)\cdots(\alpha_k + \beta_k d)]$$

where $\alpha_i \geq 0$ and $\beta_i \geq 0$ and k represents the number of stages in the carcinogenic process. The α's represent the spontaneous background occurrence for each stage and β's represent the proportionality constants for the carcinogenic process induced by the dose for each stage. The one-hit and k-hit models are special cases of the multi-stage model. The multi-stage model can be written in a somewhat simpler form as
$$P(d) = 1 - \exp[-(\gamma_0 + \gamma_1 d + \gamma_2 d^2 + \cdots + \gamma_k d^k)].$$

<u>Log-probit model</u>. The log-probit model has been used extensively in the bioassay of dichotomous responses. This model assumes that each animal has its own threshold dose below which no response occurs and above which a tumor is produced by exposure to a chemical. The log-probit model assumes that the distribution of log dose thresholds is Gaussian (Normal). Thus, the most likely log dose at which tumors occurs is the ED_{50} (effective dose which produces tumors in 50% of the animals) or average log dose at which tumors occur. The frequency of tumors decreases as log doses depart from the average log dose. The frequency of induced tumors approaches zero as doses approach zero or become extremely large. That is, there are relatively few extremely sensitive or extremely resistant animals in a population. For the log-probit model the probability of a tumor induced by an exposure to a dose d of a chemical is given by
$$P(d) = \Phi(\alpha + \beta \log_{10} d)$$
where Φ denotes the standard cummulative Gaussian (Normal) distribution. Chand and Hoel[54] showed that the log-probit dose-response is obtained when the time-to-tumor distribution is log-normal.

One of the properties of the Mantel-Bryan procedure is that it can be applied to data whether or not a statistically significant increase in tumors has been demonstrated. This is particularly important since the demonstration of statistical significance depends so strongly upon the number of animals tested. The Mantel-Bryan procedure is based upon the uncertainty of an experimental result, as measured by the upper confidence limit. Thus control is exercised even if no excess tumors are observed experimentally because the possibility exists that a larger experiment may demonstrate the carcinogenicity of the compound. Thus, the procedure does not penalize the experimenter that conducts a good experiment.

The Mantel-Bryan or Linear Extrapolation procedures lead to estimates of maximal risk or acceptable dosages based upon the uncertainty in experimental data regardless of whether or not carcinogeneity is demonstrated. For example, zero tumors in 450 animals certainly does not indicate that a compound is carcinogenic for the particular set of conditions. However, there is no guarantee that the next 450 animals tested would not contain some tumors. The upper 99% confidence limit for the proportion of tumor bearing animals in this case is 1%. For this example, if the compound were tested at 1% in the diet and a maximal risk of one in a million were desired, then the acceptable dose could be no greater than 1 ppm. This is quite a severe restriction for a compound for which carcinogenicity has not been demonstrated. Such a philosophy would unnecessarily limit many necessary and useful chemicals[32]. There is no precedent for restricting the use of a chemical on the basis of carcinogenicity until carcinogenicity has been demonstrated. However, the possibility of obtaining a false negative in an experiment often is quite high allowing the presence of carcinogenic compounds in the environment[32].

Human intake of a chemical varies among individuals and varies daily for a given individual. The simplest approach and perhaps adequate for our current state of knowledge is to calculate risks for anticipated "maximum" exposure levels. This gives additional conservatism for any prediction technique. The view is held by some that it is not necessary to attempt to protect every last individual with unusual habits, but to base predictions on average intake, for example, in an attempt to estimate the actual risk to the population[32]. Gaylor and Shapiro[32] suggest that mathematically, the proper approach to calculate the risk for the total population is to calculate the risk for a given dosage (a most difficult task) and then to multiply that risk by the proportion of the time that dosage occurs followed by integration over the distribution of dosages. Generally, the distribution of dosages for an environmental chemical in a human population would be unknown. Thus, introducing variation in consumption adds another dimension to be investigated to an already complicated problem. This gives the average risk and does not consider a segment of the population which may be at high risk.

The uses of safety factors have been discussed as a crude form of extrapolation; e.g., dividing the highest experimental no effect dose by 100 or the lowest effect level by 5000 to arrive at an acceptable dose. According to Gaylor and Shapiro[32], the shortcomings of safety factors are obvious. They are arbitrary, do not consider the magnitude of a response, and do not consider the uncertainty of a response as determined by the number of animals tested.

Other models include the logistic model of Berkson[55] and the extreme value model of Chand and Hoel[54]. While none of these models can be verified on the basis of biological arguments, some have more biological appeal than others. The multi-stage model appears to be about the most general model with the other models approximating some form of the multi-stage model according to the values of the parameters. Unfortunately, most of these models fit experimental data about equally well in the observable response rates at experimental dosage levels, but give extremely different estimated responses when extrapolated to low dosage levels to which humans may be exposed[32].

Turner[55] illustrated the mathematical forms of a spectrum of hit-theory models in which the mathematical theory between the "hits" of an agent and the critical targets was developed. Extensions were also discussed included both biological variation in the probability of a hit, and variation in the number of critical targets.

Nordling[56] and Stocks[57] later proposed theories in which a single cell can generate a tumor only after it has undergone a certain number, possibly greater than one, changes or mutations and, as such, can be termed a multievent theory of carcinogenesis. Many of the multievent models of carcinogenesis from the viewpoint of low dose kinetics have been reviewed by Crump et al[39]. Two basic assumptions were made: (1) that the cancer process is single cell in origin with possible multiple steps between initiation and complete alteration and that (2) the growth period of the completely altered cell is basically independent of the exposure level.

For direct carcinogenic processes, in which the agent or its metabolite acts at the cellular level to produce an irreversible change, it was concluded that most models of carcinogenesis will be linear at low dose levels. In addition, it was shown that if it can be assumed that the environment contains carcinogens that act in conjunction with the test agent, then all the models thus far proposed will be linear at low doses. In all of these theories, the emphasis is mainly on the stochastic nature of the changes involved in the carcinogenesis process. The role played by the carcinogen is considered to a much lesser degree. It is commonly assumed that the transitional events in the process attributable to the carcinogen occur with probabilities proportional to the level of exposure.

Brown[33] considered this a gross oversimplification of the actual process. The actual exposure is generally considered to be modified by absorption, distribution, metabolism and excretion of the chemical substance, and the effective exposure level should probably be the actual concentration of the carcinogen at, and within the target cells. Other factors that may affect the delivery of the carcinogen to intracellular compartments are membrane permeability and enzyme binding. Therefore, the "effective dose" may well be some complex function of the actual exposure along with the biochemical and physiological parameters of the host[33]. Most of these mathematical models incorporate the dose as it is actually administered in animal experiments or in human exposure. The function which relates administered dose to "effective dose", if it is not a simple case of proportionality, can have a profound effect on the dose-response relationship[33].

The various dose-response curves that have been observed may not be indicative of different carcinogenic processes once the agent has reached the target cell, but rather may indicate different functions relating the administered to the effective dose. This problem will probably relate more to chemical carcinogenesis as opposed to radiation induced cancer[33].

Even if a normal cell has been transformed to some intermediate stage in the carcinogenic process, this would not necessarily mean that cancer must occur. The possibility of cell repair or recovery, or some other response from the immune mechanisms, may be sufficient to stop or reverse the process before the final stage is reached. In addition, the death of these transformed cells may occur before the process has a chance to continue toward the eventual cancer. This is one of the major arguments in favor of the existence of a threshold[33].

However, if there is some probable chance that these recovery mechanisms may not complete their role before the occurrence of another event or transformation, then this type of threshold will not exist[33]. According the Brown[33], thresholds may be considered from two viewpoints: an "actual" threshold which is an exposure level below which any carcinogenic response attributable to the specific agent is impossible; a "practical" threshold which is an exposure level below which an attributable carcinogenic response is highly unlikely. Mantel[58] has argued that whether or not "actual" thresholds do exist is immaterial, and that one should consider the "practical" thresholds when estimating human risk, and he has suggested procedures for extrapolating the results of animal experiments performed at necessarily high exposure levels to the lower levels of human exposure. Mantel and Bryan[59] suggested that a level of risk to cancer on the order of one in 10^8 could be thought of as a "practical" threshold and they stated that efforts should be made to estimate exposure levels that would produce no greater risk than this. Using mathematical models that relate the latency period (time between initiation of exposure and appearance of cancer) to the exposure level, Jones[60] suggests that a "practical" threshold could be defined as an exposure level for which the latency period exceeds the normal lifespan.

Experimental or observational evidence for the existence of an "actual" threshold is usually presented in the form of a dose-response graph in which the percentage of animals with tumors or the average number of tumors per animal is plotted against the dose level of the carcinogen. Either the existence of those dose levels not leading to an increase in tumor incidence over controls, or the extrapolation of such curves to low dose levels, which apparently would result in no tumor increase, are cited as indications of the existence of some threshold below which no response is possible. According to Brown[33], this type of evidence is nothing more than an exercise in self-deception. In the first place, the observation of no positive responses does not guarantee that the probability of response is actually zero. From a statistical viewpoint, zero responders out of a population of size N is consistent at the 5% significance level with an actual response probability ranging between zero and approximately 3/N (e.g., when N=100 and zero responders are observed, the true probability of response may be as high as 3%).

In the second case, when an observed plot of dose against tumors is extrapolated downward to produce a no-effect level, it is assumed that the observed dose-response relationship, usually linear, will be obtained throughout the entire spectrum of dose levels and that one threshold level exists for the entire population at risk. The assumption of linearity throughout the entire dose spectrum can easily lead to erroneous conclusions when the assumption is false.[33]

Brown also states that it is much more likely that if thresholds do in fact exist, not all members of the population have the same level. The human population is a very diverse, genetically heterogeneous group, exposed in varying degrees to a large variety of toxic agents. Many different disorders may affect the frequency of mutational events within specific tissues. Disorders characterized by chromosomal instability have been found to be predisposed to malignancy. Patients with Xeroderma Pigmentosum are highly susceptible to ultraviolet induced skin cancer and they have been found to have deficient DNA repair mechanism[61]. In Bloom's syndrome there is an immune deficiency and an increased risk for leukemia and cancer of the colon[62]. There is also some evidence that the amount of an enzyme, aryl hydroxylase, which is genetically controlled, may determine the susceptibility to lung cancer from cigarette smoking[63]. This system may provide a model where the risk of mutation and subsequent malignancy following exposure to an environmental carcinogen may be genetically determined. If malignancy is the result of a series of mutational events, then there must be subpopulations at varying degrees of risks, or who have varying thresholds for the carcinogenic agent. Hence, the search for thresholds should not, according to Brown[33], be a search for one specific no-effect level, applicable to all members of the population at risk, rather the problem is one of finding many thresholds for each of the various subgroups in the population.

In addition, when considering the possibility of carcinogenic thresholds, one should keep in mind that no agent has been found to induce a type of cancer which has not already been found. It is possible, perhaps even likely, that many carcinogenic agents act by the same mechanism on the same target cells[33]. This would imply, that since there are many carcinogenic agents in our everyday environment, some additional carcinogen could act in a simple additive manner and thus any level of exposure would simply add to the already existing background. This means that for a population already being exposed to carcinogenic agents, any additional carcinogen would simply increase the expected tumor incidence in a continuous manner, no matter how low the exposure to the additional agent[33]. Hence, despite all the complexities of the mechanisms of chemical carcinogenesis, because of genetic variation among members of the population at risk, and bacause statistical analysis cannot resolve the question one way or another, Brown believes[33] that the search for an "actual" carcinogenic threshold is probably fruitless and any human exposure to a carcinogen should be considered to have some associated risk, no matter how small that risk may be.

The authorative theories of carcinogenesis that have been proposed are all stochastic in concept. They consider the probabilistic chances of the occurrence of some series of events. According to Brown[33], if one is willing to assume that these events have transition probabilities that can be affected by the carcinogen, no matter how small the concentration, and if one is willing to assume that the system of distribution and metabolism have some chance of allowing some amount of the carcinogenic agent to reach the target cells, no matter how small the chance or the amount, and if we are willing to assume that the repair and recovery systems may not do a perfect job, no matter how small the chances of this happening, then no level of carcinogenic exposure will exist that will have a zero probability of leading to a cancer[33].

Carcinogenic risk assessment involves a mixture of statistical, scientific, and public policy considerations. Concepts in current use, such as "no observed effect levels" and "virtual safety", and the problems in implementing them by means of dose-response models, particularly the probit-log dose and linear models were recently reviewed by Cornfield[35]. The upper limits to risk provided by some conservative procedures were viewed by Cornfield as inconsistent with coherent balancing of risk and benefits.

A common basis to the dose-response curves describing both carcinogenic and non-carcinogenic effects, in the opinion of Cornfield, is to be found in deactivating reactions. A simplified model in which a toxic substance is activated and deactivated in separate and simultaneous reactions was presented and the dose response curve implied by the model thence deduced. This curve has the general form of a hockey stick, with the striking part flat or nearly flat until the dose administered saturates the deactivation system, after which the probability of a response rises rapidly. Such a curve describes the Bryan-Shimkin methylcholanthrene-tumor incidence dose response curve as well as the probit log-dose model. The concept of a saturation dose viewed is relevant to risk assessments for carcinogenic and noncarcinogenic substances alike[35].

The question as to whether the carcinogenicity of chemicals is dose-dependent or whether a "threshold" dose exists below which cancer will not be induced by exposure is undoubtedly a much debated one which has engendered great controversy.

Although many toxicologists postulate the existence of a threshold dose below which no harmful effect occurs, unfortunately, it is impossible to prove the existence of a no effect level (threshold) statistically. In fact, as the incidence of tumors asymptotically approaches zero at low doses, the occurrence of tumors may be so rare as to appear that there is some dose below which tumors apparently do not occur. Gaylor and Shapiro[32] suggest that even if thresholds do exist for carcinogenic processes, a conservative approach from a risk standpoint is to assume that any dose may induce or promote carcinogenic processes with the only no effect dose being zero.

For chemical carcinogenesis, the lack of universal acceptance of dose-response is believed by Gehring and Blau[40] to result from the erroneous assumption that dose-response and threshold concepts are inextricably associated. Chemical carcinogens, regardless of chemical composition, produce a greater incidence of cancer when administered in large doses, than in small doses, and hence a dose-response curve does exist[64]. While it can be argued that for some chemicals the response is related linearly to dose rather than to the logarithm of the dose, this deviation, if true, does not preclude the establishment of a dose-response relationship[40]. Such a deviation does not alter, whatsoever, application of the dose-depending concept to the hazard assessment of chemicals, according to Gehring and Blau[40].

Arguments that there is no threshold for chemical carcinogenesis are substantive. The principal argument states that cancer is an expression of a permanent, replicable defect resulting from amplification of a defect initiated in one cell by reaction of the agent with a critical receptor. Once such a defect occurs in a cell, the cell may be dormant for years before expressing a discernible untoward effect[40]. Unlike classical responses, division of a large dose of some carcinogens into smaller repeated doses does not abolish the response. For dimethylaminoazobenzene, 4-dimethylaminostilbene and diethylnitrosamine, the total cumulative dose necessary for carcinogenesis with small daily doses is smaller than the single dose required to produce an equivalent response[64-66]. However, the size of the repeated dose can be reduced further, resulting first in an increased latency for development and, finally, no experimentally discernible response. Another argument for no threshold dose is that experiments on radiation induced cancer have not revealed a threshold within the realm of statistical reliability.

The major evidence (as summarized by Gehring and Blau[40]) in support of a threshold concept are the following arguments, a number of which have been discussed earlier in Chapter 2. (1) Chemical carcinogenesis is a multistage process involving: (a) exposure, absorption, distribution, activation, deactivation, and elimination of the chemical per se or products formed from it; (b) interaction with critical receptor sites leading to molecularly transmittable products; (c) survival and proliferation of transformed cells to clinical cancer. Interference with any of these processes may constitute a threshold. For example, promoters, which in themselves cannot initiate cancer, can enhance greatly the incidence of cancer induced by administration of an initiator. Also, the damaged receptor site may undergo repair. (2) Alteration of the physiological status may either augment or inhibit the response to a carcinogen. For example, age, sex, nutriton, population, density, hormonal state, or concomitant disease may affect the response to a carcinogen. This suggests that a precancerous status may exist or may be induced without development of cancer until the precancerous status attains some critical level or until the precancerous state can no longer be held in check by suppressive mechanisms, whatever they may be[40]. (3) As the dose of a carcinogen is decreased, the latency period for cancer development increases. This phenomen was revealed lucidly by Druckrey[65] who noted that the dose multiplied by some power of time was constant; i.e., dt_n=constant in which n=2 to 4. This relationship can imply a threshold in that multiples of a lifetime will be required for expression of cancer in response to low doses. Albert and Altshuler[66] utilized the increasing latency with decreasing dose of a carcinogen to formulate limits for unavoidable exposures to carcinogens. (4) Utilizing the relationship of dose to time-of-appearance of cancer, Jones and Grendon[57] postulated that a number of cells in close proximity require transformation to allow development of an aberrant clone of cells and ultimately cancer. This multihit hypothesis, if true, will result in a marked reduction in the incidence of cancer as the dose is decreased for the same reason that trimolecular chemical reactions become negligible as the concentrations of the reactants are decreased[40]. (5) For many chemical carcinogens, cancer occurs only when doses are given that exceed those needed to cause pathological responses, such as grossly and histologically discernible tissue damage. Some cancers develop clinically in chronically inflamed or scarred tissue, e.g., colonic cancer in patients with ulcerative colitis or regional enteritis, squamous cell carcinomas in ulcers of burn scars, squamous carcinomas of the bladder in schistosomiasis, scar carcinomas in lung, carcinomas and sarcomas arising in osteocutaneous fistulas caused by chronic osteomyelitis, and carcinoma of the stomach in autoimmune (atrophic) gastritis[68]. (6) There is a substantial and growing body of evidence that carcinogenesis is subject to immuno-surveillance, particularly cell-mediated immunity[66,69]. (7) Stress, such as administration of unrealistically large doses of chemical to laboratory animals, can enhance greatly the response to oncogenic viruses and perhaps other innate carcinogens as well[70]. There is reasonable evidence in both humans and animals, that over-nutrition, particularly excess dietary fat, is a major cause of cancer[71,72]. Malonal-dehyde, a product of peroxidative fat metabolism which is also formed spontaneously in tissues, particularly when the diet is deficient in antioxidants, has been found to be carcinogenic[73].

The TLV Committee of the American Conference of Governmental Industrial Hygienists (ACGIH) has recently proposed guidelines[74] which divided experimental carcinogens into 3 groups: those of _high_, _low_ and _intermediate_ potency.

The categorization of carcinogens is predicted on certain conditions of dosage rates according to the 3 major routes of occupational contact, viz., respiratory, dermal and gastrointestinal which elicit tumors in significant excess above that in negative control animals.

A substance of _high_ carcinogenic potency by the respiratory route must induce malignancy in dosages less than 1 mg/m^3 of inhaled air (or its equivalent in parts per million) in 6- to 7-hour daily repeated exposures throughout the animal's lifetime, or from a single, intratracheally administered dose not exceeding 1 mg of particulate of liquid, per 100 ml or less of animal minute respiratory volume.

Examples: _Bis-chloromethyl ether_, malignant nasal tumors, rats @ 0.1 ppm in 520 years. _Hexamethyl phosphoramide_, nasal squamous cell carcinoma, rats @ 0.05 ppm in 13 mos. _Be(OH)$_2$_, pulmonary adenomas, rats @ 40 µg as Be, in one year.

Substances of _low_ tumorigenic potency by the respiratory route need only to elicit tumors at dosages greater than 10 mg/m^3 (or equivalent ppm) in daily repeated exposures for 12 months, with holding period for another year, or, from intratreacheally administered dosages totaling more than 10 mg of particulate of liquid per 100 ml or more of animal minute respiratory volume.

Examples: _Beryl_ (Beryllium aluminum silicate) malignant lung tumors, rats @ 15 mg/m^3 @ 17 months. _Benzidine_, various tumors, rats, 10-20 mg/m^3 @ > 13 months.

Substances such as dimethyl sulfate and hydrazine, on present evidence might be considered experimental animal carcinogens of intermediate potency.

In order to place in proper perspective the tumorigenic findings in experimental animals in relation to practical occupational concerns for workers protection, specific dosage limits have been set for the 3 routes of administration for mice and rats above which a substance is <u>not</u> to be considered as an occupational carcinogen.

The Carcinogen Subcommittee of ACGIH[74] believes that evidence for thresholds of carcinogens takes 3 forms: (1) Evidence from epidemiologic sutides of industrial plant experience, and from well-designed carcinogenic studies in animals. (2) Indisputable biochemical, pharmacokinetic, inherent, built-in anticarcinogens and processes in our bodies. (3) Accumulated biochemical knowledge makes the threshold concept the only plausible concept.

Evidence for thresholds in carcinogenesis reportedly exhibited for BCME, 1,4-dioxane, coal tar, β-naphthylamine, hexamethyl phosphoramide, vinyl chloride, vinyl chloride plus vinylidene chloride and dimethyl sulfate is shown in Table 3. It should be noted that the 9 carcinogens shown are of widely differing structures, producing many different tumor types, possibly by different mechanisms. Thresholds appeared evident for all 3 major routes of entry irrespective of whether the carcinogens were of very high or low potency and were distinguished in 3 instances of human carcinogenesis, e.g., β-naphthylamine, dimethyl sulfate and vinyl chloride.

Advocates of the concept that safe occupational thresholds exist for carcinogens cite the fact that no <u>unique</u> molecular configuration exists for carcinogenicity and as have been noted earlier, a series of factors exist which may influence and modify the possibility that any given chemical might cause cancer in exposed workers, e.g., personal habits such as smoking and individual genetic history[74].

Flowers[75] recently described an approach to setting environmental limits for carcinogens based on the rationale that since ionizing radiation and mutagens produce comparable effects at equivalent concentrations the "rec" (radiation equivalent chemical) concept can provide a basis for setting limits.

Risk of developing cancer can be associated with the probability of an initial carcinogenic event occurring among many molecular events. This implies that an estimate or risk might be useful in establishing a carcinogen limit. For the variety of chemical and biologic agents suspected of being involved in initiation of carcinogenesis, there is a paucity of quantitative data on background exposure concentrations that might be acceptable.

For a physical agent-ionizing radiation-average background is 0.49 mrem/day, which represents the external and internal flux from a variety of sources[76]. Based on the assumption that background ionizing radiation represents a threshold of acceptable risks, it would be useful to calculate an airborne concentration of a chemical agent equivalent to radiation background.

A fundamental premise in making such a calculation is that ionizing radiation interferes with the flow of biologic information by mechanisms similar to chemical carcinogens or mutagenic agents. To translate radiation exposure into a radiation equivalent chemical (rec) dose[77], it is necessary to express ionizing radiation in chemical terms. For purposes of calculation and derivation of a useful relationship, radiation exposure can be converted to the number of ions produced per gram of tissue

TABLE 3

EVIDENCE FOR THRESHOLDS IN CARCINOGENESIS

Test Substance	Route	Species	Dose Levels Eliciting Tumors	Dose Levels Not Eliciting Tumors	Duration	References
Bis-Chloro methyl ether	Inhln	Rat	100 ppb	10 and 1 ppb	6 mos daily	1. Leong et al., 1975
1,4-Dioxane	Oral	Rat	1% H₂O	0.1 & 0.01%	2 yrs	2. Torkelson et al., 1974
	Inhln	Rat	>1000 ppm	111 ppm	2 yrs daily	3. Ibid., 1974
Coal tar	Topical	Mouse	6,400; 640; 64 mg	0.64 mg	2x/wk., 64 wks.	4. Bingham, 1974
β-Naphthylamine	Inhln & Skin	Man	>5% β in a-Form	<0.5% β a-Form	22 yrs	5. Zapp, 1975
Hexamethyl phosphoramide	Inhln	Rat	4,000; 400 ppb	50 ppb	8 mos	6. Zapp, 1975
Vinyl Chloride	Inhln	Rat	2,500; 200; 50 ppm	<50; >10 ppm '50-'59, 160 ppm average; 30-170 ppm range '60 <50 ppm, decreasing 50 10 ppm	7 mos	7. Keplinger et al., 1975
(+ Vinylidene Chloride)	Inhln	Man	>200 ppm		25 yrs	8. Kramer, Mutchler, 1971
						9. Ott et al., 1975
Dimethyl sulfate	Inhln	Rat	10; 3 ppm (est)	Unknown	>10 mos	10. Druckrey et al., 1970
	Inhln	Man	Unknown	<2-5 ppm	15 yrs	11. Pell, Dupont, 1975

by transfer of ionizing energy; the conversion of number of ions to moles of ions allows comparisons with chemical equivalents:

1. Assume mrem = mrad
   = 0.1 erg/g
   = 6.25×10^{10} ev/g

2. Ions produced each day:
   $2 \times 0.47 \times 6.25 \times 10^{10} / 2.2$ ev per pair = 2.62×10^{10} ionic pairs/g-day

3. Moles of ions for a 154 lb (70 Kg) human/day
   70,000 g $\times 2.62 \times 10^{10}$ 6.022 $\times 10^{23}$ = 2.80×10^{-7} ions-moles/day

4. Then ppt/human-day, assuming 14 cu m of air per day:
   $2.8 \times 10^{-7} \times 82.05 \times 298 \times 10^{6}$ cu m = ppt/human-day
   14 cu m = 4.9 ppt (parts per trillion)

5. For an occupational environment based on a five day week at 40 hours and inhalation of 7 cu m per day at work, then: $2 \times 10 \times 4.9$ ppt = (approximately) 0.1 ppb

Application of the (rec) concept can be shown by use of selected occupational carcinogaens. Table 4 provides estimates or proposed environmental limits for 14 carcinogens identified by NIOSH, as well as asbestos and vinyl chloride. The weight of target organs and relative yields of tumors are used as modifying factors in the estimates.

Table 4. Regulated Carcinogens[75]

Substance	Target Organs	REC Limit
2-Acetylaminofluorene	Whole body	0.1 ppb
4-Aminobiphenyl	Bladder	0.01ppb
Benzidine	Bladder	0.01ppb
3,3'-Dichlorobenzidine	Bladder and other tissues	0.01ppb
4-Dimethylamino-azobenzene	Whole body	0.1 ppb
alpha-Naphthylamine	Bladder	0.01ppb
beta-Naphthylamine	Bladder	0.01ppb
4-Nitrobiphenyl	Bladder	0.01ppb
N-Nitrosodimethylamine	Liver and kidney	0.03ppb
beta-Propiolactone	Whole body	0.1 ppb
bis(Chloromethyl)ether	Lungs	0.01ppb
Chloromethyl methyl ether	Lungs	0.01ppb
4,4'-methylene-bis(2-chloroaniline)	Bladder and other tissues	0.01ppb
Ethyleneimine	Liver and whole body	0.05ppb
Vinyl chloride	Liver	0.03ppb
Asbestos	Lungs	3 fibers/cc

4-Aminobiphenyl is regarded as being the most potent bladder or kidney carcinogen and has been assigned a potency factor of one (1.0) while other similar carcinogens have been assigned potency factors by relative yield of tumors.

Distribution of vinyl chloride and other liver carcinogens is assumed to be in liver and adipose tissues, giving a concentration of 0.25 ppb for an estimated limit. For food and water, a 154 lb. human should be allowed a maximum of 3 x 10 exp-8 moles/day of a given carcinogen with appropriate modification by factors for target organs and relative potency (tumor yields).

In Table 4, conversion of the Committee 17 data on radiation to molarity of ions produced indicates that ionizing radiation is from 1 to 10 times more potent in producing mutations than the chemical mutagens used as examples. Except for the highest radiation dose, a concentration of mutagen ranging in values of 6 ± 1.8 times the radiation induced ion concentration is required to produce comparable effects. Chemical doses used ranged from 1 million to 125 million times the radiation equivalent background while radiation exposures (x-rays) were from several hundred thousand to 5 million times background. Interpretation of the mutagenic data indicates that radiation is a more potent mutagen than typical chemical mutagens selected by Committee 17. Therefore, radiation background may be regarded as an appropriate basis for estimation of acceptable risks for exposure to carcinogenic, radiomimetic, or mutagenic agents.

Proposed environmental limits are considered by Flowers[75] to reduce the risks of initiating events as well as risks of promoting progressive developments associated with genetic and epigenetic processes leading to either cancer or expression of mutagenic changes. Use of radiation background to estimate acceptable risks is a useful concept allowing the determination of nonzero environmental limits by a defined process according to Flowers[75].

Albert et al[34] recently reviewed the rational developed by the United States Environmental Protection Agency for the assessment of carcinogenic risks.

While there can be little argument that the principal evidence that an agent is a _human_ carcinogen is provided by adequate epidemiologic data reinforced by confirmatory animal tests, it is recognized, for practial purposes, most instances that require judgements on human carcinogenicity will be based on animal studies.

The rationale for the substantial justification for using rodent bioassay systems for predicting human responses are basically predicated on the following observations: (1) of the chemical agents that are generally accepted to have produced human cancer, all but one produced carcinogenic response in rats and/or mice; (2) in _most_ tests, the cancers occur in the same organ as in humans when tested by the appropriate route of exposure; (3) in only a relative few instances have conventional bioassay tests of ingestion of inhalation by rats and mice produced false-negative results; e.g., tests with other species or routes of administration were required to produce positive results[34].

While it can be anticipated that rodent bioassay systems will produce false-positive results, no current evidence apparently exists on which to base judgements of how frequently and with what classes of agents false-positive results are likely to occur[34].

Table 5. Estimation of REC Background from Committee 17 Data*

Substance	Chemical Dose M	Airborne Equivalent**	Radiation Dose, rad***	REC Background****
Hycanthone methane sulfonate	0.0001	17.4 ppm	120 (uncorrected)	71.2 ppt
Hycanthone methane sulfonate			20 (corrected)	427.2 ppt
Ethyl methane sulfonate	0.01	1,740 ppm	4,000 (uncorrected)	213 ppt
Ethyl methane sulfonate			2,000 (corrected)	426 ppt
Ethyl methane sulfonate	0.02	348 ppm	100	1,700 ppt
Sodium nitrite	0.011	1,933 ppm	66 (uncorrected)	14,000 ppt
			6.6 (corrected)	140 ppb

*Committee 17, Council of the Environmental Mutagen Society, John Drake, Chairman Environmental Mutagenic Hazards, Science, 187, 503-514, 1975
**Airborne equivalent is the air concentration that would allow intake of an equivalent chemical dose of the mutagen as an ideal gas by a 70 kg person inhaling 10 cu m of air
***Corrected and uncorrected values are based on Committee 17 estimates of effective dose or rate factor corrections.
****Estimated REC background is the airborne equivalent concentration multiplied by 0.49 mrem for background radiation and divided by the radiation dose in mrad where mrad is assumed to be equivalent to mrem.

    The quantitative assessment of the impact of a suspect carcinogen on cancer induction in humans at unregulated levels of exposure is of critical importance to those charged with the need for assessing regulatory action[34]. It should be noted that of the half dozen examples in which quantitative comparisons can be made between animals and humans, the magnitude of carcinogenic response in the most sensitive animals tested does show a reasonable comparability of that of humans[78,79]. However, it is well recognized that substantial differences do occur in sensitivity, even among different strains in the same speices of test animals. As earlier discussed in this chapter, of equal critical importance in the quantitative risk assessment is the extrapolation of dose-response relationships from high to low levels of exposure. All the recognized instances of cancer induction in humans and animals by known chemical and physical agents have apparently involves large exposures compared with those that concern the setting of exposure standards. Hence estimates are required for the level of cancer risk at exposures significantly below those for which observable responses have been obtained.

In order to make such extrapolations, some shape to the dose-response curve must be assumed. One such possible extrapolation that has been proposed is to assume a linear non-threshold dose-response relationship to estimate the maximum response[34]. This linear pattern of response has been observed in humans who were exposed to certain forms of ionizing radiation and who developed lung cancer as a result of cigarette smoking. It is also the pattern of response observed for the induction of genetic mutations; possibly genetic damage is the fundamental derangement in cancer cells. The linear nonthreshold dose-response relationship is conservative in predicting the largest response for any given level of low-dose exposure. Such a dose-response pattern implies that a safe level of exposure is nonexistent. However, the use of several extrapolation models is appropriate to convey the range of uncertainty in these estimates[78,79].

Once detailed risk and benefit analyses are available it would appear that there is also a critical need to consider the extent of the risk, the benefits conferred by the substance, the availability of substitutes and the costs of control of the substance[34,78,79].

In considering the risks, according to Albert et al[34], it will be necessary to view the evidence for carcinogenicity in terms of a warning signal, the strength of which is a function of many factors including those relating to the quality and scope of the data, the character of the toxicological response, and the possible impact on public health. Albert et al[34] concluded that it is understood that qualifications relating to the strength of the evidence for carcinogenicity may be relevant to this consideration because of the uncertainties in our knowledge of the qualitative and quantitative similarities of human and animal responses. In all events, it is essential in making decisions about suspect carcinogens that all relevant information be taken into consideration.

A comprehensive cancer policy aimed at identifying, classifying and regulating potential cancer-causing substances in the workplace has been recently proposed by the Occupational Safety and Health Administration (OSHA) in the United States[80].

If adopted, the plan would classify hundreds of toxic substances according to their known or suspected potential for causing cancer and establish three model standards for controlling workplace exposure to them.

Category I would be reserved for confirmed carcinogens-any substance found to cause cancer in humans, or in two mammalian species of test animals or in replicative tests in the same species.

A chemical placed in that category would trigger an emergency temporary standard by OSHA, reducing workers exposure as much as practicable. At the same time, the agency would start proceedings on a permanent standard which would reduce exposure to the lowest feasible level or ban the material if OSHA determines there is a substitute.

Category II would encompass suspect carcinogens-where evidence is only suggestive or from one animal species. Substances in this category would be subject to a permanent standard to reduce workers exposure to a level low enough to prevent acute or chronic toxic effects.

Category III would be an alert area–to include other suspect substances which can't be placed in one of the first two categories because of insufficient data.

Category IV would flag attention to potential dangers--listing substances which could fall into any of the first three categories but are not presently found in U.S. workplaces.

Models will be developed for an emergency temporary standard for dealing with substances in the early stages of Category I, a permanent standard for Category I substances and a permanent standard for Category II suspect carcinogens.

The regulatory procedure goes forward automatically and inexorably once data satisfying Category I are presented to OSHA. In determining whether a material is a carcinogen, OSHA says it will make no distinction between benign and malignant tumors. The proposed regulation assumes there is no safe level of exposure to a carcinogen and would mandate lowest feasible exposure without considering economics or a ban.

An alternative to the proposed cancer policy of OSHA was proposed by the American Industrial Health Council (AIHC) and the AIHC alternative rejects the idea that short-term bacterial test can be used at all to classify substances as carcinogens[81]

The industry proposal stresses the importance of human data as opposed to data developed on laboratory animals, adds that "time to tumor" relationships and cancer causing "potency" must be taken into account in any effort to classify and regulate carcinogens. Under AIHC's proposal, substances would be classified as "known human carcinogens", "confirmed animal oncogens", or "substances for further testing".

Under the first category would be substances which on the basis of human epidemiological data, were either of "high", "intermediate", or "low" cancer potency. "Where epidemiologic evidence shows that exposure under in-use conditions has increased the age-standardized risk of developing any form of cancer by a factor of 10-fold or more", the proposal says, "the agent shall be regarded as a potent carcinogen". Where the increase is between two and 10-fold, the agent would be regarded as one of "intermediate" cancer potency, and where the increase is less than two-fold, the agent would be considered a "weak" carcinogen.

The proposal says that if human data are not adequate to assess potency, animal data may be employed under certain restrictions. The proposal also sets up criteria whereby substances would be classified as "confirmed animal oncogens" of high, intermediate, or low potency. However, AIHC would not allow substances to be classified as confirmed animal oncogens solely on their having reacted "oncogenically" where: "1. Dose via the respiratory route exceeds 1,000 mg/m^3 for the mouse and hamster or 2,000 mg/m^3 for the rat. "2. Dose by the dermal route exceeds 1,500 mg/kg for the mouse and hamster or 3,000 mg/kg for the rat. "3. Dose by the gastrointestinal route exceeds 500 mg/kg/d for a lifetime, equivalent to about 10 g total lifetime dose for the mouse and hamster, and 100 g. total lifetime dose for the rat".

Before regulating, AIHC would require positive epidemiological results of positive results "in at least two different mammalian species", where OSHA would regulate on the basis of either malignant or benign tumors in "(1) humans, or (2) two mammalian test species, or (3) a single mammalian species if those results have been replicated in the same species in another experiment, or (4) a single mammalian species if those results are supported by short-term tests". AIHC contended that "while short-term bacterial (in vitro) tests have value as screening tools, they do not predict similar human responses".[81]

While the major objective of the epidemiological method is to identify carcinogenic hazards within the environment, with a view of their control[82], epidemiology may also contribute to other aspects of environmental research. These include the determination of the socially "acceptable risk", which is of great importance relative to legislation. Further, human data represent the only objective criteria against which in vivo and in vitro laboratory screening techniques may be evaluated[3].

A major requirement in the control of cancer is to determine cost/benefit situations, for which accurate epidemiological data are necessary[3]. In making comparisons it is probably desirable to present the risk of developing a cancer within a lifetime, as this permits assessment of the risk in better perspective than if only age-specific rates are examined[3]. According to Higginson[3], it is most important to distinguish between a marked increase in the relative risk of a rare tumor affecting only a small proportion of the population, and in which the potential risk to society or the group may be quite small, and a relatively moderate increased risk of a common tumour which may be very much more important in terms not only of absolute numbers on society but also for the individual himself. Data of this type are summarized by Higginson[3] in Table 6 which is calculated on a hypothetical male population, according to statistics available from recent epidemiological studies and which can be considered as reasonably representative of the actual situation in a modern industrial state. Although the figures have been rounded off and the potentiality of certain competing risks eliminated, the effect of these more sophisticated calculations would probably not significantly affect the conclusions. Table 6 compares incidence rates per 100,000 population, the total number of cases observed and the relative risk ratio in a hypothetical population of 100 million males. The relative ratios selected are based on the average data available for smokers and non-smokers in the UK and for asbestos workers from several series[82] and from recent reports on vinyl chloride in the US. While the figures may change with the basic data used, the general principles illustrated will essentially remain the same.

The risk of dying before the age of 65 is approximately three in ten for the average male. During this period, for a non-smoker the chance of dying from lung cancer is relatively small. However, the increase in risk may be very much greater for a heavy smoker. On the other hand, workers exposed to very high levels of vinyl chloride have been stated to have an increased risk of 400 times that of the general population for developing angiosarcoma. However, since the latter is a very rare cancer, the absolute risk is still very much less than that of dying from lung cancer in the general population. A number of other occupational risks, as indicated by incidence data, are given in Table 6 as calculated by Pochin[83]. Such calculations may assist appropriate management and labour executives to place the risks in perspective for the individual workers as compared to the general population.

In the evaluation of the effects of mutagenic chemicals it is important to distinguish four distinct phases, e.g., (1) primary identification or detection of mutagenic activity; (2) verification; (3) quantification and (4) extrapolation to man[84].

Table 6. Comparison Between Incidence, Number of Cases in Population and Relative Risks[3].

Bronchogenic carcinoma in	Incidence per 10^5	Number of Cases Observed	Relative Risk Ratio
Non-smokers (say 30% population)	8	2,500	1
General population	60	60,000	7
All smokers (say 70% population)	85	57,500	10
Heavy smokers (say 25% population)	130	32,500	15
Non-smoking asbestos workers (say 20,000 persons)[a]	8	2.5	1
Smoking asbestos workers (say 20,000 persons)	560	400	70+

[a]While the incidence of lung carcinoma in non-smoking asbestos workers is not greater than that in the general population, the data are insufficient to say with certainty that lung cancer is not increased as compared with non-smokers in the general population. Mesotheliomas have been excluded.

Table 7. Estimated Rates of Fatality (or incidence) of Disease Attributed to Types of Chemical or Physical Exposure[3].

Occupation	Cause of Fatality	Rate (D/M/Y)
Workers with cutting oils- Birmingham	Cancer of the scrotum	60
Shoe industry (press and finishing rooms)	Nasal cancer	130
Printing trace workers	Cancer of the lung and bronchus	About 200
Wood machinists	Nasal cancer	700
Uranium mining	Cancer of the lung	1500
Coal carbonizers	Bronchitis and cancer of the bronchus	2800
Rubber mill workers	Cancer of the bladder	6500
Mustard gas manufacturing (Japan 1929-45)	Cancer of the bronchus	10,400
Nickel workers (before 1925)	Cancer of the lung	15,500
β-Naphthylamine	Cancer of the bladder	24,000

The problem of extrapolating human risk assessment from current mutation tests is a particularly vexing and complicated one. In the main there is a paucity of data concerning a comparison of interlaboratory results from like tests as well as elaborating the same chemical in a wide range of tests involving different species.

Differences in results might be attributable to differences in pharmacokinetics among species. While results across both laboratory and phylogenic lines can be comparable qualitatively there can be quantitative differences and these differences can have implications for risk assessment[84-86].

The genetic changes that may be induced are of different types and include (1) mutations (molecular alterations of the genes) invisible under the microscope; (2) chromosome breaks which may lead to chromosomal aberrations; (3) changes in the number of chromosomes, resulting from errors in the distribution of chromosomes at cell division, and leading to aneuploidy. Only when these changes occur in the germinal tissue, contained in the gonads, does the possibility of transmission to the offspring exist. All these genetic changes pose harmful effects in the descendents where they come to expression.

Extrapolation from experimental organisms to man presents the greatest difficulties, since chemical mutagens are characterized by great specificity of action with regard to the spectrum of genetic changes, organisms or cell types.

In Drosophila, for example, most pre-carcinogens (requiring metabolic activation), produce high levels of gene mutations, but no or relatively few chromosome breakage effects, such as translocations, dominant lethals, or chromosome loss[84]. Even potent chromosome breaking substances appear to require considerably higher concentrations to elicit chromosome breakage than mutations[84]. The same situation may well be true for mammals. Sobels[84] has stated that "since all routine mammalian assay systems (e.g., dominant, lethals, cytogenetic assays, micro-nucleic, heritable translocations) in vivo rely on the detection of chromosome breakage, they cannot be considered as diagnostic for the induction of mutations and may well generate false negatives".

While an assessment of the possible genetic hazards involved requires a quantification in terms of dose-effect curves, it is generally acknowledged that these cannot, at present, be determined in the absence of suitable fast tests for gene mutations in the intact mammal[84]. Hence at present regulatory measures in the main have to rely on confirmatory evidence from a battery of different test systems.

Systematic step-wise comparisons for different end-points of genetic damage at different concentration levels ("the parallelogram") has been suggested by Sobels[84] as a step to obtain better estimates for the induction of mutations. This approach involves extrapolations, in vitro-in vivo, that employ cytogenic damage in mammalian assay systems, to calibrate for the induction of mutations as obtained from mammalian cell lines and host-mediated assays[84].

It is clear according to Sobels[84], the additional extensive studies are required in comparative mutagenesis to define the detection capacity (both qualitatively and quantitatively) for different end-points of genetic damage (e.g., gene mutation, chromosomal aberrations, non-disjunction) and malignant transformation.

To be able to extrapolate with confidence from the results of mutagenic tests with laboratory species to man, it is necessary to demonstrate the ability to extrapolate among laboratory test systems[87].

If one asks only if a particular test system has a positive or negative response to the agents being tested, only the qualitative response of the test is extrapolated among laboratory test systems. However, in tier approaches to mutagenic tests, as for example that proposed by Bridges[88] tier three is far more demanding in that it asks for a quantitative estimation of the mutagenic risk-benefit type analysis in which the benefit to mankind is weighed against the genetic risk of using the particular material in the environment. In extrapolating from experimental organisms to man at the tier three level it is necessary to ask not only "will the agent be mutagenic in man" but also: "can the mutational response in experimental species be related in a quantitative way to response in man?"[87]

Among diverse species, we would expect the reaction of the genetic material to have greater similarity, than the metabolism of chemical agents which tends to vary among species more than does the genetic mechanism[87]. Consequently, it is important to distinguish between differences in response in the metabolism of test organisms and man and differences in response of the genetic material to chemical mutagens. Hence, comparative physiology and comparative mutagenesis must be distinguished[87].

In order to separate comparative physiology from comparative mutagenesis it is necessary to distinguish between exposure and dose. In radiation biology dose defined in terms of "rads" is the absorption of energy at the place of interest. Actually what is referred to as dose in the chemical mutation literature (for example, concentration of a chemical in the feeding media, amount and concentration injected, or length of time exposed to a gas) is really exposure of the whole organism[89]. The importance of the distinction between exposure and dose becomes apparent when one considers all the possible physiological effects and chemical interactions which could occur after exposure, for example: assimilation, circulation, transport across numerous cell membranes, the metabolic activity of the intact animal and enzymatic degradation or activation of the compounds, as well as the physiological activity of microflora in the digestive tract. Saturation of any required enzyme system may alter relationship of exposure to dose[87,89].

According to Lee[87] there is no justification in assuming linearity between exposure and dose to the germ line for chemical mutagens. The distinction between dose to the germ line and exposure of the organism is essential in developing dosimetry for comparative mutagenesis. Experimental protocols may be developed to give a linear relation between exposure and dose for a range of exposures, but this relation must be experimentally verified for each mutagen and protocol for exposure and cannot be assumed[87].

When studying chemical mutagens dose must be determined within defined germ cell stages, for extreme differences in germ cell sensitivity are well established. For example, EMS induces a high frequency of mutations in late spermatid and mature spermatozoa, while inducing a low mutation frequency in spermatogonia[90]. In contrast, chloroethyl methanesulfonate induces a higher mutation frequency in spermatogonia in postmeiotic stages[91,92]. Loveless[93] compiled a table of a number of alkylating agents that showed their varied effects on different germ cell stages. It was found that the range of response among different germ cell lines is equal to or possibly exceeds the limits of sensitivity of the genetic test[94].

It is only the germ cells in multi-cellular organisms that must be considered in the case of heritable mutations, hence, only the amount of the mutagen that penetrates to the germ line is of genetic significance.

At the most basic level, measurement of molecular dose is aimed at determining the number of chemical alterations (lesions) induced in the genetically significant target (GST) by an administered concentration of a chemical agents[95]. Once this molecular dose has been established for genetic test systems it then becomes possible to relate mutagenic response to the number of chemical alterations in the GST and to make valid comparisons between the mutagenic responses of different species to the chemical agent[95].

The chromosomes of higher animals contain most of their genetic information. Among these animals there has been a remarkable evolutionary conservation of the basic features of chromosome composition[96,97]. In contrast to that of prokaryotes the eukaryote biochemical dissymmetry ratio $(A + Y)/(G + C)$ is relatively constant. The distinguishing features of the eukaryote also include multiple chromosomes and a very well organized means of insuring fidelity of information transfer to daughter cells[98]. These eukaryote chromosomes are composed of DNA and several classes of protein. The histones, usually divided into 5 classes depending on the amino acid composition, are the most common chromosomal protein and are potentially nucleophilic molecules, having non-covalently banded amino groups at frequent intervals[99].

The selection of appropriate target molecules for determining molecular dose of chemical mutagens to the germ line has been recently described by Aaron[98]. There are many intracellular nucleophiles, but the group of molecules not likely to reflect the events in the gene are those that are known to be physically close to DNA in the genome. This purely biological consideration limits the range of selected reaction products (SRP) that can be considered. For purposes of detection, chemically stable reaction products for the estimator of the mutagen dose are required. Alkylated DNA in sperm cells of Drosophila melanogaster[89,100] and in the mouse[101] have been used for molecular dosimetry.

Another type of SRP is alkylated protein. Ehrenberg et al[102] recently proposed the use of testicular tissue dose as a correlate of genetic damage for quantitative risk estimates. This method has been criticized by Aaron[100] since it does not resolve the cellular differences in nucleophilic composition and hence is unlikely to be useful for correlations of known differences in genetic cell stage sensitivity. In addition, average alkylation level of protein in the tissue does not necessarily measure only the products in the subcellular genetic composition.

Because of the constant relationship between initially formed products in parallel irreversible reactions any stable product within the subcellular genetic compartment should correlate with the pre-mutational lesion. Furthermore, correlation between a product of nucleophilic substitution and the genetic response is not evidence that the measured SRP is the genetically significant target[100].

Molecular dosimetry is only now becoming feasible. It will undoubtedly be quite a number of years before enough data is accumulated in numerous laboratories and with a variety of species so that valid comparisons among species can be made[100].

Osterman-Golkar et al[103] reported on the use of haemoglobin of erythrocytes as a monitor of the in vivo dose after exposure to the directly alkylating agents ethyelene oxide[103] and after exposure to dimethylnitrosamine[103] and vinyl chloride[104] which both need metabolic activation. These studies indicated that, provided the reaction products with haemoglobin have a certain stability and the life-time of haemoglobin is not affected, determinations of the degree of alkylation of specific amino acids in haemoglobin might be used to integrate the dose of a chemical over a considerable period[104].

Segerback et al[105] recently reported on the quantitative determination of alkylated amino acids in haemoglobin as a measure of the dose after treatment of mice with methyl methanesulfonate (MMS). The degree of alkylation of hemoglobin exhibited a linear dependence on the quantity of MMS injected while the degree of alkylation of guanine-N-7 in DNA indicated a slight positive deviation from linearity at high doses.

The practical determination of alkylated products as a consequence of exposure to a methylating agent may be complicated by a contribution from endogenous methylations in vivo[106].

Claxton and Barry[107] consider the assessment of genetic risk for human populations to be exceptionally problematic. In lower organisms dose-response curves relating exposure to genetic damage may be generated through experiments; such dose-response curves will not be generated for human populations. While some data may be obtained from those exposed through occupational or other unwitting contact, in practice most dependence will be on extrapolations from whole mammal tests to estimate the risk posed by each mutagen[108].

A basic unit of mutagenic activity would be useful to compare the risks of various chemical mutagens. Two units have been suggested, one based on the spontaneous mutation rate and one on radiation equivalents[108]. All test systems have a spontaneous mutation rate, which is a low-level background rate of mutation occurring without the presence of a known mutagen. One proposed unit of chemical mutagenic activity would be that concentration of mutagen which produces a doubling of the spontaneous mutation rate. For mammals, this could be expressed in milligrams per kilogram of body weight.

Radiation biology has quantified doses of ionizing radiation of several types in terms of biological damage, thereby allowing comparisons between sources. The REM (radiation-equivalent-man) is a unit dose equivalence of radiation used to compare sources of ionizing radiation with respect to biological damage of a specified sort[109]. The REC (rem-equivalent-chemical) has been proposed as a unit which would allow comparison between radiation and chemical mutagens. The REC would be that dose concentration multiplied by time which produces the same amount of genetic damage as one REM of chronic irradiation[109,110].

Although the use of both the doubling concentration and the REC have serious technical drawbacks, the development of such a unit could provide a quantitative measure of mutagenic activity which would be useful in determining and comparing genetic risks of chemical mutagens.

The possibility of calculating threshold levels for a number of mutagenic substances (regardless of whether they produce point mutations, base substitutions, cross linkages in the DNA helices, chromosome breaks, or other abnormalities) was proposed by Claus[111,112]. In this study, quantitative estimates were made on molecules per cell basis for selected chemicals (e.g., trimethyl phosphate, dichlorovos, ethylmethane-sulfonate, methane sulfonate and N-methyl-N'-nitro-N-nitrosoguanidine) which have been demonstrated to be mutagenic in order to show the amounts required for a single point mutation, a base substitution or for chromosome breakage. When these quantities were compared to background mutation rates, it was concluded by Claus[111] that the burden represented by man-made chemicals at the present time is miniscule in comparison to the background levels. Additionally several orders of magnitude larger quantities from the most potent mutagens would have to be released into the environment before their effect could even be recognized.

In contrast to the above views of Claus[111,112] concerning threshold levels of mutagens, one must take note of the admonishments of Ames[113] that "it is worth emphasizing however, that one molecule of a mutagen is enough to cause a mutation and that if a large population is exposed to a "weak" mutagen it may still be a hazard to the human germ line. Since no repair system is completely effective, there may be no such thing as a completely safe dose of a mutagen". Ames[113] further states that "if a compound is a mutagen in any organism then it should not be used on humans unless there is a definitive evidence that it is neither mutagenic nor carcinogenic in animals, or unless the benefit outweights the possible risk".

References

1. Doll, R., Muir, C. S., and Waterhouse, J. A. H. (eds) "Cancer Incidence in Five Continents", Springer-Verlag, Berlin, Heidelberg, New York (1970)
2. Higginson, J., "Present Trends in Cancer Epidemiology", In Proc. 8th Canadian Cancer Conference, Honey Harbour, ONtario (1969) 40-75
3. Higginson, J., Importance of Environmental Factors in Cancer, Inserm Symposia Series, No. 52 (1976) 15-24; IARC Scientific Publ. No. 13, International Agency for Research on Cancer, Lyon (1976)
4. Higginson, J., and Muir, C. S., Epidemiology, In Cancer Medicine, (eds) Holland, J. F., and Frei, E., Lea & Febiger, Philadephia, PA, (1973) 241-306
5. Case, R. A. M., Hosker, M. E., McDonald, D. B. and Pearson, J. T., Tumors of the urinary bladder in workmen engaged in the manufacture and use of certain dyestuff intermediates in the British Chemical Industry, I. The role of aniline, benzidine, alpha-naphthylamine and beta-naphthylamine, Brit. J. Ind. Med. 11 (1954) 75-104
6. Haenszel, W., Cancer mortality among the foreign born in the United States, J. Natl. Cancer Inst., 26 (1961) 37-132
7. Haenszel, W., and Kurihara, M., Studies of Japanese migrants, I. Mortality from cancer and other diseases among Japanese in the United States, J. Natl. Cancer Inst., 40 (1968) 43-68
8. Shubik, P., and Clayson, D. B., Application of the results of carcinogen bioassays to man, Inserm Symposia Series Vol. 52 (1976) 241-252; IARC Scientific Publ. No. 13, International Agency for Research on Cancer, Lyon (1976)
9. Hammond, E. C., The epidemiological approach to the etiology of cancer, Cancer 35 (1975) 652-654
   Janerich, D. T., and Lawrence, C. E., Epidemiological strategies for identifying carcinogens, Mutation Res., 33 (1975) 55-63
11. Miller, R. W., Genetics of human cancer: An epidemiologists view, In Genetics of Human Cancer, (eds) Mulvihill, J. J., Miller, R. W., and Fraumeni, J. F., Jr., Raven Press, New York (1977) 481-482
12. Creech, J. L., Jr. and Johnson, M. N., Angiosarcoma of liver in the manufacture of polyvinyl chloride, J. Occup. Med., 16 (1974) 150-151
13. Selikoff, I. J., Recent perspectives in occupational cancer, Ambio, 4/1 (1975) 14-17
14. IARC, IARC Monograph Programme on the Evaluation of the Carcinogenic Risk of Chemicals to Humans, IARC Internal Tech. Rept. No. 77/002, International Agency for Research on Cancer, Lyon, October 1977, pp. 15-17
15. Tomatis, L., The value of long-term testing for implementation of primary prevention In "Origins of Human Cancer" (eds) Hiatt, H. H., Watson, J. D., and Winsten, J. A., Cold Spring Harbor, New York, Cold Spring Harbor Laboratory (1977)
16. Tomatis, L., Agthe, C., Bartsch, H., Huff, J., Montesano, R., Saracci, R., Walker, E., and Wilbourn, Evaluation of the carcinogenicity of chemicals: A review of the Monograph Program of the International Agency for Research on Cancer (1971-1977), Cancer Res., 38 (1978) 377-885
17. Rall, D. P., Species differences in carcinogenesis testing, In "Origins of Human Cancer", (eds) Hiatt, H. H., Watson, J. D., and Winsten, J. A., Cold Spring Harbor, New York, Cold Spring Harbor Laboratory (1977)
18. Health and Welfare of Canada "The Testing of Chemicals for Carcinogenicity, Mutagenicity and Teratogenicity", Ottawa, Canada, Sept. 1973

19. Weinhouse, S., Problems in the assessment of huma risk of carcinogenicity by chemicals, In "Origins of Human Cancer" (eds) Hiatt, H. W., Watson, J. D., and Winsten, J. A., Cold Spring Harbor, New York, Cold Spring Harbor Laboratory (1977)
20. Roe, F. J. C., and Tucker, M. J., Recent developments in the design of carcinogenicity tests on laboratory animals, Proceedings of the European Society for the Study of Drug Toxicity and Their Significance in Man, Vol. XV, Zurich, June (1973) 171-177
21. IARC Monographs on the Evaluation of Carcinogenic Risk of Chemicals to Man, Volume 1, International Agency for Research on Cancer, 1972, Lyon France
22. IARC Monographs on the Evaluation of Carcinogenic Risk of Chemicals to Man, Volume 10, International Agency for Research on Cancer, 1976, Lyon, France
23. Upton, A. C., Comparative observations on radiation carcinogenesis in man and animals. Carcinogenesis. A Broad Critique: 631-675, The Williams and Wilkins Company, Baltimore, 1967
24. Schneiderman, M. A., Mantel, N., and Brown, C. C., From mouse to man- or how to get from the laboratory to Park Avenue and 59th Stree, Annals of the New York Academy of Sciences, 1975, 246, 237-248
25. Mantel, N., Bohidar, N. R., Brown, C. C., Ciminera, J. L. and Tukey, J. W., An improved Mantel-Bryan procedure for "safety" testing of carcinogens, Cancer Res., 25 (1975) 865-872
26. Druckrey, H., Pharmacological approach to carcinogenesis, In Ciba Foundation Symposium on Carcinogenesis: Mechanisms of Action, (eds) Wolstenholme and O'Connor, Churchill, London (1959) p. 110
27. Berenblum, I., and Shubik, P., New quantitative approach to the study of stages of chemical carcinogenesis in the mouse's skin, Brit. J. Cancer, 1 (1947) 383-391
28. Berenblum, I., and Shubik, P., The persistence of latent tumour cells induced in the mouse's skin by a single application of 9,10-dimethyl-1,2-benzanthracene, Brit. J. Cancer, 3 (1949) 384-386
29. D'Aguanno, W., Interpretation of test results in terms of carcinogenic potential to the test animal: The regulatory point of view, In Carcinogenesis Testing of Chemicals, (ed) Golberg, G., CRC Press, Cleveland, Ohio, (1973) pp. 41-44
30. Dinman, D. B., "Non-concept" of "no-threshold": Chemicals in the environment, Science, 175 (1972) 495
31. Friedman, L., Problems of evaluating the health significance of the chemicals in foods, Pharmacology and the Future of Man, Proceedings of the Fifth International Congress on Pharmacology, San Francisco, (1972), Vol. 2, Karger, Basel, (1973) p. 30
32. Gaylor, D. W., and Shapiro, R. E., Extrapolation and risk estimation for carcinogenesis, In "Advances in Modern Toxicology" (eds) Mehlman, M. A., Blumenthal, H., and Shapiro, R. E., Hemisphere Press, Washington, DC (1977) in press
33. Brown, C. C., Mathematical aspects of dose-response studies in carcinogenesis- The concepts of thresholds, Oncology, 33 (1976) 62-65
34. Albert, R. E., Train, R. E., and Anderson, E., Rational developed by the Environmental Protection Agency for the assessment of carcinogenic risks, J. Natl. Cancer Inst., 58 (1977) 1537-1541
35. Cornfield, J., Carcinogenic risk assessment, Science, 198 (1977) 693-699
36. Newill, V. A., Methodologies of risk assessment, Ann. NY Acad. Sci., 271 (1976) 413-417

37. Albert, R. E., and Altshuler, B., Asessment of environmental carcinogen risks in terms of life shortening, Env. Hlth. Persp., 13 (1976) 91-94
38. Hoel, D. G., Statistical extrapolation methods for estimating risks from animal data, Ann. NY Acad. Sci., 271 (1976) 418-420
39. Crump, K. S., Hoel, D. G., Langley, C. H., and Peto, H., Fundamental carcinogenic processes and their implications for low dose risk assessment, Cancer Res., 36 (1976) 2973-2979
40. Gehring, P. J., and Blau, G. E., Mechanisms of carcinogenesis: Dose reponse, J. Env. Pathol. Toxicol., 1 (1977) 163-179
41. Hoel, D. G., Gaylor, D. W., Kirschstein, R. L., Saffiotti, U., and Schneiderman, M. A., Estimation of risks of irreversible delayed toxicity, J. Toxicol. Environ. Hlth., 1 (1975) 133-151
42. Hoel, D. G., Biometry branch: Summary statements, Env. Hlth. Persp., 20 (1977) 217-219
43. Blum, H. F., (ed.) Carcinogenesis by Ultraviolet Light, Princeton University, Press, Princeton, NJ (1959)
44. Druckrey, H., Quantitative aspects of chemical carcinogenesis, In Potential Carcinogenic Hazards from Drugs (Evaluation of Risks), R. Truhaut, (ed.) UICC Monograph Series, Vol. 7; 60-78, Springer-Verlag, New York, NY (1967)
45. Albert, R. E., and Altshuler, B., Considerations relating to the formation of limits for unavoidable population exposure to environmental carcinogens, In Radionuclide Carcinogenesis, J. E. Ballour et al. (ed) 233-253, AEC Symposium Series, Conf-72050, NTIS, Springfield, VA (1973)
46. Peto, R., Lee, P. N., and Paige, W. S., Statistical analysis of the bioassay of continuous carcinogens, Brit. J. Cancer, 26 (1972) 258-261
47. Peto, R., and Lee, P. N., Weibull distributions for continuous-carcinogenesis experiments, Biometrics, 29 (1973) 457-470
48. Day, T. D., Carcinogenic action of cigarette smoke condensate on mouse skin, Brit. J. Cancer, 21 (1967) 56-81
49. Pike, M. C., A method of analysis of a certain class of experiments in carcinogenesis, Biometrics, 22 (1966) 142-161
50. Arley, N. & Iversen, N., On the mechanism of experimental carcinogenesis, Acta. Path. Microbiol. Scand., 31 (1952) 164-171
51. Armitage, P., and Doll, R., Stochastic models for carcinogenesis, Proc. Fourth Berkely Symp. Mathematical Statistics and Probability, Vol. 4; 19-38, University of California Press, Berkeley, CA (1961)
52. Neyman, J., and Scott, E. L., Statistical aspect of the problem of carcinogenesis, Proc. Fifth Berkeley Symp. Mathematical Statistics and Probability, Vol. 4, 745-776, University of California Press, Berkeley, CA (1965)
53. Iverson, S., and Arley, N., On the mechanism of experimental carcinogens, Acta. Pathol. Microbiol. Scand., 27 (1950) 773-803
54. Chand, N., and Hoel, D. G., A comparison of models for determining safe levels of environmental agents, In "Reliability and Biometry: Statistical analysis of life length" (eds) Proschan, F., and Serfling, Society for Industrial and Applied Math, Philadelphia, PA (1973)
55. Turner, M. E., Some classes of hit-theory models, Math. Biosc., 23, (1975) 219-235
56. Nordling, C. O., A new theory on the cancer inducing mechanisms, Brit. J. Cancer, 7 (1953) 68-72
57. Stocks, P., A study of the age curve for cancer of the stomach in connection with a theory of the cancer producing mechanism, Brit. J. Cancer, 7 (1953) 407-417

58. Mantel, N., The concept of threshold in carcinogenesis, Clin. Pharmac. Ther., 4 (1963) 104-109
59. Mantel, N., and Bryan, W. R., "Safety" testing of carcinogenic agents, J. Natl. Cancer Inst., 27 (1961) 455-470
60. Jones, H., Dose-effect relationships in carcinogenesis and the matter of threshold in carcinogenesis; NIEHS Conference on Problems of Extrapolating the Results of Laboratory Animal Data to Man and of Extrapolating the Results from High Dose Level Experiments to Low Dose Level Experiments, Pinehurst, North Carolina, March 10-12 (1976)
61. Robbins, J. H., Kraemer, K. H., and Lutzner, M. A., et al., Xeroderma pigmentosum, An inherited disease with sun sensitivity, multiple cutaneous neoplasms, and abnormal DNA repair, Ann. Intern. Med., 80 (1974) 221-248
62. German, J., Genes which increase chromosomal instability in somatic cells and predispose to cancer, Prog. Med. Genet., 8 (1972) 61-101
63. Kellerman, G., Shaw, C. R., and Luyten-Kellerman, M., Aryl hydrocarbon hydroxylase inducibility and bronchogenic carcinoma, New Engl. J. Med., 289 (1973) 934-937
64. Weisberger, J. H., Chemical Carcinogenesis, In "Toxicology" eds. Casarett, L. J., and Doull, J., MacMillan, New York (1975) pp. 333-378
65. Druckrey, H., Quantitative aspects in chemical carcinogenesis, In Potential Carcinogenic Hazards from Drugs, Evaluation of Risks, eds. Truhaut, R., UICC Monograph Series Vol. 7, Springer-Verlag (1967) pp. 60-78
66. Albert, R. E., and Altshuler, B., Considerations relating to the formulations of limits for unavoidable population exposures to environmental carcinogens, In Radionuclide Carcinogenesis, Proceedings of the 12th Annual Hanford Biology Symposium at Richland, Washington, (1973) 234-253
67. Jones, H. B., and Grendon, A., Environmental factors in the origin of cancer and estimation of possible hazard to man, Food Cosmet. Toxicol., 13 (1975) 251-269
68. Laroye, G. J., How efficient is immunological surveillance against cancer and why does it fail? Lancet, i (1974) 1097-1100
69. Roe, F. J. C., and Tucker, M. J., Recent developments in the design of carcinogenicity tests on laboratory animals, In Experimental Model Systems in Toxicology and their Significance for the Study of Drug Toxicity, Vol. 15, New York, American Elsevier Publ. Co., (1974) pp. 171-177
70. Riley, V., Mouse mammary tumors, Alteration of incidence as apparent function of stress, Science, 189 (1975) 465-467
71. Weisburger, J. H., Environmental cancer, J. Occup. Med., 18 (1976) 245-252
72. Wynder, E. L., Nutrition and cancer, Fed. Proc., 35 (1976) 1309-1315
73. Shamberger, R. J., Anderone, T. L., and Willis, C. E., Antioxidants and cancer IV. Initiating activity of malonaldehyde as a carcinogen, J. Natl. Cancer Inst., 53 (1974) 1771-1773
74. Stokinger, H. E., The case for carcinogen TLV's continues strong, Occup. Hlth. Safety, March/April (1970) 54-60
75. Flowers, E., Proposal: A non-zero approach to environmental limits for carcinogenesis, Occup. Hlth. Safety, May/June (1977) 35-39
76. Morgan, K. Z., Permissable exposure to ionizing radiation, Science, 139 (1963) 565-581
77. Committee 17 (John Drake, Chairman) Council of the Environmental Mutagen Society, Environmental Mutagenic Hazards, Science, 187 (1975) 503-514
78. Executive Committee of the National Academy of Sciences: An assessment of present and alternative technologies, Vol. 1., In Contemporary Pest Control Practices and Prospects: A Report. Washington, DC., Natl. Acad. Sci. (1975) 66-82

79. National Cancer Advisory Board Subcommittee on Environmental Carcinogenesis: General criteria for assessing the evidence for carcinogenicity of chemical substances: A Report. J. Natl. Cancer Inst., 58 (1977) 461-465
80. Anon, OSHA proposes sweeping plan to curb workplace carcinogens, Chem. Ecology Nov. (1977) 5
81. Anon, Industry alternative to OSHA Cancer Policy rejects in vitro test entirely, Pesticide & Toxic News, 6 (8) (1978) 7-9
82. Higginson, J., Developments in cancer prevention through environmental control. In Advances in Tumour Prevention, Detection and Characterization, Volume 2, Cancer Detection and Prevention, C. Maltoni (ed), pp. 3-18, Excerpta Medica, Amsterdam
82a. Biological effects of asbestos, P. Bogovski, J. C. Gilson, V. Timbrell, and J. C. Wagner, (eds) IARC Scientific Publication No. 8, International Agency for Research on Cancer, Lyon, France
83. Pochin, E.E., Occupational and other fatality areas, Community Health, 6 (1974) 2-13
84. Sobels, F. H., Some problems associated with the testing for environmental mutagens and a perspective for studies in "comparative mutagenesis", Mutation Res., 46 (1977) 245-260
85. Anon, Can human risk assessment flow from current mutation tests? Experts differ Pesticide & Toxic Chem. News, 5 (51) (1977) 26-27
86. NIEHS, Conference on Comparative Mutagenesis Workshop, Research Triangle Park, NC, October 31-November 4, 1977
87. Lee, W. R., Molecular Dosimetry of Chemical Mutagens. Determination of molecular dose to the germ line, Mutation Res., 38 (1976) 311-316
88. Bridges, B. A., The three-tier approach to mutagenicity screening and the concept of radiation-equivalent dose, Mutation Res., 26 (1974) 335-340
89. Lee, W. R., Comparison of the mutagenic effects of chemicals and ionizing radiation using Drosophila melanogaster test systems, In: Nygaard, O. F., Adler, H. I., and Sinclair, W. K. (eds), Radiation Research, Biomedical Chemical and Physical Perspectives, Proceedings of the 5th International Congress of Radiation Research, Academic Press, New York (1975) pp. 976-983
90. Jenkins, J. B., Mutagenesis at a complex locus in Drosophila with monofunction alkylating agent, ethylmethanesulfonate, Genetics, 57 (1967) 783-793
91. Fahmy, O. G., and Fahmy, M. J., Mutagenicity of 2-chloroethyl methanesulfonate in Drosophila melanogaster, Nature, 177 (1956) 996-997
92. Matthew, C., The nature of delayed mutation after treatment with chloroethyl methanesulfonate and other alkylating agents, Mutation Res., 1 (1964) 163-172
93. Loveless, A., Genetic and allied effects of alkylating agents, Pennsylvania State Univ. Press (1966) p. 270
94. Ehling, U. H., Cumming, R. B., and Malling, H. V., Induction of dominant lethal mutations by alkylating agents in male mice, Mutation Res., 5 (1968) 417-428
95. Sega, G. A., Molecular dosimetry of chemical mutagens: Measurement of molecular dose and DNA repair in mammalian germ lines, Mutation Res., 38 (1976) 317-326
96. Delange, R. J., and Smith, E. L., Histones: structure and function, Ann. Rev. Biochem., 40 (1971) 279-814
97. Taylor, J. H., The structure and duplication of chromosomes, In: "Genetic Organization" Vol. I, eds., Caspari, E. W., and Ravin, A. W., Academic Press, New York (1969)
98. Aaron, C. S., Molecular dosimetry of chemical mutagens: Selection of appropriate target molecules for determining molecular dose to the germ line, Mutation Res. 38 (1976) 303-310

99. Delange, R. J., and Smith, E. L., The structure of histones, Acc. of Chem. Res. 5 (1972) 368-373
100. Aaron, C. S., Lee, W. R., Janca, F. C., and Seamster, P. M., Ethylation of DNA in the germ line of male Drosophila melanogaster, Genetics, 74 (1973) 1
101. Sega, G. A., Cumming, R. B., and Walton, M. F., Dosimetry studies on the ethylation of mouse sperm DNA after exposure to (^{3}H)-ethylmethanesulfonate; Mutation Res., 24 (1974) 317-333
102. Ehrenberg, L., Hiesche, K. D., Osterman-Golkar, S., and Wennberg, I., Evaluation of genetic risks of alkylating agents; tissue doses in the mouse from air contaminated with ethylene oxide, Mutation Res., 24 (1974) 83-103
103. Osterman-Golkar, S., Ehrenberg, L., Segerback, D., and Hallstrom, I., Evaluation of genetic risk of alkylating agents, II. Haemoglobin as a dose monitor, Mutation Res., 34 (1976) 1-10
104. Osterman-Golkar, S., Hultmark, D., Segerback, D., Calleman, C. J., Gothe, R., Ehrenberg, L., and Wachtmeister, C. A., Alkylation of DNA and proteins in mice exposed to vinyl chloride, Biochem. Biophys. Res. Commun., 7 (1977) 259-266
105. Segerback, D., Calleman, C. J., Ehrenberg, L., Logroth, G. and Osterman-Golkar, S., Evaluation of genetic risks of alkylation agents. IV. Quantitative determination of alkylated amine acids in haemoglobin as a measure of the dose after treatment of mice with methylmethanesulfonate, Mutation Res., 49 (1978) 71-82
106. Paik, W. K., and Kim, S., Protein methylation, enzymatic methylation of proteins after translations may take part in control of biological activities of proteins, Science, 17 (1971) 114-119
107. Claxton, L. D., and Barry, P. Z., Chemical Mutagenesis: An emerging issue for public health, Am. J. Publ. Hlth., 67 (1977) 1037-1042
108. Committee 17, Environmental Mutagen Society, Environmental Mutagenic Hazards, Science, 187 (1975) 503-514
109. Advisory Committee on the Biological Effects of Ionizing Radiation. The Effects on Populations of Exposure to Low Levels of Ionizing Radiation. National Academy of Sciences, National Reserach Council, Washington, DC (1972)
110. Bridges, B. A., The three-tier approach to mutagenicity screening and the concept of radiation-equivalent dose, Mutation Res., 26 (1974) 335-340
111. Claus, G., Bolander, K., and Krisko, I., Man-made chemical mutagens in the natural environment: An evaluation of hazards, Studia Biophysica, 50 (1975) 123-126
112. Claus, G., Krisko, I., and Bolander, K., Chemical carcinogens in the environment and in the human diet: Can a threshold be established? Food Cosmet. Tox., 12 (1974) 737-746
113. Ames, B. N., The detection of chemical mutagens with enteric bacteria, In "Chemical Mutagens", Vol. 1 (ed) Hollaender, A., Plenum Press, New York (1971) 267-282

POTENTIAL INDUSTRIAL CARCINOGENS AND MUTAGENS

Class	Chemical Name & Synonym	CAS#	Carcinogenicity	Mutagenicity - Bacteria	Yeast	Neurospora	Drosophila	Mammalian Cells	Human Cells	Dominant Lethal	Host Mediated	Production Quantities [a]
I. Alkylating Agents												
J. Halogenated Alkanols												
ClCH$_2$CH$_2$OH	2-Chloroethanol (Ethylene-chlorohydrin)	107-07-3	−	+	0	0	0	0	0	0	0	>1000 lbs (1974)
K. Halogenated Ethers												
ClCH$_2$OCH$_3$	Methylchloromethylether* [b] (CMME)	107-30-2	+	+	0	0	0	0	0	0	0	>1000 lbs (1974)
ClCH$_2$OCH$_2$Cl	Bis(chloromethyl)ether* [b] (BCME)	432-88-1	+	+	0	0	0	0	0	0	0	<1000 lbs (1973)
Cl(CH$_2$)$_2$O(CH$_2$)$_2$Cl	Bis(2-chloroethyl)ether [b]	111444	±	±	+	0	+	0	0	0	0	<1000 lbs (1974)
L. Aldehydes												
HCHO	Formaldehyde	50-00-0	±	±	0	+	+	0	0	0	0	5765 x 10^6 lbs (1974)
CH$_3$CHO	Acetaldehyde	75-07-0	0	0	0	0	+	0	0	0	0	1670 x 10^6 lbs (1976)
CH$_2$=CHCHO	Acrolein	10202-8	0	+	0	0	+	0	0	+	0	61 x 10^6 lbs (1974)

POTENTIAL INDUSTRIAL CARCINOGENS AND MUTAGENS

Class	Chemical Name & Synonym	CAS#	Carcinogenicity	Mutagenicity - Bacteria	Yeast	Neurospora	Drosophila	Mammalian Cells	Human Cells	Dominant Lethal	Host Mediated	Production Quantities [a]
II. Acylating Agents												
(CH$_3$)$_2$N-C(O)-Cl	Dimethylcarbamoyl Chloride [b] (DMNG)	79-44-7	+	+	0	0	0	0	0	0	0	<1000 lbs (1974)
(C$_2$H$_5$)$_2$N-C(O)-Cl	Diethylcarbamoyl Chloride	88-10-8	0	+	+	0	0	0	0	0	0	1.5 x 10^4 lbs (1974)
C$_6$H$_5$-C(O)-Cl	Benzoyl Chloride	98-88-4	+	+	0	0	0	0	0	0	0	15 x 10^6 lbs (1972)
CH$_2$=C=O	Ketene	46-35-14	0	0	0	−	+	0	0	0	0	

POTENTIAL INDUSTRIAL CARCINOGENS AND MUTAGENS

Class	Chemical Name & Synonym	CAS#	Carcinogenicity	Mutagenicity									Production Quantities [a]
				Bacteria	Yeast	Neurospora	Drosophila	Mammalian Cells	Human Cells	Dominant Lethal	Host Mediated	
III. Peroxides												
$(CH_3)_3COOC(CH_3)_3$	Di-tert, butylperoxide	110-05-4	0	0	0	+	0	0	0	0	0	3×10^6 lbs (1974)
$(CH_3)_3COOH$	Tert, butylperoxide	75-91-2	0	+	0	+	+	0	0	0	0	>1000 lbs
$C_6H_5C(CH_3)_2OOH$	Cumene Hydroperoxide	80-15-9	0	+	0	+	0	0	0	0	0	3062×10^6 lbs (1974)
$HOOC(CH_2)_2COOH$	Succinic Acid Peroxide	3504130	0	+	0	0	0	0	0	0	0	>1000 lbs (1974)
CH_3COOOH	Peracetic Acid (Peroxy acetic acid; acetylhydreperoxide)	79-21-0	0	+	0	0	0	0	0	0	0	
H_2O_2	Hydrogen Peroxide	772-28-41	0	±	+	+	−	0	0	0	0	1.9×10^5 lbs (1974)

POTENTIAL INDUSTRIAL CARCINOGENS AND MUTAGENS

Class	Chemical Name & Synonym	CAS#	Carcinogenicity	Mutagenicity									Production Quantities [a]
				Bacteria	Yeast	Neurospora	Drosophila	Mammalian Cells	Human Cells	Dominant Lethal	Host Mediated	
IV. Halogenated Unsat'd & Sat'd Hydrocarbons & Aromatic Derivatives												
A. Unsat'd Hydrocarbons												
$CH_2=CHCl$	Vinyl Chloride (VCM)*[b]	75-01-4	+	+	+	−	+	0	+	0	+	5621×10^6 lbs (1974)
$CH_2=CCl_2$	Vinylidene Chloride (1,1-dichloro-ethylene)[b]	75-35-4	±	+	0	0	0	0	0	−	0	60×10^6 lbs (1974)
$ClCH=CCl_2$	Trichloroethylene[b]	79-01-4	±	+	+	0	0	0	0	0	+	610×10^6 lbs (1976)
$CH_2=\underset{Cl}{C}-CH=CH_2$	Chloroprene (2-chloro-1,3-butadiene)*[b]	126-99-8	+	+	0	0	0	0	+	0	0	349×10^6 lbs (1975)
$H-\underset{H}{\overset{Cl}{C}}=\underset{HCl}{\overset{H}{C}}-C-H$	Trans-1,4-dichlorobutene (1,4-dichloro-2-butene)[b]	764-41-0	+	+	+	0	0	0	0	0	0	
$Cl_2C=CCl_2$	Tetrachloroethylene (perchloroethylene)	127-18-4	±	±	0	0	0	0	0	0	+	1210×10^6 (1976)
$Cl_2\text{-}\underset{}{C}=\underset{}{C}\text{-}CCl_2$ (Cl Cl)	Hexachlorobutadine (Perchlorobutadinene) (HCBD)	87-68-3	±	+	0	0	0	0	0	0	0	10×10^6 lbs (1974)[c]
$CH_2=CH\text{-}CH_2Cl$	Allyl Chloride	590-21-6	−	±	−	0	0	0	0	0	0	300×10^6 lbs (1976)

POTENTIAL INDUSTRIAL CARCINOGENS AND MUTAGENS

Class	Chemical Name & Synonym	CAS#	Carcinogenicity	Mutagenicity - Bacteria	Yeast	Neurospora	Drosophila	Mammalian Cells	Human Cells	Dominant Lethal	Host Mediated	Production Quantites [a]
IV. Halogenated Unsat'd & Sat'd Hydrocarbons & Aromatic Derivatives												
Aryl Derivatives												
CH_2Cl (benzene ring)	Benzyl Chloride	100-44-7	+	+	+	0	0	0	0	0	0	80 x 10^6 lbs (1972)
Hexachlorobenzene structure	Hexachlorobenzene	118-74-1	+	0	+	0	0	0	0	—	0	8.5 x 10^6 lbs (1974)
Polyaromatics												
$(Cl)_n$–biphenyl–$(Cl)_n$	Polychlorinated Biphenyls[b]	1336363	±	±	0	0	0	0	—	—	0	40 x 10^6 lbs (1974)

POTENTIAL INDUSTRIAL CARCINOGENS AND MUTAGENS

Class	Chemical Name & Synonym	CAS#	Carcinogenicity	Mutagenicity - Bacteria	Yeast	Neurospora	Drosophila	Mammalian Cells	Human Cells	Dominant Lethal	Host Mediated	Production Quantites [a]
IV. Halogenated Unsat'd & Sat'd Hydrocarbons & Aromatic Derivatives												
B. Sat'd Hydrocarbons												
$CHCl_3$	Chloroform[b]	67-66-3	+	—	0	0	0	0	0	0	0	300 x 10^6 lbs (1974)
CH_2Cl_2	Methylene Chloride	75-09-2	0	+	—	0	0	0	0	0	0	1000 x 10^6 lbs (1976)
CCl_4	Carbon Tetrachloride[b]	56-23-5	+	—	0	0	0	0	0	0	0	1000 x 10^6 lbs (1974)
CH_3Cl	Methyl Chloride	74-87-3	0	+	0	0	0	0	0	0	0	493 x 10^6 lbs (1974)
CH_3I	Methyl Iodide	74-88-4	+	+	+	0	0	0	0	0	0	19,000 lbs (1975)
$ClCH_2CH_2Cl$	1,2-ethylene dichloride	107-06-2	0	+	0	0	+	0	0	0	0	9165 x 10^6 lbs (1974)
$BrCH_2CH_2Br$	1,2-ethylene dibromide	106-93-4	+	+	+	+	+	0	0	0	0	332 x 10^6 lbs (1974)
$CH_3CHClCH_2Cl$	1-2-Dichloropropane	13063-43-9	0	+	0	0	0	0	0	0	0	84.2 x 10^6 lbs (1974)
$Cl_2CH.CH.Cl_2$	1,1,2,2-Tetrachloroethane	79-34-5	+	+	0	0	0	0	0	0	0	1000 lbs (1975)
CH_3CCl_3	1,1,1-Trichloroethane (Methyl Chloroform)	71-55-6	—	+	0	0	0	0	0	0	0	340 x 10^6 lbs (1976)
CCl_3CCl_3	Hexachloroethane	67-72-1	+	0	0	0	0	0	0	0	0	1000 lbs (1975)

POTENTIAL INDUSTRIAL CARCINOGENS AND MUTAGENS

Class	Chemical Name & Synonym	CAS#	Carcinogenicity	Mutagenicity - Bacteria	Yeast	Neurospora	Drosophila	Mammalian Cells	Human Cells	Dominant Lethal	Host Mediated	Production Quantities [a]
V. Hydrazines, Hydroxylamines, Carbamates												
A. Hydrazines												
H_2N-NH_2	Hydrazine [b]	302-01-2	+	+	O	O	+	O	O	−	+	3.1×10^6 lbs (1971)
$(CH_3)_2N-NH_2$	1,1-Dimethylhydrazine [b] (UDMH)	57-14-7	+	O	O	O	O	O	O	O	O	$<1.1 \times 10^6$ lbs (1973)
$CH_3-NH-NHCH_3$	1,2-Dimethylhydrazine [b] (SDMH)	54-07-3	+	O	O	O	O	O	O	O	O	<1000 lbs (1974)
NH_2NHCNH_2 || O	Hydrazine Carboxamide [b] (semicarbazide)	57-56-7	O	O	O	O	O	+	O	O	O	>1000 lbs (1971)

POTENTIAL INDUSTRIAL CARCINOGENS AND MUTAGENS

Class	Chemical Name & Synonym	CAS#	Carcinogenicity	Mutagenicity - Bacteria	Yeast	Neurospora	Drosophila	Mammalian Cells	Human Cells	Dominant Lethal	Host Mediated	Production Quantities [a]
V. Hydrazines, Hydroxylamines, Carbamates												
B. Hydroxylamines												
NH_2OH	Hydroxylamine	7803498	O	+	+	+	O	+	+	O	O	<1000 lbs (1974)
CH_3-N-OH	N-Methylhydroxylamine	593771	O	+	O	O	+	O	O	O	O	<1000 lbs (1974)
H_2N-OCH_3	O-Methylhydroxylamine	67-62-9	O	+	O	+	O	+	O	O	O	<1000 lbs (1974)
C. Carbamates												
$H_2N-C-OC_2H_5$ || O	Ethyl Carbamate (Urethan) [b]	51-79-6	+	+	+		+	+	O		O	1×10^5 lbs (1972)

POTENTIAL INDUSTRIAL CARCINOGENS AND MUTAGENS

Class	Chemical Name & Synonym	CAS#	Carcin-ogenicity	Mutagenicity								Production Quantites [a]
				Bacteria	Yeast	Neurospora	Drosophila	Mammalian Cells	Human Cells	Dominant Lethal	Host Mediated	
VI. Acetamides, Thio-Acetamides and Thioureas												
A. Acetamides $H_2N-C(=O)-CH_3$	Acetamide	60355	+	—	o	o	o	o	o	o	o	
B. Thioacetamides $H_2N-C(=S)-CH_3$	Thioacetamide	62555	+	—	o	o	o	o	o	o	o	
C. Thioureas $NH_2-C(=S)-NH_2$	Thiourea	62566	+	—	o	o	o	o	o	o	o	
Ethylenethiourea structure	Ethylenethiourea (2-Imidazolidinethiourea)	96457	+	—	o	o	—	—	o	—	—	

POTENTIAL INDUSTRIAL CARCINOGENS AND MUTAGENS

Class	Chemical Name & Synonym	CAS#	Carcin-ogenicity	Mutagenicity								Production Quantites [a]
				Bacteria	Yeast	Neurospora	Drosophila	Mammalian Cells	Human Cells	Dominant Lethal	Host Mediated	
VII. Nitrosamines $(CH_3)_2 N-NO$	Dimethylnitrosoamine[b] (DMN)	62-75-9	+	±	±	±	+	+	0	0	0	< 1000 lbs (1976)
$(C_2H_5)_2 N-NO$	Diethylnitrosamine (DEN)[b]	55-18-5	+	±	±	±	+	+	0	0	0	< 1000 lbs (1974)
VIII. Aromatic Amines $H_2N-C_6H_4-C_6H_4-NH_2$	Benzidine*[b]	92-87-5	+	+	0	0	+	0	0	0	0	1.5 x 10⁶ lbs (1972)
3,3'-Dichlorobenzidine structure	3,3'-Dichlorobenzidine[b]	91-94-1	+	+	0	0	0	0	0	0	0	4.6 x 10⁶ lbs (1972)
2-Aminobiphenyl structure	2-Aminobiphenyl[b]	90415	+	+	0	0	0	0	0	0	0	< 1000 lbs (1974)

POTENTIAL INDUSTRIAL CARCINOGENS AND MUTAGENS

Class	Chemical Name & Synonym	CAS#	Carcin-ogenicity	Mutagenicity							Production Quantites [a]
				Bacteria	Yeast	Neurospora	Drosophila	Mammalian Cells	Human Cells	Dominant Lethal / Host Mediated	
VIII. Aromatic Amines (Cont.'d)											
	4-Aminobiphenyl*[b]	92-67-1	+	+	0	0	0	0	0	0 0	<1000 lbs (1974)
	1-Naphthylamine (α-naphthylamine)[b]	134-32-7	±	+	0	0	0	0	0	0 0	7 x 10^6 lbs (1974)
	2-Naphthylamine (β-naphthylamine)*[b]	91-59-8	+	+	0	0	0	0	0	0 0	<1000 lbs (1974)
	4,4'-Methylene Bis (2-chloroaniline) (MOCA)[b]	101-14-4	±	+	0	0	0	0	0	0 0	7.7 x 10^6 lbs (1972)
	4,4'-Methylene Bis (2-methylaniline)[b]	180552	+	0	0	0	0	0	0	0 0	<1000 lbs (1974)

POTENTIAL INDUSTRIAL CARCINOGENS AND MUTAGENS

Class	Chemical Name & Synonym	CAS#	Carcin-ogenicity	Mutagenicity							Production Quantites [a]
				Bacteria	Yeast	Neurospora	Drosophila	Mammalian Cells	Human Cells	Dominant Lethal / Host Mediated	
IX. Azo Dyes											
	Azobenzene[b]	103-33-3	±	+	0	0	−	0	0	0 0	<1000 lbs (1974)
	paraAmino Azo Benzene[b]	60-09-3	+	+	0	0	−	−	0	0 0	3.3 x 10^5 lbs (1974)
	p-Dimethylamino Azo Benzene (DAB)[b]	60-11-7	+	+	0	0	±	0	0	0 0	1 x 10^4 lbs (1971)
	ortho-Amino Azo Toluene (o-AT)[b]	97-56-3	+	+	0	+	0	+	0	0 0	4.5 x 10^5 lbs (1973)

POTENTIAL INDUSTRIAL CARCINOGENS AND MUTAGENS

Class	Chemical Name & Synonym	CAS#	Carcinogenicity	Mutagenicity								Production Quantites a/
				Bacteria	Yeast	Neurospora	Drosophila	Mammalian Cells	Human Cells	Dominant Lethal	Host Mediated	
X. **Heterocyclic Aromatic Amines**												
	Quinoline	91-22-5	+	+	0	0	0	0	0	0	0	
	8-Hydroxyquinoline	184-24-3	±	+	0	0	−	0	0	0	+	
XI. **Nitrofurans**												
	Nitrofuran	609392	0	+	0	0	0	0	0	0	0	<1000 lbs (1974)
	N-(4-(5-nitro-2-furyl)-2-thiazolyl)acetamide) (NFTA)	531-82-8	0	+	0	0	0	0	0	0	0	<1000 lbs (1974)
	N-(4-(5-nitro-2-furyl)-2-thiazolyl)formamide (FANFT)	24554265	+	+	+	0	0	0	0	0	0	<1000 lbs (1974)

POTENTIAL INDUSTRIAL CARCINOGENS AND MUTAGENS

Class	Chemical Name & Synonym	CAS#	Carcinogenicity	Mutagenicity								Production Quantites a/
				Bacteria	Yeast	Neurospora	Drosophila	Mammalian Cells	Human Cells	Dominant Lethal	Host Mediated	
XII. **Anthraquinones**												
	1,2-dihydroxy-9,10-anthraquinone (alizarin)	74-48-0	0	+	0	0	0	0	0	0	0	
	1,4-dihydroxy-9,10-anthraquinone (Quiniznrin)	81-64-1	0	+	0	0	0	0	0	0	0	
	1,2,3-Trihydroxy-9,10-anthraquinone (Anthragallol)	602-64-2	0	+	0	0	0	0	0	0	0	
	1,2,4-Trihydroxy-9,10-anthraquinone (Purpurin)	81-54-9	0	+	0	0	0	0	0	0	0	
	1,4-Diamino-9,10-anthraquinone	128-95-0	0	+	0	0	0	0	0	0	0	

POTENTIAL INDUSTRIAL CARCINOGENS AND MUTAGENS

Class	Chemical Name & Synonym	CAS#	Carcinogenicity	Mutagenicity - Bacteria	Yeast	Neurospora	Drosophila	Mammalian Cells	Human Cells	Dominant Lethal	Host Mediated	Production Quantites [a]
XIII. Quinones & Polyhydric Phenols												
A. Quinones	para-Quinone (1,4-Benzoquinone)	106-51-4	+	0	0	−	0	0	0		0	
B. Polyhydric Phenols	1,4-Dihydroxybenzene (Hydroquinone)	123-31-9	+	+	0	0	0	0	0	0	0	
	1,2-Dihydroxybenzene (Catechol)	120-80-9	+	0	0	0	0	0	0	0	0	
XIV. Aromatic Hydrocarbons	Benzene*[b]	74-43-2	±	0	0	0	0	+	0	0	0	1.4 billion gallons (1976)

POTENTIAL INDUSTRIAL CARCINOGENS AND MUTAGENS

Class	Chemical Name & Synonym	CAS#	Carcinogenicity	Mutagenicity - Bacteria	Yeast	Neurospora	Drosophila	Mammalian Cells	Human Cells	Dominant Lethal	Host Mediated	Production Quantites [a]
XV. Halogenated Benzenes	o-Dichlorobenzene	95-50-1	±	0	0	0	0	0	0	0	0	153 x 10^6 lbs (1976)
	p-Dichlorobenzene	100-46-7	+	0	0	0	0	0	0	0	0	161 x 10^6 lbs (1976)
	Hexachlorobenzene (HCB)	118-74-1	+	0	+	0	0	0	0	−	0	5 x 10^6 lbs (1974)[c]
XVI. Nitroaromatics & Aliphatics												
A. Nitroaromatics	2,4-Dinitrotoluene	121-14-2	0	+	0	0	0	±	0	−	0	
	2,4,6-Trinitrotoluene	118-96-7	0	+	0	0	0	0	0	0	0	

POTENTIAL INDUSTRIAL CARCINOGENS AND MUTAGENS

Class	Chemical Name & Synonym	CAS#	Carcinogenicity	Mutagenicity								Production Quantites [a]
				Bacteria	Yeast	Neurospora	Drosophila	Mammalian Cells	Human Cells	Dominant Lethal	Host Mediated	
XVI. Nitroaromatics & Aliphatics												
B. Nitroalkanes												
$CH_3-CH(NO_2)-CH_3$	2-Nitropropane	79-46-9	+	0	0	0	0	0	0	0	0	30×10^6 lbs (1976)
XVII. Phosphoramides and Phosphoniumsalts												
A. Phosphoramides												
$(N(CH_3)_2)_3P=O$	Hexamethyl phosphoramide (Tris (dimethylamino) phosphine oxide, HMPA)	680-31-9	+	0	0	0	0	±	0	0	0	
B. Phosphonium Salts												
$[HOH_2C-P^+(CH_2OH)_3] X^-$	Terakis (Hydroxymethyl) Phosphoniumchloride (THPC)		+	0	0	0	0	0	0	0	0	

POTENTIAL INDUSTRIAL CARCINOGENS AND MUTAGENS

Class	Chemical Name & Synonym	CAS#	Carcinogenicity	Mutagenicity								Production Quantites [a]
				Bacteria	Yeast	Neurospora	Drosophila	Mammalian Cells	Human Cells	Dominant Lethal	Host Mediated	
XVIII. Unsaturated Nitriles												
$CH_2=CHCN$	Acrylonitrile (vinyl cyanide, 2-Propenenitrile)	167-13-1	+	+	0	+	−	0	0	0		1670×10^6 lbs (1976)
XIX. Azides												
$Na^+ N=N=N^-$	Sodium azide	266-28-228	0	0	0	−	0	0	0	0		< 1000 lbs (1974)
XX. Cyclic Ethers												
(1,4-dioxane ring structure)	1,4-Dioxane [b]	123911	+	0	0	0	0	0	0	0		163×10^6 lbs (1973) [c]

+ Reported positive in the literature

- Reported negative in the literature

0 Not tested, unreported, or unknown

* Human carcinogen

a Data from Stanford Research Institute (SRI); Chemical ProducersIndex; Chemical Marketing Reports; Chemical Week; Chem. Eng. News; U.S. International Trade Commission Reports

b Reviewed in IARC (International Agency for Research on Cancer) Monographs on the Evaluation of Carcinogenic Rsk of Chemicals to Man

c Produced as Hex Wastes from the manufacture of Perchloroethylene, Trichloroethylene, Carbon Tetrachloride and Chlorine

CHAPTER 5

ALKYLATING AGENTS-EPOXIDES AND LACTONES

A. Epoxides

The epoxides (oxiranes, cyclic ethers) include a number of very reactive reagents that are exceptions to the generalization that most ethers are resistant to cleavage.

Because of the strain energy of oxiranes, they react with acidic reagents even more rapidly than acyclic ethers do, producing β-substituted alcohols. The direction of opening of an oxirane in the S_N2 and acid-catalyzed processes differs. The less highly substituted carbon (sterically more accessible) is the site of the attack in the S_N2 process, whereas the more highly substituted carbon (more stable carbonium ion) is the site of attack in the acid-catalyzed process. Due to their reactivity, a number of epoxides have broad general utility.

1. **Ethylene Oxide** (H_2C-CH_2; oxirane; 1,2-epoxyethane; ETO) possesses a three-membered ring which is highly strained and readily opens under mild conditions; e.g., even in the unprotonated form it reacts with nucleophiles to undergo S_N2 reactions. The ease of this reaction is ascribed to the bond angle strain of the 3-membered ring (estimated to be about 27k cal/mode) a strain that is relieved in the course of the ring-opening displacement reaction, the bond angle strain providing a driving force for the reaction[1].

Ethylene oxide is considered unlikely to persist chemically unaltered, due to its high reactivity. The half-life of ethylene oxide in water is reported to be 76 hrs at 37°C and 6 months at 4°C. It reacts with water yielding ethylene glycol; with hydrogen halides to produce ethylene halohydrins; with alcohols and phenols to produce ethylene glycol ethers; with amines to produce ethanol amines; with acids to produce ethylene glycol esters and with sulphydryl compounds to produce thioethers. The most frequently found degradation products are ethylene chlorohydrin and ethylene glycol[2,3].

Ethylene oxide is produced in the U.S. by the catalytic oxidation of ethylene with air or oxygen. The commercial importance of ethylene oxide which is used in enormous quantities, as cited earlier, lies in its extreme reactivity and readiness to form other important compounds.

The scope of the annual production of ethylene oxide can be gleaned from the following: the current U.S. production is approximately 2,000 million kg; European production in 1972 was estimated at 865 million kg; and Japanese production in 1974 was 415 million kg[4].

The consumption pattern for ethylene oxide in the U.S. in 1974 was as follows (%): ethylene glycol, 59; acylic monionic surface-active agents, 8; glycol ethers, 7; ethanolamines, 6; diethylene glycol, 5; cyclic nonionic surface active agents, 5;

polyethylene glycol, 3; triethylene glycol, 2; polyether polyols, 2; and others (e.g., choline and choline chloride, ethylene chlorohydrin, hydroxyethyl starch, aryl-ethanolamines, acetal copolymer resins, and cationic surface-active agents), 3. Consumption of ethylene oxide for these uses is expected to increase an average of 4.7-5.2% per year to 1980[3]. The largest use for ethylene oxide is an intermediate in the production of ethylene glycol which is used mainly in anti-freeze products, and as an intermediate for polyethylene terephthalate polyester fiber and film production.

In 1975, approximately 100,000 pounds of ethylene oxide (approximately 0.1% of total production) was used in fumigation and sterilization of a broad spectrum of products including: (a) in the treatment of scientific and medical equipment and supplies made of glass, metals, plastics, rubber or textiles, drugs, leather, motor oil, clothing, furs, furniture, books and transportation equipment (jet aircraft, buses, railroad passenger cars) and (b) as a fungicidal fumigant in the post harvest treatment and sterilization of whole spices, copra, black walnut meats, cocoa, and flour, dried egg powder, dried fruits and dehydrated vegetables. The fumigant and sterilization uses of ethylene oxide is expected to grow at an average annual rate of 3.5-5% to 1980[3].

Ethylene oxide use in hospitals for "cold" sterilization of medical equipment has resulted in some instances of ethylene oxide concentrations exceeding the Occupational Safety and Health Administration (OSHA) short term exposure level of 75 ppm by a factor of 10. Of the 8,300 private hospitals in the U.S., approximately 7,500 use ethylene oxide sterilizers. It has been estimated that about 72,000 people are potentially directly exposed to ethylene oxide in U.S. hospitals, with probably another 30,000 casually exposed[5].

There appears to be two major populations at risk to the potential adverse effects of ethylene oxide. These are (a) individuals using this agent in the operation of sterilization equipment on a routine basis and/or who spend most of their time in the immediate area of the operation of this equipment and (b) individuals who use ethylene oxide as a fumigant on a routine basis and/or often enter storage chambers shortly after fumigation[2,6].

NIOSH recommended in Sept. 1977 that occupational exposure to ethylene oxide be limited to 59 ppm (90 mg/m^3) for a time-weighted average over an 8 hr exposure period, and that workers not be exposed to ethylene oxide at a concentration greater than 75 ppm (135 mg/m^3) as a ceiling concentration. These recommendations pertain to ethylene oxide use as a sterilant in medical facilities which accounts for about 0.5 million kg or 0.02% of all ethylene oxide produced in the U.S.[7]. The current MAC for ethylene oxide in the USSR is 1 mg/m^3 [4].

The use of ethylene oxide as a fumigant and its transformation into ethylene chlorohydrin have been reviewed by FAO/WHO[8]. It is noted that ethylene oxide reacts with inorganic chloride in foods to form ethylene chlorohydrin[9,10].

Ethylene oxide has been found in low concentrations in untreated cigarettes but the levels increased with longer treatment time and/or higher concentration of ethylene oxide. Similar levels were found in the smoke from commercial cigarettes from various countries (e.g., U.S., U.K., Jspan, Austria, Federal Republic of Germany)[11]. Commercial charcoal filters remove ethylene oxide from cigarette smoke[11].

There are few published carcinogenic studies on ethylene oxide[2,4], although a 2 year study is now in progress[2]. In limited studies, no carcinogenic effect was found when ethylene oxide was tested in ICR/Ha Swiss mice by skin application[12] and rats by subcutaneous injection[4,13].

Ethylene oxide reacts with DNA, primarily at the N-7 position of guanosine, forming N-7-hydroxyethylguanine[14]. Ethylene (1,2-^{3}H) oxide alkylated protein fractions taken from different organs of mice exposed to air containing 1.15 ppm of the labelled agent. The highest activity was found in lung followed by liver, kidney, spleen and testis[15].

Treatment with a 9.5 mM ethanol solution of ethylene oxide for 1 hr at 25°C produced reverse mutations in S. typhimurium TA 1535[16]. Reverse mutations were also produced in Neurospora crassa at the adenine locus by aqueous solutions of ethylene oxide[17-19]. Ethylene oxide induces recessive lethals[20-22], translocations[23,24] and minute mutations[25] in Drosophila melanogaster and is mutagenic in barley[26-28] and Aspergillus nidulans[29].

Exposure of male Long-Evans rats for 4 hours to 1.83 g/m^3 (1000 ppm) ethylene oxide produced dominant lethal mutations[30], while chromosome aberrations were found in bone-marrow cells of male rats of the same strain exposed to 0.45 g/m^3 (250 ppm) ethylene oxide for 7 hours/day for 3 days[30]. Ethylene oxide induces chromosome aberrations in mammalian somatic cells[31,32].

Strekalova found a significant increase in metaphase chromosome aberrations in bone marrow cells from rats given a single i.p. injection of ethylene oxide in water[31]. Embree et al[30] cited similar results using inhaled ethylene oxide.

It should be restressed that ethylene oxide readily converts to the glycol in an aqueous medium and to ethylene chlorohydrin if chloride ions are present[33]. Ethylene chlorohydrin is mutagenic in S. typhimurium plate assay[34,35].

Ehrenberg et al[15] recently described the evaluation of genetic risks of ethylene oxide. The degree of alkylation was used to determine tissue doses D_c; e.g., the concentration of free alkylating agent, integrated over time, in resting male mice exposed for 1-2 hr to air containing 1-35 ppm ethylene oxide. The exposure doses were 0.03-2% of the LD$_{50}$. The results agree with an absorption of all ethylene oxide in alveolar ventilation, a rapid distribution to all organs, and a rapid detoxication and excretion, e.g., biological half-life of about 9 minutes. D_t was found to be proportional to the exposure dose within the range studied. In most organs, including the tests, the D_t was about 0.05 μM hr per ppm hr of exposure; the degree of alkylation of DNA agreed with expectation from the doses determined.

On the basis of dose-effect curves of ethylene oxide and X-rays in barley, or tissue dose of ethylene oxide in man of 1 mM·hr may be provisionally set equal to 80 rad of low LET radiation. According to Ehrenberg et al[15] allowing for the difference in alveolar ventilation between mouse and man, this would mean that epoxide operators working at 5 ppm ethylene oxide for 40 hrs/week receive a weekly gonad dose of ethylene oxide amounting to abut 4 "rad-equivalents".

2. __Propylene Oxide__ (1,2-epoxypropane; $H_2C-\overset{H}{\underset{O}{C}}-CH_3$) (less reactive than ethylene oxide) is produced by two processes, viz., (1) chlorohydrin process from 1-chloro-2-propanol and $Ca(OH)_2$, and (2) a peroxidation process based on the oxidation of isobutane to tert.butyl alcohol and tert.butyl hydroperoxide, the latter after separation, is used to oxidize propylene to propylene oxide[36].

The consumption pattern for propylene oxide in the U.S. in 1975 was as follows (in %): polyether polyols for use in the production of polyurethan resins; 59; propylene glycol, 21; polyether polyols for the production of surface-active agents; hydraulic brake fluids, lubricants for rubber molds and textile fibers, heat transfer fluids, metal working fluids, and compression lubricants, 7; dipropylene glycol, 3; glycol ethers, 2; synthetic glycerin 1.8; and miscellaneous uses including the preparation of isopropanol amines, propylene carbonate, hydroxypropyl cellulose, and hydroxypropyl starch, 6.2. Small amounts of propylene oxide are used as a stabilizer for nitrocellulose lacquers and as a fumigant for a spectrum of materials ranging from plastic medical instruments to foodstuffs[3,27].

In the U.S., propylene oxide is registered as an insecticidal and fungicidal fumigant for cocoa, gums, processed spices, starch and processed nut meats (except peanuts) when such fulk foods are to be further processed into a final food. (The residue tolerance for products so treated is 300 mg/kg, expressed as propylene oxide)[38]. Use of propylene oxide (8 parts) and carbon dioxide (92 parts) mixture for fumigation of the above materials for periods not to exceed 48 hrs at temperature not to exceed 125°F (with the same residue tolerance) has been recently approved by EPA[39]. It is estimated that the U.S. consumption of propylene oxide will increase at an average rate of 9-10.5% from 1975 to 1980[3].

In 1973, European propylene oxide capacity was estimated to be 903 million kg; while the Japanese total output in 1974 was 131 million kg[37].

Permissible levels of propylene oxide in the working environment have been established in various countries[40]. The threshold limit value in the U.S. is 240 mg/m^3 (100 ppm); the maximum allowable concentration in the USSR is 1 mg/m^3.

Several studies have been reported on the persistence of propylene oxide in treated food[37]. Following fumigation of a broad spectrum of foods with 2:8 propylene oxide-carbon dioxide for 2-4 hrs at 50°C, levels of a propylene oxide residue ranging from 1000-2500 mg/kg were found in nutmeg, pepper and coriander white dried fish, wheat and soyabean flours contained the lowest amounts of residue[41].

In a very limited study, propylene oxide was carcinogenic in rats producing local sarcomas following subcutaneous injection of 1500 mg/kg of propylene oxide in Arachis oil (total dose)[42]. Propylene oxide induced recessive lethal mutations in __Drosophila melanogaster__[21,22,43]. Reverse mutations were induced in __Neurospora__ crassa after treatment with 0.5 M propylene oxide in water for 15 minutes[44]. It was also recently reported to be an active mutagen when tested with a gradient plate technique with __S. typhimurium__ and __E. coli__ tester strains[45].

Propylene oxide reacts with DNA at neutral pH to yield N-7-(2-hydroxypropyl) guanine and N-3(2-hydroxypropyl)adenine[46].

3. **1,2-Butylene Oxide** (1,2-epoxybutane; propyl oxirane; ethylethylene oxide; 1-butene oxide; $CH_3CH_2\underset{\underset{O}{\smile}}{CH-CH_2}$) can be prepared by the epoxidation of 1-butene with peroxyacetic acid[47], or via the catalytic reaction with molecular oxygen in the liquid phase[48]. It undergoes the usual reactions of epoxides with compounds having labile hydrogen atoms such as water, alcohols, polyols, phenols, thiols, ammonia, amines and acids. The principal use 1,2-epoxybutane has been as a corrosion inhibitor or acid scavenger in chlorinated solvents such as trichloroethylene and methyl chloroform at levels of 3-8%[3,47]. Inclusion of about 0.25 to 0.5 of 1% of 1,2-epoxybutane during preparation of vinyl chloride and co-polymer has served to minimize both container corrosion and metal pick-up[47]. 1,2-Epoxybutane can be polymerized or copolymerized with other alkylene oxides to yield polyethers[47,49] and polyurethans[50,51].

1,2-Epoxybutane was found to be non-carcinogenic when applied as a 10% solution in acetone to the skin of ICR/Ha female Swiss mice[52]. It was reported to be weakly mutagenic when tested in the *Salmonella typhimurium*/microsome test[53]. In recent studies by Cline et al[45], 1,2-epoxybutane was shown to be a strongly mutagenic in WP2 uvrA⁻ and TA 100 and weakly positive in WP2 and TA 1535.

4. **Epichlorohydrin** (1-chloro-2,3-epoxypropane; chloropropylene oxide; $\underset{\underset{O}{\smile}}{CH_2-CH}-CH_2Cl$) is produced by the high temperature chlorohydrination of allyl chloride (obtained by chlorination of propylene)[54], and is extensively used as an intermediate for the manufacture of synthetic glycerins, epoxy resins, (e.g., via reaction with Bisphenol-A), elastomers, and in the preparation of pharmaceuticals, textile coatings, cleaning agents, glycidyl ethers, paper sizing agents, ion-exchange resins, surface active agents, corrosion inhibitors, inks and dyes[54] and as a solvent for resins, gums, cellulose and paints. The U.S. consumption pattern in 1973 for refined epichlorohydrin was as follows[3] (%): manufacture of epoxy resins, 72; elastomers, 3; glycidyl ethers and modified epoxy resins, 3; wet-strength resins, 3; water-treatment resins, 2.8; surfactants, 2.2; and miscellaneous applications, 6. The use of epichlorohydrin is expected to increase annual (%) as follows: epoxy resins, 8-9; elastomers, 8-9; and for all other uses, 4-5[3]. Production of epichlorohydrin by 2 companies in the U.S. in 1973 was about 165.5 million kg, while in 1974 the Japanese production was about 41 million kg[55]. Epichlorohydrin is also produced by one or more companies in the following countries: France, Federal Republic of Germany, the Netherlands, U.K., German Democratic Republic, Czechoslovakia, and USSR[54].

Reactions of epichlorohydrin with active hydrogen compounds (e.g., alcohols, amines) normally occur at the more reactive epoxide site, although reactions involving the chlorine are known to occur[55].

Epichlorohydrin has recently been reported to produce squamous cell carcinomas of the nasal epithelium in rats following inhalation at levels of 100 ppm for 6 hours/day[56]. Epichlorohydrin has been previously shown to induce local sarcomas in ICR/Ha Swiss mice following subcutaneous injection[56], and was active as an initiator in a two-stage skin carcinogenesis study in mice[57].

Epichlorohydrin (without metabolic activation) at concentrations of 1-50 mM per 1 hr, induced reverse mutations in *S. typhimurium* G46 and TA 100 tester strains[58].

The mutagenic activity with TA 1535 tester strain was markedly reduced in the presence of liver homogenates[59]. Using the gradient plate technique, Cline et al[45] found epichlorohydrin (without metabolic activation) mutagenic in TA 1535, TA 100, E. coli WP2 and WP2 uvrA⁻ and with liver activation in S. typhimurium G46, TA 1538 and TA 98. In earlier studies, epichlorohydrin produced reverse mutations in E. coli strain B/r[60] and in Neurospora crassa when tested as a 0.15 M aqueous solution[44]; recessive lethal mutations in Drosophila melanogaster[21]. It was also mutagenic in Klebsiella pneumoniae[61].

Doses of 50 and 100 mg/kg of epichlorohydrin after 3 hours increased the frequency of reverse mutations using S. typhimurium strains G46, TA 100 and TA 1950 in ICR female mice in a host-mediated assay[58].

Mutagenic activity (as determined with the TA 1535 strain of S. typhimurium) was detected in the urine of mice after oral administration of 200-400 mg/kg epichlorohydrin[59]. Although an initial evaluation of mutagenic activity (utilizing the above system) in the urine of 2 industrial workers exposed to a concentration in excess of 25 ppm was regarded as borderline, additional mutagenic testing revealed more definitive evidence of activity, with the active compound appearing as a conjugate[59].

Epichlorohydrin induced dose-dependent chromosome abnormalities in bone marrow of ICR mice injected i.p. to a single dose of 5-40 mg/kg, 150 mg/kg, repeated doses of 5 x 1-10 mg/kg, p.o. in a single dose of 20 or 40 mg/kg or by repeated dose at 5 x 4-20 mg/kg[62]. Human peripheral lymphocytes exposed to $10^{-5} - 10^{-7}$ epichlorohydrin in vitro during the last 24 hours of cultivation showed chromosomal aberrations. It was 4-5 times less mutagenic thatn the polyfunctional mutagenic agent TEPA when tested analogously[63]; the epichlorohydrin induced changes were mainly classified as chromatid and isochromatid breaks and exchanges. These results demonstrated the ability of epichlorohydrin to induce gene and chromosome mutations in somatic cells. The finding of no changes in gametic cells was suggested to be the result of biotransformation changes of epichlorohydrin into forms which then cannot react gametic cells in a concentration capable of inducing dominant lethal effects[63].

A recent prospective cytogenetic study in Czeckoslovakia conducted on 35 workers occupationally exposed to epichlorohydrin disclosed that percentage of cells with chromosomal aberrations in blood samples (peripheral lymphocytes) of workers was 1.37 before exposure, 1.91 after the first year and 2.69 after the second year of exposure[64]. The difference between the percentages of aberrant cells before and after two years of occupational exposure was considered highly significant ($P<0.0001$). Particularly significant were the increase of chromatid and chromosomal breaks after exposure simultaneously with an increased number of breaks per 100 cells[38].

NIOSH estimated that approximately 50,000 U.S. workers are occupationally exposed to epichlorohydrin. Approximately 550 million pounds of epichlorohydrin were produced in the US in 1975[65].

A study covering 864 employees in Shell's Deer Park, Texas and Norco, Louisiana plants who worked with epichlorohydrin for at least 6 months between 1948 and 1965 (and followed up to 1976) showed that the mortality rates and causes of death were not significantly different from those of other people living in the same states. Among some workers respiratory cancer was somewhat higher than average. Total deaths from all forms of cancer among the exposed workers were less than expected and none of the differences between reported deaths and expected deaths were statistically significant[66].

5. <u>Glycidol</u> (2,3-epoxy-1-propanol; $CH_2CH\text{-}CH_2OH$ with O bridge) is widely employed in textile finishings as water repellant finishes, and as intermediates in production of glycerol and glycidyl esters, esters and amines of industrial utility. It is mutagenic in <u>Drosophila</u>[21], <u>Neurospora</u>[21], and <u>S. typhimurium</u> tester strains TA 98 (for frame-shift mutagens) and TA 100 (for base pair substitution mutagens)[67], both with and without rat liver microsomal extract (RME) (although less effective in the presence of RME) and in <u>Klebsiella pneumoniae</u> auxotroph[61].

6. <u>Glycidaldehyde</u> (2,3-epoxy-1-propanol; $CH_2\text{-}CHCHO$ with O bridge) is prepared from acrolein by the action of hydrogen peroxide or sodium hypochloride[68]. It has been used as a cross-linking agent, vapor-phase disinfectant and suggested synthetic intermediate[68].

Undiluted glycidaldehyde is stable at room temperature, but a 30% aqueous solution lost 94% of the epoxide after 2 months at room temperature. Both the epoxide and aldehyde are very reactive, entering into typical reactions associated with these sites[68].

Glycidaldehyde is carcinogenic in ICR/Ha Swiss mice following skin application[69] or subcutaneous injection[70] and in Sprague-Dawley rats following its subcutaneous administration[71,72], producing malignant tumors at the stie of application in both species.

Glycidaldehyde produces base-pair mutations in <u>S. typhimurium</u> TA 1535[53], TA 1000[53] and TA 100[67] tester strains (on a molar basis glycidaldehyde was about 20 to 50 times more potent in producing mutations than glycidol in TA 100[67]). Glycidaldehyde induces reverse base-pair mutations in <u>Saccharomyces cerevisiae</u> strain S211[72] and petite cytoplasmic mutations by strain N123 of <u>S. cerevisiae</u>[72]. It produces base-pair transitions (primarily A-T to G-C), frame-shift mutations and some deletions in bacteriophage T4[73,74] and is mutagenic in <u>Klebsiella pneumoniae</u>[61].

7. <u>Styrene Oxide</u> (1,2-epoxyethylbenzene; epoxystyrene; phenyl oxirane; phenylethylene oxide; $C_6H_5\text{-}CH\text{-}CH_2$ with O bridge) is produced either via the epoxidation of styrene with peroxyacetic acid or by the chlorohydrin route from α-phenylβ-iodoethanol and KOH[75]. Styrene oxide is produced in the U.S. by one company with annual production estimated to be 450-900 thousand kg[76]. In Japan, one company produced an estimated 1.8 million kg in 1976[76].

Styrene oxide is used primarily as a reactive diluent in epoxy resins and as an intermediate in the preparation of agricultural chemicals, cosmetics, surface coatings, and in the treatment of textiles and fibers[75,76]. Styrene oxide may be used as a catalyst and cross-linking agent for epoxy resins in coatings having a capacity of 1,000 gallons or more when such containers are intended for repeated use in contact with alcoholic beverages containing up to 8% of alcohol by volume according to a recent FDA ruling[77].

In Japan, styrene oxide is used as a raw material for production of perfume[76].

Styrene oxide has been detected as a by-product in commercial samples of styrene chlorohydrin[78]; in effluent water from latex manufacturing plants in Louisville, Kentucky and from chemical manufacturing plants in Louisville and Memphis, Tennessee[79]. It has also been detected as a volatile component of a Burley tobacco concentrate[80].

Styrene oxide has been tested by skin application in C3H and Swiss ICR/Ha mice with no significant increase in the incidence of skin tumors observed[81,82].

Styrene oxide induces reverse mutations in S. typhimurium strains TA 1535[53,83-85] and TA 100[83,84] without metabolic activation, (producing base-pair substitutions). Previous recent mutagenic analyses with styrene oxide on strains of S. typhimurium (TA 1537 and TA 1538) sensitive to frame-shift producing agents were negative[86]. Styles and Anderson[87] reported metabolic activation studies with styrene oxide (as well as benz(a)pyrene) using S. typhimurium and P388 mouse lymphoma cells in vitro employing S-9 mixes derived from rat, mouse, guinea pig, hamster, rabbit and human livers. No potency correlation could be shown between the potential carcinogenicity/mutagenicity in mammals because the result was dependent upon the balance between activation and detoxification. It was also noted that in both test systems, the response obtained for a particular species could be varied by the presence or absence of induction, the species of origin and the fraction of the S-9 mix.

A dose-dependent increase of forward mutations in S. pombe and gene conversion in S. cerevisiae was obtained by treatment with styrene oxide[88]. Styrene oxide (100 mg/kg) increased the frequency of gene conversion in S. cerevisiae but not forward mutations in S. pombe in the host-mediated assay using male Swiss albino mice[88]. Styrene oxide induced a dose-dependent increase of forward mutations in mammalian somatic cells in cultase (azaguanine-resistant mutants in V79 Chinese hamster cells)[88]. In this latter case styrene oxide was more active than ethyl methane sulfonate.

It should be stressed that styrene oxide is a metabolite in the proposed transformation of styrene to hippuric acid in man and animals[89-92], viz.,

$$R-CH=CH_2 \rightarrow R-CH-CH_2 \rightarrow R-CH-CH_2 \rightarrow R-CH-COOH$$
$$\underset{O}{\diagdown\diagup}\underset{OH\ OH}{}\underset{}{OH}$$

Styrene   Styrene oxide   Styrene glycol   Mandelic acid

$R=C_6H_5$

$$R-\underset{O}{\overset{\|}{C}}-NHCH_2COOH \leftarrow R-COOH \leftarrow R-CH_2OH \leftarrow R-\underset{O}{\overset{\|}{C}}-COOH$$

Hippuric acid   Benzoic acid   Benzyl alcohol   Phenyl glyoxylic acid

For example, as postulated by Leibman[92], styrene is metabolically converted to styrene oxide, and subsequently to styrene glycol by microsomal mixed function oxidases and microsomal epoxide hydrase from the liver, kidneys, intestine, lungs, and skin from several mammals[93]. The principal metabolites which have been detected in the urine of factory workers exposed to styrene vapor on the job, or volunteers exposed under controlled conditions for 4-60 ppm styrene vapor for 2 hours are mandelic acid and phenylglyoxylic acid[90,91].

In the U.S. styrene is principally produced by the catalytic dehydrogenation of ethylbenzene (made by alkylation of benzene or recovery from refinery streams). In another process, styrene is produced as a coproduct with propylene oxide[91]. In 1976, 13 U.S. companies produced 2864 million kg; total West European companies produced about 1090 million kg of styrene. Worldwide production of styrene is estimated to have been 7000 million kg[94].

The principal areas of application of an estimated 87% of the styrene consumed in the U.S. in 1976 (excluding exports) was in the production of plastics and resins (polystyrene resins) (61%); acrylonitrile-butadiene-styrene terpolymer (ABS) and styrene-acrylonitrile copolymer (SAN) resins (11%); styrene-butadiene copolymer resins (8%); unsaturated polyesters (7%); styrene-butadiene rubber (SBR) (11%). Miscellaneous applications consumed the remaining 2%[94].

An estimated 62% of the styrene produced world wide in 1974 was consumed in the manufacture of polystyrene (including expandable polystyrene), in ABS and SAN resins (13%), in styrene-butadiene rubber (SBR) (17%) and in other applications (8%).

The current U.S. OSHA 8-hour time weighted average for styrene is 100 ppm in the workplace air in any 8 hr workshift of a 40 hr work week. Exposure to styrene may not exceed a ceiling concentration limit of 200 ppm except for a time period of 5 minutes in any 3 hour period when the concentration may be as high as 300 ppm[95].

Work environment hygiene standards (8 hr time weighted averages) for styrene in other countries are as follows[40]: The Federal Republic of Germany, 420 mg/m^3; Sweden, 210 mg/m^3 (50 ppm); the German Democratic Republic, 200 mg/m^3; the MAC of styrene in the USSR is 5 mg/m^3.

Styrene has been detected in concentrations of 0.6 to 1.2 µg/g in the subcutaneous fat tissue of some U.S. styrene polymerization workers[96].

Styrene has been detected in the workplace air at levels of 0.2 to 136 ppm during the fabrication of glass-reinforced plastic pipe and at 25 to 144 ppm (time-weighted average) during the production of plastic boat components[94,97]; in the curing of tires in the U.S. at concentrations of 61-146 ppb[94] and in the production of fiber-reinforced laminated plastics in Czeckoslovakia[98].

Styrene has been detected in the Kanawka River in West Virginia[99], in the Scheldt River in the Netherlands at a concentration of 1 µg/l[99], in effluent discharged from petroleum refining (31 µg/l), chemical (30 µg/l), rubber (3 µg/l; 2.6 µg/l) and textile manufacturing plants in the U.S.[79,94,99].

Styrene has also been detected in U.S. finished drinking water at concentrations less than 1 µg/l[100] and in commercial charcoal-filtered drinking water in New Orleans, Louisiana[101].

Styrene has been detected in concentrations of 18.0 µg/cigarette in the smoke of U.S. domestic brand cigarettes[102] and with 6 aromatic hydrocarbons, in the side stream smoke of American-blend cigarettes[103].

Information is scant regarding the carcinogenicity of styrene. Styrene vapor at concentrations of 600 and 1000 ppm administered to female rats, 5 days/week for almost 2 years produced "a possible increase in the frequency rate of tumors in the lymphoid or hematopoietic classification". Male rats had no lesions attributable to this regimen[104]. The Manufacturing Chemists Association (MCA) in transmitting this finding to several U.S. Federal Agencies noted that the conclusions become "questionable" because of the "complications posed by pneumonia". The Trade Association also noted that long term studies on styrene being conducted by the National Cancer Institute, Litton Biometics for MCA and by Professor C. Maltoni in Italy have not reported "any relationship between treatment with styrene and the occurrence of tumors". MCA also noted that the animals in the MCA studies were exposed to levels 6 to 10 times OSHA's occupational standard for styrene, e.g., 100 ppm averaged over 8 hours[104].

Aspects of the mutagenicity of styrene is somewhat conflicting. For example, styrene, in the presence of a 9000 g supernatant from livers of rats pretreated with Clophen C or Aroclor 1254 induced reverse mutations in S. typhimurium TA 1535 and TA 100, while it was not mutagenic to TA 1537, TA 1538 and TA 98[84,105]. Styrene was non-mutagenic in a spot test with various strains of S. typhimurium without metabolic activation[83]. Styrene was also non-mutagenic in spot tests and plate incorporation assays in strains TA 1535, 1537, 1538, 98 and 100 at concentrations up to 1 mg/plate even in the presence of liver metabolizing systems[85].

Styrene was non-mutagenic when tested on forward mutation and gene-conversion systems of yeast (S. pombe and S. cerevisiae respectively). However, it was mutagenic only in a host-mediated assay (using male Swiss albino mice) with yeast (S. cerevisiae) when tested at very high doses (1000 mg/kg), but did not increase the forward mutation frequency in S. pombe[88].

Styrene did not increase the frequency of mutation to 8-azaguanine resistance in cultured V79 Chinese hasmter cells in the absence of metabolic activation[88].

Following exposure of male rats through inhalation to 300 ppm styrene for 6 hr/day, 5 days/week (2-11 weeks), an increase in the frequency of chromosomal aberrations in bone-marrow cells (8 to 12% in the exposed group compared to 1 to 6% in the control) was observed between 9 and 11 weeks)[106].

8. **Diglycidyl Resorcinol Ether** [1,3-bis(2,3-epoxypropoxybenzene); meta-bis-(2,3-epoxypropoxy)-benzene; resorcinol diglycidyl ether; RDGE;

is produced by the reaction of an excess of epichlorohydrin with resorcinol in alkaline solution[107]. The major areas of application of diglycidyl resorcinol ether are as a liquid epoxy resin, as a reactive diluent in the production of other epoxy resins[107,108] and to cure polysulfide rubber[107].

Although there are two producers of this epoxy derivative in the U.S., the quantity produced is not known. An estimated 5 thousand kg were produced in 1975 by one company in Japan[107].

No information is available concerning potential populations at risk, levels of exposure or occurrence of the material in the environment.

In preliminary communications, diglycidyl resorcinol ether induced skin tumors in C57Bl mice following thrice weekly skin painting[107,109] and induced local tumors in Long-Evans rats following subcutaneous administration[107,109]. However, no skin tumors were found in 8-week old Swiss ICR/Ha mice which received thrice weekly skin paintings of a 1% solution of diglycidyl resorcinol ether in benzene (approximately 0.1 ml solution per application) for life[69,107].

9. Miscellaneous Epoxides

Although the eight epoxides considered above are of principal concern based on their potential carcinogenicity and/or mutagenicity, production quantities and use patterns, it is important to note that there are a number of additional industrial epoxides of potential toxicological significance. However, very much less is known of their toxic properties. For example, 21-25 million pounds of alkyl (predominantly C_{12} and C_{14}) glycidyl ethers were consumed in the U.S. in 1975 for the captive production of alkyl glyceryl ether sulfonates which are used primarily as shampoos, light-duty liquid detergents, and combination soap-syndet toilet bar soap[3]. Additional recommended uses include: (a) as reactive diluents for epoxy resins sytems: (b) as stabilizers for PVC resins and (c) as stabilizers for chlorinated paraffins and other halogenated products[3]. It has been estimated that the annual growth rate for use of these alkyl glycidyl ethers in the manufacture of ether sulfonates in 8-9% to 1982[3].

A variety of epoxidized esters are employed as plasticizers, primarily for PVC resins, where they also function as heat and light stabilizers (synergistically with barium-cadmium-zinc stabilizers)[3]. The major epoxy esters used in the U.S. in 1974 (in millions of pounds) were: epoxidized soya oils, 127; epoxidized linseed oil, 5.5; octyl (n-octyl and 2-ethylhexyl) epoxy tallates, 14.9; and epoxidized toll oils, octyl epoxy stearates and other epoxidized esters, 6.5.

1-Epoxyethyl-3,4-epoxycyclohexane (1,2-epoxy-4-(epoxyethyl)cyclohexane; 4-vinylcyclohexenediepoxide; vinylcyclohexenedioxide; CH-CH$_2$; CAS # 106-87-6) can be prepared by the action of peracetic epoxidation of 4-vinylcyclohexane[110,111]. It has been used as a reactive diluent for other diepoxides and for epoxy resins derived from bisphenol-A and epichlorohydrin. Other recommended uses for 1-epoxyethyl-3,4-epoxycyclohexane include: (a) as a chemical intermediate for condensation with dicarboxylic acids, (b) as a monomer for the preparation of polyglycols containing unreacted epoxy groups and (c) for homopolymerization to a 3-dimensional resin[111,112]. 1-Epoxyethyl-3,4-epoxycyclohexane produced squamous-cell skin carcinomas in Swiss ICR/Ha mice following application of 0.1 ml of a 10% solution in benzene thrice weekly[82].

Diepoxybutane (butadiene diepoxide; 2,2'-bioxirane; 1,1'-bi(ethylene oxide); butadiene dioxide; dioxybutane; erythritol anhydride; CH_2CH-CH-CH_2) exists in four forms (e.g., DL-, D-, L-, and meso). Diepoxybutane has been prepared via the reaction of 1,4-dichloro-2,3-butanediol or a 2,3-dihalogeno-1,4-butanediol with sodium hydroxide. The DL- form has been prepared from 1,4-dibromo-2-butene and the meso- form from 1,4-dihydroxy-2-butene or 3,4-epoxy-1-butene, while the L-form is made from (2S,3S)-threitol-1,4-bis-methanesulphonate via (2S,3S)-1,2-epoxy-3,4-butanediol-4-methanesulphonate[113].

Diepoxybutane has been employed in some countries to a limited extent for the curing of polymers, cross-linking textile fibers and in the prevention of microbial spoilage[113,114]. In addition several patents have been issued on its use in polymer, paper, and textile treatments and as a chemical intermediate. D,L- and meso-1,2: 3,4-diepoxybutane are carcinogenic in mice by skin application producing squamous-cell skin carcinomas[70]. L-1,2:3,4-diepoxybutane is carcinogenic in mice by i.p. injection[113,115] and the D,L-racemate produced local sarcomas in mice and rats by subcutaneous injection[70,113].

Diepoxybutane induced reverse mutations in S. typhimurium TA 1535[53], in S. pombe[116], and in E. coli B and B/r strains after treatment for 1 hr at 37°C with 0.01 M and 0.02 M aqueous solutions respectively[117] and in Neurospora crassa after treatment with a 0.2 M solution[17]. Diepoxybutane produces mitotic gene conversions in S. cerevisiae strain D4 after 5 hrs treatment with a 0.005 M solution[118]. In Drosophila melanogaster, diepoxybutane produced sex-linked recessive lethal mutations, visible mutations, semi-lethal mutations[119], translocations[120] and minute mutations[121].

Huang and Rader[122] recently investigated the association between mutagenicity and adduct formation of 1,2,7,8-diepoxyoctane (DEO) and 1,2,5,6-diepoxycycloctane (DECO) structural analogs of diepoxybutane. The mutagenicity of DEO and DECO was investigated using diploid Chinese hamster lung cells, with 6-thioguanine resistance as the marker. DEO was found to be genetically active, while DECO was totally inactive. The difference between the two compounds in chemical and genetic activities was attributed to their molecular conformations and the resulting differential flexibilities and adduct-forming abilities. Association between mutagenicity and adduct formation was claimed by the authors to be conclusive.

An additional industrial epoxide of potential carcinogenic and/or mutagenic concern is phenylglycidyl ether (PGE; $C_6H_5OCH_2\text{-}CHCH_2$) made from the reaction of phenol and epichlorohydrin. Commercial products can contain phenol and diglycidylether as trace impurities. This epoxide reacts rapidly with hydroxyl and amine groups and has been used to improve polymer processing[123].

No data are available as to the carcinogenicity or mutagenicity of phenyl glycidyl ether.

B. Lactones

Lactones constitute a class of highly reactive compounds that possess a broad spectrum of current and suggested industrial uses including: wood processing, protective coatings and impregnation of textiles, modification of flax cellulose, urethan foam manufacture, intermediates in the preparation of insecticides, plasticizers and medicinals[124,125].

1. β-Propiolactone ($CH_2\text{-}CH_2$; BPL; β-hydroxypropionic acid lactone; hydracrylic acid; β-lactone) is by far the most important lactone produced commercially. It is prepared by the condensation of ketene with formaldehyde in the presence of zinc or aluminum chloride[124]. (It should be noted that both formaldehyde and ketene have been found mutagenic[125-128] and are further reviewed in the text). Commercial

grade BPL (97%) can contain trace quantities of the reactants. Samples of the common commercial produce have also been found to contain impurities including: acrylic acid, acrylic anhydride, acetic acid and acetic anhydride[124]. Industrially, BPL has been used mainly as an intermediate in the production of acrylic acid and esters.

During the past 2 decades, the most extensive use of BPL has been in plastic polymerization processes. As of 1974, two U.S. companies had a combined capacity of approximately 40 million kg/year of acrylic acid and esters based on BPL[124]. Only one company in the U.S. produced BPL til 1974[124,129]. BPL is known to have been produced elsewhere each by one company (e.g., Federal Republic of Germany and Canada), the production capacity of these companies, as well as that produced formally in the U.S. in not definitively known[124]. However, the statement has been made that "BPL production in the U.S. has grown from over 500 million pounds in 1957 to over 50 billion pounds by 1974"[130]. This figure appears excessive by most reckoning of high volume synthetic organic chemical production in the U.S.

BPL has also been used as an intermediate in the synthesis of propionic compounds and as a polymerization terminator for plastics. Minor applications of BPL have included its use in the chemical modification of natural products such as starches, proteins and oils, in the dairy industry[131] and in biological research[132,133]. Because of its potent reactivity with biological molecules, BPL has also been reported as being used in the sterilization of blood plasma, tissue grafts, surgical instruments and enzymes and as a vapor disinfectants in enclosed spaces[124]. Such uses are now believed to have been discontinued in several countries[124].

The use of BPL has been restricted because of its potential carcinogenic effects for man[134]. BPL is now listed with 13 other chemicals as being subject to an Emergency Temporary Standard on certain carcinogens under an order made by the Occupational Safety and Health Administration, Department of Labor on April 26, 1973[135].

The very high chemical reactivity of BPL is due to the presence of a strained four-membered lactone ring. It is a nucleophilic alkylating agent which reacts readily with acetate, halogen, thiocyanate, thiosulphate, hydroxyl and sulphydryl ions[125,130,136-139].
It reacts very rapidly with nucleophilic centers such as proteins, particularly sulfur containing amino acids, and nucleic acids[130]. The predominant alkylated base in DNA and RNA exposed to BPL is 7-(2-carboxyethyl)guanine[130,136].

Adenine is alkylated to a lesser extent at the 3-position[140]. BPL acts as an alkylating agent by undergoing ring-opening at the O-CO bonds with unionized molecules and at the CH_2-O bonds with ionized molecules[130,141]. BPL is relatively soluble in water (37% v/v.) and undergoes hydrolysis (50% for 3 hr at 35°C) of the ring structure to produce relatively non-toxic beta-hydroxypropionic acid[139,142].

The carcinogenic and mutagenic properties of BPL have been recently reviewed by Brusick[130]. Most tumors induced by BPL (the most active molecule among a group carcinogenic lactones) are sarcomas and are located at the site of injection. BPL is carcinogenic in the mouse by skin application, subcutaneous or i.p. application and in the rat by subcutaneous injection[124,143,144], while oral administration in the rat produced squamous cell carcinomas of the forestomach[70,124]. These results combined with previous tumor distribution observations illustrate the inability of BPL to reach systemic organs when administered by the typical routes of exposure.

The potential environmental carcinogenic problems associated with BPL has been reviewed by Hueper[134] and suggest that BPL would pose the highest risk as a skin carcinogen in man[130,134].

The molecular interaction that determines the carcinogenic and mutagenic properties of BPL is its ability to interact with DNA and more specifically, with the base guanine[130,136,145] to form 7-(2-carboxyethyl)guanine, as previously cited[131,137,145]. High concentrations of BPL produce interstitial cross-links suggesting that BPL as well as other monofunctional alkylating agents (e.g., nitrous acid and methyl methane sulfonates) are capable of producing DNA alterations[130,145].

Several comparative mutagenesis studies with genetically defined reverse mutation systems in E. coli[146,147] and S. typhimurium[49,130,148-150] in which the nucleotide sequences of the mutant strains were known, confirmed earlier indications that BPL induces predominantly base-pair substitution mutations. Unlike data from other microorganisms there is no evidence for BPL-induced deletions in bacteria[130].

The genetic effects of BPL have probably been studied more intensively in fungi than in any other organism[130]. BPL has been shown to revert ade$^-$ and inos$^-$ mutants[151] and induce auxotrophic mutants[152] of Neurospora, induce mitotic gene conversion in S. cerevisiae strain D4[153,154] and auxotrophic mutants and respiration deficient mutants in a haploid strain of S. cerevisiae[155]. The latter study indicated that BPL induces predominantly base-pair substitution mutations with a number of deletions and a small fraction of frameshift mutations[130,155].

Prakash and Sherman[156] in a study of the reversion of iso-1-cytochrome C mutants of S. cerevisiae demonstrated that a predominance of GC to AT transitions were induced by BPL.

BPL has also been found mutagenic in bacteriophages and animal viruses[73,130,157], inducing primarily guanine//cytosine//thymine base pair transitions in phage T4[73]. This type of mispairing was suggested by Corbett et al[73] to be the most probable cause of BPL induced mutagenesis.

The in vitro induction of chromosome aberrations in rat leukocytes by treatment with 2.5×10^{-4}M was demonstrated by Dean[158]. Clive[159] demonstrated the mutagenicity of BPL in mouse lymphoma cells (L5178) strains that are heterozygous at the thymidine kinase (TK) locus and detect TA$^{+/-}$ → TK^{-1-} forward mutation using a BUdR containing selective medium[160]. In a comparison of the ability of BPL to induce chromosome and chromatid aberrations in cultured Chinese hamster ovary fibroblasts with the frequency of sister chromatid exchanges (SCE), Perry and Evans[161] found at concentrations of 3×10^{-5}M and 3×10^{-4}M, BPL produced only a few chromosome or chromatid aberrations in the 20 mitoses examined whereas the number of SCE in 20 cells was 425 and 1,668 at the respective concentrations.

Chromosome aberrations were not induced in rats (100 mg/kg i.p.) up to 96 hr after treatment[158] nor did 3-10 mg/kg i.p. induce dominant lethality in ICR/Ha Swiss mice[62].

Transplacental induction of malignant transformation in hamster embryo cells following i.p. injection of 2.5-3.0 mg/kg BPL to pregnant hamsters has been reported by DiPaolo et al[164].

BPL also induced chromosomal aberrations in Vicia faba[163] and Allium cepa[164].

Because the biological half-life of BPL in vivo is probably too short for substantial quantities of the alkylated form to reach target sites, it is difficult to assess the impact of BPL or humans[130]. Brusick[130] summarized that risk assessments using normal methods of exposure such as drinking water, feed or i.p. or i.m. injection would be expected to show little or no mutagenic effect in acute or chronic studies.

2. β-Butyrolactone (3-hydroxybutanoic acid; β-lactone; 3-hydroxybutyric acid lactone; 2-methyl-2-oxetanone; $CH_3-CH-O\overset{CH_2-C=O}{|}$) can be prepared by the catalytic hydrogen-atom of diketene. In Japan, it was made from ketene and acetaldehyde until 1973, during which year 30 thousand kg were produced[165].

No data are available concerning the use patterns, occurrence and potential populations at risk from β-butyrolactone.

β-Butyrolactone is carcinogenic in ICR/Ha Swiss mice by skin application[52,69] and in Sprague-Dawley rats by subcutaneous injection[70] and when administered orally produced squamous-cell carcinoma of the forestomach[70].

β-Butyrolactone inactivates transforming DNA from Bacillus substilis[166].

3. γ-Butyrolactone (butyric acid lactone; 1,4-butanolide; 4-butyrolactone; γ-hydroxybutyric acid lactone; γBL; $\overset{CH_2CH_2}{\underset{\underset{O}{C-CH_2}}{O}}$) is produced commercially in the U.S. by the dehydrogenation of 1,4-butanediol over a copper catalyst at 200-250°C and in Japan by the hydrogenation of maleic anhydride[168,169]. Since 1971, approximately 1.5 million kg (year of γ-butyrolactone) have been made by two Japanese companies, one of which may have stopped production recently. Most of the production is used to manufacture N-methyl-2-pyrrolidone[171].

Commercial production of γ-butyrolactone was first reported in the U.S. in 1953 and one company was producing it as of 1976[169]. U.S. consumption of γ-butyrolactone in 1974 is estimated to have been approximately 14 million kg. γ-Butyrolactone is used principally as: (1) a chemical intermediate in the production of 2-pyrrolidone; (2) as an intermediate for other organic chemicals and (3) as a solvent. The U.S. consumption of 2-pyrrolidone in 1973 is estimated to have been about 7 million kg, used primarily as intermediates for vinyl pyrrolidone in the manufacture of homo- and copolymers[169]. These polymers are employed as film formers in hair sprays, as clarifying agents in beer and wine and as blood expanders. Less than 50 thousand kg of 2-pyrrolidone were used in 1974 in the production of nylon 4 fibers[169].

U.S. consumption in 1974 of the most important of N-methyl-2-pyrrolidone, another important derivative of γ-butyrolactone, is estimated to have been approximately 5 million kg. It is used primarily as a solvent (e.g., in the extraction of butadiene and as a polymer solvent)[169].

γ-Butyrolactone is also used as a solvent for many polymers including: polyacrylonitrile, polyvinylchloride, polystyrene, polyamides, polyvinylcarbazole, and cellulose (e.g., in the textile industry as a spinning and coagulating solvent for polyacrylonitrile), as a selective solvent for acetylene[169] in the petroleum industry and as an intermediate in the production of the herbicide 4-(2,4-dichlorophenoxy)butyric acid[170].

γ-Butyrolactone has been reported in beer and apple brandy of 2 mg/l[171] and 5-31 mg/l[172] respectively. It has also been detected in wine[173], vinegar[174], cooked meats[175], coffee[176], tomatoes[177] and in tobacco smoke condensate[178].

γ-Butyrolactone was found to be non-carcinogenic when tested in C_3H ad XVII/G mice by oral administration[172], in Swiss ICR/Ha mice by skin application[69], in Swiss-Webster mice by subcutaneous administration[179], in male albino rats by oral administration[165], and in male Wistar rats subcutaneously[180]. No data are available on the mutagenicity of γ-butyrolactone.

References

1. Gutsche, C. D., and Pasto, D. J., Fundamentals of organic chemistry, Prentice-Hall, Englewood Cliffs, NJ (1975) p. 296
2. EPA, Environmental Protection Agency Position Document for the Rebuttable presumption against Registration of Pesticide Products Containing Ethylene Oxide, Chem. Reg. Reptr., 1 (4) (1978) 1650-1657
3. EPA, A study of industrial data on candidate chemicals for testing, Contract #68-01,4109, Environmental Protection Agency, Washington, DC, Feb. 17 (1977)
4. IARC, Vol. 11, Cadmium, nickel, some epoxides, miscellaneous industrial chemicals and general considerations on volatile anaesthetics, International Agency for Research on Cancer, Lyon, 1976, pp. 157-167
5. Anon, Ethylene oxide use in hospitals as health threat, controls recommended, Pesticides & Chemical News, 5 (1977) 27
6. Anon, Ethylene oxide, RPAR Draft Position Document Detailed, Pesticide & Toxic Chemicals News, 5 (1977) 27
7. Anon, Ethylene oxide exposure limit, Toxic Material News, 4 (34) (1977) 212
8. FAO/WHO, 1971 Evaluations of Some Pesticide Residues in Food. Report of the 1971 Joint Meeting of the FAO Working Party of Experts on Pesticide Residues and the WHO Expert Committee on Pesticide Residues, WHO Pest. Res. Series, No. 1 (1971) 28-288
9. Ragelis, E. P., Fisher, B. S., and Klimeck, B. A., Note on determination of chlorohydrins in foods fumigated with ethylene oxide and with propyelen oxide, J. Assoc. Off. Anal. Chem., 49 (1966) 963-965
10. Wesley, F., Rourke, B., and Darbishire, O., The formation of persistent toxic chlorohydrins in foodstuffs by fumigation with ethylene oxide and with propylene oxide, J. Food Sci., 30 (1965) 1037-1042
11. Muramatsu, M., Obi, Y., Shimada, Y., Takahashi, K., and Nishida, K., Ethylene oxide in cigarette smoke, Rep. Centr. Res. Lab. Jap. Monop. Manuf., 110 (1968) 217-222
12. Van Duuren, B. L., Orris, L., and Nelson, N., Carcinogenicity of epoxides, lactones, and peroxy compounds, II, J. Natl. Cancer Inst., 35 (1965) 707-717
13. Walpole, A. L., Carcinogenic action of alkylating agents, Ann. NY Acad. Sci. 68 (1958) 750-761
14. Brookes, P., and Lawley, P. D., The alkylation of guanosine and guanylic acid, J. Chem. Soc. (1961) 3923-3928
15. Ehrenberg, L., Hiesche, K. D., Osterman-Golkar, S., and Wennberg, I., Effects of genetic risks of alkylating agents: Tissue doses in the mouse from air contaminated with ethylene oxide, Mutation Res., 24 (1974) 83-103
16. Rannug, U., Göthe, R., and Wachtmeister, C. A., The mutagenicity of chloroethylene oxide, chloroacetaldehyde, 2-chloroethanol and chloroacetic acid, conceivable metabolites of vinyl chloride, Chem. Biol. Interactions (1976) in press
17. Kolmark, G., and Westergaard, M., Further studies on chemically induced reversions at the adenine locus of Neurospora, Hereditas, 39 (1953) 209-224
18. Kolmark, H. G., and Kolbey, F. J., Kinetic studies of mutation induction by epoxides in Neurospora crassa, Mol. Gen. Genetics, 401 (1968) 89-98
19. Kilbey, B. J., and Kolmark, H. G., A mutagenic after-effect associated with ethylene oxide in Neurospora crassa, Mol. Gen. Genetics, 101 (1968) 185-188
20. Bird, M. J., Chemical production of mutations in Drosophila: Comparison of techniques, J. Genet., 50 (1952) 480-485

21. Rapoport, I. A., Dejstvie akisi etilena, glitsida i glikoles na gennye mutatsii (Action of ethylene oxide glycides and glycols on genetic mutations), Dokl. Acad. Nauk. SSR, 60 (1948) 469-472
22. Rapoport, I. A., Alkylation of gene molecule, Dokl. Acad. Nauk. SSR, 59 (1948) 1183-1186
23. Watson, W. A. F., Further evidence on an essential difference between the genetic effects of mono- and bifunctional alkylating agents, Mutation Res., 3 (1966) 455-457
24. Nakao, Y., and Auerbach, C., Test of possible correlation between cross-linking and chromosome breaking abilities of chemical mutagens, Z. Verebungsl., 92 (1961) 457-461
25. Fahmy, O. G., and Fahny, M. J., Gene elimination in carcinogenesis: Reinterpretation of the somatic mutation theory, Cancer Res., 30 (1970) 195-205
26. Sulouska, K., Lindgren, D., Eriksson, G., and Ehrenberg, L., The mutagenic effect of low concentrations of ethyleneoxide in air, Hereditas, 62 (1969) 264-266
27. Lindgren, D., and Sulouska, K., The mutagenic effect of low concentrations of ethylene oxide in air, Hereditas, 63 (Suppl 1) (1969) 13
28. Ehrenberg, L., Chemical mutagenesis: Bochemical and chemical points of view on the mechanism of action, Abh. Deut. Akad. Wiss. Berlin Kl. Med., 1 (1960) 124-136
29. Morpurgo, G., Induction of mitotic crossing-over in Aspergillus nidulans by bifunctional alkylating agents, Genetics, 49 (1963) 1259-1263
30. Embree, J. W., Lyon, J. P., Hine, C. H., The mutagenic potential of ethylene oxide using the dominant-lethal assay in rats, Toxicol. Appl. Pharmacol., 40 (1977) 261-267
31. Strekalova, E. E., Mutagenic action of ethylene oxide on mammals, Toksikol. Nov. Prom. Khimschesk. Veshchestv., 12 (1971) 72-78
32. Fomenko, V. N., and Strekalova, E. E., Mutagenic action of some industrial poisons as a function of concentration and exposure time, Toksikol. Nov. Promysclen. Khimschesk. Veshchestv., 13 (1973) 51-57
33. Willson, J. E., Ethylene oxide sterilant residues, Bull. Parent. Drug Assoc. 24 (1970) 226-234
34. Rosenkranz, H S., and Wlodkowski, T. J., Mutagenicity of ethylene chlorohydrin, A degradation product present in food-stuffs exposed to ethylene oxide, J. Agr. Food Chem., 22 (1974) 407-409
35. Rosenkranz, H. S., and Carr, S., 2-Haloethanols, Mutagenicity and relativity with DNA, Mutation Res., 26 (1974) 367-370
36. Lowenheim, F. A., and Moran, M. K., Propylene oxide, in Faith, Keyes, and Clark's Industrial Chemicals, 4th ed., John Wiley & Sons, New York (1975) pp. 692-697
37. IARC, Monographs on the Evaluation of Carcinogenic Risk of Chemicals to Man, Vol. 11, Lyon (1976) pp. 191-199
38. U.S. Code of Federal Regulations, Propylene oxide, Food and Drugs, Title 21, Part 123.380, U.S. Govt. Printing Office, Washington, DC (1975) p. 677
39. Anon, Propylene oxide-carbon dioxide fumigation up to 40 hours approved, Pesticide & Toxic Chem. News, 6 (1) (1977) 11
40. Winell, M. A., An international comparison of hygienic standards for chemicals in the work environment, Ambio, 4 (1975) 34-36
41. Oguma, T., Hosogai, Y., Fujii, S., and Kawashiro, I., Gaseous anti-microbial agents. I. Determination of propylene oxide residues in foods, J. Food Hyg. Soc. Japan, 9 (1968) 395-398

42. Walpole, A. L., Carcinogenic action of alkylating agents, Ann. NY Acad. Sci. 68 (1958) 750-761
43. Schalet, A., Drosophila Info. Service, 28 (2) (1954) 155
44. Kolmark, G., and Giles, N. H., Comparative studies on monoepoxides as inducers of reverse mutations in Neurospora, Genetics, 40 (1955) 890-902
45. Cline, J. C., Thompson, C. Z., and McMahon, R. E., A convenient technique for the detection of volatile liquid mutagens in ten tester strains, 9th Annual Meeting of Environmental Mutagen Society, San Francisco, March 9-13 (1978) 68-69
46. Lawley, P. D., and Jarman, M., Alkylation by propylene oxide of deoxyribonucleic acid, adenine, guanosine and deoxyguanylic acid, Biochem. J., 126 (1972) 893-900
47. Lapkin, M., Epoxides, In Kirk-Othmer Encyclopedia of Chemical Technology, Vol. 8, 2nd ed., Wiley and Sons, New York (1965) pp. 263-293
48. Bocard, C., Gadelle, C., Mimoun, H., and Seree De Roch, I., Catalytic epoxidation of propylene by molecular oxygen, French Patent, 2,044,007, 19 Feb., 1971, Chem. Abstr., 75 (1971) 151660Q
49. Milgrom, J., Telomeric ethers with hydroxy or thio end group, Ger. Offen. 1,803,383, 12 June (1969) Chem. Abstr., 71 (1969) P71182N
50. Beitchman, B.D., Cross-linked isocyanurate-polyurethan elastomers and plastics prepared with triethylenediamine-cocatalyst combinations, Rubber Age, 98 (1966) 65-72
51. Jefferson Chem. Co., Polyurethan from poly(butyleneoxide) polyols, Belgian Patent, 670,367, 30 March (1966) Chem. Abstr., 65 (1967) 15620H
52. Van Duuren, B. L., Langseth, L., Goldschmidt, B. M., and Orris, L., Carcinogenicity of epoxides, lactones, and peroxy compounds, VI. Structure and carcinogenic activity, J. Natl. Cancer Inst., 39 (1967) 1217-1225
53. McCann, J., Choi, E., Yamasaki, and Ames, B. N., Detection of carcinogens as mutagens in the Salmonella/microsome test: Assay of 300 chemicals, Proc. Natl. Acad. Sci., 12 (1975) 5135-5139
54. Lowenheim, F. A., and Moran, M. K., Epichlorohydrin, In Faith, Keyes and Clarks Industrial Chemicals, 4th ed., John Wiley & Sons, New York (1975) pp. 335-338
55. IARC, Monographs on the Evaluation of Carcinogenic Risk, Vol. 11, Lyon (1976) 131-139
56. Anon, Epichlorohydrin causes nose cancers in rats, NYU's Nelson Reports, Pesticide & Toxic News, 5 (19) (1977) 27-28
57. Van Duuren, B. L., Goldschmidt, B. N., Katz, C., Siedman, I., and Paul, J. S., Carcinogenic activity of alkylating agents, J. Natl. Cancer Inst., 53 (1974) 695-700
58. Sram, Cerna, M., and Kucerova, M., The genetic risk of epichlorohydrin as related to the occupational exposure, Biol. Zbl., 95 (1976) 451-462
59. Kilian, D. J., Pullin, T. G., Conner, T. H., Legator, M. S., and Edwards, H. N., Mutagenicity of epichlorohydrin in the bacterial assay system: Evaluation by direct in vitro activity and in vivo activity of urine from exposed humans and mice, Presented at 8th Annual Meeting of Environmental Mutagen Society, Colorado Springs, Colo. Feb. 13-17 (1977) p. 35
60. Strauss, B., and Okubo, S., Protein synthesis and the induction of mutations by E. coli by alkylating agents, J. Bact., 79 (1960) 464-473
61. Voogd, C. E., Mutagenic action of epoxy compounds and several alcohols, Mutation Res., 21 (1973) 52-53
62. Epstein, S. S., Arnold, E., Andrea, J., Bass, W., and Bishop, Y., Detection of chemical mutagens by the dominant lethal assay in the mouse, Toxicol. Appl. Pharmacol., 23 (1972) 288-325

63. Kucerova, M., Polivakova, Z., Sram, R., and Matousek, V., Mutagenic effect of epichlorohydrin. I. Testing on human lymphocytes in vitro in comparison with TEPA, Mutation Res., 34 (1976) 271-278
64. Kucerova, M., Zhurokov, V. S., Polivkova, Z., and Ivanova, J. E., Mutagenic effect of epichlorohydrin. II. Analysis of chromosomal aberrations in lymphocytes of persons occupationally exposed to epichlorohydrin, Mutation Res., 48 (1977) 355-360
65. Anon, NIOSH recommends new epichlorohydrin standard, revision on allyl chloride, Toxic Materials News, 3 (1976) 154
66. Anon, Epichlorohydrin. Shell Chemical Company Announces Results of Mortality Study of Exposed Workers, Chem. Reg. Reptr., 1 (1977) 988-989
67. Wade, M., and Moyer, J., The mutagenicity of epoxides, Fed. Proc., 35 (1976) 404
68. IARC, Monographs on the Evaluation of Carcinogenic Risk, Vol. 11, Lyon (1976) pp. 176-181
69. Van Duuren, B. L., Orris, L., and Nelson, N., Carcinogenicity of epoxides, lactones, and peroxy compounds, II., J. Natl. Cancer Inst., 35 (1965) 707-717
70. Van Duuren, B. L., Langseth, L., Orris, L., Teebor, G., Nelson, N., and Kuschner, M., Carcinogenicity of epoxides, lactones and peroxy compounds, IV. Tumor response in epithelial and connective tissue in mice and rats, J. Natl. Cancer Inst., 37 (1966) 825-834
71. Van Duuren, B. L., Langseth, L., Orris, L., Baden, M., and Kushner, M., Carcinogenicity of epoxides, lactones and peroxy compounds, V. Subcutaneous injection in rats, J. Natl. Cancer Inst., 39 (1967) 1213-1216
72. Izard, M. C., Recherches sur les effects mutagenes de l'acroleine et de ses deux epoxydes le glycidol et le glycidal sur S. cerevisiae, C.R. Acad. Sci. Ser. D., 276 (1973) 3037-3040
73. Corbett, T. H., Heidelberger, C., and Dove, W. F., Determination of the mutagenic activity to bacteriophage T4 of carcinogenic and noncarcinogenic compounds, Mol. Pharmacol., 6 (1970) 667-679
74. Corbett, T. H., Dove, W. F., and Heidelberger, C., Attempts to correlated carcinogenic with mutagenic activity using bacteriophage, Int. Cancer Congr. Abstr., 10 (1970) 61-62; 10th Int. Cancer Congr., Houston, TX May 22-29 (1970)
75. IARC, Monographs on the Evaluation of Carcinogenic Risk of Chemicals to Man, Vol. 11 Lyon (1976) 201-208
76. IARC, Monographs on the Evaluation of Carcinogenic Risk of Chemicals to Man, Vol. 17 Lyon (1978) in press
77. U.S. Code of Federal Regulations, Title 21, Food and Drugs, part 175.300, Washington, DC (1977) p. 454
78. Dolgopolov, V.D., and Lishcheta, L. I., Qualitative determination of by-products in commercial samples of styrene chlorohydrin, Khim. Farm. Zh., 5 (1971) 55-56
79. Shackelford, W. M., and Keith, L. H., Frequency of Organic Compounds Identified in Water, EPA No. 600/4-76-062, U.S. Environmental Protection Agency, Athens, GA (1976) p. 214
80. Demole, E., and Berthet, D., A chemical study of Burley tobacco flavor (Nicotiana Tabacum L.) I. Volatile to medium-volatile constituents (b.p. $\leq 84°/0.001$ Torr), Helv. Chim. Acta, 55 (1972) 1866-1882
81. Weil, C. S., Condra, N., Huhn, C., and Striegel, J. A., Experimental carcinogenicity and acute toxicity of representative epoxides, J. Am. Ind. Hyg. Assoc. 24 (1963) 305-325

82. Van Duuren, B. L., Nelson, N., Orris, L., Palmes, E. D., and Schmidt, F. L., Carcinogenicity of epoxides, lactones and peroxy compounds, J. Natl. Cancer Inst., 31 (1963) 41-55
83. Milvy, P., and Garro, A. J., Mutagenic activity of styrene oxide (1,2-epoxy ethyl benzene) a presumed styrene metabolite, Mutation Res., 40 (1976) 15-18
84. Vainio, H., Paakkonen, R., Ronnholm, K., Raunio, V., and Pelkonen, O., A study on the mutagenic activity of styrene and styrene oxide, Scand. J. Work. Env. Hlth., 3 (1976) 147-151
85. Stoltz, D. R., and Withey, R. J., Mutagenicity testing of styrene and styrene epoxide in Salmonella typhimurium, Bull. Env. Contam. Toxicol., 17 (1977) 739-742
86. Glatt, H. R., Oesch, F., Frigerio, A., and Garattini, S., Epoxides metabolically produced from some known carcinogens and from some clinically used drugs, I. Difference in mutagenicity, Int. J. Cancer, 16 (1975) 787-797
87. Styles, J. A., and Anderson, D., Metabolic activation studies with benz(a)-pyrene and styrene oxide using Salmonella typhimurium and mouse lymphoma cells, 9th Annual Meeting Environmental Mutagen Society, San Francisco, March 9-13 (1978) 55-56
88. Loprieno, N., Abbondandolo, A., Barale, R., Baroncelli, S., Bonatti, S., Bronzetti, G., Cammellini, A., Corsi, C., Corti, G., Frezza, D., Leporini, C., Mazzaccaro, A., Nieri, R., Rosellini, D., and Rossi, A., Mutagenicity of industrial compounds: Styrene and its possible metabolite styrene oxide, Mutation Res., 40 (1976) 317-324
89. Abbondandolo, A., Bonatti, S., Colella, C., Corti, G., Matteucci, F., Mazzaccaro, A., and Rainaldi, G., A comparative study of different protocols for mutagenesis assays using the 8-azaguanine resistance system in Chinese hamster cultured cells, Mutation Res., 37 (1976) 293-306
90. Ohtsuji, H., and Ikeda, M., The metabolism of styrene in the rat and the stimulatory effect of phenobarbital, Toxicol. Appl. Pharmacol., 18 (1971) 321-328
91. Ikeda, M., and Imamura, Evaluation of hippuric, phenylglyoxylic and mandelic acids on urine as indicators of styrene exposure, Int. Arch. Arbeitsmed., 32 (1974) 93-101
92. Leibman, K. C., Metabolism and toxicity of styrene, Env. Hlth. Persp., 11 (1975) 115-119
93. Oesch, F., Mammalian epoxide hydrase: Inducible enzymes catalyzing the inactivation of carcinogenic and cytotoxic metabolites derived from aromatic and olefinic compounds, Xenobiotica, 3 (1973) 305-340
94. IARC, Monographs on the Evaluation of Carcinogenic Risk of Chemicals to Man, Vol. 17, Lyon (1978) in press
95. U.S. Occupational Safety and Health Administration, Occupational Safety and Health Standards Subpart Z-Toxic and Hazardous Substances Code of Federal Regulations, Title 29, Chapter XVII, Section 1910.1000, Bureau of National Affairs, Washington, DC (1976) p. 31:8304
96. Wolff, M. S., Daum, S. M., Lorimer, W. V., Selikoff, I. J., and Aubrey, B. B., Styrene and related hydrocarbons in subcutaneous fat from polymerization workers, J. Toxicol. Env. Hlth., 2 (1977) 997-1005
97. NIOSH, Health Hazard Evaluation Determination. Reinell Boats, Inc. Poplar Bluff MO, Rept. 75-150-378, U.S. Dept. HEW, National Institute for Occupational Safety and Health, Cincinatti (1977) pp. 1, 10-14, 21-22, 24-26
98. Cakrtova, E., and Vanecek, M., Evaluation of the styrene exposure in the production of glass fiber-reinforced laminated plastics, Prac. Lek., 26 (1974) 370-374

99. Eurocop-Cost, "A Comparehensive List of Polluting Substances Which have been Identified in Various Fresh Waters, Effluent Discharges, Aquatic Animals, and Plants, and Bottom Sediments", 2nd ed., EUCO/MDU/73/76, Commission of the European Communities, (1976) p. 101
100. Environmental Protection Agency, Preliminary Assessment of Suspected Carcinogens in Drinking Water, U.S. Environmental Protection Agency, Washington, DC. p. II-7
101. Dowty, B. J., Carlisle, D. R., and Laseter, J. L., New Orlean's drinking water sources tested by gas-chromatography-mass spectrometry, Env. Sci. Technol., 9 (1975) 762-765
102. Baggett, M. S., Morie, G. P., Simmons, M. W., and Lewis, J. S., Quantitative determination of semi-volatile compounds in cigarette smoke, J. Chromatog., 97 (1974) 79-82
103. Jermini, C., Weber, A., and Grandjean, E., Quantitative determination of various gas-phase components of the side-stream smoke of cigarettes in the room air as a contribution to the problem of passive smoking, Int. Arch. Occup. Environ. Hlth., 36 (1976) 169-181
104. Anon, MCA find that styrene produces tumors in rats but cast doubt on finding, Pesticide & Toxic Chem. News, 6 (13) (1978) 7-8
105. deMeester, C., Poncelet, F., Roberfroid, M., Rondelet, J., and Mercier, M., Mutagenicity of styrene and styrene oxide, Mutation Res., 56 (1977) 147-152
106. Meretoja, T., Vainio, H., Sorsa, M., and Härkönen, H., Occupational styrene exposure and chromosomal aberrations, Mutation Res., 56 (1977) 193-197
107. IARC, Monographs on the Evaluation of Carcinogenic Risk of Chemicals to Man, Vol. 11, Lyon (1976) pp. 125-129
108. Lee, H., and Neville, K., "Handbook of Epoxy Resins", McGraw-Hill, San Francisco, (1967) p. 4-59
109. McCammon, C. J., Kotin, P., and Falk, H. L., The carcinogenic potency of certain diepoxides, Proc. Amer. Ass. Cancer Res., 2 (1957) 229-230
110. Wallace, J. G., Epoxidation, In Kirk-Othmer Encyclopedia of Chemical Technology, Vol. 8, 2nd ed., Wiley and Sons, New York (1964) pp. 249-265
111. IARC, Monographs on the Evaluation of Carcinogenic Risk of Chemicals to Man, Vol. 11, Lyon (1976) 141-145
112. Anon, Bakelite epoxy resiner 1-4206, Technical Bulletin, #4-1456, Union Carbide Co., New York (1964)
113. IARC, Monographs on the Evaluation of Carcinogenic Risk of Chemicals to Man, Vol. 11, International Agency for Research on Cancer, Lyon (1976) 115-123
114. Stecher, P. G., (ed) "The Merck Index", 8th ed., Merck & Co., Rahway, NJ (1968) p. 418
115. Shimkin, M. B., Weisburger, J. H., Weisburger, E. K., Bugareff, N., Suntzeff, V., Bioassay of 29 alkylating chemicals by the pulmonary-tumor response in strain A mice, J. Natl. Cancer Inst., 36 (1966) 915-935
116. Clarke, C. H., and Loprieno, N., The influence of genetic background on the induction of methionine reversions by diepoxybutane in S. pombe, Microbiol. Genet. Bull., 22 (1965) 11-12
117. Glover, S. W., A comparative study of induced reversions in E. coli. In "Genetic Studies with Bacteria", Carnegie Institution of Washington, Publ. No. 612, Washington, DC (1956) pp. 121-136
118. Zimmermann, F. K., and Vig, B. K., Mutagen specificity in the induction of mitotic crossing over in S. cerevisiae, Mol. Gen. Genet., 139 (1975) 255-268
119. Bird, M. J., and Fahmy, O. G., Cytogenetic analysis of the action of carcinogens and tumor inhibitors in Drosophila melanogaster I. 1:2,3:4-Diepoxylentane, Proc. Roy Soc. Ser. B., 140B (1953) 556-578

120. Watson, W. A. F., Studies on a recombination-deficient mutation of Drosophila. II. Response to X-rays and alkylating agents, Mutation Res., 14 (1972) 299-307
121. Fahmy, O. G., and Fahmy, M. J., Gene elimination in carcinogenesis: reinterpretation of the somatic mutation theory, Cancer Res., 30 (1970) 195-205
122. Huang, S. L., and Rader, D. N., The association between mutagenicity and adduct formation of 1,2,7,8-diepoxyoctane and 1,2,5,6-diepoxycyclooctane, 9th Annual Environmental Mutagen Society Meeting, San Francisco, March 9-13 (1978) 42
123. Terrill, J. B., and Lee, K. P., The inhalation toxicity of phenylglycidyl ether. I. 90-day inhalation study, Toxicol. Appl. Pharmacol., 42 (1977) 263-269
124. IARC, Monographs on the Evaluation of Carcinogenic Risk of Chemicals to Man, Vol. 4, International Agency for Research on Cancer, Lyon (1974) 259-269
125. Fishbein, L., Flamm, W. G., and Falk, H. L., "Chemical Mutagens", Academic Press, New York (1970) pp. 215-216
126. Rapoport, I. A., Acetylation of gene proteins and mutations, Dokl. Akad. Nauk. SSSR, 58 (1947) 119
127. Auerbach, C., Drosophila tests in pharmacology, Nature, 210 (1966) 104
128. Khishim, A. F. E., The requirement of adenylic acid for formaldehyde mutagenesis, Mutation Res., 1 (1964) 202
129. Hyatt, J. C., U.S. mulls rules for handling of chemicals that can lead to cancer in plant workers, Wall Street J. April 11 (1973) p. 36
130. Brusick, D. J., The genetic properties of beta-proprolactone, Mutation Res., 39 (1977) 241-256
131. Murata, A., and Kanegawa, T., Effect of beta-proprolactone on Lactobacillus cassei phage J1 system, J. Gen. Appl. Microbiol., 19 (1973) 467-480
132. Wainberg, M. A., Hjorth, R. N., and Howe, C., Effect of beta-propiolactone on Sendai virus, Appl. Microbiol., 22 (1971) 618-621
133. Thoma, R. W., Use of mutagens in the improvement of production strains of microorganisms, Folia Microbiol., 16 (1971) 197-204
134. Hueper, W. C., Environmental carcinogenesis in man and animals, Ann. NY Acad. Sci., 108 (1963) 963-1038
135. U.S. Government, Occupational Safety and Health Standard, Federal Register 38 No. 85 (1973) 10929
136. Boutwell, R. L., Colburn, N. H., and Muckerman, C. C., In vivo reactions of β-propiolactone, Ann. NY Acad. Sci., 163 (1969) 751
137. Fishbein, L., Degradation and residues of alkylating agents, Ann. NY Acad. Sci. 163 (1969) 869
138. Brookes, P., Covalent interaction of carcinogens with DNA, Life Sci., 16 (1975) 331-344
139. Dickens, F., Carcinogenic lactones and related substances, Brit. Med. Bull., 20 (1964) 96-101
140. Lawley, P. D., and Brookes, P., Further studies on the alkylation of nucleic acids and their constituent nucleotides, Biochem. J., 89 (1963) 127-138
141. Zaugg, H. E., β-Lactones, Organic Reactions, 8 (1954) 305-363
142. Long, F. A., and Purchase, M., The kinetics of hydrolysis of β-propiolactone in acid, neutral and basic solutions, J. Am. Chem. Soc., 72 (1950) 3267
143. Van Duuren, B. L., Carcinogenic epoxides, lactones, and haloethers and their mode of action, Ann. NY Acad. Sci., 163 (1969) 633
144. Dickens, F., and Jones, H. E. H., Further studies on the carcinogenic action of certain lactones and related substances in the rat and mouse, Brit. J. Cancer, 19 (1965) 392

145. Kubrinski, H., and Szybalski, E. H., Intermolecular linking and fragmentation of DNA by beta-propiolactone, a monoalkylating carcinogen, Chem. Biol. Interactions, 10 (1975) 41-55
146. Dean, B. J., The mutagenic effects of organophosphorus pesticides on microorganisms, Arch. Toxikol., 30 (1972) 67-74
147. Mukai, F., and Troll, W., The mutagenicity and initiating activity of some aromatic amine metabolites, Ann. NY Acad. Sci., 163 (1969) 828-836
148. Brusick, D. J., and Zeiger, E., A comparison of chemically induced reversion patterns of S. typhimurium and S. cerevisiae mutants using in vitro plate tests, Mutation Res., 14 (1972) 271-275
149. Ames, B. N., and Whitefield, H. S., Jr., Frameshift mutagenesis in Salmonella, Cold Spring Harbor Symp. on Quant. Biol., 31 (1966) 221-225
150. Chen, C. C., Speck, W. T., and Rosenkranz, H. S., Mutagenicity testing with Salmonella typhimurium strains, II. Effect of unusual phenotypes on the mutagenic response, Mutation Res., 28 (1975) 31-36
151. Smith, H. H., and Srb, A. M., Induction of mutations with β-propiolactone, Science, 114 (1951) 490-492
152. Schlottfeld, C. S., Induction of mutation in Neurospora crassa with beta-propiolactone, PH. D. Thesis, Cornell University, Ithaca, New York (1953) 123
153. Zimmermann, F. K., Induction of mitotic gene conversion by mutagens, Mutation Res., 11 (1971) 327-337
154. Brusick, D. J., and Mayer, V. W., New developments in mutagenicity screening techniques with yeast, Environ. Hlth. Persp., 6 (1973) 83-96
155. Brusick, D. J., The mutagenic activity of beta-propiolactone in Saccharmocyes cerevisiae, Mutation Res., 15 (1972) 425-434
156. Prakash, L., and Sherman, F., Mutagenic specificity: Reversion of iso-1-cytochrome C mutants of yeast, J. Mol. Biol., 79 (1963) 65-82
157. Fukuda, S., and Yamamoto, N., Effect of beta-propiolactone on bacteriophage and Salmonella typhimurium, Cancer Res., 30 (1970) 830-833
158. Dean, B. J., Chemical-induced chromosome damage, Lab. Animals, 3 (1969) 157-174
159. Clive, D., Personal communication, described in ref. 130.
160. Clive, D., and Spector, J. F. S., Laboratory procedure for assessing specific locus mutations at the TK locus in cultured L5178Y mouse lymphoma cells, Mutation Res., 31 (1975) 17-29
161. Perry, P., and Evans, H. J., Cytological detection of mutagen-carcinogen exposure by sister chromatid exchange, Nature, 258 (1975) 121-125
162. DiPaolo, A., Nelson, R. C., Donovan, P. J., and Evans, C. H., Host-mediated in vivo/in vitro assay for chemical carcinogenesis, Arch. Pathol., 95 (1973) 380-385
163. Swanson, C. P., and Merz, T., Factors influencing the effect of β-propiolactone on chromosomes of Vicia faba, Science, 129 (1959) 1364
164. Smith, H. H., and Lofty, T. A., Effects of β-propiolactone and ceepryn on chromosomes of Vicia and Allum, Am. J. Botany, 42 (1955) 750
165. IARC, Monographs on the Evaluation of Carcinogenic Risk of Chemicals to Man, Vol. 11, International Agency for Research on Cancer, Lyon (1976) 225-229
166. Melzer, M. S., Effect of carcinogens andother compounds on deoxyribonuclease, Biochim. Biophys. Acta., 138 (1967) 613-616
167. Stecher, P. G., (ed) "The Merck Index", 8th ed., Merck & Co., Rahway, NJ (1968) p. 184
168. Minoda, S., and Mijajima, M., Make δ-BL and THF from maleic, Hydrocarbon Processing, Nov (1970) 146-178

169. IARC, Monographs on the Evaluation of Carcinogenic Risk of Chemicals to Man, Vol. 11, International Agency for Research on Cancer, Lyon (1976) 231-239
170. Anon, "Herbicide Handbook", 3rd ed., Weed Science Society of America, Champaign, Illinois (1974) pp. 126-128
171. Spence, L. R., Palamand, S. R., and Hardwick, W. D., Identification of C_4 and C_5 lactones in beer, Tech. Quart. Master Brew. Ass. Amer., 10 (1973) 127-129
172. Rudali, G., Apiou, F., Boyland, E., and Castegnaro, M., A propos de l'action cancerigene de la δ-butyrolactone chez les souris, CR Acad. Sci. (Paris) 282 (1976) 799-802
173. Webb, A. D. Gayon, P. R., and Boidron, J. N., Composition-d'une essence extraite d'un vin de V. vinifera (Variete cabernet-sauvignon), Bull. Soc. Chim. Fr., 6 (1964) 1415-1420
174. Kahn, J. H., Nickol, G. B., and Conner, H. B., Identification of volatile components in vinegars by gas chromatography-mass spectrometry, J. Agr. Food Chem., 20 (1972) 214-218
175. Liebich, H. M., Douglas, D. R., Zlatkis, A., Muggler-Chavan, F., and Donzel, A., Volatile components in roast beef, J. Agr. Food Chem., 20 (1972) 96-99
176. Gianturco, M. A., Giammarino, A. S., and Griedel, P., Volatile constituents of coffee, Nature, 210 (1966) 1358
177. Johnson, A. E., Nursten, H. E., and Williams, A. A., Vegetable volatiles: A survey of components identified. II. Chem. & Ind., Oct. 23 (1971) 1212-1224
178. Neurath, G., Dunger, M., and Kustermann, I., Untersuchung der "semi-volatiles" des cigarettenrauches, Beitr. Tabakforsch., 6 (1971) 12-20
179. Swern, D., Wieder, R., McDonough, M., Meranze, D. R., and Shimkin, M. B., Investigation of fatty acids and derivatives for carcinogenic activity, Cancer Res. 30 (1970) 1037-1046
180. Dickens, F., and Jones, H. E. H., Carcinogenic activity of a series of reactive lactones and related substances, Brit. J. Cancer, 15 (1961) 85-100

CHAPTER 6

AZIRIDINES, ALIPHATIC SULFURIC ACID ESTERS, SULTONES, DIAZO ALKANES AND ARYLALKYLTRIAZENES

A. Aziridines

Aziridines are extremely reactive alkylating agents which can undergo two major types of reactions[1,2]: (1) ring-preserving reactions in which an aziridine (e.g., ethyleneimine) acts as a secondary amine reacting with many organic functional groups containing an active hydrogen, undergo replacement reactions of the hydrogen atom by nucleophilic attack at one of the methylene groups and (2) ring-opening reactions similar to those undergone by ethylene oxide. Aziridines, because of their dual functionality and high degree of reactivity, exhibit actual or potential utility in a broad range of applications[2] including: (1) textiles: crease proofing, dyeing and printing, flame proofing, water-proofing, shrink proofing, form stabilization and stiffening; (2) adhesives and binders; (3) petroleum products and synthetic fuels; (4) coatings; (5) agricultural chemicals; (6) ion-exchange resins; (7) curing and vulcanizing polymers; (8) surfactants; (9) paper and printing; (10) antimicrobials; (11) flocculants and (12) chemotherapeutics.

1. <u>Aziridine</u> (ethyleneimine; azacyclopropane; dimethylenimine; dihydro-1-H-aziridine; $\begin{smallmatrix}CH_2\\CH_2\end{smallmatrix}\!\!>\!\!NH$) can be prepared commercially by the reaction of ammonia and ethylene dichloride or by the cyclization of 2-aminoethyl hydrogen sulfate[1].

Currently there is only one U.S. manufacture and its annual production is estimated to be about 1.3 million kg[3]. No current information is available on the amounts of aziridine produced in Europe. One producer in the Federal Republic of Germany suspended sales of this chemical and all derivatives except polyethyleneimine in 1973[1]. Production of aziridine in Japan during the period 1970-1974 is estimated to have been 600,000 kg. Currently there is only one producer[1].

Aziridine is used principally as a chemical intermediate to provide one or more aziridinyl-(CH_2CH_2-N-) or ethyleneimino-(-CH_2CH_2NH-) substituents.

Approximately 50% of the aziridine produced in the U.S. is polymerized to polyethylenimine which contains less than 1 ppm residual monomer. Polyethylenimine is used principally as a flocculant in water treatment and in the textile and paper industries where it is used as a wet-strength additive, as an adhesion promoter in various coating applications, and in the textile industry (to improve dyeing and printing for waterproofing and to impart antistatic properties).

Of the remaining 50% of the aziridine produced, most is used as a chemical intermediate in drug, cosmetic and dye manufacture, in the production of N-2-hydroxyethylenimine (2-aziridinyl ethanol) and of triethylenemelamine, and as a intermediate and monomer for oil additive compounds, ion-exchange resins, coating resins, adhesives, polymer stabilizers and surfactants[1-4].

In Japan, aziridine is believed to be used mostly for the production of polyethylenimine with lesser quantities used to produce taurine and as an intermediate in the production of pesticides and dyestuffs[1].

Data are not available on the numbers of individuals employed in the production of aziridine or in its use applications.

Aziridine is included in the U.S. Occupational Safety and Health Administration's (OSHA) list of "occupational carcinogens". The exposure of an employee to aziridine during a 40-hour week can not exceed an 8-hour time weighted average of 1 mg/m^3 (0.5 ppm)[5]. The maximum exposure to aziridine in the Federal Republic of Germany and the German Democratic Republic is limited to 1 mg/m^3 (0.5 ppm) and in the USSR to 0.02 mg/m^3 [6].

Aziridine is carcinogenic in two strains of mice (C56Bl/6 x C3H/Anf)F1 and (C56Bl/6xAKRF1) following its oral administration producing an increased incidence of liver-cell and pulmonary tumors[1,7].

Aziridine induces both transmissible translocations and sex-linked recessive lethal mutations in Drosophila melanogaster[8,9]; specific locus mutations in silkworms (Bombyx Mori)[10]. Aziridine also produces leaky mutants, mutants with polarized and non-polarized complementation patterns, and non-complementing mutants and multilocus deletions in Neurospora crassa[11] and induces mitotic recombination[12] and gene conversion in Saccharomyces cerevisiae[13].

Aziridine induces chromosome aberrations in cultured human cells[14], mouse embryonic skin cultures[15] and Crocker mouse Sarcoma 188[15]. When rabbits were inseminated with spermatozoa which had been treated with aziridine in vitro, only 40% of embryos were found to be viable relative to the number of corpora lutea in comparison to 78% in controls[16]. Aziridine has been reported to possess teratogenic activity in the rat fetus[17].

2. **2-Methylaziridine** (propylenimine; 2-methylazacyclopropane; methylethylenimine; $H\text{-}C\overset{CH_3}{\underset{H_2C}{\diagup\!\!\!\!\diagdown}}NH$) can be prepared commercially via the reaction of 1,2-dichloropropane with ammonia at elevated temperatures or by the reaction of sodium hydroxide with 2-chloropropylamine hydrochloride (formed by the addition of hydrogen chloride to 1-amino-2-propanol)[1].

Currently, 2-methylaziridine is produced in the U.S. by one manufacturer and production data are not available.

Reactions of 2-methylaziridine fall into two general categories, e.g., as a secondary amine yielding N-substituted propylenimines in which the ring is intact, and as a cyclic amine involving ring-opening reactions[1].

2-Methylaziridine is apparently used in the U.S. exclusively as an intermediate since there is no available evidence of its use in the monomeric form[1]. It is used principally in the modification of latex surface coating resins to improve adhesion. Polymers modified with 2-methylaziridine or its derivatives have been used in the adhesive, textile and paper industries, because of the enhanced bonding of imines to cellulose derivatives. 2-Methylaziridine has also been employed to modify dyes for specific adhesion to cellulose. Derivatives of 2-methylaziridine have been used in photography, gelatins, synthetic resins, as modifiers for viscosity control in the oil additive industry and as flocculants in petroleum refining, and in imine derivatives for use in agricultural chemicals[1]. It is believed to be used in Japan primarily in the treatment of paper[1].

OSHA's standards for occupational exposure to skin contaminants require that an employee's exposure to 2-methylaziridine not exceed 5 mg/m^3 (2 ppm) in air[5]. A similar threshold limit value has been established in the Federal Republic of Germany.

2-Methylaziridine has been reported to be a powerful carcinogen affecting a wide range of organs in the rat when administered orally[18]. For example, brain tumors (gliomas) and squamous cell carcinomas of the ear duct have been found in both sexes; disseminated granulocytic leukemia in males and a number of multiple mammary tumors (some metastasizing to the lung) were found in females at the end of 60 weeks following twice weekly 10 and 20 mg/kg oral administrations.

2-Methylaziridine (as well as aziridine) have been shown to be mutagenic in the Salmonella/microsome test system[19].

3. 2-(1-Aziridinyl)ethanol (β-hydroxy-1-ethyl aziridine; N-(2-hydroxyethyl)-aziridine; aziridine ethanol; $\mathrm{\begin{smallmatrix}H_2C\\H_2C\end{smallmatrix}\!\!>\!\!N\text{-}CH_2CH_2OH}$) can be prepared by the addition of aziridine to ethylene oxide[1]. (It should be noted that both reactive intermediates are carcinogenic and mutagenic).

It has been produced commercially in the U.S. since about 1965 and production by one company in 1973 was about 45,000 kg. It is reported to be used commercially in the modification of latex polymers for coatings, textile resins and starches, as well as in the preparation of modified cellulose products such as paper, wood fibers and fabrics[1]. 2-(1-Aziridinyl)-ethanol is carcinogenic in mice producing malignant tumors at the site of its injection[20] and has been reported to induce sex-linked recessive lethals in Drosophila melanogaster[21].

B. Aliphatic Sulfuric Acid Esters

Alkyl sulfates such as dimethyl- and diethyl sulfates are very reactive alkylating agents that have been extensively employed both in industry and the laboratory for converting active-hydrogen compounds such as phenols, amines and thiols to the corresponding methyl and ethyl derivatives[22].

1. Dimethyl Sulfate (dimethyl monosulfate; methyl sulfate; sulfuric acid; dimethyl ester; DMS; $H_3CO\overset{\overset{O}{\|}}{\underset{\underset{O}{\|}}{S}}\text{-}OCH_3$) has been produced commercially for at least 50 years, principally by the continuous reaction of dimethyl ether with sulfur trioxide[22].

While no information is available on world commercial production of dimethylsulfate it is known that in the U.S. one company manufactures the chemical, one producer was reported in the U.K. in 1970, 2 in the Federal Republic of Germany in 1967 and 1 in Italy in 1969.

Dimethyl sulfate has been used extensively as a methylating agent both in industry and the laboratory[2,22]. Its utility includes the methylation of cellulose, preparation of alkyl lead compounds, preparation of alkyl ethers of starch, solvent for the extraction of aromatic hydrocarbons, curing agent for furyl alcohol resins and the polymerization of olefins[2]. DMS has been employed commercially for the preparation of quaternary ammonium methosulfate salts (via its reaction with the respective tertiary amine)[22] (For this reaction dimethyl sulfate must be used as the alkylating agent). Included in this group are six cationic surfactants: dimethyl dioctadecylammonium methosulfate; (3-lauramidopropyl)trimethyl ammonium methosulfate; (3-oleamideopropyl)-trimethyl ammonium methosulfate; the methosulfate of a stearic acid-diethanolamine condensate; the methosulfate of N-(2-hydroxyethyl)-N,N',N'-tris(2-hydroxypropyl)-ethylenediamine distearate; and the methosulfate of N,N2,N',N'-tetrakis(2-hydroxypropyl)ethylene-diaminedioleate. U.S. production of eight compounds of this type amounted to at least 500 kg each in 1970[22].

Dimethyl sulfate has also been used for the preparation of anticholinergic agents (e.g., diphemanil methyl sulfate and hexocyclium methyl sulfate, and the parasympathomimetic agent, neostigmine methyl sulfate[22].

There are no data on the numbers of workers employed in the production of dimethylsulfate (or diethylsulfate) and their use applications.

Dimethyl sulfate has been listed among the industrial substances suspect of carcinogenic potential for man. The current TLV in the U.S. is 1 ppm[23].

Dimethyl sulfate has been shown to be carcinogenic in the rat (the only species tested) by inhalation[22,24], subcutaneous injection[24] and following pre-natal exposure[24]. It is carcinogenic to the rat in a single-dose exposure[22,24]. The possibility of carcinogenicity of dimethyl sulfate in man occupationally exposed for 11 years has been raised[25], however, good epidemiological evidence is unavailable to confirm this[22,25].

Dimethyl sulfate is mutagenic in S. typhimurium TA 1535[26], E. coli[27], Drosophila[27,28], Neurospora[29,30], S. pombe[31] and in the host-mediated assay with S. typhimurium TA 1950 as genetic indicator system[26]. In the latter system the lowest effective dose of dimethyl sulfate was 2500 μ mole/kg. Dimethyl sulfate also induced chromosome breakage in plant material[32].

2. <u>Diethyl Sulfate</u> (diethylmonosulfate; ethyl sulfate; sulfuric acid; diethyl ester, DES; $H_5C_2O-\overset{\overset{O}{\|}}{\underset{\underset{O}{\|}}{S}}-OC_2H_5$) similarly to DMS has been produced for at least 50 years,

and has been prepared principally by absorbing ethylene in concentrated sulfuric acid or by the action of fuming sulfuric acid on ethyl ether or ethanol[22].

Diethyl sulfate has been employed in a variety of ethylation processes in a number of commercial areas and organic synthesis including[2,22]: finishing of cellulosic yarns, etherification of starch, stabilization of organophosphorus insecticides, as a catalyst in olefin polymerization and acrolein-pentaerythritol resin formation. Diethyl sulfate has been used as the ethylating agent for the commercial preparation of a number of cationic surfactants including: (2-aminoethyl) ethyl (hydrogenated tallow alkyl) (2-hydroxyethyl) ammonium ethosulphate; 1-ethyl-2-(8-heptadecenyl)-1-(2-hydroxyethyl)-2-imidazolinium ethosulphate; N-ethyl-N-hexadecyl-morpholinium ethosulphate; N-ethyl-N-(soybean oil alkyl) morpholinium ethosulphate; ethyl dimethyl (mixed alkyl) ammonium ethosulphate; and triethyl octadecyl ammonium ethosulphate. U.S. production of 6 compounds of this type amounted to at least 500 kg each in 1970[22].

Diethyl sulfate is carcinogenic in the rat (the only species tested) following subcutaneous administration and pre-natal exposure[22,24]. The evidence for carcinogenicity of diethyl sulfate in the rat following oral administration is inconclusive. Opposite results have been published for the carcinogenic potential of structurally related chemicals differing in their alkylating properties. For the dialkyl sulfates, the less reactive diethyl sulfate was found to be a stronger carcinogen than dimethyl sulfate[24].

Diethyl sulfate is mutagenic in S. typhimurium TA 1535[26], E. coli[33-35], Drosophila[27,28,36,37], bacteriophage T-2[38], Neurospora[29], S. pombe[31,39] and Aspergillus nidulans[40], and in the host-mediated assay with S. typhimurium TA 1950 and 2500 μ mole/kg (the lowest effective dose)[26].

3. Cyclic Aliphatic Sulfuric Acid Esters

Similarly to dialkyl sulfates, alkyl alkanesulfonates and alkane sultones, cyclic aliphatic sulfuric acid esters are monofunctional alkylating agents which react with neutral or ionic nucleophiles, Y or Y-, along the general pathway (1) or (2)

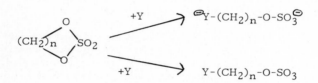

Members of this class, 1,2-ethylene sulfate (ESF, I), 1,3-butylene sulfate (BSF, II) and 1,3-propylene sulfate (PSF, III) are used to introduce sulfoxyalkyl groups, e.g., into nitrogen heterocycles[41], especially into cyanine dyes[42,43].

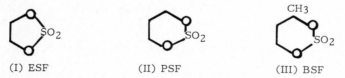

(I) ESF   (II) PSF   (III) BSF

No information is available on the amounts of cyclic aliphatic sulfuric acid esters produced, the number of workers involved in their production or their use applications.

1,2-Ethylene sulfate (glycol sulfate) has recently been shown to be a carcinogenic alkylating agent inducing local malignant tumors when administered s.c. to ICR/Ha mice[44]. Although no similar studies or long-term carcinogenesis studies have been reported for the cyclic sulfates PSF and BSF, their carcinogenic potential must also be taken into account[26,45]. It is interesting to note that glycol sulfite ($\begin{smallmatrix} H_2C-O \\ H_2C-O \end{smallmatrix} S=O$) did not exhibit carcinogenic activity when administered s.c. to mice. Van Duuren et al[44] in comparing the relative activities of glycol sulfate (ESF) and glycol sulfite suggested the activity of ESF occurred because the sulfate when reacting with nucleophiles in neutral media undergoes C-O-alkyl scission while the sulfite under comparable conditions undergoes S-O scission.

The cyclic sulfates ESF, PSF and BSF revert the tester strain TA 1535 of S. typhimurium[26] in vitro indicating their ability to induce base substitutions. Compared with alkyl sulfates and sultones, the mutagenic activity in the plate test was 1,3-propane sulfone > 1,3-propylene sulfate > 1,3-butylene sulfate > 1,4-butane sulfone > 1,2-ethylene sulfate > diethyl sulfate > dimethyl sulfate[26].

Dose-response studies in the host-mediated assay with tester strains TA 1950 of S. typhimurium as genetic indicator system revealed a linear dose dependency of mutagenic activity. The lowest effective dose (LED) (μ mole/kg) for the cyclic aliphatic sulfates were: 1,3-propylene sulfate, 10; 1,3-butylene sulfate, 50 and 1,2-ethylene sulfate which was only weakly mutagenic, 500০[26].

C. Sultones

Sultones such as the 1,3-propane-, and 1,4-butane derivatives are being increasingly employed industrially to introduce the sulphopropyl and sulphobutyl groups ($-CH_2CH_2CH_2SO_3^-$ and $-CH_2CH_2CH_2CH_2SO_3^-$ respectively) into polymer chains containing nucleophilic centers in order to enhance water solubility and confer an anionic character[1].

1. 1,3-Propane Sultone (3-hydroxy-1-propane sulphonic acid sultone; 1,2-oxathiolane-2,2-dioxide; $\begin{smallmatrix} CH_2\text{-}CH_2 \\ H_2C \quad\, O \quad O \\ \diagdown S \diagup \\ O \end{smallmatrix}$) the simplest sultone is produced commercially via the dehydration of gamma-hydroxy-propanesulphonic acid, which is prepared from sodium hydroxypropanesulphonate. This sodium salt is initially prepared by the addition of sodium bisulphite to allyl alcohol[1].

1,3-Propane sultone is a monofunctional alkylating agent which reacts with nucleophiles, Y-, along the general pathway (1) or (2) as follows: [46]

$$Y^- + \begin{matrix} CH_2\text{-}CH_2 \\ | \quad\quad\quad\, \diagdown SO_2 \\ CH_2\text{-}O \diagup \end{matrix} \rightarrow Y\text{-}CH_2CH_2CH_2SO_2O^- \quad\quad (1)$$

$$Y + \begin{matrix} CH_2\text{-}CH_2 \\ | \quad\quad\quad\, \diagdown SO_2 \\ CH_2\text{-}O \diagup \end{matrix} \rightarrow Y^+\text{-}CH_2CH_2CH_2SO_2O^- \quad\quad (2)$$

A large number of sulphopropylated products and their potential uses[1,47,48] include: (a) derivatives of amines, alcohols, phenols, mercaptans, sulphides and amides useful as detergents, wetting agents, lathering agents and bacteriostats; (b) soluble starches used in the textile industry; (c) solubilized cellulose, which was reported to have soil-suspending properties; (d) dyes, (e) an antistatic additive for polyamide fibers; (f) cation-exchange resins (prepared by condensing the sulphonic acid product derived from phenol and propane sultone which formaldehyde); and (g) phosphorus-containing sulphonic acids (produced from organic phosphines, neutral esters of trivalent phosphorous acids, and phosphorous and phosphoric triamides), useful as insecticides, fungicides, surfactants and vulcanization accelerators.[1]

The production of 1,3-propane sultone in the U.S. was estimated at 1000 pounds in 1973, but the only producer ceased production in 1975[3]. No information is available of 1,3-propane sultone being produced elsewhere.

1,3-Propane sultone is carcinogenic in the rat when administered orally, intravenously or by pre-natal exposure, and exhibits a local carcinogenic effect in the mouse and the rat when given subcutaneously[1,18,20,49,50]. An aqueous solution of 1,3-propane sultone given to Charles River CD rats by gavage twice weekly at doses of 56 mg/kg body wgt for 32 weeks and 28 mg/kg body wgt for 60 weeks produced an incidence of gliomas of 29/52 and 27/52, respectively in rats killed at 60 weeks[18]. The incidence of gliomas was similar at both dose levels in both sexes. In addition several rats had leukemia, ear duct tumors and adenocarcinomas of the small intestine[18].

The homolog, 1,4-butane sultone is chemically far less reactive[51,52] and displays only a low carcinogenic potency[49]. (Less than 1,000 pounds of 1,4-butane sultone was produced in the U.S. in 1977[3]).

1,3-Propane sultone has been classified as a potent mutagen toward Schizesaccaromyces pombe[31] while the mutagenic effectiveness of 1,4-butane sultone towards S. pombe was found to be much lower[31,52]. 1,3-Propane sultone is mutagenic in S. typhimurium TA 1535 and S. cerevisiae D3 in vitro and in the host-mediated assay[53].

Both 1,3-propane-, and 1,4-butane sultones are active in the transformation of Golden Syrian hamster embryo cells in vitro[54].

1,3-Propane sultone at 3Mat 30°C for 4 hr, is a powerful mutagen in higher plants, e.g., barley (Hordeum vulgare)[55]. It also causes mutations in Proteus mirabilis[56], Pisum sativum[57] and in some non-tuberous species of Solanum[58], besides producing chromosomal aberrations in a variety of plant systems[59-63].

As a typical alkylating agent[49,52,64], the biological activity of 1,3-propane sultone may be attributed to its reaction with genetic material. As a bifunctional agent, it is likely to introduce cross-links between DNA strands thus interfering with normal replication of DNA and inhibiting its synthesis.

Recently, 1,3-propane sultone induced mutagenic activity in S. pombe was reported to be at par with that of methyl- and ethyl methane sulfonate (MMS and EMS) when expressed in terms of alkylating events at specific nucleophilicity[52].

D. Diazoalkanes

Diazoalkanes represent an extremely reactive class of alkylating agents. Diazoalkanes are sufficiently basic to abstract protons from many compounds containing acidic hydrogens, the rate of protonation increasing as the acidity of the proton donor increases. The protonated diazoalkane (an alkyldiazonium ion) is exceedingly unstable, losing molecular nitrogen yielding a carbonium ion which then becomes affixed to whatever nucleophile is available. Hence the overall reaction is a replacement of the nitrogen of the diazoalkane by the hydrogen and accompanying nucleophilic portion of the protic compound[65].

1. Diazomethane (azimethylene) is a resonance hybrid of the three forms: $H_2\overset{\ominus}{C}-\overset{\oplus}{N}\equiv N \rightleftharpoons H_2C=N=N \rightleftharpoons H_2C-N=N$. Because of its toxicity and its explosive nature, diazomethane is not manufactured for distribution and sale. When used as a methylating agent in the laboratory it is produced and used in situ. Most laboratory preparations of diazomethane during the late 1940's and the 1950's were probably made from N-methyl-N'-nitro-N-nitrosoguanidine. However, in most countries in recent years, it is believed to be made from N-nitroso-methyl-p-toluenesulphonamide[66].

OSHA's health standards for air contaminants require that an employee's exposure to diazomethane does not exceed an 8-hour time-weighted age of 0.2 ppm (0.4 mg/m^3) in the work place during any 8-hour workshift for a 40 hour work week[67].

Diazomethane is a powerful methylating agent for acidic compounds such as carboxylic acids, phenols, and enols, and, as a consequence, is both an important laboratory reagent and has industrial utility (with acids, diazomethane yields esters and with enols it gives O- alkylation)[66]. When heated, irradiated with light of the appropriate wavelength, or exposed to certain copper-containing catalysts, diazomethane loses molecular nitrogen and forms carbene, via., $CH_2N_2 \rightarrow CH_2: + N_2$. Carbenes are exceedingly reactive species which, for example, can add to alkenes to form cyclopropanes. Carbenes can react with the electrons of a carbon-hydrogen bond to "insert" the carbon of the carbene between carbon and hydrogen, e.g., transforming -CH to -C-CH$_3$[65].

Diazomethane can react with many biological molecules, especially nucleic acids and their constituents. For example, its action on DNA includes methylation at several positions on the bases and the deoxyribose moiety as well as structural alterations that result in lower resistance to alkaline hydrolysis and altered hyperchromicity[67-71]. Acid hydrolysis of diazomethane-treated DNA yields 7-methylguanine and 3-methylguanine, while acid hydrolysis of diazomethane-treated RNA yields 7-methylguanine, 1-methyladenine and 1-methylcytosine[70].

Diazomethane has been found to be carcinogenic in limited studies in mice and rats, the only species tested[72,73]. In A/2G mice it increased the incidence of lung tumors following skin application[72], while exposure to the gas induced lung tumors in rats[72]. The role of diazomethane as the active agent responsible for the carcinogenic action of many compounds (e.g., nitroso derivatives) has been noted[74-76].

The mutagenicity of diazomethane in Drosophila[77], Neurospora[77-79] and Saccharomyces cerevisiae[80,81] have been described. The implication of diazomethane in the mutagenesis of nitroso compounds has also been cited[82-85].

The carcinogenicity of diazoacetic ester (ethyl diazoacetic ester (ethyl diazoacetate) (DAAE) N=N=CHCOOC$_2$H$_5$ in Wistar rats, Syrian hamsters and Swiss mice has recently been reported[86]. DAAE is mainly locally effective as most tumors were of the digestive tract or skin following oral and percutaneous administration. However, a systemic effect does occur as shown by the increased incidence of lung adenomas in mice following both oral and percutaneous administration[86].

It is also germane to consider aryl and heterocyclic diazo compounds as a class as potential environmental electrophiles[87]. For example, 4-aminoimidazole-5-carboxamide, a component of human urine derived from the de novo purine biosynthetic pathway, was recently reported to undergo in vivo diazotization in rats following its sequential administration with $NaNO_2$[88]. The diazotization product 4-diazoimidazole-5-carboxamide undergoes intramolecular cyclization to azahypoxanthine (Figure 1). 4-Diazoimidazole-5-carboxamide exhibited dose-related mutagenicity in S. typhimurium TA 100 strain and represents a potent electrophilic reactant similar to the proposed ultimate carcinogenic forms of arylalkylnitrosamines[89] and arylnitrosamides[90] (e.g., aryldiazonium ions).

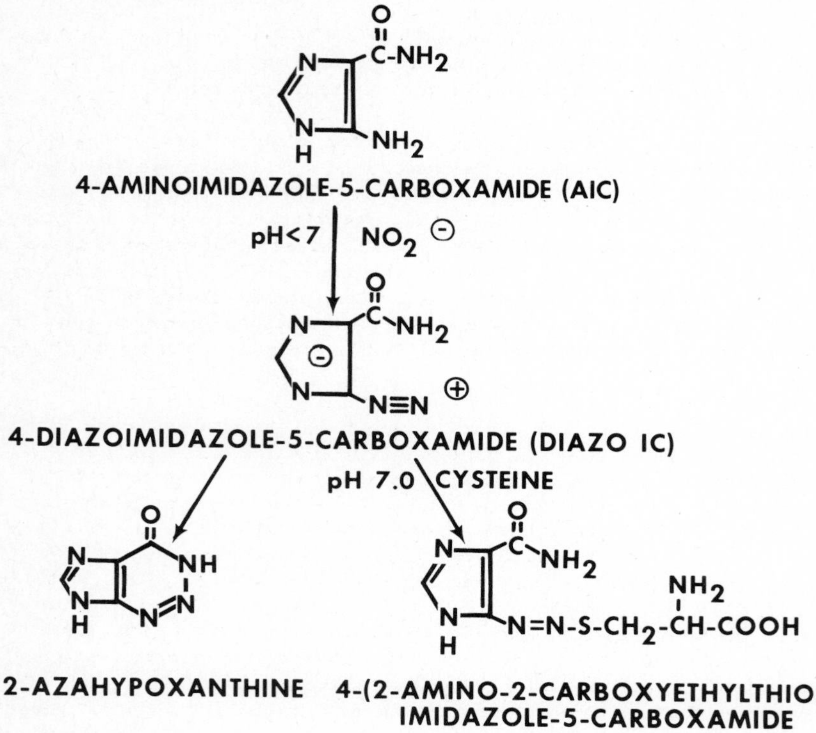

4-AMINOIMIDAZOLE-5-CARBOXAMIDE (AIC)

pH<7 | NO_2^-

4-DIAZOIMIDAZOLE-5-CARBOXAMIDE (DIAZO IC)

pH 7.0 | CYSTEINE

2-AZAHYPOXANTHINE   **4-(2-AMINO-2-CARBOXYETHYLTHIOAZO)-IMIDAZOLE-5-CARBOXAMIDE**

FIGURE 1. Formation of 4-diazoimidazole-5-carboxamide by the diazotization of 4-aminoimidazole-5-carboxamide, its reactions with sulfhydryl compounds, and its internal cyclization to 2-azahypoxanthine.[87]

E. Aryldialkyltriazenes

Triazenes of the general formula $X-\emptyset-N=N-N(CH_3)_2$ (x=substituent); $\emptyset$=phenyl or a heterocyclic residue are industrial intermediates, as well as anti-neoplastic agents and have been patented as rodent repellants and herbicides.

The carcinogenicity of a number of aryldialkyltriazenes has been described[91-93]. The target organs of the carcinogenic activity of the majority of triazenes are the kidney, the central nervous system and the brain[91], and less frequently, the heart[93].

The mutagenicity of a number of these agents has been reported in S. typhimurium (metabolically activated)[94], Drosophila melanogaster[95,96], Neurospora crassa[97], S. cerevisiae[95,98,99].

1-Phenyl-3,3-dimethyltriazene (PDT) causes gene conversion in S. cerevisiae both as a direct mutagen and in the host mediated assay in the mouse where it may be metabolized[98,100]. In Drosophila, PDT is strongly mutagenic inducing sex-linked recessive lethals in sperm, spermatids, spermatocytes, and mature and immature oocytes. There are, however, few events such as chromosome loss and 2/3 translocations[100]. PDT has only very slight effects on human lymphocyte chromosomes in vitro[100]. The very low aberration frequency suggests that the enzymes essential for the activation of triazenes are not present in lymphocytes or that the slight effects observed could be due to spontaneously forming hydrolytic products of PDT[101].

Significant increases in the number of aberrations of both the isolocus and chromatid types were found in rat lymphocytes cultured in vitro after treatment in vivo with single applications of 35 or 50 mg/kg PDT[101]. These positive results could imply the existence of an active circulating metabolite of PDT detectable by the chromosomal damage it causes to rat lymphocytes.

The observed carcinogenicity, mutagenicity and toxicity[102] are suggested to be dependent on at least two molecular mechanisms[95] (Figure 2). One mechanism involves non-enzymic cleavage of the diazoamino side chain liberating arenediazonium cations. In the other mechanism, the major metabolic pathway is an enzymic oxidative monodealkylation yielding the corresponding monoalkyltriazenes, with subsequent hydrolysis

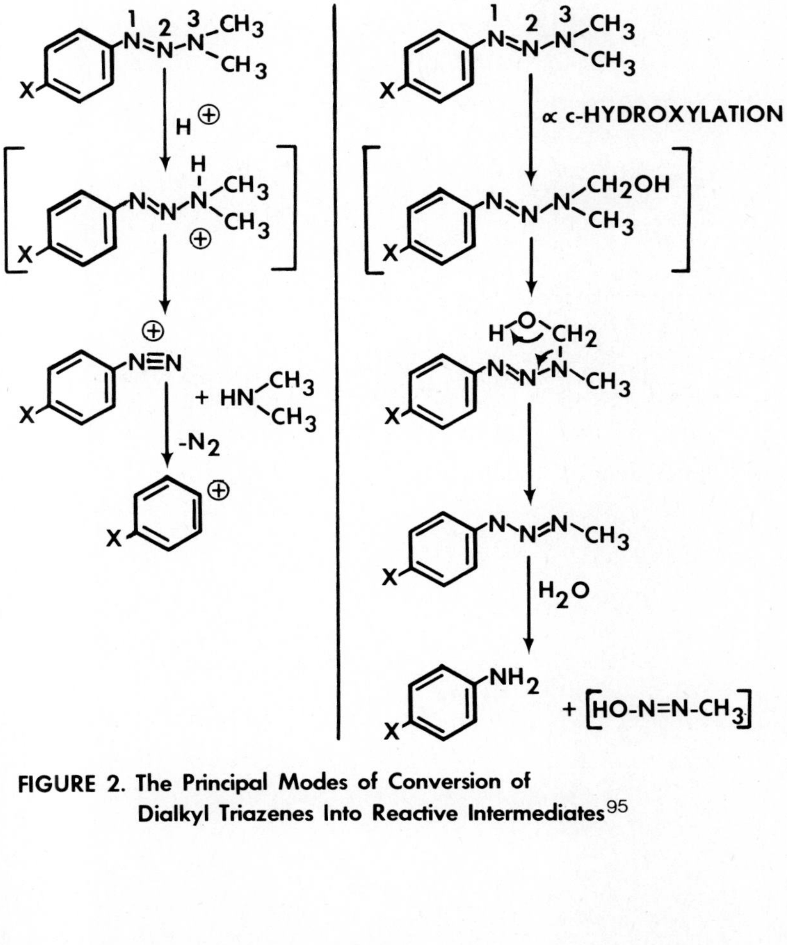

FIGURE 2. The Principal Modes of Conversion of Dialkyl Triazenes Into Reactive Intermediates[95]

yielding alkylating reactants[94,103] (e.g., methylating species similar to those formed from alkylnitrosoureas). A more recent report[104] described the formation of modified anilines during the catabolic degradation of the carcinogen (1-(4-chlorophenyl)-3,3-dimethyl triazene (4Cl-PDMT). Following subcutaneous administration of 1-(4-chloro-[U-^{14}C]phenyl)-3,3-dimethyltriazene, 4-chloro-2-hydroxyaniline (15.1%) and 4-chloroaniline (5.2%) were the most abundant metabolites arising by in vivo fission of the diazoamino group. The structures and distribution of 4-hydroxyaniline (less than 0.1%), 4-chloro-3-hydroxyaniline (0.7%) and 3-chloro-4-hydroxyaniline (8.2%) suggest that these metabolites are derived from a common 3,4-epoxy intermediate and arise either by elimination of chlorine, by opening of the epoxide ring, or by an intramolecular hydroxylation-induced chlorine migration respectively. The identification of 3-chloro-4-hydroxyaniline amongst the metabolites of 4-Cl-PDMT is of special interest according to the authors since it represents the first case of a catabolic degradation of a chemical carcinogen accompanied by the chlorine walk (N.I.H. shift).

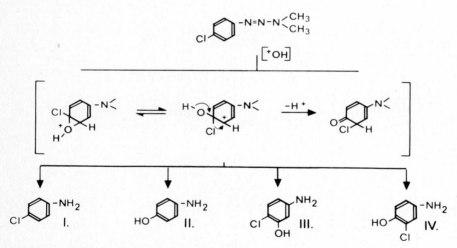

Fig. 3. Formation of modified anilines from a 3,4-epoxy intermediate[104]

References

1. IARC, Monographs on the Evaluation of Carcinogenic Risk of Chemicals to Man, Vol. 9, International Agency for Research on Cancer, Lyon (1975) 37-46, 47-49, 61-65
2. Fishbein, L., Flamm, W. G., and Falk, H. L., "Chemical Mutagens", Academic Press, New York (1970) pp. 143-145, 216-217, 253-258
3. Stanford Research Institute, A Study of Industrial Data on Candidate Chemicals for Testing, EPA-560/5-77-066; Menlo Park, CA, August (1977) pp. 4-29 to 4-31
4. Hawley, G. G., "The Condensed Chemical Dictionary" 8th ed., Van Nostrand/Reinhold, New York (1971) p. 367
5. U.S. Code of Federal Regulations, Occupational Safety and Health Standards, Title 29, part 1910.23, Government Printing Office, Washington, DC (1973) p. 21
6. Winell, M., An international comparison of hygienic standards for chemicals in the work environment, Ambio, 4 (1975) 34-36
7. Innes, J. R. M., Ulland, B. M., Valerio, M. G., Petrucelli, L., Fishbein, L., Hart, E. R., Pallotta, A. J., Bates, R. R., Falk, H. L., Gart, J. J., Klein, M., Mitchell, I., and Peters, J., Bioassay of pesticides and industrial chemicals for tumorigenicity in mice: A preliminary note, J. Natl. Cancer Inst., 42 (1969) 1101
8. Alexander, M. L., and Glanges, E., Genetic damage induced by ethylenimine, Proc. Nat. Acad. Sci., 53 (1965) 282
9. Sram, R. J., The effect of storage on the frequency of translocations in Drosophila melanogaster, Mutation Res., 9 (1970) 243
10. Inagaki, E., and Oster, I. I., Changes in the mutational response of silkworm spermatozoa exposed to mono- and polyfunctional alkylating agents following storage, Mutation Res., 7 (1969) 425
11. Ong, T. M., and deSerres, F. J., Mutagenic activity of ethylenimine in Neurospora Crassa, Mutation Res., 18 (1973) 251
12. Zimmermann, F. K., and VonLaer, U., Induction of mitotic recombination with ethylenimine in Saccharomyces cerevisiae, Mutation Res., 4 (1967) 377
13. Zimmermann, F. K., Induction of mitotic gene conversion by mutagens, Mutation Res., 11 (1971) 327
14. Chang, T. H., and Elequin, F. T., Induction of chromosome aberrations in cultured human cells by ethylenimine and its relation to cell cycle, Mutation Res., 4 (1967) 83
15. Biesele, J.J., Philips, F. S., Thiersch, J. B., Burchenal, J. H., Buckley, S. M., and Stock, C. C., Chromosome alteration and tumor inhibition by nitrogen mustards: The hypothesis of cross-linking alkylation, Nature, 166 (1950) 1112-1113
16. Nuzhdin, N. I., and Nizhnik, G. V., Fertilization and embryonic development of rabbits after treatment of spermatozoa in vitro with chemical mutagens, Dokl. Akad. Nauk. SSSR Otd. Biol., 181 (1968) 419
17. Murphy, M. L., DelMoro, A., and Lacon, C., The comparative effects of five polyfunctional alkylating agents on the rat fetus with additional notes on the chick embryo, Ann. NY Acad. Sci., 68 (1958) 762
18. Ulland, B., Finkelstein, M., Weisburger, E. K., Rice, J. M., and Weisburger, J. H., Carcinogenicity of industrial chemical propylene imine and propane sultone, Nature, 230 (1971) 460
19. McCann, J., Choi, E., Yamasaki, E., and Ames, B. N., Detection of carcinogens as mutagens in the Salmonella/microsome test: Assay of 300 chemicals, Proc. Nat. Acad. Sci., 72 (1975) 5135

20. Van Duuren, B. L., Melchionne, S., Blair, R., Goldschmidt, B. M., and Katz, C., Carcinogenicity of isoesters and epoxides and lactones: Aziridine ethanol, propane sultone and related compounds, J. Natl. Cancer Int., 46 (1971) 143
21. Filippova, L. M., Pan'shin, O. A., and Kostyankovskii, R. G., Chemical mutagens, IV. Mutagenic activity of germinal systems, Genetika, 3 (1967) 134
22. IARC, Monographs on the Evaluation of Carcinogenic Risk of Chemicals to Man, Vol. 4, International Agency for Research on Cancer, Lyon (1974) pp. 271-276; 277-281
23. ACGIH, Threshold Limit Values for Chemical Substances in Workroom Air Adopted by ACGIH for 1976, American Conference of Governmental Industrial Hygienists, Cincinatti, Ohio (1976)
24. Druckrey, H., Druse, H., Preussmann, R. Ivankovic, S., and Landschütz, C., Cancerogene alkylierende substanzen. III. Alkylhalogenide, -sulfate-, sulfonate und ringgespannte heterocyclen, Z. Krebsforsch., 74 (1970) 241
25. Druckrey, H.,Preussmann, R., Nashed, N., and Ivankovic, S., Carcinogenic alkylierende substanzen. I. Dimethylsufate, carcinogene wirkung an ratten und wahrscheinliche ursache von berufskrebs, Z. Krebsforsch., 68 (1966) 103
26. Braun, R., Fischer, G. W., and Schöneich, J., The mutagenicity and DNA-damaging activity of cyclic aliphatic sulfuric acid esters, Chem. Biol. Interactions, 19 (1977) 241-252
27. Alderson, T., Ethylation versus methylation in mutation of E. coli and Drosophila, Nature, 203 (1964) 1404
28. Pelecanos, M., and Alderson, T., The mutagenic action of diethylsulfate in Drosophila melanogaster I. The dose-mutagenic response to larval and adult feeding Mutation Res., 1 (1964) 173
29. Malling, H. V., Identification of the genetic alterations in nitrous acid-induced AD-3 mutants of Neurospora crassa, Mutation Res., 2 (1965) 320
30. Kolmark, G., Mutagenic properties of esters of inorganic acids investigated by the Neurospora back mutation test, Ser. Physiol., 26 (1956) 205-220
31. Heslot, H., Etude quantitative de reversions biochmiques induites chez la levure schizosaccharomyces pombe par des radiations et des substances radiomimetiques, Abhandl. Deut. Akad. Wiss., Berlinkl. Med. (1962) 193
32. Loveless, A., and Ross, W. C. J., Chromosome alteration and tumour inhibition by nitrogen mustards: The hypothesis of cross-linking alkylation, Nature, 166 (1950) 1113
33. Strauss, B., and Okubo, S., Protein synthesis and the induction of mutations by E. coli by alkylating agents, J. Bact., 79 (1960) 464
34. Alderson, T., Brit. Empire Cancer Campaign Ann. Rept. (1963) p. 416
35. Zamenhof, S., Leidy, G., Hahn, E., and Alexander, H. E., Inactivation and unstabilization of the transforming principle by mutagenic agents, J. Bact., 72 (1956) 1
36. Alderson, T., and Pelecanos, M., The mutagenic activity of diethylsulfate in Drosophila melanogaster, II. The sensitivity of the immature (Larval) and adult testis, Mutation Res., 1 (1964) 182
37. Pelacanos, M., Induction of cross-overs, autosomal recessive lethal mutations, and reciprocal translocations in Drosophila after treatment with dimethyl, Nature 210 (1965) 1294
38. Loveless, A., The influence of radiomimetic substances on DNA synthesis and function studied in E. coli phage systems, III. Mutation of T2-bacteriophage as a consequence of alkylation in vitro: The uniqueness of ethylation, Proc. Roy. Soc., B150 (1959) 497

39. Heslot, H., S. Pombe: A new organism for the study of chemical mutagenesis, Abhandl. Deut. Akad. Wiss. Berlin. Kl. Med., 1 (1960) 98-105
40. Alderson, T., and Clark, A. M., Interlocus specificity for chemical mutagens in Aspergillus nidulans, Nature, 210 (1966) 593
41. Lichtenberger, J., and Lichtenberger, R., Sur les sulfates de diols, Bull. Soc. Chim. (Fr.) (1948) 1002
42. Brunken, J., Glöckner, G., and Poppe, E. J., Uber cyclische sulfate und ihre verwendung zur herstellung von sulfatebetainen, Veroeff. Wiss. Photo-lab. AGFA 9 (1961) 61
43. Poppe, E.-J., Sensibilisierungsfarbstoffe mit sauren substituenten, Z. Wiss. Photogr., Photophys. Photochem., 63 (1969) 149
44. Van Duuren, B. L., Goldschmidt, B. M., Katz, C., Seidman, I., and Paul, J. S., Carcinogenic activity of alkylating agents, J. Natl. Cancer Inst., 53 (1974) 695-700
45. Schöneich, J., Safety evaluation based on microbial assay procedures, Mutation Res., 41 (1976) 89
46. Goldschmidt, B. M., Frenkel, K., and Van Duuren, B. L., The reaction of propane sultone with guanosine, adenosine and related compounds, J. Heterocyclic Chem., 11 (1974) 719
47. Fischer, F. F., Propane sultone, Int. Eng. Chem., 56 (1964) 41
48. Bremond, J., Propane sulton, propriétés et emplois, Rev. Prod. Chim., Sept. 15 (1964) 433
49. Druckrey, H., Krause, H., Preussmann, R., Ivankovic, S., Landschütz, C., and Jimmy, J., Cancerogene allylierende substanzen, IV. 1,3-Propanesulton und 1,4-butansulton, Z. Krebsforsch., 75 (1970) 69-84
50. Druckrey, H., Krause, H., and Pruessmann, R., Propane sultone, A potent carcinogen, Naturwiss, 55 (1968) 449
51. Nilsson, T., "Nagra sultoner deras framstallninoch hydrolys", Ph. D. Thesis Lund Unviersity, Lund Sweden, 1946
52. Osterm-Golkar, S., and Wachtmeister, C. A., On the reaction kinetics in water of 1,3-propane sultone and 1,4-butane sultone: A comparison of reaction rates and mutagenic activity of some alkylating agents, Chem. Biol. Interactions, 14 (1976) 195
53. Simmon, V. F., In vitro mutagenicity assays with Saccharomyces cerevisiae D3, J. Natl. Cancer Inst. (1978) in press
54. Dunkel, V. C., Wolff, J. S. and Pienta, R. J., In vitro transformation as a presumptive test for detecting chemical carcinogens, Cancer Bulletin, 29 (1977) 167-174
55. Singh, C., and Kaul, B. L., Propane sultone: A powerful mutagen for barley, Mutation Res., 56 (1978) 355-357
56. Bohme, H., Sensitization of Proteus mirabilis to the lethal action of ethyl methanesulfonate by pretreatment with manganous chloride, Biochem. Biophys. Res. Commun., 6 (1961) 108-111
57. Kak, S. N., and Kaul, B. L., Role of Mn+ ions on the modification of mutagenic activity of some alkylating agents, Cytologia, 38 (1973) 577-585
58. Zutshi, U., and Kaul, B. L., Polyploidy and sensitivity to alkylating mutagens, Radiat. Bot., 15 (1975) 59-68
59. Kaul, B. L., The production of chromosomal aberrations by propane sultone in Vicia faba, Mutation Res., 7 (1969) 339-347
60. Kaul, B. L., and Kak, S. N., Influence of Mn+ ions on the radiomimetic activity of some alkylating mutagens, The Nucleus, 15 (1972) 96-101

61. Kaul, B. L., and Kak, S. N., Influence of post-irradiation treatment with some chemicals on the recovery of chromosomal aberrations in barley, Egypt. J. Genet. Cytol., 2 (1973) 76-83
63. Singh, C., and Kaul, B. L., Propane sultone mutagenesis in higher plants, II. Influence of some pre- and post-treatment conditions, Indian J. Agr. Res., 7 (1973) 147-152
64. Loveless, A., Genetic and Allied Effects of Alkylating Agents", Butterworths, London (1966)
65. Gutsche, C. D., and Pasto, D. J., "Fundamentals of Organic Chemistry", Prentice-Hall, Englewood Cliffs, NJ (1975) p. 693
66. IARC, Monographs on the Evaluation of Carcinogenic Risk of Chemicals to Man, Vol. 7, International Agency for Research on Cancer, Lyon (1974) 223-230
67. U.S. Code of Federal Regulations, Air Contaminants, Title 29, Par. 1910.93, Government Printing Office, Washington, DC (1974)
68. Friedman, O. M., On a chemical degradation of deoxyribosenucleic acid, Biochem. Biphys. Acta., 23 (1957) 215
69. Friedman, O. M., Mahapatra, G. N., Dash, B., and Stevenson, R., Studies on the action of diazomethane on deoxyribonucleic acid, Biochim. Biphys. Acta. 103 (1965) 286-297
70. Kriek, E., and Emmelot, P., Methylation of deoxyribonucleic acid by diazomethane Biochim. Biophys. Acta., 91 (1964) 59-66
71. Holy, A., and Scheit, K. H., On the methylation of dinucleoside phosphates with diazomethane, Biochem. Biophys. Res., 123 (1966) 430
72. Schoental, R., and Magee, P. N., Induction of squamous carcinoma of the lung and of the stomach and oesophagus by diazomethane and N-methyl-N-nitroso-urethane respectively, Brit. J. Cancer, 16 (1962) 92-100
73. Schoental, R., Carcinogenic action of diazomethane and of nitroso-N-methyl urethane, Nature, 188 (1960) 420
74. Rose, F. L., In "The Evaluation of Drug Toxicity" (ed) Walpole, A. L., and Spinks, A., Little, Brown, Boston, Mass. (1958) p. 116
75. Mizrahi, I. J., and Emmelot, P., The effect of cysteine on the metabolic changes produced by two carcinogenic N-nitroso dialkylamines in rat liver, Cancer Res. 22 (1962) 339
76. Druckrey, H., Carcinogenicity and chemical structure of nitrosamines, Acta. Unio Intern. Contra. Cancrum., 19 (1963) 510
77. Rapoport, I. A., Alkylation of gene molecule, Dokl. Akad. Nauk. SSSR, 59 (1948) 1183
78. Gensen, K. A., Kolmark, G., and Westergaard, M., Back-mutations in Neurospora Crassa adenine locus induced by diazomethane, Hereditas, 35 (1949) 521-527
79. Jensen, K. A., Kirk, I., Kolmark, G., and Westergaard, M., Chemically induced mutations in Neurospora, Cold Spring Harbor Symp Quant. Biol., 16 (1951) 245-261
80. Marquardt, H., Zimmermann, F. K., and Schwaier, R., Nitrosamide als mutagene agentien, Naturwissenschaften, 50 (5063) 625
81. Marquardt, H., Schwaier, R., and Zimmermann, F., Nicht-mutagenitat von nitro-saminen bei Neurospora crassa, Naturwissenschaften, 50 (1963) 135
82. Zimmermann, F. K., Schwaier, R., and Von Laer, U., The influence of pH on the mutagenicity in yeast of N-methylnitrosamides and nitrous acid, Z. Vererbungslehre, 97 (1965) 68-71
83. Schwaier, R., Vergleichende mutations versuche mit sieben nitrosamiden im ruckmutationstest an hefen, Z. Verebungslehre, 97 (1965(55-67

84. Schwaier, R., Zimmermann, F. K., and Von Laer, U., The effect of temperature on the mutation induction in yeast by N-alkylnitrosamides and nitrous acid, Z. Vererbungslehre, 97 (1965) 72-74
85. Marquardt, H., Zimmermann, F. K., and Schwaier, R., The action of carcinogenic nitrosamines and nitrosamides on the adenine-G-45 reverse mutations system of S. cerevisiae, Z. Verebungslehre, 95 (1964) 82-96
86. Love, L. A., Pelfrene, A. F., and Garcia, H. G., Carcinogenicity of diazoacetic acids (DAAE), Anat. Rec., 189 (1977) 547-548
87. Lower, G. M., Jr., Lanphear, S. P., Johnson, B. M., and Bryan, G. T., Aryl and heterocyclic diazo compounds as potential environmental electrophiles, J. Toxicol. Env. Hlth., 2 (1977) 1095-1107
88. Lijinsky, W., and Ross, A. E., Alkylation of rat liver nucleic acids not related to carcinogenesis by N-nitrosamines, J. Natl. Cancer Inst., 49 (1972) 1239-1249
89. Goodall, C. M., Lijinski, W., and Tomatis, L., Toxicity and oncogenicity of nitrosomethylaniline and nitrosomethylcyclohexylamine, Toxicol. Appl. Pharmacol. 17 (1970) 426-432
90. Perussmann, R., Druckrey, H., and Bucheler, J., Carcinogene wirkung von phenyl-nitroso-harnstoff, Z. Krebsforsch., 71 (1968) 63-65
91. Preussmann, R., Ivankovic, S., Landschutz, C., Gimmy, J., Flohr, E., and Griesbach, U., Carcinogene wirkung von 13 aryldialkyltriazenen an BD-ratten, Z. Krebsforsch., 81 (1974) 285-310
92. Druckrey, H., Ivankovic, S., and Preussmann, R., Neurotrope carcinogene wirkung von phenyl-dimethyltriazen an ratten, Naturwiss, 54 (1967) 171
93. Preussmann, R., Druckrey, H., Ivankovic, S., and Von Hodenberg, A., Chemical structure and carcinogenicity of aliphatic hydrazo, azo and azoxy compounds and of triazenes, potential in vivo alkylating agents, Ann. NY Acad. Sci., 163 (1969) 697-716
94. Malaveille, C., Kolar, G. F., and Bartsch, H., Rat and mouse tissue-mediated mutagenicity of ring-substituted 3,3-dimethyl-1-phenyltriazenes in Salmonella typhimurium, Mutation Res., 36 (1976) 1
95. Kolar, G. F., Fahrig, R., and Vogel, E., Structure activity dependence in some noval ring-substituted 3,3-dimethyl-1-phenyl triazenes. Genetic effect in D. melanogaster and S. cerevisiae, Chem. Biol. Interactions, 9 (1974) 365
96. Vogel, E., Chemische konstitution und mutagene wirkung VI. Induction dominanter und rezessiv-geschlechtsgebundener lethal mutationen durch aryldialkyltriazene bei drosophila melanogaster, Mutation Res., 11 (1971) 379
97. Ong, T. M., and DeSerres, F. J., The mutagenicity of 1-phenyl-3,3-dimethyl-triazene and 1-phenyl-3-monomethyltriazene in Neurospora crassa, Mutation Res. 13 (1971) 276
98. Fahrig, R., Metabolic activation of aryldialkyltriazenes in the mouse: Induction of mitotic gene conversion in Saccharomyces cerevisiae in the host mediate assay, Mutation Res., 13 (1971) 436-439
99. Siebert, D., and Kolar, G. F., Induction of mitotic gene conversion by 3,3-dimethyl-phenyltriazene, 1-(3-hydroxyphenyl)-3,3-dimethyltriazene in Saccharomyces cerevisiae, Mutation Res., 18 (1973) 267
100. Vogel, E., Fahrig, R., and Obe, G., Triazenes, a new group of indirect mutagens Comparative investigations of the genetic effects of different aryldialkyltriazenes using Salmonella cerevisiae, the host mediated assay, Drosophila melanogaster, and human chromosomes in vitro, Mutation Res., 21 (1973) 123-136
101. Newton, M. F., Bahner, B., and Lilly, L. J., Chromosomal aberrations in rat lymphocytes treated in vivo with 1-phenyl-3,3-dimethyltriazene and N-nitroso-

morpholine. A further report on a possible method for carcinogenic screening, Mutation Res., 56 (1977) 39-46
102. Andrysova, A., Rambousek, V., Jirasek, J., Zverina, V., Matrka, M., and Marhold, J., 1-Aryl-3,3-dialkyltriazene compounds. Toxicity of parasubstituted 1-phenyl-3,3-dialkyltriazene compounds, Physiol. Bohenoslov. (1972) 63
103. Preussmann, R., Von Hodenberg, A., and Hengy, A., Mechanisms of carcinogens with 1-aryl-3,3-dialkyltriazenes. Enzymatic dealkylation by rat liver microsomal fraction in vitro, Biochem. Pharmacol., 18 (1969) 1
104. Kolar, G. F., and Schlesiger, J., Biotransformation of 1-(4-chlorophenyl)-3,3-dimethyltriazene into 3-chloro-4-hydroxyaniline, Cancer Letters, 1 (1975) 43-47

CHAPTER 7

PHOSPHORIC ACID ESTERS

Phosphoric acid (HO-P(=O)-OH), as a tribasic acid, can form mono-, di-, and triesters
$\overset{\text{O}}{\underset{\text{OH}}{\text{HO-P-OH}}}$
with a broad spectrum of alcohols, thiols and phenols, a number of the resultant reactive organophosphates have utility as alkylating agents, intermediates in chemical synthesis and as organophosphorus insecticides[1] (e.g., Dichlorvos, Parathion, Malathion, Diazinon).

The common structural element of all organophosphates is =P-O-C-, with both phosphorus and carbon being electrophilic sites. Alkylation (e.g., methylation or ethylation) can occur as a result of nucleophilic attack on the carbon atom with subsequent cleavage of the C-O bond. Alternatively, a nucleophile can preferentially attack the phosphorus atom and undergo phosphorylation. The type and rate of reaction with a given nucleophile depends to a major extent on its nature, as in the presence of several nucleophiles such as occur in competitive reactions in a living cell[1].

1. <u>Trimethylphosphate</u> [$(CH_3O)_3P=O$; phosphoric acid-trimethyl ester, TMP] is the simplest trialkyl ester of phosphoric acid and has been mainly employed as a methylating agent, in the preparation of organophosphorus insecticides [e.g., Dichlorvos (DDVP) $(CH_3O)_2P(=O)(OCH-CCl_2)$ via reaction with chloral][2], as a low-cost gasoline additive[3,4] and as a catalyst for polyester manufacture[5,6].

Trimethylphosphate is known to alkylate <u>E. coli</u>, phage T_4B[7], to cause chromosome breaks in bone marrow cells of rats[8,9], or cultured human lymphocytes[10], and to produce mutations in bacteria[11-14], including <u>Salmonella typhimurium</u> tester strains with R factor plasmids[14], <u>Neurospora</u>[15], <u>Drosophila</u>[16] and mice[17-19].

A number of additional phosphate esters have suggested utility in diverse industrial purposes including: catalysts for curing resins, chemical intermediates, solvents, gasoline and lubricant additives, anti-foaming antioxidants and flotation agents[20].

2. <u>Triethylphosphate</u> [$(C_2H_5O)_3P=O$, phosphoric acid, triethyl ester, TEP] is used to impart flame-resistance in polyesters[21], as a heat stabilizer for neoprene rubber[22] and as a plasticizer for injection moldable bisphenol-based polyesters[23]. Triethylphosphate has been shown to be mutagenic in <u>Drosophila</u>[13].

3. <u>Tris(2,3-dibromopropyl)phosphate</u> (Tris-BP, Tris; [$(BrCH_2CHCH_2O)_5P=O$] can be manufactured by the addition of bromine to triallyl phosphate or by the esterification reaction of phosphorus oxychloride with 2,3-dibromopropanol[24]. Commercial preparations of Tris can be obtained in two grades, viz., HV (high in volatiles) and LV (low

in volatiles). A typical LV sample has been reported to contain the following impurities[25]: 0.05 1,2-dibromo-3-chloropropane ($BrCH_2CHBrCH_2Cl$); 0.05 1,2,3-tribromopropane ($BrCH_2CHBrCH_2Br$); and 0.20% 2,3-dibromopropanol ($BrCH_2CHBrCH_2OH$).

Currently, about 300 million pounds of flame-retardant chemicals are being produced mainly for use in fabrics, plastics and carpets. Approximately two-thirds of this amount are inorganic derivatives such as alumina trihydrate and antimony oxide while the remaining one-third are large numbers of brominated and chlorinated organic derivatives[26-29].

Tris has been used principally as an additive to impact flame retardants to synthetic fibers and fabrics such as polyester, acetate and triacetate and acrylics. About 65% of the 10 million pounds of Tris produced annually in the U.S. by 6 manufacturers are applied to fabrics used for childrens fabrics (e.g., polyester, acetate and triacetate) to impact flame retardance to meet the standards promulgated under the Flammable Fabrics Act[24,30]. Tris is added to fabrics used for childrens' garments to the extent of 5-10% by weight.

As a result of concern about the mutagenic[31,32] and carcinogenic[32,34] properties of Tris, the manufacturers of this type of sleepwear ceased using Tris-treated fabrics in January, 1977 and the Consumer Product Safety Commission in the U.S. banned the sale of Tris-treated sleepwear in April, 1977[24,30].

Tris has also been used in significant quantities to impart flame retardancy to a broad spectrum of other synthetic polymers including: rigid and flexible polyurethane foams, cellulose nitrate surface coatings, polystyrene, polyvinyl chloride phenolics, paper coatings and rubber. Although the use of Tris to date has not been banned in these applications in the U.S., two major suppliers of Tris were reported to be dropping production and clearly the production of Tris is expected to decrease rapidly[30].

It is not certain to what extent Tris and/or its by products or degradation products can reach the environment from its production or use applications. Flame retardants are added in the final mixing steps of polymer production processes or during fiber spinning processes and there are hundreds of production locations in the U.S. alone[24].

A significant portion of the total (approximately 10%) is estimated to reach the environment from textile finishing plants and laundries while most of the remainder is postulated to eventually end up on solid wastes (e.g., manufacturing waste and used clothing)[35]. The U.S. Consumer Product Safety Commission (CPSC) has estimated that repeated washing of Tris treated fabrics removes 20-85% of the Tris and hence wash water containing unknown amounts of Tris then becomes a potential pollutant of natural water systems[36]. The population affected by water effluents is not known[24].

It is difficult to estimate the population at risk due to the production of Tris per se as well as those in contact or exposed to with fibers and polymers containing Tris[24]. CPSC maintains that a large population may be affected by production and use of fibers and polymers containing Tris. At the time of the CPSC ban in April, 1977, about 40% of all children's sleepwear in the retail pipeline (about 20 million garments) contained Tris, and approximately 10-20% of these garments had already been sold to consumers. In addition it was estimated that about 7 million square yards of Tris-treated fabrics was then in inventory with fabric and garment producers[36]. Public exposure to other Tris-treated products has not been reported.

CPSC considered cancer incidence projections from three studies, in assessing the risk from Tris-treated apparel. For example, it was estimated in the study of Hooper and Ames that for one year exposure to Tris-treated garments, 1.7% of the children would develop cancer, or 17,000 cases/million. While CPSC's estimates ranged from 25 to 5,100 cases per million, the Environmental Defense Fund predicted incidences as high as 6,000 per million male population[36]. Preliminary results from the National Cancer Institute demonstrated the carcinogenicity of Tris fed in the diet at levels of 50 and 100 ppm to mice and 500 and 1000 ppm to rats of both sexes. The test mice developed tumors in the liver, kidney, lung and stomach and the test rats developed kidney tumors[33,34].

The carcinogenicity of an impurity in commercial grades of Tris, e.g., 1,2-dibromo-3-chloropropane (DBCP) should also be noted. This compound caused a high incidence of squamous carcinoma of the stomach in both rats and mice as early as 10 weeks after initiation of feeding by oral intubation[37,38].

Tris is mutagenic[25,31,32] to histidine-requiring strains of Salmonella typhimurium (Ames' Salmonella/microsome test)[39]. For example, Prival et al[31] reported Tris mutagenic to S. typhimurium strains TA 1535 and TA 100, but not TA 1538 indicating that Tris induces mutations of the base-pair substitution type. On a quantitative basis, no significant activity was found among 9 different commercial samples, including high and low volatile materials from 5 different supplies. Highly purified samples of Tris containing 0.029% 1,2,3,-tribromopropane and less than 0.002% each of 1,2-dibromo-3-chloropropane had approximately the same mutagenic activity as the commercial samples.[31] Each of the 3 contaminants displayed some mutagenic activity, but insufficient to account for the mutagenicity of Tris when the level of these compounds in Tris was taken into account[31].

Extracts of fabrics treated with Tris were also found capable of inducing mutations in TA 1535 and TA 100 strains of S. typhimurium[31].

Tris has been found to induce heritable mutations (sex-linked recessive lethals) in Drosophila melanogaster[40]. It is also mutagenic with and without activation in mouse lymphoma cells[41], increases sister chromatid exchange six-fold in V-79 cells in culture[41], but is negative in the bone-marrow cytogenetic test[41]. Tris induces unscheduled DNA synthesis[42] and lesions in DNA in human cells in culture[43]. These findings which indicate a specific action of Tris on cellular DNA suggest that this agent is potentially capable of inducing genetic effects not only in bacteria but in eukaryotic cells as well and hence adds a further dimension of concern.

Additionally, the risk of reproductive effects on children from Tris-treated pajamas has been cited[44]. 1,2-Dibromo-3-chloropropane (DBCP) an impurity present in Tris in amounts of 0.05% is closely related to 2,3-dibromopropane, a metabolite of Tris. DBCP is a mutagen, a carcinogen and is known to cause testicular atrophy in animals[45]. It causes sterility in animals and has recently been suggested to cause human sterility in workers engaged in its production[44].

The risk of reproductive effects on children from Tris-treated pajamas is amplified because the scrotum is about 20 times more permeable to chemicals than other skin.

It was recently reported that the dermal application of Tris (neat) at a dose of 1 ml/kg (2.27 g/kg) to the clipped backs of albino rabbits, once each week for 3 months resulted in testicular atrophy and chronic interstitial nephritis in males. Females did not show any adverse effects[45].

In a subchronic 90-day study in which groups of young adult Osborne-Mendel rats were given daily gavages of Tris or its primary metabolite 2,3-dibromopropanol at doses of 25, 100 or 250 mg/kg in propylene glycol for 13 weeks, a prominent lesion was present in the kidney of all Tris treated animals. There was an increased incidence and severity of chronic nephritis with associated regenerative epithelium, hypertrophy and dysplasia of renal tubular epithelial cell nuclei. The complex of changes was more severe with the high doses and among males[46].

References

1. Wild, D., Mutagenicity Studies on Organophosphorus Insecticides, Mutation Res., 32 (1975) 133
2. British Crop Protection Council, "Pesticide Manual", (ed) Martin, H., Worcester, (1968) p. 152
3. Ehrenberg, L., Osterman-Golkar, S., Singh, D., and Lundquist, U., On the Reaction Kinetics and Mutagenic Activity of Methylating and β-Halogenoethylating Gasoline Additives, Radiation Botany, 15 (1974) 185
4. Kerley, R. V., and Flet, A. E., Fuels for Automotive Engines, U.S. Patent 3,807,974, April 30, (1974), Chem. Abstr., 81 (1974) 172852T
5. Watanabe, T., Ichikawa, H., Yokouch, R., Manufacture of Polyesters for Film-molding Use, Japan Pat. 73 42, 712, Dec. 7 (1973) Chem. Abstr., 81 (1974) 136948V
6. Murayama, K., and Yamadera, R., Heat-Stable Polyesters, Japan Kokai, 73, 102, 194, Dec. 22 (1973), Chem. Abstr., 81 (1974) 14004P
7. Gumanov, G. D., and Gumanov, L. L., Mutagenic Action of Alkyl Phosphates on T_4B Phage and Alkylation of Phage DNA, Dokl. Akad. Nauk., SSSR, 198, (1972) 1442-1444
8. Adler, I. D., Ramarao, G., and Epstein, S. S., In Vivo Cytogenetic Effects of Trimethylphosphate and of Tepa on Bone Marrow Cells of Male Rats, Mutation Res., 13 (1971) 263
9. Legator, M. S., Palmer, K. A., and Adler, I. D., A Collaborative Study of In Vivo Cytogentic Analysis. I. Interpretation of Slide Preparations, Toxicol. Appl. Pharmacol., 24 (1973) 337
10. Söderman, G., Chromosome Breaking Effect of Gasoline Additive in Cultured Human Lymphocytes, Hereditas, 71 (1972) 335
11. MacPhee, D. G., Salmonella typhimurium His G46 (R-Utrecht): Possible use in Screening Mutagens and Carcinogens, Appl. Microbiol., 26 (1973) 1004
12. Voogd, C. E., Jacobs, J. J. J. A. A., Vanderstel, J. J., On the Mutagenic Action of Dichlorvos, Mutation Res., 16 (1972) 413
13. Dyer, K. F., and Hanna, P. J., Comparative mutagenic activity and toxicity of triethylphosphate and dichlorvos in bacteria and Drosophila, Mutation Res. 21 (1973) 175
14. McCann, J., Spingarn, N. E., Kobori, J., and Ames, B. N., Detection of Carcinogens as Mutagens: Bacterial Tester Strains with R Factor-Plasmids, Proc. Natl. Acad. Sci., 72 (1975) 979
15. Kolmark, G., quoted in Epstein, S. S., Base, W., Arnold, E., and Bishop, Y., Mutagenicity of Trimethylphosphate in Mice, Science, 168 (1970) 584
16. Dyer, K. F., and Hanna, P. J., Mutagenic and Antifertility Activity of Trimethylphosphate in Drosophila Melanogaster, Mutation Res., 16 (1972) 327
17. Dean, B. J., and Thorpe, E., Studies with dichlorovs vapor in dominant lethal mutation tests on mice, Arch. Toxicol., 30 (1972) 51
18. Epstein, S. S., Trimethylphosphate (TMP), EMS Newsletter, 2 (1969) 33
19. Epstein, S. S., Bass, W., Arnold, E., and Bishop Y., Mutagenicity of Trimethylphosphate in Mice, Science (1970) 584
20. Hanna, P. J., and Dyer, K. F., Mutagenicity of Organophosphorus Compounds in Bacteria and Drosophila, Mutation Res., 28 (1975) 405-420
21. Byrd, S. M., Jr., Flame-Retardant Polyesters: Two Approaches, Proc. Annual Conf., Reinf. Plast. Compos. Inst. Soc. Plast. Ind., 29 (1974) 23D, 9 pp., Chem. Abstr., 81 (1974) 1064993G

22. Takahashi, H., Taketsume, M., and Nakata, M., Chloroprene Rubber Composition, Japan Patent, 74 00, 981, Jan 10 (1974) Chem. Abstr., 81 (1974) 79138E
23. Sakata, H., Okamoto, T., Hasegawa, H., and Nagata, K., Polyester Compositions With Improved Workability, Japan Kokai, 74 34, 546, April 2 (1974), Chem. Abstr., 81 (1974) 137062
24. Environmental Protection Agency, Status Assessment of Toxic Chemicals, 14. Tris (2,3-Dibromopropylphosphate), Industrial Environmental Research Laboratory, Cincinatti, Ohio, Sept. 6 (1977)
25. Kerst, A. F., Toxicology of Tris(2,3-Dibromopropyl)phosphate, J. Fire Flamm, Fire Retardant Chem. (suppl), 1 (1974) 205
26. Pearce, E. M., and Liepins, R., Flame Retardants, Env. Hlth. Persp., 11 (1975) 59-69
27. Hutzinger, O., Sundstrom, G., and Safe, S., Environmental chemistry of flame retardants, I. Introduction and principles, Chemosphere, 1 (1976) 3-10
28. Anon, For chemicals and additives, a new splash in R & D, Mod. Plastics, 52 (9) (1975) 41
29. WHO, Health Hazards from New Environmental Pollutants, Tech. Rept. Series No. 586 (1976) 52-63 World Health Organization, Geneva
30. Stanford Research Institute, A Study of Industrial Data on Candidate Chemicals for Testing, EPA Rept. 560/5-77-066; Menlo Park, CA, August (1977) pp. 4-44 to 4-49
31. Prival, M. J., McCoy, E. C., Gutter, B., and Rosenkranz, H. S., Tris(2,3-dibromopropyl)phosphate: Mutagenicity of a widely used flame retardant, Science 195 (1977) 76-78
32. Blum, A., and Ames, B. N., Flame-retardant additives as possible cancer hazards, Science, 195 (1977) 17-23
33. National Cancer Institute, Results of bioassay with Tris, National Institutes of Health (1977)
34. NIOSH, Children's apparel containing Tris, Part 8, Federal Register, 42 (18) April 8 (1977)
35. EPA, Summary characterizations of selected chemicals of near-term interest, EPA, Rept. 560/4-76-004, Office of Toxic Substances, Environmental Protection Agency, Washington, DC April (1976) pp. 53-54
36. Consumer Product Safety Commission, Tris and Fabrics, Yarn or Fiber Containing Tris, Federal Register, 42 (105) June 1 (1977)
37. Olson, W. A., Haberman, R. T., Weisburger, E. K., Ward, J. M., and Weisburger, J. H., Induction of stomach cancer in rats and mice by halogenated aliphatic fumigants, J. Natl. Cancer Inst., 51 (1973) 1993
38. Powers, M. B., Voelker, W., Page, N. P., Weisburger, E. K., and Kraybill, H. F., Carcinogenicity of ethylene dibromide (EDB) and 1,2-dibromo-3-chloro-propane (DBCP) Toxicol. Appl. Pharmacol., 33 (1975) 171-172
39. McCann, J., Choi, E., Yamasaki, E., and Ames, B. N., Detection of carcinogens as mutagens in Salmonella/microsome test: Assay of 300 chemicals, Proc. Natl. Acad. Sci. (USA) 72 (1975) 5135
40. Valencia, R., Drosophila mutagenicity tests of saccharin, Tris, PtCl4 and other compounds, 9th Annual Meeting of Environmental Mutagen Society, San Francisco, CA, March 9-13 (1978) p. 64
41. Matheson, D., Tris(2,3-dibromopropyl)phosphate: A flame retardant chemical, Presented at 9th Annual Meeting of Environmental Mutagen Society, San Francisco, CA, March 9-13 (1978)

42. Stich, H. F., cited by Blum, A., and Ames, B. N., in reference 32
43. Gutter, B., and Rosenkranz, H. S., The flame retardant Tris(2,3-dibromopropyl)-phosphate: Alteration of human cellular DNA, Mutation Res., 56 (1977) 89-96
44. Anon, Tris may cause sterility in boys, Berkeley researchers tell safety unit, Chem. Reg. Reptr., 1 (1977) 894-895
45. Torkelson, T. R., Sadek, S. E., Rowe, V. K., Kodama, J. K., Anderson, H. H., Loquvam, G. S., and Hine, C. H., Toxicological investigations of 1,2-dibromo-3-chloropropane, Toxicol. Appl. Pharmacol., 3 (1961) 545-559
46. Osterberg, R. E., Bierbower, G. W., and Hehir, R. M., Renal and testicular damage following dermal application of the flame retardant Tris(2,3-dibromopropyl)phosphate, J. Toxicol. Env. Hlth., 3 (1977) 979-987
47. Osterberg, R. E., Bierbower, G. W., Ulsamer, A. H., Porter, W. K., Jr., and Jones, S., Abstr. of 17th Annual Meeting, Society of Toxicology, San Francisco, CA, March 12-16 (1978) 71,73.

CHAPTER 8

ALDEHYDES

The carbonyl group (as typified in aldehydes and ketones) occurs in many substances of both biological and commercial importance. Compounds such as aldehydes have been employed in a large number of organic syntheses and reactions. For example, as a result of the polarization of the carbonyl group, $\diagup\!\!\!\!\diagdown C = O \longleftrightarrow \overset{+}{C} - \overset{-}{O}$, aldehydes (and ketones) have a marked tendency to add nucleophilic species (Lewis bases) to the carbonyl carbon, followed by the addition of an electrophilic species (Lewis acids) to the carbonyl oxygen, the reactions are classified as 1,2 nucleophilic additions, via.:

$$\diagup\!\!\!\!\diagdown C = O + Nu^- \rightleftharpoons \diagup\!\!\!\!\diagdown C \diagdown\!\!\!\!\!{}^{O^-}_{Nu} \overset{E^+}{\rightleftharpoons} \diagup\!\!\!\!\diagdown C \diagdown\!\!\!\!\!{}^{O-E}_{Nu}$$

which can involve carbon or nitrogen nucleophilics.

1. **Formaldehyde** ($H_2C=O$), the simplest of the aldehydes is considerably more reactive than its higher homologs. The chemical stability associated with enol-keto tautomerism in the higher aldehydes is lacking in formaldehyde where the carbonyl groups is attacked directly to two hydrogens. Formaldehyde is produced by the oxidation of methanol.

Formaldehyde is used extensively as a reactant in a broad spectrum of commercial processes because of its high chemical reactivity and good thermal stability[1,2]. These reactions can be arranged in three major categories, via., 1) self-polymerization reactions; 2) oxidation-reduction reactions and addition or condensation reactions with a large number of organic and inorganic compounds.

Since pure formaldehyde is a gas at ordinary temperatures and hence cannot be readily handled, it is marketed chiefly in the form of aqueous solutions containing 37% to 50% formaldehyde be weight. In aqueous solution formaldehyde exists entirely in the hydrated form $H_2C\diagdown\!\!\!\!\!{}^{OH}_{OH}$.

Formaldehyde has a marked tendency to react with itself to form linear polymers (designated as paraformaldehyde) or a cyclic trimer (designated as trioxane); i.e.,

$$HO(CH_2O)_xH \xleftarrow{H_2O} \overset{H}{\underset{H}{\diagdown}}C=O \xrightarrow{H^+, \Delta} \text{trioxane ring}$$

Paraformaldehyde   Trioxane

In general, the major chemical reactions of formaldehyde with other compounds involve the formation of methylol (—CH$_2$OH) or methylene derivatives. Other typical reactions include alkoxy-, amido-, amino-, cyano-, halo-, sulfo-, and thiocyano-methylations. For example, reactions with amides and carbamates yield methylol derivatives, e.g., methylolureas and methylol carbamates which are used in the treatment of textiles (for crease-resistance; crush proof, flame resistance and shrinkproof fabrics). Reaction of formaldehyde with hydrochloric acid can yield the carcinogenic bischloromethyl ether (BCME).

Aldol condensations are important in the synthesis of β-hydroxycarbonyl compounds which can be used in further synthesis, e.g., pentaerythritol production. Methylol derivatives are highly reactive species which can be polymerized to yield methylene or ether bridges, e.g., phenolic resins. Condensation of formaldehyde with ammonia yields hexamethylenetetramine which undergoes many reactions including decomposition into formaldehyde and ammonia, and nitramine formation upon nitration.

The major uses of formaldehyde and its polymers are in the synthetic resin industry (e.g., in the production of urea-formaldehyde-, phenolic-, polyacetal-, and melamine-formaldehyde resins) and in the manufacture of pentaerythritol and hexamethylenetetramine. Production levels are currently approximately 6000 million pounds annually on a 37% basis)[2]. Over 50% of the formaldehyde produced is used in the manufacture of resins (e.g., urea-formaldehyde and phenol-formaldehyde).

Pentaerythritol is used mainly in alkyd surface coating resins, rosin and tall oil resins, varnishes, pharmaceuticals, plasticizers and insecticides[2], as well as in the preparation of butanediol, acetols (polyformaldehydes) and slow-release fertilizers. Butanediol is used to make tetrahydrofuran and polybutyleneterephthalate (PBT).

Formaldehyde is employed in a number of minor applications in agriculture, paper, textile and dyestuffs manufacture, medicine, analysis, etc. (Table 1).

Feldman[3] has recently reviewed the reactions of formaldehyde with various components of biological importance (e.g., nucleic acids and nucleoproteins). The genetic and cytogenetical effects of formaldehyde and related compounds have recently been surveyed by Auerbach et al[4].

The mutagenicity of formaldehyde has been described most extensively for Drosophila[5-13] (with hydrogenperoxide)[14] and established for Neurospora cassida (also with hydrogen peroxide)[15-17] and E. coli[17-20].

Formaldehyde effects on E. coli B/r in a special mutant lacking a DNA polymerase (pol A$^-$) and, therefore, a repair deficient strain were elaborated by Rosenkranz[19]. Formaldehyde treatment of pol A$^+$ and pol A$^-$ strains showed differential toxicity, determined by the "zone of inhibition" surrounding a formaldehyde-soaked disc placed on the surface of the growth agar. There was a preferential inhibition of growth in the pol A$^-$ strain, indicating that some repair capability may affect the survival of formaldehyde treated bacteria.

In the above studies, Rosenkranz[19] also described the interaction of known carcinogens (e.g., methyl methanesulfonate and N-hydroxylaminofluorene) with both the pol A$^+$ and pol A$^-$ strains of E. coli and concluded that "in view of the present

TABLE 1

Minor Uses of Formaldehyde and Its Products

Agriculture

1. Treatment of bulbs, seeds and roots to destroy microorganisms.
2. Soil disinfectant.
3. Prevention of rot and infections during crop storage.
4. Treatment of animal feed grains.
5. Chemotherapeutic agent for fish.

Dyes

1. Manufacture of intermediate for production of rosaniline dyes.
2. Preparation of phenyl glycine, an intermediate in the manufacture of indigo dyes.
3. Used to prepare formaldehydesulfoxylates which are stripping agents.

Metals Industries

1. Pickling agent additive to prevent corrosion of metals by H_2S.
2. Preparation of silver mirrors.
3. Hexamethylenetetramine is used to produce nitrilotriacetic acid and formaldehyde to produce ethylenediaminetetracetic acid. These compounds are excellent metal sequestering agents.

Paper

Formaldehyde is used to improve the wet-strength, water shrink, and grease resistance of paper, coated papers and paper products.

Photography

1. Used in film to harden and insolubilize the gelatin and reduce silver salts.
2. Photographic development.

Rubber

1. Prevent putrefaction of latex rubber.
2. Vulcanize and modify natural and synthetic rubber.
3. Hexamethylenetetramine is used as a rubber accelerator.
4. Synthesis of tetraphenylmethylenediamine, a rubber antioxidant.

Hydrocarbon Products

1. Prevent bacterial action from destroying drilling fluids or muds.
2. Remove sulfur compounds from hydrocarbons.
3. Stabilize gasoline fuels to prevent gum formation.
4. Modify fuel characteristics of hydrocarbons.

(Continued on p. 146)

TABLE 1 (continued)

Leather

Tanning agent for white washable leathers.

Solvents and Plasticizers, Surface Active Compound

1. Synthesis of ethylene glycol.
2. Synthesis of formals.
3. Synthesis of methylene derivatives.
4. Synthesis of surface active compounds.

Starch

Formaldehyde is used to modify the properties of starch, by formation of acetals and hemiacetals.

Textiles

Modification of natural and synthetic fibers to make them crease, crush and flame resistant and shrink-proof.

Wood

Used as an ingredient in wood preservatives.

Concrete and Plaster

Formaldehyde is used as an additive agent to concrete to render it impermeable to liquids and grease.

Cosmetics and Deodorants

Formaldehyde is utilized in deodorants, foot antiperspirants and germicidal soaps.

Disinfectants and Fumigants

Formaldehyde is employed to destroy bacteria, fungi, molds, and yeasts in houses, barns, chicken coops, hospitals, etc.

Medicine

1. Treatment of athlete's foots and ring worm.
2. Hexamethylenetetramine is used as a urinary antiseptic.
3. Conversion of toxins to toxoids.
4. Synthesis of Vitamin A.
5. Urea-formaldehyde is used as a mechanical ion exchange resin.

Analysis

Small quantities are used in various analytical techniques.

findings and because the procedure used seems to be quite reliable for detecting carcinogens, it would seem that continued use of formaldehyde requires reevaluation and monitoring as exposure to even low levels of this substance might be deleterious especially if it occurs over prolonged periods of time, a situation which probably increases the chance of carcinogenesis".

Formaldehyde is also mutagenic in E. coli B/r strains which were altered in another repair function, Hcr⁻. (This strain lacks the ability to reactivate phage containing UV-induced thymine dimers because it lacks an excision function.) Strains of E. coli B/r which were Hcr⁻ showed more mutation to streptomycin resistance or to tryptophan independence than did the repair competent Hcr⁺ strain. Ultraviolet inactivation of Hcr⁻ strains was enhanced by treatment with formaldehyde, possibly indicating some effect of formaldehyde on the repair function[20].

Formaldehyde has been found to combine with RNA or its constituent nucleotides[21-23], with the formation of hydroxymethylamide ($HOCH_2CONH_2$) and hydroxymethylamine (NH_2CH_2OH) by hydroxymethylation of amido ($-CONH_2$) and amino groups respectively.

Formaldehyde has been found to combine more readily with single stranded polynucleotides such as replicating DNA[24] or synthetic poly A[25]. The reaction products may also include condensation products of adenosine such as methylene bis AMP. The possibility of formation of these compounds in vivo has led to the postulation that adenine dimers may be found in polynucleotides in situ or may be erroneously incorporated into polynucleotides[23,24].

An alternative mechanism of action for formaldehyde involving the formation of peroxidation products by autooxidation of formaldehyde or by its reaction with other molecules to form free radicals has been proposed[17]. The synergism between hydrogen peroxide and formaldehyde in producing mutations in Neurospora has been described[17] The combination of formaldehyde and H_2O_2 was found to be differentially mutagenic at two loci, adenine and inositol utilization. These two loci showed divergent dose response curves when similarly treated with formaldehyde and H_2O_2. This was taken as evidence for a mutagenic peroxidation product.

According to Auerbach et al[4], the elucidation of formaldehyde action at the molecular level is confused and to some extent it will probably remain so because a variety of mechanisms are involved in different organisms, under different conditions, and for the production of different endpoints.

A number of carcinogenic studies of formaldehyde per se[25-28] as well as that of hexamethylene tetramine[29,30] (an agent known to release formaldehyde) have been reported. To date the assessment of the carcinogenicity based on these studies would appear to be equivocal.

NIOSH has recently recommended that exposure to formaldehyde be limited to a ceiling concentration of 1.2 mg/m^3 (1 ppm) determined on the basis of a 30-minute sampling period. The recommendation includes all substances that yield or generate formaldehyde as gas or in solution. NIOSH estimates that about 8000 employees are potentially exposed to formaldehyde[31]. The major sources of exposure to formaldehyde occurs in (1) industrial applications involving its use in chemical synthesis, textiles, acrolein in various systems. Acrolein has been found mutagenic in Drosophila melanogaster[63], in S. typhimurium strains TA 1538 and TA 98[64] and non-mutagenic in the dominant-lethal assay in ICR/HA Swiss mice when tested at 1.5 and 2.2 mg/kg[65].

Information on absorption, distribution, excretion and metabolism of acrolein is scant. Egle[66] reported that in dogs 81-84% of inhaled acrolein is retained in the upper respiratory tract. Rats metabolized at least 11% of a subcutaneous dose of acrolein to N-acetyl-S-(3-hydroxypropyl)-2-cysteine[67].

The current TLV for acrolein in the U.S. is 0.25 mg/m^3 (0.1 ppm) in terms of an 8 hour time weighted average in the workplace air in any 8 hour work shift of a 40 hour work week[68].

The work environment hygiene standard (in terms of an 8-hour time weighted average) for acrolein is 0.25 mg/m^3 in the Federal Republic of Germany, the German Democratic Republic, and Sweden, and 0.5 mg/m^3 in Czechoslovadia. The MAC of acrolein in the USSR is 0.7 mg/m^3[69].

3. Acetaldehyde

Acetaldehyde (ethanol; acetic aldehyde; CH3CHO) is employed primarily as an intermediate in the production of paraldehydes, acetic acid, acetic anhydride, pentaerythritol, butyl alcohol, butyraldehyde, chloral, 2-ethyl hexanol and other aldol products; peroxy acetic acid, cellulose acetate, vinyl acetate resins and pyridine derivatives[70]. Lesser uses of acetaldehyde include the production of aniline dyes; thermosetting resins from the condensation products with phenol and urea; the preparation of Schiff bases (via reaction with aliphatic and aromatic amines) which are used as accelerators and antioxidants in the rubber industry. Acetaldehyde has been used as a preservative for fruit and fish, as a denaturant for alcohol, in fuel compositions, for hardening gelatin, glue and casein products, for the prevention of mold growth on leather, and as a solvent in the rubber, tanning and paper industries[70,71].

Aldehyde is manufactured in the hydration of acetylene, the oxidation or dehydrogenation of ethanol, or the oxidation of saturated hydrocarbons or ethylene[70]. Acetaldehyde is a highly reactive compound exhibiting the general reactions of aldehydes, e.g., undergoing numerous condensation addition and polymerization reactions. In addition, acetaldehyde is readily oxidized to peroxy compounds such as peroxy acetic acid, acetic anhydride, or acetic acid, the principal product(s) dependent on the specific oxidation conditions employed.

It is the product of most hydrocarbon oxidations; it is a normal intermediate product inthe respiration of higher plants; it occurs in traces in all ripe fruits and may form in wine andother alcoholic beverages after exposure to air. Acetaldehyde is an intermediate product in the metabolism of sugars in the body and hence occurs in traces in blood. It has been reported in fresh leaf tobacco[72] as well as in tobacco smoke[73,74] and in automobile and diesel exhaust[75,76].

Information as to the mutagenicity of aldehydes (with the exception of formaldehyde) is scant. Acetaldehyde has been found mutagenic in Drosophila[63].

The chromosome breaking ability of acetaldehyde in Vicia faba root tips was found to be much higher than that of its precursor ethyl alcohol at concentrations ranging from 5×10^{-3} to 5×10^{-2}M. This effect was strongly influenced by temperature, being maximal at 12°C and less at higher temperatures[77]. Chromosomal aberrations were preferentially localized in the heterochromatic segments of the genome as is the case after exposure to many alkylating agents but not after ionizing radiations[77].

References

1. Kirk-Othmer, "Encyclopedia of Chemical Technology" 2nd ed., Vol. 10, Wiley-Interscience, New York (1966) p. 77.
2. Environmental Protection Agency, "Investigation of Selected Potential Environmental Contaminants-Formaldehyde, Final Report", Office of Toxic Substances, Environmental Protection Agency, August, 1976
3. Feldman, M. Y., Reaction of nucleic acids and nucleo proteins with formaldehyde, Progr. Nucleic Acids Res. Mol. Biol., 13 (1975) 1-49
4. Auerbach, C., Moustchen-Dahmen, M., and Moutschen, J., Genetic and cytogenetical effects of formaldehyde and related compounds, Mutation Res., 39 (1977) 317-362
5. Alderson, T., Chemically Induced Delayed Germinal Mutation in Drosophila, Nature, 207 (1965) 164
6. Auerbach, C., Mutation Tests on Drosophila Melanogaster with Aqueous Solutions of Formaldehyde, Am. Naturalist, 86 (1952) 330-332
7. Khishin, A. F. E., The Requirement of Adenylic Acid for Formaldehyde Mutagenesis Mutation Res., 1 (1964) 202
8. Burdette, W. J., Tumor Incidence and Lethal Mutation Rate in a Tumor Strain of Drosophila Treated with Formaldehyde, Cancer Res., 11 (1951) 555
9. Auerbach, C., Analysis of the Mutagenic Action of Formaldehyde Food. III Conditions influencing the effectiveness of the treatment, Z. Vererbungslehre, 81 (1956) 627-647
10. Auerbach, C., Drosophila Tests in Pharmacology, Nature, 210 (1966) 104
11. Rapoport, I. A., Carbonyl Compounds and the Chemical Mechanisms of Mutations, C.R. Acad. Sci. USSR, 54 (1946) 65
12. Kaplan, W. D., Formaldehyde as a Mutagen in Drosophila, Science, 108 (1948)
13. Auerbach, C., The Mutagenic Mode of Action of Formalin, Science, 110 (1949) 119
14. Sobels, F. H., Mutation Tests with a Formaldehyde-Hydrogen Peroxide Mixture in Drosophila, Am. Naturalist, 88 (1954) 109-112
15. Dickey, F. H., Cleland, G. H., and Lotz, C., The Role of Organic Peroxides in the Induction of Mutations, Proc. Natl. Acad. Sci., 35 (1949) 581
16. Jensen, K. A., Kirk, I., Kolmark, G., and Westergaard, M., Cold Spring Symp. Quant. Biol., 16 (1951) 245
17. Auerbach, C., and Ramsey, D., Analysis of a Case of Mutagen Specificity in Neurospora Crassa, Mol. Gen. Genetics, 103 (1968) 72
18. Demerec, M., Bertani, G., and Flint, J., Chemicals for Mutagenic Action on E. coli, Am. Naturalist, 85 (1951) 119
19. Rosenkranz, H. S., Formaldehyde as a Possible Carcinogen, Bull. Env. Contam. Toxicol., 8 (1972) 242
20. Nishioka, H., Lethal and Mutagenic Action of Formaldehyde, Hcr$^+$ and Hcr$^-$ Strains, E. Coli, Mutation Res., 17 (1973) 261
21. Hoard, D. E., The Applicability of Formol Titration to the Problem of End-Group Determinations in Polynucleotides. A Preliminary Investigation, Biochem. Biophys. Acta., 40 (1960) 62
22. Haselkorn, R., and Doty, P., The Reaction of Formaldehyde with Polynucleotides, J. Biol. Chem., 236 (1961) 2730
23. Alderson, T., Significance of Ribonucleic Acid in the Mechanism of Formaldehyde Induced Mutagenesis, Nature, 185 (1960) 904
24. Voronina, E. N., Study of the Spectrum of Mutations Caused by Formaldehyde in E. Coli K-12, In Different Periods of a Synchronized Lag Period, Sov. Genet., 7 (1971) 788

25. Fillippova, L. M., Pan'shin, O. A., and Koslyankovskii, F. K., Chemical Mutagens, Genetika, 3 (1967) 135
26. Watanabe, F., Matsunaga, T., Soejima, T., and Iwata, Y., Study on the Carcinogenicity of Aldehyde, 1st. Report, Experimentally Produced Rat Sarcomas by Repeated Injections of Aqueous Solution of Formaldehyde, Gann., 45 (1954) 451
27. Watanabe, F., and Sugimoto, S., Study on the Carcinogenicity of Aldehyde, 2nd Report, Seven Cases of Transplantable Sarcomas of Rats Appearing in the Area of Repeated Subcutaneous Injections of Urotropin, Gann., 46 (1955) 365
28. Horton, A. W., Tye, R., and Stemmer, K. L., Experimental Carcinogenesis of the Lung, Inhalation of Gaseous Formaldehyde on an Aerosol Tar By C3H Mice, J. Natl. Cancer Inst., 30 (1963) 31
29. Della Porta, G., Colnagi, M. I., and Parmiani, G., Non-Carcinogenicity of Hexamethylenetetramine in Mice and Rats, Food Cosmet. Toxicol., 6 (1968) 707
30. Brendel, R., Untersuchungen an ratten zur vertraglickeit von hexamethylene tetramin, Arznei-Forsch., 14 (1964) 51
31. Anon, NIOSH proposes 1.2 mg standard for formaldehyde, Occup. Hlth. Safety Letter, 6 (24), Jan. 8 (1977) p. 6
32. Schaal, G. E., Make acrolein from propylene, Hydrocarbon Processing, Sept. (1973) pp. 218-220
33. Anon, Degussa brings in new acrolein capacity, European Chem. News, July 16 (1971) p. 10
34. IARC, Monographs on the Evaluation of Carcinogenic Risk of Chemicals to Man, Vol. 16, International Agency for Research on Cancer, Lyon (1978) in press
35. Guest, H. R., Kiff, B W., and Stansbury, H. A., Jr., Acrolein and derivatives In Kirk, R. E. and Othmer D. F., eds. "Encyclopedia of Chemical Technology", 2nd ed., Vol. 1, Wiley & Sons, New York (1963) 255-274
36. Weschler, J. R., Epoxy resins, In Kirk, R. E., and Othmer, D. F., eds, "Encyclopedia of Chemical Technology", 2nd ed., Vol. 8, Wiley & Sons, New York (1965) 299-300
37. Anon, UCC finishes first phase in Louisiana expansion, Chem. Marketing Reptr., Oct. 6 (1976) p. 35
38. Windholz, M. "The Merck Index", 9th ed., Merck & Co., Inc. Rahway, NJ (1976) p.17
39. U.S. Code of Federal Regulations, Title 21, Food and Drugs, Parts 172.892 and 176.300, Washington, DC (1977) pp. 420 and 494
40. Izard, C., and Libermann, C., Acrolein, Mutation Res., 47 (1978) 115-138
41. Packer, J., and Vaughan, J., "A Modern Approach to Organic Chemistry", Clarendon, Oxford (1958)
42. Champeix, J., and Catilina, P., "Les Intoxications par L'acroleine", Masson, Paris (1967)
43. Plotnikova, M., Estimation des taux d'acroleine pollutant l'atmosphere, Gig. Sanit., 22 (6) (1957)
44. Protsenko, G. A., Danilov, V. I., Timchenko, A. N., Nenartovich, A. V., Trubiko, V. I., and Savchenkov, V. A., Working conditions when metals to which a primer has been applied and welded evaluted from the heath and hygiene aspect, A Vtomat. Svarka, 26 (1973) 65-68
45. Masek, V., Aldehydes in the air in coal and pitch coking plants, Staub-Reinhalt. Luft., 32 (1972) 335-336
46. Volkova, Z. A., and Bagdinov, Z. M., Industrial hygiene problems in vulcanization processes of rubber production, Gig. Sanit., 34 (1969) 33-40
47. Doorgeest, T., Paint and air pollution, Tno Nieuws, 25 (1970) 37-42

48. Smythe, R. J., and Karasek, F. W., Analysis of diesel engine exhausts for low-molecular-weight carbonyl compounds, J. Chromatog., 86 (1973) 228-231
49. Hoshika, Y., and Takata, Y., Gas chromatographic separation of carbonyl compounds as their 2,4-dinitrophenyl hydrazones using glass capillary columns, J. Chromatog., 120 (1976) 379-389
50. Testa, A., and Joigny, C., Gas-layer chromatographic estimation of acrolein and other α,β/unsaturated compounds from the gas phase of cigarette smoke, Ann. Dir. Etud. Equip., Seita (Serv. Exploit. Ind. Tab. Allumettes), Sec. 1, 10 (1972) 67-81
51. Rathkamp, G., Tso, T. C., and Hoffman, D., Chemical studies on tobacco smoke XX. Smoke analysis of cigarettes made from bright tobaccos differing in variety and stalk positions, Beitr. Tabakforsch., 7 (1973) 179-189
52. Wynder, E. L., Goodman, D. A., and Hoffmann, D., Ciliatoxic components in cigarette smoke, III. In vitro comparison of different smoke components, Cancer 18 (1965) 1562-1658
53. Skog, E., A toxicological investigation of lower aliphatic aldehydes, Acta Pharmacol. (Copenhagen) 6 (1950) 229-318
54. Bouley, G., Dubreuil, A., Godin, J., and Boudene, C., Effects, chez le rat, d'une faible dose d'acroleine en halee en countinue, J. Eur. Toxicol., 8 (1975) 291-297
55. Dahlgren, S. E., Dalen, H., and Dalham, T., Ultrastructural observations on chemically induced inflammation in guinea pig trachea, Virchows Arch. B., 11 (1972) 211-233
56. Holmberg, B., and Malmfors, T., Cytotoxicity of some organic solvents, Env. Res. 7 (1974) 183
57. Koerker, R. L., Berlin, A. J., and Schneider, F. H., The cytotoxicity of short-chain alcohols and aldehydes in cultured neuroblastoma cells, Toxicol. Appl. Pharmacol., 37 (1976) 281-288
58. Munsch, N., deRecondo, A. M., and Erayssinnet, C., Effects of acrolein on DNA synthesis in vitro, FEBS Letter, 30 (1973) 286-290
59. Zollner, H., Inhibition of some mitochondrial functions by acrolein and methyl vinyl ketone, Biochem. Pharmacol., 22 (1973) 1171-1178
60. Feron, V. J., and Kruysse, A., Effects of exposure to acrolien vapor in hamsters simultaneously treated with benzo(a)pyrene or diethylnitrosamine, J. Toxicol. Env. Hlth., 3 (1977) 379-394
61. Izard, C., Recherches sur les effets mutagenes de l'acrolein et de ses deux epoxydes: Le glycidol et le glycidal, sur Saccharomyces cerevisiae, CR Acad. Sci. Ser. D, 276 (1973) 3037-3040
62. Van Duuren, B. L., Orris, L., and Nelson, N., Carcinogenicity of epoxides, lactones, and peroxy compounds, II. J. Natl. Cancer Inst., 35 (1965) 707-717
63. Rapoport, I. A., Mutations under the influence of unsaturated aldehydes, Dokl. Akad. Nauk. SSSR, 61 (1948) 713-715
64. Bignami, M., Cardamone, G., Comba, P., Ortali, V. A., Morpurgo, G., and Carere, A., Relationship between chemical structure and mutagenic activity in some pesticides: The use of Salmonella typhimurium and Aspergillus nidulans, Mutation Res., 46 (1977) 243-244
65. Epstein, S. S., Arnold, E., Andrea, J., Bass, W., and Bishop, Y., Detection of chemical mutagens by the dominant lethal assay in the mouse, Toxicol. Appl. Pharmacol., 23 (1972) 288-325
66. Egle, J. L., Retention of inhaled formaldehyde, propionaldehyde and acrolein in the dog, Arch. Environ. Hlth., 25 (1972) 119-124

67. Kaye, C. M., Biosynthesis of mercapturic acids from alkyl alcohol, alkyl esters and acrolein, Biochem. J., 134 (1973) 1093-1101
68. U. S. Occupational Safety and Health Administration, Occupational Safety and Health Standards, Subpart 2-Toxic and Hazardous Substances, Code of Federal Regulations, Title 29, Chapter XVII, Section 1910.1000, Washington, DC (1976) p. 31:8302
69. Winell, M., An international comparison of hygienic standards for chemicals in the work environment, Ambio, 4 (1) (1975) 34-36
70. Hayes, E. R., Acetaldehyde, In Kirk-Othmer Encyclopedia of Chemical Technology, 2nd ed., Vol. 1, Interscience Publishers, New York (1963) pp. 75-95
71. Merck & Co., "The Merck Index", 9th ed., Merck & Co., Inc., Rahway, NJ (1976) p. 35
72. Shaw, W. G. J., Stephens, R. L., and Weybrew, J. A., Carbonyl constituents in the volatile oils from flue-cured tobacco, Tobacco Sci., 4 (1960) 179
73. Irby, R. M., Harlow, E. S., Cigarette Smoke, I. Determination of vapor constituents, Tobacco, 148 (1959) 21
74. Johnstone, R. A. W., and Plimmer, J. R., The chemical constituents of tobacco smoke, Chem. Revs., 59 (1959) 885
75. Linnel, R. H., and Scott, W. E., Diesel exhaust analysis, Arch. Environ. Hlth., 5 (1962) 616
76. Ellis, C. F., Kendall, R. F., and Eccleston, B. H., Identification of oxygenates in automobile exhausts by combined gas-liquid chromatography and I.R. techniques, Anal. Chem., 37 (1965) 511
77. Rieger, R., and Michaelis, A., Chromosomen aberrationen nach einwirkung von acetaldehyd auf primär wurzeln von Vicia faba, Biol. Zentralbl., 79 (1960) 1-5

CHAPTER 9

ACYLATING AGENTS

Acylating agents (e.g., acid chlorides or anhydrides) are another class of highly reactive electrophilic agents (via the initial formation of a complex with a Lewis acid catalyst such as aluminum chloride). Acylation and alkylation processes are closely related, for example, in ther activity toward arenes. The reaction effectively introduces an acyl group, RCO-, into an aromatic ring and the product is an aryl ketone (or an aldehyde if the acid chloride is formylchloride). It can be assumed that these substances can interfere with normal biological reactions because of their high chemical reactivity to biochemical substances.

1. **Dimethylcarbamoylchloride** ($(CH_3)_2N-\overset{O}{\underset{\parallel}{C}}-Cl$; DMCC) is prepared via the reaction of phosgene and trimethylamine and is used primarily inthe preparation of pharmaceuticals, e.g., neostigmine bromide, neostigmine methylsulfate and pyridostigmine bromide, agents used in the treatment of myasthenia gravis and secondarily as an intermediate in the synthesis of carbamates which are used as pesticides, drugs and industrial intermediates in the synthesis of dyes and unsymmetrical dimethylhydrazine (a rocket fuel)[1,2]. Dimethylcarbamaylchloride has also been found at levels of up to 6 ppm during production of phthaloylchlorides[1-3].

The carcinogenic potential (high incidence and short latency period) of dimethylcarbamoyl chloride by inhalation in rats has recently been reported[2]. Eighty-nine of 93 rats exposed by inhalation to 1 ppm DMCC developed squamous cell carcinomac of the nose within 200 days. The carcinogenic potential of DMCC was first reported in 1972[4] in a preliminary note, and in 1974[5] describing a high incidence of skin tumors and subcutaneous sarcomas, along with some papillary tumors of the lung in ICR/Ha Swiss mice following applications of DMCC to skin by both subcutaneous and intraperitoneal injection.

Dimethylcarbamoyl chloride has been shown to be mutagenic in *Salmonella typhimurium* TA 100 and TA 98 containing an R factor (plasmids carrying antibiotic resistance genes)[6] and two *E. coli* strains (WP2 and WP25)[2].

2. **Diethylcarbamoyl chloride** ($(C_2H_5)_2N-\overset{O}{\underset{\parallel}{C}}-Cl$; DECC) is used commercially primarily in the synthesis of anthelmintic diethylcarbamazine citrate[2]. DECC has recently been found to be less mutagenic than DMCC in *E. coli* strains WP2 and WP25[2].

3. **Benzoyl Chloride** ($C_6H_5-\overset{O}{\underset{\parallel}{C}}-Cl$) It should be noted that other acylating agents, e.g., benzoyl chloride, phthaloyl chloride are widely employed as chemical intermediates. A recent report has cited the incidence of lung cancer among benzoyl chloride manufacturing

workers in Japan.[7] The manufacturing process involved the initial chlorination of toluene to benzotrichloride with subsequent hydrolysis or reaction of the intermediate benzotrichloride with benzoic acid to yield benzoyl chloride. The sequence of reactions is as follows:

$$C_6H_5-CH_3 \xrightarrow{3Cl_2} C_6H_5-CCl_3 + 3HCl \xrightarrow[C_6H_5COOH]{H_2O} C_6H_5-\underset{O}{\overset{\|}{C}}-Cl$$

Minor reaction products in the original chlorination step were found to be benzyl chloride ($C_6H_5CH_2Cl$) and benzal chloride ($C_6H_5CHCl_2$). Benzyl chloride has been shown to be carcinogenic in rats[8] and mutagenic in Salmonella typhimurium TA 100 and TA 98 tester strains with R-factor phasmids[6]. The high reactivity of both benzoyl chloride and benzotrichloride used as reagents to introduce benzoyl ($C_6H_5CO\cdot$) and benzenyl ($C_6H_5C\cdot$) radicals respectively suggest the potential of these reagents for carcinogenic and/or mutagenic activity.

4. Ketene ($CH_2=C=O$; ethenone) is a highly reactive acylating agent formed by pyrolysis of virtually any compound containing an acetyl group, e.g., acetone. Ketene is widely used in organic synthesis for the acylation of acids, hydroxy compounds, aromatic hydrocarbons (Friedel-Crafts acylation), amines etc.[9]. The major areas ot utility of ketene include: the manufacture of acetic anhydride and the dimerization to diketene which is an important intermediate for the preparation of dihydroacetic acid, acetoacetic esters, acetoacetanilide, N,N-dialkylacetoacetamides, and cellulose esters which are used in the manufacture of fine chemicals, drugs, dyes, and insecticides.

Ketene has utility as a rodenticide[10], in textile finishing[11], in the acetylation of viscose rayon fiber[12], as an additive for noncorrosive hydrocarbon fuels[13], and in the acetylation of wood for improved water resistance[14].

Ketene has been found mutagenic in Drosophila[15], but non-mutagenic in Neurospora[16].

References

1. Anon, Dupont Takes Steps to Protect Workers Against DMCC, Occup. Hlth. Safety Letter, $\underline{6}$ (1976) 7
2. Finklea, J. F., Dimethylcarbamoyl Chloride (DMCC), National Institute of Occupational Safety and Health, Current Intelligence Bulletin, July 7, 1976
3. E.I. dupont de Nemours & Co., Dimethylcarbamoyl Chloride, Advisory Letter to NIOSH, June 23, 1976
4. Van Duuren, B. J., Dimethyl Carbamyl Chloride, A Multi Potential Carcinogen, J. Natl. Cancer Inst., $\underline{48}$ (1972) 1539
5. Van Duuren, B. J., Goldschmidt, B. M., Katz, C., Seidman, I., and Paul, T.S., Carcinogenic Activity of Alkylating Agents, J. Natl. Cancer Inst., $\underline{53}$ (1974) 695
6. McCann, J., Spingarn, N. E., Kobori, J., and Ames, B. N., Detection of Carcinogens as Mutagens: Bacterial Tester Strains with R Factor Plasmids, Proc. Natl. Acad. Sci., $\underline{72}$ (1975) 979
7. Sakabe, H., Matsushita, H., Koshi, S., Cancer Among Benzoyl Chloride Manufacturing Workers, Ann. N.Y. Acad. Sci., $\underline{271}$ (1976) 67
8. Druckrey, H., Kruse, H., Preussmann, R., Ivankovic, S., and Landschütz, C., Cancerogene Alkylierende Substanzen, III. Alkylhalogenide,-Sulfate-, Sulfonate Und Ringgespante Heterocyclen, Z. Krebsforsch., $\underline{74}$ (1970) 241
9. Lacey, R. N., In "Advances in Organic Chem. Methods Results", (eds, Raphael, R. A., Taylor, E. C., Wynberg, H.) Wiley Interscience, New York(1960), p. 213
10. Farbenfabriken Bayer Akt., British Patent 862,866 (1961), Chem. Abstr., $\underline{55}$, 15823b (1961)
11. Gagliardi, D. D., Symposium on Textile Finishing, I. Chemical Finishing of Cellulose, Am. Dyestuff Reptr., $\underline{50}$ (1961) 34
12. Takigawa, M., and Kanda, I., Toho Reiyon Kenkyu Hokoku $\underline{3}$, 30 (1956); Chem. Abstr., $\underline{53}$ (1959) 7602
13. Fields, J. E., and Zopf, Jr., G. W., U.S. Patent 2,291,843 (1960)
14. Karlsons, I., Svalbe, K., Ozolina, I., and Sterni, S., USSR Patent 391,924, 27 July (1973), Chem. Abstr., $\underline{81}$ (1974) 39278U
15. Rapoport, I. A.,Alkylation of gene molecule, Dokl. Akad. Nauk. SSSR, $\underline{59}$ (1948) 1183-1186
16. Jensen, K. A., Kirk, I., Kolmark, and Westergaard, M., Chemically induced mutations in Neurospora, Cold Spring Harbor Symp. Quant. Biol., $\underline{16}$ (1951) 245-261

CHAPTER 10

PEROXIDES

The utility of a variety of organic peroxides in a broad spectrum of commercial and laboratory polymerization reactions is well established. The commercial organic peroxides are highly reactive sources of free radicals: RO:OR → RO· + ·OR. organic peroxides are employed predominantly in the polymer industry (e.g., plastics, resins, rubbers, elastomers, etc.) and are used in applications including the following: (a) initiators for the free-radical polymerizations and/or co-polymerizations of vinyl and diene monomers, (b) curing agents for resins and elastomers and (c) cross-linking agents for polyolefins. Miscellaneous uses of organic peroxides in the polymer industry include: vulcanization of natural and butadiene rubbers; curing polyurethanes and adhesives; preparation of graft copolymers; cross-linking polyethylenes and ethylene-containing co-polymers; and as flame retardant synergists for polystyrene[1].

1. <u>Di-tert.butyl peroxide</u> [$(CH_3)_3COOC(CH_3)_3$; bis(1,1-dimethyl ethyl)peroxide] is used extensively in organic synthesis as a free radical catalyst, as a source of reactive methyl radicals; as an initiator for vinyl monomer polymerizations and co-polymerizations of ethylene, styrene, vinyl acetate, and acrylics and as a curing agent for thermoset polyesters, styrenated alkyds and oils, and silicone rubbers, as a vulcanization agent for rubber[2], in lubricating oil manufacture[3], for cross-linking of fire-resistant polybutadiene moldings[4] and for cross-linking of high density polyethylene[5].

Di-tert-butyl peroxide is mutagenic in <u>Neurospora</u>[6] but has been reported to be inactive toward transforming-DNA[7].

2. <u>tert-Butyl hydroperoxide</u> [$(CH_3)_3COOH$; 1,1-dimethyl ethyl hydroperoxide] is used as an initiator for vinyl monomer polymerizations and copolymerizations with styrene, vinyl acetate, acrylics, acrylamide, unsaturated polyesters[4,8,9] and as a curing agent for thermoset polyesters.

The mutagenic effect of tert-butyl hydroperoxide in <u>Drosophila</u>[10,11], <u>E. coli</u>[12], and <u>Neurospora</u>[13] and its induction of chromosome aberrations in <u>Vicia faba</u>[14,15] and <u>Oenothera</u>[16] has been described.

3. <u>Cumene hydroperoxide</u> [$C_6H_5C(CH_3)_2OOH$; 1-methyl-1-phenyl ethyl hydroperoxide] is employed as an initiator for vinyl monomer polymerizations and co-polymerizations with styrene, acrylics, butadiene-styrene, cross-leaked foamed polyesters[1,17,18], as a curing agent for thermoset polyesters, styrenated alkyds and oils, acrylic monomers,[1,19] and as a promoter for oxidation of hydrocarbons[20]. Cumeme hydroperoxide is mutagenic in <u>E. coli</u>[12] and <u>Neurospora</u>[6,7].

4. Succinic acid peroxide [HOOC-CH$_2$-CH$_2$-$\overset{\text{—}}{\text{C}}$-O-OH] is used as an initiator for vinyl monomer polymerizations and copolymerizations with ethylene and fluorolefins. It is mutagenic in E. coli[12,20], and has been found to inactivate T2 phage[21], and transforming DNA of H. influenzae[22].

5. Peracetic acid (peroxyacetic acid; acetyl hydroperoxide; ethaneperoxic acid; CH$_3$COOOH) is generally prepared from acetaldehyde via autoxidation or from acetic acid and hydrogen peroxide[23]. Organic peroxyacids are the most powerful oxidizing agents of all organic peroxides[23].

The commercial form of peroxyacetic acid is usually in 40% acetic acid, and is primarily employed as a bleaching and epoxidizing agent. The range of utility of peracetic acid includes: (1) the bleaching of textiles and paper-pulp; (2) as a catalyst for polymerization of aminopropionitrile; (3) as a co-catalyst for a sterospecific polymerization of aldehydes; (4) as a fungicide; (5) disinfectant for rooms and medical machines; (6) sterilizing agent for blood serum for tissue culture; (7) as an oxidizing and hydroxylating reagent in organic synthesis, and (8) as a bactericide in food industries.

Peracetic acid (as well as hydrogen peroxide) have been found mutagenic in S. typhimurium inducing mainly deletions with H$_2$O$_2$ being the more effluent agent[24]. For lower concentrations, cells are protected against these peroxides by superoxide dismutase and catalase. A 500 fold ratio between the concentrations of H$_2$O$_2$ and peracetic acid producing a similar biological effect indicated that cells were more efficiently protected against hydrogen peroxide. This was believed to result from the decomposition of peracetic acid via O$_2$ radical liberation; thence subsequently converted by superoxide dismutase into H$_2$O$_2$ which is further decomposed by catalase. Once the cell protection is overcome (e.g., at levels of 5g/l H$_2$O$_2$ and 20 mg/l peracetic acid) the differences in survival suggest that induced genetic damage by H$_2$O$_2$ would be partly repaired, whereas that induced by peracetic acid would not[24].

6. Hydrogen Peroxide. It is also of importance to consider hydrogen peroxide per se in terms of its utility, sources, and mutagenic activity.

Hydrogen peroxide undergoes a variety of reactions:

Decomposition	$2 H_2O_2 \rightarrow 2 H_2O + O_2$	(1)
Molecular addition	$H_2O_2 + Y \rightarrow Y \cdot H_2O_2$ (2)	(2)
Substitution	$H_2O_2 + RX \rightarrow ROOH + HX$	(3)
or	$H_2O_2 + 2 RX \rightarrow ROOR + 2 HX$	(4)
H$_2$O$_2$ as oxidizing agent	$H_2O_2 + Z \rightarrow ZO + H_2O$	(5)
H$_2$O$_2$ as reducing agent	$H_2O_2 + W \rightarrow WH_2 + O_2$	(6)

In entering into these reactions, hydrogen peroxide may either react as a molecule or it may first ionize or be dissociated into free radicals. The largest commercial use for H$_2$O$_2$ is in the bleaching of cotton textiles and wood and chemical (Kraft and sulfite) pulps. (The rate of bleaching appears directly related to alkalinity, with the active species assumed to be the perhydroxyl anion OOH$^-$.) The next largest use of H$_2$O$_2$ is in the oxidation of a variety of important organic compounds. For example, soybean oil, linseed oil, and related unsaturated esters are converted to the epoxides for use as plasticizers and stabilizers for polyvinyl chloride. Other important commercial processes include the hydroxylations of olefinic compounds, the synthesis of glycerol from propylene, the conversion of tertiary amines to corresponding amine oxides, and the conversion of thiols to disulfides and sulfides to sulfoxides and sulfones. In the textile field, H_2O_2 is used to oxidize vat and sulfur dye.

In addition, many organic peroxides are made from hydrogen peroxide. These include peroxy acids such as peroxyacetic acid; hydroperoxides such as tert-butyl hydroperoxides; diacyl peroxides such as benzoyl peroxide; ketone derivatives such as methylethylketone peroxide. Other uses of hydrogen peroxide include its use as a blowing agent for the preparation of foam rubber, plastics, and elastomers, bleaching, conditioning, or sterilization of starch, flour, tobacco, paper, and fabric[25] and as a component in hypergolic fuels[26]. Solutions of 3% to 6% H_2O_2 are employed for germicidal and cosmetic (bleaching) use, although concentrations as high as 30% H_2O_2 have been used in dentistry[27]. Concentrations of 35% and 50% H_2O_2 are used for most industrial applications.

Hydrogen peroxide has been found mutagenic in E. coli[28,29], Staphylococcus aurcus[30-32], and Neurospora[6,33,34] (including mixtures of hydrogen peroxide and acetone, and hydrogen peroxide and formaldehyde)[33]. The inactivation of transforming DNA by hydrogen peroxide[7,35-37], as well as by peroxide-producing agents (e.g., compounds which contain a free N-OH group as hydroxylamine, N-methylhydroxylamine, hydroxyurea, hydroxyurethan and hydrazines on exposure to oxygen) has been described[1,37]. Hydrogen peroxide has induced chromosome aberrations in strains of ascites tumors in mice[38] and in Vicia faba[39].

Hydrogen peroxide has also been found to be non-mutagenic in bacteria[40,41] and Drosophila[42]. Recent studies by Thacker and Parker[43] on the induction of mutation in yeast by hydrogen peroxide suggest that it is inefficient in the induction of nuclear gene mutation. This could be because radical action produces certain types of lesions, leading to inactivation and mitochondrial genome mutation, but not to point mutational changes.

There is general agreement that the effects of hydrogen peroxide, (as well as hydrogen peroxide-producing agents) on DNA are caused by the free radicals they generate.

Hydrogen peroxide (H_2O_2) decomposes into two ·OH radicals in response to UV irradiation or spontaneously at elevated temperatures. It also gives rise to HOO· radicals in the presence of reduced transition metals (e.g., Fe^{++}, Cu^+). These radicals can then react with organic molecules to produce relatively more stable organic peroxy radicals and organic peroxides which may later decompose again into free radicals. This process of "radical-exchange" sustains the effectiveness of short-lived radicals such as ·OH, HOO·, and H· and gives them an opportunity to reach the genome where they can exert their effect[1].

7. <u>Benzoyl Peroxide</u> (dibenzoyl peroxide, benzoyl superoxide; lucidol; benoxyl;) is prepared by the interaction of benzoyl chloride and a cooled solution of sodium peroxide. It is primarily employed as an oxidizer in bleaching oils, flour, etc. and as a catalyst in the plastics industry where it serves

as an initiator of free radicals for polymerization reactions[44], viz.,

$$\text{PhC(O)O-OC(O)Ph} \xrightarrow{70-80°} 2\, \text{PhC(O)O}\cdot \longrightarrow 2\, \text{Ph}\cdot + 2CO_2$$

While benzoyl peroxide has not been found to be carcinogenic or mutagenic in limited studies[45] thus far, its inclusion here based on its reactivity and relation to mutagenic peroxides considered above, appears warranted.

NIOSH estimates that 25,000 employees in the U.S. are exposed or potentially exposed to benzoyl peroxide, with fewer than 1000 involved in manufacture[9]. Primary irritation of the skin and mucous membranes and sensitization have occurred in individuals as a result of flour bleached with, or resins treated with benzoyl peroxide[45]. The current standard for occupational exposure to airborne benzoyl peroxide in the U.S. is limited to a time-weighted average concentration of 5 mg/m^3[45].

References

1. Fishbein, L., Flamm, W. G., and Falk, H. L., "Chemical Mutagens", Academic Press, New York, (1970) pp. 269-273
2. Maeda, I., Aoshima, M., Low-Temperature Vulcanization of Chlorinated Rubbers, Ger. Offen., 2,401,375, August 14 (1974), Chem. Abstr., 82 (1975) P183704
3. Bitter, J. G. A., Ladeur, P., Maas, R. J., and Bernard, J. D. J., Lubricating Oils with High Viscosity Index, Ger. Offen. 2,361,653, June 20 (1974), Chem. Abstr., 82 (1975) P46240Z
4. Musashi, A., Yamazaki, M., and Hiruta, M., Hardenable Flame Proof Polybutadiene Composition Ger. Offen. 2,403,639, August 15 (1974), Chem. Abstr., 82 (1975) P179950A
5. Proskurnina, N. G., Akutin, M. S., and Budnitskii, Y. M., Cross-Linking of High-Density Polyethylene by Peroxides on Zeolites, Plast. Massy, 12 (1974) 50, Chem. Abstr., 82 (1975) 140928B
6. Jensen, K. A., Kirk, I., Kolmark, G., and Westergaard, M., Chemically Induced Mutations in Neurospora, Cold Springs Harbor Symp. Quant. Biol., 16 (1951) 245-261
7. Latarjet, R., Rebeyrotte, N., and Demerseman, P., In "Organic Peroxides in Radiobiology", (ed. Haissinsky, M.) Pergamon Press, Oxford (1958) p. 61
8. Meyer, H., Schmid, D., Schwarzer, H., and Twittenhoff, H. J., Setting of Unsaturated Polyester Resins, U.S. Patent 3,787,527, Jan. 22 (1974), Chem. Abstr., 81 (1974) 50578N
9. Ulbricht, J., Thanh, V. N., Effect of Redox Systems on the Degree of Polymerization and in Homogeneity of PVC, Plaste Kaut., 21 (1974) 186, Chem. Abstr., 81 (1974) 37902
10. Altenberg, L. S., The Production of Mutations in Drosophila by Tert.Butyl Hydroperoxide, Proc. Natl. Acad. Sci., 40 (1940) 1037
11. Altenberg, L. S., The Effect of Photoreactivating Light on the Mutation Rate Induced in Drosophila by Tert.Butyl Hydroperoxide, Genetics, 43 (1958) 662
12. Chevallier, M. R., and Luzatti, D., The Specific Mutagenic Action of 3 Organic Peroxides on Reversal Mutations of 2 Loci in E. Coli, Compt. Rend., 250 (1960) 1572-1574
13. Dickey, F. H., Cleland, G. H., and Lotz, C., The Role of Organic Peroxides in the Induction of Mutations, Proc. Natl. Acad. Sci., 35 (1949) 581
14. Revell, S. H., Chromosome Breakage by X-rays and Radiomimetic Substances in Vicia, Heredity, 6 (1953) Suppl., 107-124
15. Loveless, A., Qualitative Aspects of the Chemistry and Biology of Radiomimetic (Mutagenic) Substances, Nature, 167 (1951) 338
16. Oehlkers, F., Chromosome Breaks Induced by Chemicals, Heredity Suppl., 6 (1953) 95-105
17. Doyle, E. N., Thermosetting Unsaturated Polyester Foam Products, U. S. Patent 3,823,099, July 9 (1974) Chem. Abstr., 81 (1974) P137087A
18. Gruzbarg, K. A., and Barboi, W. M., Copolymerization for Cross linking of Oligomeric Polyesters with Vinyl Monomers in Presence of Various Initiators, Teknol. Svoistva Polim. Mater Radiats. Otverzhdeniya. (1971) pp. 34-46, Chem. Abstr., 81 (1974) 14061E
19. Toi, Y., Shinguryo, H., Nakano, S., and Tsuchida, R., Polymerizing Acrylic Monomers Absorbed in Organic Materials, Japan Kokai, 74 16,791, Feb. 14 (1974) Chem. Abstr., 81 (1974) P50704A

20. Luzzati, D., and Chevallier, M. R., Comparison of the Lethal and Mutagenic Action of an Organic Peroxide and of Radiations on E. Coli, Ann. Inst. Pasteur., 93 (1957) 366
21. Latarjet, R., Ciba Foundation. Symp. Ionizing Radiations Cell Metab. (1957) p. 275
22. Wieland, H., and Wingler, A., Mechanism of oxidation processes. V. Oxidation of aldehydes, Ann. Chem., 431 (1923) 301
23. Mageli, O. L., and Sheppard, C. S., Peroxides and Peroxy Compounds, Organic, In Kirk-Othmer Encyclopedia of Chemical Technology, 2nd ed., Vol. 14, Interscience, New York (1967) pp. 794-820
24. Agnet, M., Dorange, J. L., and Dupuy, P., Mutagenicity of Peracetic Acid on Salmonella Typhimurium, Mut. Res., 38 (1976) 119
25. Young, J. H., U. S. Patent 2,777,749 (1957); Chem. Abstr., 51 (1957) 5442g
26. Ayers, A. L., and Scott, C. R., U. S. Patent 2,874,535 (1959); Chem. Abstr. 53, (1959) 9621b
27. Ludewig, R., Z. Deut. Zahnaerztl., 55 (1960) 444; Chem. Abstr., 54 (1960) 10138d
28. Demerec, M., Bertani, G., and Flint, J., Chemicals for Mutagenic Action on E. Coli Am. Naturalist, 85 (1951) 119
29. Iyer, V. N., and Szybalski, W., Two Simple Methods for the Detection of Chemical Mutagens, Appl. Microbiol., 6 (1958) 23
30. Wyss, O., Stone, W. S., and Clark, J. B., The Production of Mutations in Staphylococcus Audreus by Chemical Treatment of the Substrate, J. Bacteriol., 54 (1947) 767
31. Wyss, O., Clark, J. B., Haas, F., and Stone, W. S., The Role of Peroxide in the Biological Effects of Irradiated Broth, J. Bacteriol., 56 (1948) 51
32. Haas, F. L., Clark, J. B., Wyss, O., and Stone, W. S., Mutations and Mutagenic Agents in Bacteria, Am. Naturalist., 74 (1950) 261
33. Dickey, F. H., Cleland, G. H., and Lotz, C., The Role of Peroxides in the Induction of Mutations, Proc. Natl. Acad. Sci. US, 35 (1949) 581
34. Wagner, R. P., Haddox, C. R., Fuerst, R., and Stone, W. S., The Effect of Irradiated Medium, Cyanide and Peroxide on the Mutation Rate in Neurospora, Genetics, 35 (1950) 237
35. Luzzati, D., Schweitz, H., Bach, M. L., and Chevallier, M. R., Action of Succinic Peroxide on Deoxyribonucleic Acid (DNA), J. Chim. Phys., 58 (1961) 1021
36. Zamenhof, A., Alexander, H. E., and Leidy, G., Studies on the Chemistry of the Transforming Activity, I. Resistance to Physical and Chemical Agents, J. Exptl. Med., 98 (1953) 373
37. Freese, E. B., Gerson, J., Taber, H., Rhaese, H. J., and Freese, E. E., Inactivating DNA Alterations by Peroxides and Peroxide Inducing Agents, Mutation Res., 4 (1967) 517
38. Schöneich, J., The Induction of Chromosomal Aberrations by Hydrogen Peroxide in Strains of Ascites Tumors in Mice, Mutation Res., 4 (1967) 385
39. Lilly, L. J., and Thoday, J. M., Effects of Cyanide on the Roots of Vicia Faba, Nature, 177 (1956) 338
40. Doudney, C. O., Peroxide Effects on Survival and Mutation Induction in UV Light Exposed and Photo-reactivated Bacteria, Mutation Res., 6 (1968) 345-353
41. Kimball, R. F., Hearon, J. Z., and Gaither, N., Tests for a Role of H_2O_2 in X-ray Mutagenesis. II. Attempts to Induce Mutation by Peroxide, Radiation Res., 3 (1955) 435-443
42. Sobels, F. H., Peroxides and the Induction of Mutations by X-Rays, Ultraviolet, and Formaldehyde, Radiation Res., Suppl. 3 (1963) 171-183

43. Thacker, J., Parker, W. F., The Induction of Mutation in Yeast by Hydrogen Peroxide, <u>Mutation Res.</u>, <u>38</u> (1976) 43-52
44. Roberts, J. D., and Caserio, M. C., "Modern Organic Chemistry", W. A. Benjamin, Inc., New York (1967) p. 721
45. Anon, 5 mg. standard for benzoyl peroxide recommended, <u>Occup. Hlth. Safety Letter</u>, <u>7</u> (1) June 8 (1977) p. 7

CHAPTER 11

HALOGENATED UNSATURATED HYDROCARBONS

The halogenated aliphatic hydrocarbons represent one of the most important categories of industrial chemicals from a consideration of volume, use categories, environmental and toxicological considerations and hence most importantly, potential population risk.

In recent years, there has been recognized concern over the environmental and toxicological effects of a spectrum of halogenated hydrocarbons, primarily the organo chlorine insecticides and related derivatives, e.g., DDT, dieldrin, Mirex, Kepone, polychlorinated biphenyls (PCBs), polybrominated biphenyls (PBBs), dibromochloropropane (DBCP) and chlorinated dioxins. This concern has now been extended to practically all of the major commercial halogenated hydrocarbons, numerous members of which have extensive utility as solvents, aerosol propellants, degreasing agents, dry-cleaning fluids, refrigerants, flame-retardants, synthetic feedstocks, cutting fluids and in the production of textiles and plastics, etc., and hence are manufactured on a large scale.

1. Vinyl chloride (chloroethylene; ethylene monochloride; chloroethene; VCM; $CH_2=CHCl$) is used in enormous quantities primarily (ca. 96-97%) for the production of homo-polymer (for PVC production) and co-polymer resins (e.g., Saran and other plastics). Lesser quantities of vinyl chloride are used in the production of 1,1,1-trichloroethane (methyl chloroform), as an additive in specialty coatings and as a component of certain propellant mixtures. It is important to note the VCM production processes because of the halocarbon precursors and intermediates as well as the nature of the potential carcinogenic and mutagenic trace impurities. VCM monomer production processes employ one of the following: (1) the acetylene plus hydrogen reaction; (2) the direct chlorination of ethylene and dehydrochlorination and (3) the balanced direct and oxychlorination of ethylene and dehydrochlorination. The overall processes differ primarily in the manner in which the intermediate ethylene dichloride is produced. The bulk of VCM is produced by process (3) above[1-3] (over 95% of VCM produced in the U.S. was made from ethylene)[3]. A typical commercial product can contain the following impurities in mg/kg[2]: unsaturated hydrocarbons, 10; acetaldehyde, 2; dichloro compounds, 16; water, 15; HCl, 2; non volatiles, 200; iron, 0.4; phenol as a stabilizer, 25-50; and trace amounts of organic impurities including: acetylene, 1,3-butadiene, methyl chloride, vinylidene chloride and vinylacetate[1]. VCM is generally supplied as a liquid under pressure and currently most VCM is not inhibited for shipping[3].

As will be discussed in more detail later, ethylene dichloride is mutagenic in Drosophila[4,5], in S. typhimurium TA 1530, TA 1535 and TA 100 tester strains (without metabolic activation)[6,7] and in E. coli (DNA polymerase deficient pol A⁻ strain[8].

The growth patterns of VCM per se as well as that of its primary end product polyvinylchloride (PVC) plastic resin have been well documented[1-3,8-12]. The total world production of VCM in 1971 was 7,059 million kg and in various areas was estimated as follows (in millions of kg)[2]: Western Europe (2,497); Eastern Europe (817); U.S. (1,969); Japan (1,275); and other areas (499). VCM production in the U.S. in 1974 exceeded 2.6 billion kg (about one-third Western World's output) with the annual growth rate expected to exceed 10% per year through the 1980's[1]. (In 1976, nine U.S. companies reported the production of 2,580 million kg of VCM[9]).

Countries producing VCM, listed in decreasing order of estimated annual production in recent years and the numbers of producers in each year are as follows: Japan (15), the Federal Republic of Germany (5), Italy (3), France (4), Belgium (3), the United Kingdom (4), The Netherlands (2), Brazil (3), Spain (3), Turkey (3), Taiwan (2), Argentina (2), Sweden (1), Mexico (1), USSR (2), South Korea (1), Finland (1), Czeckoslovakia (1), Venezuela (1), Egypt (1), Thailand (1), Rumania (1), Chile (1), Greece (1) and Australia (1)[2,13].

About 96% of the 2,274 million kg of VCM consumed in the U.S. in 1976 was used for the production of vinyl chloride homopolymer and copolymer resins[3]. The remainder was used (essentially by one company internally) as a co-monomer with vinylidene chloride in the production of resins, and in the production of methylchloroform[3]. The consumption patterns of vinyl chloride monomer in Western Europe and Japan are believed to be similar to that in the U.S.[3]. The world production of PVC in 1975 is estimated to be 9-10 million tons.

The major market for PVC resins is in the production of plastic pipe and conduit. Other important uses include: in floor coverings, consumer goods, electrical applications, and in transportation[3].

The total world-wide employment in the VCM and PVC industries is over 70,000 workers. Those employed in industries using PVC as a basic element are believed to number in the millions[11].

PVC is produced (in U.S.) via 4 major processes (in % total production) as follows:[1] (1) suspension polymerization, 78; (2) emulsion polymerization, 12; (3) bulk polymerization, 6; and (4) solution, 4.

The current U.S. Occupational Safety and Health Administration (OSHA) health standards for exposure to air contaminants requires that an employee's exposure to vinyl chloride does not exceed an 8-hour time weighted average of 1 ppm in the workplace air in any 8-hour workshift of a 40 hour work week. During any work shift an employees exposure may not exceed a ceiling concentration limit of 5 ppm averaged over any period of 15 minutes or less[3].

IARC[3] has summarized the work environment hygiene standards for exposure to vinyl chloride as reported by Bertram[14] in terms of time weighted averages for 8-hour (8-hr) and 15-minute (15-m) (a ceiling concentration) time period unless otherwise stated for the following countries: Belgium-200 ppm max; Canada-10 ppm (8-h) and 25 ppm (15-m); Finland-5 ppm (8-hr) and 10 ppm (10 min. period); France-no limits; Federal Republic of Germany-5 ppm max. technical standard, with the average concentration not exceeding 15 ppm over the period of 1 hr; United Kingdom-25 ppm (8-h) and 50 ppm (15-m); Italy-50 ppm (8-h) (this is expected to change to 25 ppm (8-h) and 15 ppm (15-m)); Japan-expected to be 10 ppm; The Netherlands-10 ppm (8-h); Norway-1 ppm (8-h) and 5 ppm (15-m); Sweden-1 ppm (8-h and 5 ppm (15-m); Switzerland-100 ppm (this is expected to change to 10 ppm (8-h)) and USSR-12 ppm.

The hazard of vinyl chloride was originally believed to primarily concern workers employed in the conversion of VCM to PVC who may receive a particularly high exposure of VCM in certain operations (e.g., cleaning of polymerization kettles) or a long-term exposure to relatively low concentrations in air of VCM at different factory sites. Much larger populations are now believed to be potentially at risk including: (1) producers of VCM, (2) people living in close proximity to VCM- or PVC producing industries, (3) users of VCM as propellant in aerosol sprays; (4) persons in contact with resins made from VCM; (5) consumers of food and beverage products containing leachable amounts of unreacted VCM from PVC packaged materials and (6) ingestion of water containing unreacted VCM leached from PVC pipes.

Gaseous vinyl chloride is emitted at both vinyl chloride and PVC resin plants and is distributed into the atmosphere surrounding the emissions source in patterns that depend on the amount of vinyl chloride released, the nature of the plant area from which it is released and the meterological conditions[1]. It is estimated by the U.S. Environmental Protection Agency that the total vinyl chloride escaping to the atmosphere in the United States exceeded 100 million kg per year prior to 1975[1]. Vinyl chloride loss from the average VC plant is estimated to be about 0.45 kg/100 kg of VCM produced[1]. Based on limited data, ambient concentrations of vinyl chloride exceeded 1 ppm (2560 μg/m^3) less than 10% of the time in residential areas located in the vicinity of plants producing VC or PVC. The maximum concentration of vinyl chloride found in ambient air was 33.0 ppm (84,480 μg/m^3) at a distance of 0.5 km from the center of the plant[1]. The average concentration of VCM in air around the PVC plants prior to 1975 was 17 ppb[15]. VCM has been determined in the air at levels of 3.1-1250 ppb in the Houston, Texas area (where an estimated 40% of the U.S. production capacity of PVC is located)[16].

The U.S. Environmental Protection Agency recently proposed new rules to reduce the national emission standard for vinyl chloride from 10 ppm to 5 ppm in order to effect a reduction of vinyl chloride emissions by one-half within 3 years of the actual rulemaking. Based on new average-sized plants, this would result in hourly emissions of 5.1 kg from an ethylene dichloride-vinyl chloride plant instead of 10.3 kg; 9 kg from a dispersion PVC plant instead of 17.5 kg and 13.5 kg from a suspension PVC resin plant instead of 16 kg[17].

Early occupational exposure studies have revealed a wide range of VCM concentrations dependent on the manufacturing processes involved[18]. The air concentration of VCM in a polymerization reaction prior to ventilation has been reported[18] to be 7800 mg/m^3 (3000 ppm) and range from 1560-2600 mg/m^3 (600-1000 ppm) in a polymerization reactor after washing[19].

Concentrations of VCM in the working atmospheres in some plants producing PVC have been reported in the ranges of 100-800 mg/m^3 (40-312 ppm) with peaks up to 87,300 mg/m^3 (34,000 ppm)[20].

Additional concentrations of vinyl chloride have been reported for workplace air including: (1) 0.15 to 0.35 ppm in air in three English cable factories[21]; (2) greater than 75 ppm in air in a Yugoslavian PVC manufacturing plant[22]; and (3) greater than 113.6 mg/m^3 in air in a Russian synthetic leather plant[23].

In 1974, Heath[24] estimated that 20,000 U.S. workers in the past and to 1974 had been exposed to VCM in manufacturing plants. Baretta[25] in 1969 reported that on a time-weighted average, the concentrations of VCM to which coagulator operators are exposed ranges from 130-650 mg/m^3 (50-250 ppm). A more recent survey by NIOSH of 3 VCM plants reported that the time-weighted-average exposure to VCM ranged from 0.07-26.46 ppm[3,26].

The concentration of residual VCM monomer in PVC powder that is fabricated into final products is also an important determinant of VCM in the ppm range. The entrapped concentration is dependent upon the production process and can range from 0.1 to 5.8 thousand ppm, which can be liberated during fabrication, particularly when heated[1]. PVC leaving certain plants may contain 200-400 ppm VCM, on delivery to the customer, the level of VCM is about 250 ppm, and after processing, levels of 0.5-20 ppm are reached, depending on the method of fabrication[27]. However, new processing methods developed since 1974, leave as little as 1 to 2 ppm residual VCM in VCM resins[28]. Residual VCM in commercial food grade resins have been reduced by special techniques to 115 ppb (wt/wt) for resin and less than 0.048 ppb (wt/wt) for compound and sheet PVC[29].

VCM has been detected in effluents discharged by chemical and latex manufacturing plants and in raw water in the U.S.[30]. The U.S. Environmental Protection Agency estimated in 1974 that about 12.3 kg/day of VCM were discharged in the wastewater effluent from two VCM plants in the Long Beach, California area[3].

Vinyl chloride has been found in municipal water supplies in the United States[1,14] in representative samples of the nation's community drinking water supplies that chlorinate their water and represent a wide variety of raw water sources, treatment techniques and geographical locations. The highest concentration of vinyl chloride detected in finished drinking water in the U.S. was 10.0 μg/l[31]. The sources of the vinyl chloride found in the Miami, Florida and Philadelphia water supplies (5.6 and 0.27 μg/l respectively have not been identified[1].

Available results indicate that migration of vinyl chloride from rigid PVC water pipes does occur, and that it is a linear function of the residual vinyl chloride level in the pipe itself[1]. Only limited data are available on vinyl chloride emissions from the incineration of plastics. The quantities of vinyl chloride and combustion products varied as a function of temperature as well as with the type of plastics and their polymers[1]

It is believed that vinyl chloride should disappear significantly in its transport over long distances, however, in the immediate vicinity of emission sources, vinyl chloride can be considered a stable pollutant[1]. While no mechanism is presently known for the removal of vinyl chloride from the air at night, biological sinks such as microbial removal in soil may be of significance in depletion of vinyl chloride over a long time period. However, such sinks would not be expected to be important in terms of urban scale transport of vinyl chloride[1].

In studies performed in a laboratory model ecosystem it was reported that vinyl chloride, despite its lipophilicity, is so volatile (vapor pressure in 2660 mm at 25°C) that it does not bioaccumulate or transfer appreciably through food chains, at least at ordinary temperatures[32].

Vinyl chloride has been found in a relatively small number of products packaged in PVC containers (e.g., in edible oils in a concentration of 0.05 to 2 ppm[33] and in butter and margarine[34]).

The U.S. Treasury Department in 1973 banned the use of PVC for the packaging of alcoholic beverages as a result of FDA reports indicating the presence of VCM at levels up to 20 mg/kg in some alcoholic beverages packaged in this material[35].

Vinyl chloride has been found in domestic and foreign cigarettes and little cigars in concentrations of 15-17 nanograms/cigarette[36].

New automobile interiors have been found to contain vinyl chloride in concentrations of 0.4 to 1.2 ppm[3,37].

Experimental as well as human evidence of the carcinogenicity of vinyl chloride has been reviewed by Bartsch and Montesano[10] and IARC[2,3]. The results to date show that vinyl chloride administered by inhalation is carcinogenic in rats, mice and hamster, producing angiosarcoma of the liver as well as of other tissues in all three species. In addition, in rats vinyl chloride induces tumors of the Zymbal glands and of the skin, nephroblastomas, hepatomas and neuroblastomas, in mice tumors of the lung and skin were also observed[38-48]. The lowest reported effective dose which is carcinogenic in rats and mice is 50 ppm[10,44-48]. Two angiosarcomas of the subcutaneous tissue and a carcinoma of the Zymbal glands were observed in the offspring of mothers exposed during pregnancy to inhalation of vinyl chloride at 10,000 and 5000 ppm[3,10]. It is noteworthy that these experiments not only indicate the carcinogenicity of vinyl chloride but also demonstrate that the type of tumor induced, angiosarcoma of the liver, was the same as that observed in workers exposed to VCM[2,3,10].

The relationship between exposure to VCM and development of angiosarcoma of the liver was first reported by Creech and Johnson[49] who cited three cases among workers employed in a single vinyl chloride polymerization facility in the U.S. These men had cleaned reactor vessels as part of their employment. At present, a total of 17 cases of angiosarcoma of the liver have been reported among workers exposed to VCM in the U.S. Extensive world wide epidemiological studies to date have indicated about 50 cases of angiosarcomas of the liver associated with VCM exposure among workers employed in the manufacture of PVC resins (Table 1)[2,3,10,12,49-52]. Most of the tumors were observed in workers employed in the cleaning of autoclaves where the concentration of VCM is high (e.g., ranging from 500 to 2000 or more ppm in the air). It should be noted however that angiosarcomas of the liver were also found in people exposed to much lower concentrations of VCM (e.g., in a worker in the production of PVC bags who had been exposed to VCM for a total of 3 years and in an accountant employed for 10 years in a vinyl fabric factory)[10]. It has also been suggested that at least two other types of cancer, carcinoma of the brain and of the lung, appear with increased frequency in workers exposed to VCM[10,54,55].

Infante et al[52] cited a significant excess of mortality from cancer of the lung and brain in addition to cancer of the liver, among workers occupationally exposed to VCM. The risk of dying from cancer of the lymphatic and hematopoietic system also appears to increase with an increase in latency. A study of cancer mortality among populations residing proximate to VCM polymerization facilities also demonstrated an increased risk of dying from CNS and lymphatic cancer[52]. However, it was noted by Infante et

TABLE I
CASES OF ANGIOSARCOMA OF THE LIVER AMONG WORKERS EXPOSED TO VCM OR PVC[10]

Country	Number of cases	Age at diagnosis diagnosis	Years from first exposure to	Total exposure in years	Occupation
USA	15	48(36-61)[a]	20(12-30)[a]	17(4-30)[a]	Polymerization workers
UK	2	71	26	20	-,-
		37	8	3.5	-,-
Norway	1	56	22	21	-,-
Sweden	1	43	19	18	-,-
Federal Republic of Germany	5	40	14	12	-,-
		38	12	12	-,-
		44	17	17	-,-
		49	17	11	-,-
		43	13	12	-,-
France	2	43	19	19	-,-
		54	15	12	-,-
Italy	2	43	15	6	-,-
		55	22	21	-,-
Rumania	1[b]	27	11	5.5	-,-
Czechoslovakia	2	37	14	14	-,-
		45	15	15	-,-
Canada	9		no data available		-,-
Yugoslavia	4		no data available		-,-
Japan	1		no data available		-,-
UK	1	55	24	11	Pouring PVC oil onto fabric base
Sweden	1	61	27	23	VCM production worker
Federal Republic of Germany	1	43	14	14	Filling pesticide cans with VCM as propellant
Italy	1	36	6	3	Production of PVC bags

[a]Average with, in parenthesis, minimal and maximal values.

[b]This case was diagnosed as hepatocarcinoma associated with cholangiosarcoma.

al[52] that although these findings raise cause for concern about out-plant emissions of VCM, without further study these cancers cannot be unequivocally interpreted as being related to out-plant exposure to VCM.

The risk of cancer mortality among residents 45 years of age and older in three communities having vinyl chloride polymerization facilities was studied by Infante[56]. Among males the death rate from CNS tumor was higher than that for the state as a whole. No excess mortality from leukemia and aleukemia or from lymphoma was found.

dene chloride is polymerized to plastic resins at 12 facilities and the resins are then fabricated into plastics at 60 to 75 plants throughout the country[120]. Estimates of the number of workers engaged in the preparation of polymers and co-polymers of vinylidene chloride and vinyl chloride (e.g., the preparation of Saran wrap) are not available, nor are data available at present on workers exposed to only vinylidene chloride during their working lifetimes.

The American Conference of Governmental Industrial Hygienists (ACGIH) recommends that an employee's exposure to vinylidene chloride does not exceed an eight-hour time weighted average of 10 ppm (40 mg/m^3) in the workplace air in any eight-hour work shift of a forty-hour work week. During any fifteen minute period, the ACGIH proposes an absolute ceiling concentration limit of 29 ppm (80 mg/m^3) provided the daily threshold limit value (in terms of 8-hour time weighted average values) is not exceeded[123].

There is a paucity of data concerning worker exposure to vinylidene chloride. Worker exposure has generally not been monitored in the past-according to EPA[120]. Tests have shown that 20,000 ppm can easily be reached in the proximity of a spill. In some cases past worker exposure to vinylidene chloride may have exceeded those of vinyl chloride which were measured at 300-1000 ppm before OSHA limits were imposed.

Workers involved in manufacturing factilities using vinylidene chloride in polymerization processes such as the production of PVC have been reported to be exposed to vinylidene chloride in amounts of less than 5 ppm and most frequently to trace levels[124,125]. Levels of 2 ppm of vinylidene chloride have also been detected as a contaminant of submarine and spacecraft atmospheres[119,126].

A substantial amount of vinylidene chloride appears to be vented to the atmosphere during its production, polymerization and fabrication. Emissions of vinylidene chloride in the U.S. in 1974 have been estimated at 1.5 million kg from monomer synthesis operations. This was reduced to 277 thousand kg by new control technology in late 1975. Vinylidene chloride losses from polymer synthesis operations and polymer fabrication opeations were 308 thousand kg and 13.8 thousand kg respectively[127].

To estimate the population at risk due to vinylidene chloride emissions, EPA[120] assumed that the populations of the cities and countries which the production and use of vinylidene chloride exist are at risk. Using a total U.S. population figure of 212 million, the population at risk due to these producers and major users of vinylidene chloride (>$1000 annually or >1000 lbs annually) was estimated to be about 4.7 percent to the U.S. population or approximately 10 million people. Other sources of exposure to the population include the other user facilities located throughout the country and those involved in transportation. Based on solubility data it is estimated by EPA that all vinylidene chloride in wastewater is released to the atmosphere and thus probably little exposure occurs through the use of water[120].

Vinylidene chloride has been detected in effluent discharged by chemical and latex manufacturing plants in the U.S.[119] and in effluent discharged from chemical manufacturing plants in the Netherlands at a concentration of 32 µg/l[128]. Vinylidene chloride has also been identified in the U.S. in well, river, and raw water[129]. It has also been found in finished drinking water in the U.S. where the highest reported concentration was 0.1 µg/l[130].

Vinylidene chloride has also been found as an impurity in vinyl chloride monomer[131], and trichloroethylene[132] and at a level of 0.011% in commercial chloroprene[133].

As much as 25 percent of the vinylidene chloride used in any given Saran production run has been estimated to be disposed of in landfill (primarily in polymerized form) although there are no estimates of the levels of unreacted monomer[120].

Commercial household and industrial Saran wrap have been analyzed for residual vinylidene chloride monomer. Six rolls of household film had monomer concentrations ranging from 6.5 to 10.4 ppm with an average of 8.8 ppm with no significant differences found in samples taken from the beginning (outside) or the end (inside) of each roll. Levels of monomer ranging from 10.8 to 26.2 ppm were found in the industrial film with levels increasing from the beginning to the end of the roll[119,134].

Although the widespread use of vinylidene polymers as food wraps could result in the release of unreacted monomer into the food chain[121,135], and vinylidene chloride copolymers containing a minimum of 85% vinylidene chloride have been approved for use with irradiated foods[136], information is scant as to the migration of unreacted monomers from these sources either into food or via disposal of the polymeric material per se. One report states that no more than 10 ppm of unreacted vinylidene chloride is contained in Dow's product Saran Wrap and that within detectable limits, no more than 10 ppb could get into food, even under severe conditions of use[137].

An investigation of the cancer risk among a cohort of 138 workers exposed to vinylidene chloride (where vinyl chloride was not used as a copolymer) revealed no findings statistically related or individually attributed to vinylidene chloride exposure[124]. Fifty-five people had less than 15 years since first exposure and only 5 deaths were observed; 27 workers were lost to follow up but were considered alive in the analysis[124].

The health effects of vinylidene chloride have been reviewed by Haley[138] and IARC[119] and the U.S. Environmental Protection Agency[139].

Aspects of the reported carcinogenicity of vinylidene chloride appear conflicting and indicate sex, species, and strain specificity. Twenty-four of 150 Swiss male mice exposed to 25 ppm of vinylidene chloride in air for 4 hrs daily, 4-5 weeks for 52 weeks, developed adenocarcinoma of the kidney (compared to 1 out of 150 females). No such tumors occurred in mice exposed to 10 ppm for 52 weeks or in controls[140,141], nor in BALB/C, C56BL or C$_3$H mice or Sprague-Dawley rats and hamsters similarly exposed to vinylidene chloride[142].

An increased incidence of mammary fibroadenomas and carcinomas was reported in female Sprague-Dawley rats exposed to 10, 25, 50, 100 and 150 ppm vinylidene chloride in air for 4 hrs/day, 4-5 days a week for 52 weeks and observed for up to 82 weeks. No dose-response relation was found and in one rat treated with 100 ppm, vinylidene chloride, one Zymbal gland carcinoma was observed[141]. Viola[143] reported that male and female Wistar rats exposed to 100 ppm of vinylidene chloride by inhalation developed abdominal lymphomas and subcutaneous fibromas.

Two year studies at Dow Chemical Co., involving both vinylidene chloride administered in the drinking water (60, 100 and 120 ppm) and repeated inhalation (10 or 40 ppm 6 hours/day; 5 days/week; after 5 weeks, 75 ppm for up to 18 months) to male and female Sprague-Dawley rats have been carried out[145-146] and indicated no dose-related clinical differences or cumulative mortality differences or findings of neoplasia. Reproduction studies with vinylidene chloride administered to Sprague Dawley rats by inhalation or ingestion in the drinking water showed the compound to be neither a teratogen or mutagen or one adversely affecting reproductivity[145]. The vinylidene chloride (99.5%) tested in Dow studies contained trace amounts (ppm) of the following impurities: vinyl bromide, 4; vinyl chloride, 3-50; trans-1,2-dichloroethylene, 138-1300; cis-1,2-dichloroethylene, 0.013-0.16%; 1,1,1-trichloroethane, 0.03; and 1,1,2-trichloroethane[145].

Winston et al[147] and Lee[148] reported the only tumor in CD-rats exposed to 55 ppm vinylidene chloride for 9 months was a subcutaneous hemangiosarcoma of the skin in one of the rats tested. Hemangiosarcoma of the liver and lung in CD rats exposed for 9-12 months to 250 or 1000 ppm of vinylidene chloride have been reported. In addition there were no lesions in the testes or accessory organs indicative of a treatment related effect on reproductive performance[66].

In a preliminary report by Maltoni et al[141] of an ongoing study involving Sprague-Dawley rats administered 5, 10 or 20 mg/kg body weight, vinylidene chloride by stomach tube once daily, 4-5 days/week for 52 weeks, 1 carcinoma of the Zymbal gland was observed in a rat treated with a 10 mg/kg dose. At the time of reporting, the rats had been observed for 93 weeks after the start of treatment.

In most recent studies at IARC[119], vinylidene chloride when given by the oral route, induced malignant and benign liver tumors in C57Bl mice of both sexes and gastric tumors in female mice[119] and or oral cavity tumors and of meningiomas in male BDIV rats.

Maltoni et al[141] reported in a study still in progress that no tumors had occurred at 74 weeks among a group of 30 male and 30 female Chinese hamsters, 28 weeks of age exposed to 25 ppm vinylidene chloride in air for 4 hrs/day, 4-5 days/week for 52 weeks.

Vinylidene chloride in air (2% and 20%) produced reverse mutations in S. typhimurium TA 1530 and TA 100 in the presence of 9,000 g supernatant from mouse and rat liver, lung and kidney[149]. The lower mutagenic response observed with a concentration of 20% vinylidene chloride may have resulted from an inhibitory action of vinylidene chloride and/or its metabolite(s) on the microsomal enzymes responsible for its metabolic activation. It was postulated by Bartsch et al[149] that 1,1-dichloroethylene oxide (in analogy with chloroethylene oxide the suggested primary metabolite of vinyl chloride) may be a primary reactive metabolite of vinylidene chloride. It is also considered possible that partial dechlorination of vinylidene chloride by microsomal enzymes results in vinyl chloride and its metabolic products[149].

Vinylidene chloride in solution induced reverse mutations in E. coli K12 in the presence of 9,000 g supernatant from mouse liver[78].

Vinylidene chloride was not mutagenic in the dominant lethal test in male CD-1 mice exposed by inhalation to 10, 30, 50[150] and 55[151] ppm for 6 hrs/day for 5 days/week.

No chromosomal aberrations have been found in Sprague-Dawley rats exposed to 75 ppm vinylidene chloride 6 hrs/day, 5 days/week for 26 weeks[145].

Biotransformation mechanisms have been proposed for vinylidene chloride by Haley[138] and Hathaway[152] (Figure 6) analogous to that of vinyl chloride involving the initial formation of a chloroethylene oxide; e.g., 1,1-dichloroethylene oxide. Of the compounds shown in Figure 6, chloroacetic acid, thiodiglycollic acid, thioglycollic acid, dithioglycollic acid and lactam compounds have already been isolated from the urine of vinylidene chloride treated animals[152].

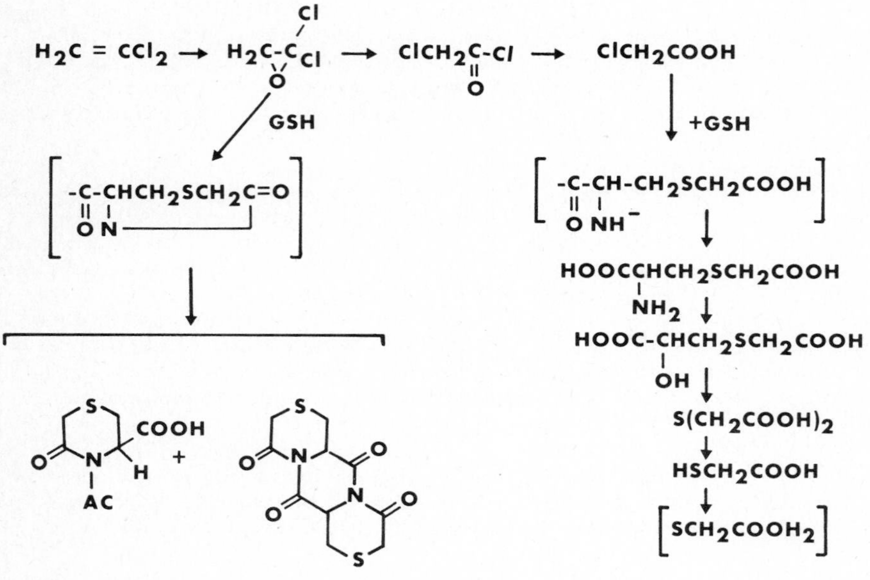
H2C = CCl2
Cl
H2C-C Cl
O
ClCH2C-Cl
O
ClCH2COOH
GSH
+GSH
-C-CHCH2SCH2C=O
O N
-C-CH-CH2SCH2COOH
O NH⁻
HOOCCHCH2SCH2COOH
NH2
HOOC-CHCH2SCH2COOH
OH
S(CH2COOH)2
HSCH2COOH
[SCH2COOH2]
S
COOH
N
H
AC
+
S
N
O
N
O
S

Fig. 6 Proposed mammalian metabolism of vinylidene chloride[152]

3. <u>Trichloroethylene</u> (1-chloro-2,2-dichloroethylene; 1,1,2-trichloroethylene; trichloroethene; acetylene trichloride; TCE; $ClCH{=}CCl_2$) may be produced from acetylene or ethylene. Although the acetylene process (involving chlorination to 1,1,2,2-tetrachloroethane, which is then dehydrochlorinated) has been the dominant method in the past, only 8 percent of the reported U.S. capacity relied on this process. Trichloroethylene is produced primarily by the chlorination and dehydrochlorination of ethylene dichloride. U.S. production of trichloroethylene in 1974 amounted to approximately 193 million kg[153,154]. This represents a 30 percent decrease from the record annual production of 1970[154], and was due primarily to legislation restricting the use and emissions of trichloroethylene and to the closing of three acetylene-based and one ethylene-based plants. Five U.S. companies produced 98 million kg of trichloroethylene

during the first 9 months of 1975[155]. In Japan in 1974, four companies produced 90 million kg, compared to 112 million kg in 1970[153]. It was forecast that the world market for trichloroethylene during 1975 would be about 680 million kg[156].

Approximately 90% of trichloroethylene consumed in the U.S. (345 million pounds in 1974) is for vapor degreasing and cold cleaning of fabricated metal parts. Because of its implications in smog production in the U.S. and resultant legislation restrictive its use, it is expected that during the next five years consumption of trichloroethylene for metal cleaning will decline at an average rate of 3% and be most probably replaced by 1,1,1-trichloroethane and perchloroethylene[157].

Six percent (25 million pounds) of the trichloroethylene production is used as a chain terminator for polyvinyl chloride production[153,154]. Additional areas of utility include: as an extract in food processing (e.g., for decaffeinated coffee), as a chemical intermediate; as a solvent in the textile industry and research laboratories; as an ingredient in printing inks, lacquers, varnishes and adhesives, and in the dry-cleaning of fabrics. A pharmaceutical grade of trichloroethylene is used as a general anesthetic in surgical, dental and obstetrical procedures.

Largely because of its solvent properties, trichloroethylene is incorporated in a number of consumer products (e.g., cleansers for automobiles, buffing solution, spot remover, rug cleaner, disinfectant and deodorant)[158].

The threshold limit value for trichloroethylene in the U.S. is 0.535 mg/m^3 (100 ppm). OSHA is currently in the process of approving the Criteria Document for trichloroethylene which recommends that a TLV expressed as a time-weighted average exposure for an 8-hr workday continue at 100 ppm and that the present ceiling be reduced from 1.07 mg/l (200 ppm) to 0.80 mg/l (150 ppm)[159,160]. The maximum allowable concentration in the USSR is 10 mg/m^3 and the MAC in several European countries has been set at 0.273 mg/l (50 ppm) or even lower[161].

The number of U.S. workers exposed to trichloroethylene has been estimated to be about 283 thousand[158] (Table 2). Levels of 1076-43,000 mg/m^3 (200-8000 ppm) of trichloroethylene have been reported in a small U.S. factory[162].

Emissions of commercial organic solvent vapors into the atmosphere have been increasing dramatically in the last decade[163]. The loss of trichloroethylene and perchloroethylene to the global environment in 1973 was estimated to be each over 1 million tons[163]. Trichloroethylene emissions can occur principally from three sources, production, transportation and consumption. Estimated emissions from trichloroethylene production are 57.0 lbs emitted/ton produced[154,164]. The quantity of trichloroethylene discharged from domestic transport is very difficult to evaluate but it is believed that emissions occur almost inevitably from loading and transfer operations as well as accidental spills[165].

The major sources of emissions resulting from trichloroethylene consumption can be attributed to its use as a solvent in open top vapor degreasers[154]. The average emission rate for an open top vapor degreaser in the U.S. is 110 tons/year[166]. Assuming that 55% of the vapor degreasing operations in 1974 used trichloroethylene, the total national emissions would have been approximately 121,000 tons or roughly 70% of the total amount of trichloroethylene used in metal cleaning operations[154,167].

TABLE 2. Occupational Exposure – Estimated Number of Workers Exposed to Trichloroethylene in Industry.

Industry	Estimated Number Exposed*
Agricultural services	124
Oil and gas extraction	267
Ordnance	57
Food products	2,502
Textile mill products	1,014
Apparel/textile products	858
Lumber products	72
Furniture mfg.	162
Paper products mfg.	2,240
Printing trades	2,876
Chemical mfg.	9,552
Petroleum products	713
Rubber/plastics mfg.	4,985
Leather products	725
Stone/clay products	2,685
Primary steel mfg.	11,672
Metal fabrication	11,709
Machinery mfg.	7,481
Electrical equipment	66,727
Transportation equipment	54,174
Instrument mfg.	4,815
Miscellaneous mfg.	1,516
Trucking/warehousing	642
Air transportation	23
Communication	5,560
Wholesale trade	·3,327
Automotive dealer	223
Furniture stores	597
Banking	2,391
Personal services	583
Misc. business services	27,759
Auto repair	5,246
Misc. repair	17,198
Amusement services	7,987
Mechanical services	20,053
Misc. unclassified	4,138
ESTIMATED TOTAL	282,653

*Projections based on preliminary data obtained from the National Occupational Hazard Survey, Hazard Surveillance Branch, Office of Occupational Health Surveillance and Biometrics, NIOSH (does not include anesthetic use or use in tradename products).

Cold cleaners are another type of metal degreaser which can contribute to trichloroethylene emissions. The average emission rate for a cold cleaner is about 0.33 tons/year[166] and compared to vapor degreasers, a minor amount of trichloroethylene is used in this type of operation.

Concentrations of trichloroethylene vapor in degreasing units have been reported to range from 20-500 ppm (at head height above the bath) with the highest levels being over the baths which relied on manual removal of articles[168], and between 150 and 250 ppm in degreasing rooms per se[169]. Concentrations of vapor in a dial assembly workshop in a Japanese factory ranged from below 135 mg/m^3 (25 ppm) to over 538 mg/m^3 (100 ppm)[169].

Concentrations of anesthetics (including trichloroethylene) in operating rooms to which surgeons and nurses were exposed varied from 1.6-554 mg/m^3 (0.3 to 103 ppm)[170]. It is estimated that about 5000 medical, dental, and hospital personnel are routinely exposed to trichloroethylene[158].

Typical concentrations of trichloroethylene at 8 locations in 5 U.S. states in 1974 ranged from 970 ng/m^3 (180 ppt) in urban areas to less than 110 ng/m^3 (20 ppt) in rural areas[171]. Less than 30 ng/m^3 (5 ppt) were found in air samples in rural Pullman, Washington, from December, 1974 to February, 1975[172]. Trichloroethylene has been recently detected in air over 3 New Jersey industrialized cities[173]. Concentrations of trichloroethylene in air samples taken at 5 land stations ranged from 2-28 ng/m^3 (0.5-5 ppt) at 11 sea stations from 1-22 ng/m^3 (0.2-4 ppt) and over the northeast Atlantic Ocean from 5-11 ng/m^3 (1.2 ppt)[174].

Slightly enhanced levels of trichloroethylene following chlorination of water at sewage treatment plants have been found in the U.S.[175,176]. Trichloroethylene has been found in the organic constituents of Mississippi River water (before and after treatment) and in the organic constituents of commercial deionized charcoal filtered water[177]. Trichloroethylene concentrations of 54 kg/day of 1.2 ng/l in average raw wastewater flow have resulted from a decaffeination process used in the manufacture of soluble (instant) coffee in California[178].

Trichloroethylene has been found in foodstuffs such as dairy products, meats, oils and fats, beverages and fruits and vegetables in levels ranging from 0.02 μg/kg in wine to 60 μg/kg in packaged tea[163]. Trace levels of trichloroethylene have also been found in edible oils after extraction[179]. The use of trichloroethylene for caffeine has recently been discontinued in the U.S. and the FDA in 1977 announced that it intends to ban the use of trichloroethylene in foods, drugs and cosmetics. In the latter category, it had been used as a topical anesthetic in some cosmetics[180,181].

The National Cancer Institute (NCI) in the U.S. has recently issued a "state of concern" alert, warning producers, users, and regulatory agencies that trichloroethylene administered by gastric intubation to B6C3F mice induced predominantly hepatocellular carcinomas with some metastases to the lungs, e.g., 30 of 98 (30.6%) of the mice given the low dose (1200 mg/kg and 900 mg/kg for male and female respectively) and 41 or 95 (43.2%) of the mice given the higher dose (2400 mg/kg and 1800 mg/kg for male and female respectively). Only one of 40 (2.5%) control mice developed these carcinomas[158,182].

No hepatocellular carcinomas were observed in both sexes of Osborne-Mendel rats administered trichloroethylene at levels of 1.0 or 0.5 g/kg by gastric intubation 5 times weekly for an unspecific period[158].

No liver lesions or hepatomas were found in NLC mice given oral doses by gavage or 0.1 ml of a 40% solution of trichloroethylene in oil twice weekly for an unspecified period[183].

Henschler et al[184] recently postulated that the carcinogenic effect of the technical sample of trichloroethylene used in the bioassay experiments[185] is most probably predominantly, if not exclusively, due to the epoxides[186] which are added to some but by no means to all brands of trichloroethylene. Technical grade trichloroethylene contains several impurities which must be stabilized for use as a degreasing agent, by antioxidants[153,184] (e.g, amines at levels of 0.001 to 0.2%) or combinations of epoxides and esters (0.2 to 2% total)[153]. Gas chromatographic-mass spectrometric analysis of the trichloroethylene sample performed by Henschler et al[184] indicated a proportion of identified contaminants amounting to 0.65% including strong monofunctional alkylating agents of known mutagenicity and/or carcinogenicity (e.g., epichlorohydrin and 1,2-epoxybutane.

Trichloroethylene (3.3 mM) in the presence of a metabolic activating microsomal system induced reverse mutations in E. coli strain K12[78]. It has also been shown to induce frameshift as well as base substitution mutation in S. cerevisiae strain XV185-^{14}C in the presence of mice liver homogenate[187].

Trichloroethylene exhibited dose-dependent mutagenicity at concentrations of 0.01, 0.1 and 1 mg/ml when tested in S. typhimurium strains TA 1535 and 1538 without metabolic activation[188]. In the host-mediated assay (with female ICR mice) using strains TA 1950, TA 1951 and TA 1952, there was a significant increase in the number of revertants at trichloroethylene and perchloroethylene doses on the level of LD_{50} and $\frac{1}{2}LD_{50}$.

Cytogenetic analysis of mice bone marrow cells performed after single and repeated i.p. applications (5 applications in one-day intervals) after 6, 24 and 48 hr. following the last application showed no significant increase in chromosomal aberrations following treatment by either trichloroethylene or perchloroethylene[188].

No adverse effects on embryonal or fetal development were observed following exposure of pregnant mice and rats to levels of 0.34 mg/l (65 ppm) or 1.605 mg/l (300 ppm) of trichloroethylene[189].

The metabolism of trichloroethylene has been reviewed by IARC[153], Kelley[190] Henschler[191], Bonse and Henschler[192] and Van Duuren[193].

There are three major metabolic transformations: (1) oxidation to chloral hydrate which takes place in the microsomal fraction of liver cells, (2) reduction to trichloroethanol and trichloroethylene is metabolized to trichloroethanol and trichloroacetic acid in rats[194] and dogs[195]. The earlier suggestion of Powell[196] in 1945 that formation of these metabolites implies rearrangement of the transient trichloroethylene oxide intermediate into chloral has been confirmed by a number of findings including: (1) the identification of chloral as a trichloroethylene metabolite in vitro[197,198] and (2) by study of the rearrangement of the oxides belonging to a series of chlorinated ethylenes[191,192]. The formation of trichloroethylene oxide (2,2,3-trichloro-oxirane) from trichloroethylene in aerobic incubates of liver microsomes and NADPH was confirmed by spectral investigation of the cytochrome P-450 complex[199].

Epoxides are now recognized as obligatory intermediates in the metabolism of olefins by hepatic microsomal mixed-function oxidases[200]. The metabolism of trichloroethylene through the oxirane includes different reactive molecular species. The participation of radical intermediates during opening of the oxirane ring and chlorine migration is possible[192].

Trichloropropane oxide has recently been reported to augment the microsomal dependent covalent binding of trichloroethylene to DNA or protein, to a greater degree in male than female mice. A correlation between such binding and liver tumor induction by trichloroethylene in B6C3F1 mice (but not in Osborne-Mendel rats) was suggested[193,201,202]. The irreversible binding of ^{14}C-labeled trichloroethylene to mice hepatic protein in vivo and in vitro has also recently been reported[203].

4. Tetrachloroethylene

Tetrachloroethylene (perchloroethylene; $Cl_2C=CCl_2$) is prepared primarily via two processes:[203] (1) the Huels method whereby direct chlorination of ethylene yields 70% perchloroethylene, 20% carbon tetrachloride and 10% other chlorinated products and (2) hydrocarbons such as methane, ethane or propane are simultaneously chlorinated and pyrolyzed to yield over 95% perchloroethylene plus CCl_4 and HCl.

The world-production of perchloroethylene in 1972 was 680 million kg; its growth rate estimated at 7%/year with a total world production estimated at 1100 million kg for 1980[204]. The consumption pattern for perchloroethylene in the United States in 1974 is estimated to have been as follows: textile and dry cleaning industries, 69%; metal cleaning, 16%; chemical intermediate (e.g., preparation of trichloroacetic acid in some fluorocarbons), 12%; and miscellaneous 3%. Perchloroethylene is used as a solvent in the manufacture of rubber solutions, paint removers, printing inks, and solvent soaps, as a solvent for fats, oils, silicones and sulfur and as a heat-transfer medium[203].

Some 500,000 workers currently are at risk of exposure to perchloroethylene according to NIOSH. It was also noted that over 20,000 dry cleaning establishments and a large number of other industries manufacture or use perchloroethylene[205].

The current TLV for perchloroethylene in the U.S. is 670 mg/m^3 (100 ppm). OSHA is attempting to reduce this standard to 50 ppm[206].

Work environment hygiene standards (all in terms of 8 hour time weighted averages) for perchloroethylene reported by Winell[207] are as follows (mg/m^3): Federal Republic of Germany, 670; the German Democratic Republic, 300; Sweden, 200; Czeckoslovakia, 250. The MAC for perchloroethylene in the USSR is 10 mg/m^3.

Depending on its source strength, meteorological dilution, sunlight intensity, and the presence of other trace constituents, perchloroethylene or its predominant product, phosgene, accordingly may or may not be observed in non-urban areas[171]. It was estimated that an ambient concentration of 10 ppb perchloroethylene observed in New York City should lead to the formation of 12 ppb phosgene[171] (TLV=100 ppb). In 1974, at 8 locations in 5 industrial states, the concentrations of perchloroethylene ranged from 1.2 ppb in the urban area to less than 0.02 ppb in rural areas. Perchloroethylene was measured at concentrations exceeding 0.06 ppb at least 50% of the time at all locations[171]. Pearson and McConnell[208] reported that city atmospheres contain from less than 0.1 to up to 10 ppb (<0.68-68 μg/m^3) of perchloroethylene. The air in the center of Munich was found to be contaminated with 0.88 ppb (6μg/m^3) of perchloroethylene in 1975, while the suburbs of Munich had an ambient concentration of 0.59 ppb (4 μg/m^3)[209]. Populations living in urban atmospheres polluted by 6 μg/m^3 of perchloroethylene were calculated to breathe in daily levels of about 90 μg[210].

Persons consuming three liters of tap water contaminated by 28.3 ppb of the solvent as described by Giger and Molnar[211], would ingest 85 μg perchloroethylene/day[210].

Chlorination at sewage treatment plants has resulted in slightly enhanced levels of perchloroethylene in water[212,176]. Similarly to trichloroethylene, perchloroethylene (5 μg/l) has been found in the organic constituents of Mississippi River water and in the organic constituents of commercial deionized charcoal-filtered water[177].

Pearson and McConnell[208] reported the concentration of perchloroethylene in animals to range from 0.5 to 50 ppb. The levels in tissues of animals at the upper end of food chains were elevated 100 times the environmental concentrations at most. This was considered a weak bioaccumulation potential as compared to other halogen compounds. The accumulation coefficient is the quotient from the uptake rate and clearance. Neeley et al[213] found an accumulation ratio of perchloroethylene of 39.6 ± 5.5. The respective value of hexachlorobenzene (HCB) is 7880.

Perchloroethylene has very recently reported to be carcinogenic in NCI studies, producing liver hepatocellular tumors in B6C3F1 hybrid male and female mice when tested at MTD and ½ MTD dose levels in corn oil solution by gavage[214,215]. No carcinogenic activity was observed in analogously treated Osborne-Mendel rats of both sexes[214].

Perchloroethylene has not been found to be carcinogenic in inhalation studies with rabbits, mice[216], rats, guinea pigs and monkeys. Perchloroethylene showed dose-dependent mutagenicity at concentrations of 0.01, 0.1 and 1 mg/ml in S. typhimurium TA 100 without metabolic activation[188].

Perchloroethylene (0.6 mM) as well as the cis- and trans-isomers of 1,2-dichloroethylene were found to be non-mutagenic when tested in the metabolizing in vitro system with E. coli K12[78,217]. The mutagenicity of vinyl chloride, vinylidene chloride, trichloroethylene, in the above test system was attributed to their initially forming unstable oxiranes, whereas halocarbons such as perchloroethylene and cis- and trans-1,2-dichloroethylene which form much more stable oxiranes were non-mutagenic[78,192,217].

The metabolic formation of trichloroacetic acid can be explained by the primary formation of the oxirane and subsequent rearrangement to trichloroacetyl chloride and its subsequent hydrolysis.

In a recent review, Henscher[217] contrasted the metabolism and mutagenicity of chlorinated olefins. In this class of molecules, the electron-withdrawal effect dominates over the mesomeric donator effect of the involved carbon atom hence decreasing the electron density in the double bond which in turn results in a chemical stabilization against electrophilic attack[192]. It is optimal in perchloroethylene as has been demonstrated by the reactivity of chlorinated ethylenes with ozone[218].

As has been previously cited[108,192,217], the chlorinated ethylenes may undergo a variety of reactions, e.g., (1) reaction with nucleophilic cellular macromolecules under alkylation; (2) conjugation with low molecular nucleophiles (mainly glutathione) both enzymatically and non-enzymatically; (3) hydrolysis to diols, with and without the catalytic action of enzymes such as epoxide hydrase; and (4) intramolecular rearrangement. The latter reaction according to Henschler[217] represents a deactivation mechanism and is of considerable importance for the potential of acute toxicity as well as of carcinogenicity and mutagenicity of the different members of the series of chlorinated ethylenes.

5. Chloroprene

Chloroprene (2-chlorobutadiene; 2-chloro-1,3-butadiene; beta-chloroprene; $CH_2=C-CH=CH_2$) is the monomer for neoprene, the specialty rubber. It is prepared by two major routes: (1) the dimerization of acetylene to monovinylacetylene and addition of hydrogen chloride and (2) the chlorination of butadiene to a mixture of dichlorobutenes, from which 3,4-dichloro-1-butene is isolated and then is subjected to dehydrochlorination[219,220]. The latter method is believed to be the basis of the current U.S. production of chloroprene. In this procedure, butadiene is first reacted with chlorine to yield a mixture of dichlorobutene isomers from which the 3,4-dichlorobutene-1 isomer is isolated and then reacted with caustic to form chloroprene. 1,4-Dichlorobutene-2, the other isomer, can either be isomerized to 3,4-dichlorobutene-1 for additional chloroprene production or it can be utilized for in the production of adiponitrile[220].

A typical specification for chloroprene made from butadiene is as follows: chloroprene, 98.5% min., 1-chlorobutadiene, 1.0% max., aldehydes (as acetaldehydes), 0.2% max., 3,4-dichlorobutene-1, 0.01% max., dimers, 0.01% max., peroxides, 1 ppm max., and no detectable amount of vinyl acetylene[221].

In 1976, two U.S. companies produced an estimated 164 million kg of chloroprene, while three Japanese companies produced a total of 80 million kg (70% of chloroprene is based on acetylene and 30% on butadiene). The total Western European production in 1977 amounted to an estimated 100 million kg of chloroprene. The total world production of chloroprene in 1977 is estimated to have been 300 million kg[220].

Chloroprene is extremely reactive, e.g., it can polymerize spontaneously at room temperatures, the process being catalyzed by light, peroxides, and other free radical initiators. It can also react with oxygen to form polymeric peroxides and because of its instability, flammability and toxicity, chloroprene has no end product uses. It is used exclusively and without isolation in the production of neoprene elastomers[222].

Neoprene, as obtained by emulsion polymerization of chloroprene, consists mainly of transpolychloroprene. There are two main classes, the sulfur modified type and the non-sulfur modified type, indicating the differences in polymerization techniques[222,223]

About 100 million kg of neoprene were consumed in the U.S. in 1976 with the following consumption pattern: the production of industrial and automotive rubber goods, 63%; wire and cable applications, 13%; construction applications, 10%; adhesive applications, 8% and miscellaneous uses, 6%[220].

An estimated 2,500 workers are currently exposed to chloroprene in the U.S.[222,224].

The TLV for chloroprene in the U.S. is 25 ppm (90 mg/m^3). In August, 1977, NIOSH recommended that occupational exposure to chloroprene be limited to a maximum concentration of 1 ppm (3.6 mg/m^3) in air determined as a ceiling for a 15 minute period during a 40-hour work week[225].

Work environment hygiene standards (all in terms of 8-hour time weighted averages) for chloroprene reported by Winell[207] are as follows (in mg/m^3): Federal Republic of Germany, 90; the German Democratic Republic, 10; Sweden, 90 and Czeckoslovakia, 50. The MAC of chloroprene in the USSR is 2 mg/m^3.

Chloroprene has been detected as an impurity in commercial vinyl chloride in Italy[226], Japan[227] and in acrylonitrile in the USSR[228].

Chloroprene concentrations of 14.5-53.4 mg/m^3 have been reported in the air inside a Russian neoprene rubber plant; 0.2-1.57 mg/m^3 500 meters from the plant and 0.12-0.38 mg/m^3 7000 meters from the plant[229]. In another neoprene rubber plant in the USSR, the chloroprene concentration in air in the immediate vicinity was 28.45 mg/m^3; 0.727 mg/m^3 500 meters from the plant, and 0.199 mg/m^3 7000 meters away[230]. Workers in a Russian shoe factory have been reported to be often exposed to chloroprene concentrations of 20-25 mg/m^3[231].

Chloroprene which has been used since 1930 in the manufacture of synthetic rubber has recently been suggested to be responsible for the increased incidence of skin and lung cancer in exposed workers in the USSR[232,233]. During the period 1956-1970 epidemiological studies of industrial workers in the Yerevan region revealed 137 cases of skin cancer among approximately 25,000 workers over 25 years of age. Each of the 25,000 workers was classified as belonging to one of the following groups: I. Never worked in industrial plants; II. Persons working in non-chemical industries; III Persons working in chemical industries but not exposed to chloroprene or its derivatives; IV. Persons working in industries using chloroprene derivatives and V. Persons with extended work experience in chloroprene production. The incidence of skin cancer within these groups in order from I to V revealed a striking gradient, e.g., 0.12%, 0.40%, 0.66%, 1.60% and 3.0%. The study also indicated a gradient in the average age of the skin cancer cases and the average duration of employment, with workers exposed to chloroprene or the derivatives showing the lowest values. The chloroprene workers who developed skin cancer had an average age of 59.6 years and an average duration of employment of 9.5 years[233].

During the same period, 87 cases of lung cancer were identified among 19,979 workers in the same region. The group exposed to chloroprene or its derivatives had the highest incidence of lung cancer (1.16%). These workers' average age was 44.5 years with an average duration of employment of 8.7 years. Of the 34 cases of lung cancer in this group, 18 were among persons having a direct and prolonged exposure to chloroprene monomer, the remaining 16 involved individuals exposed to chloroprene latexes[232]. The frequency ratio of primary lung cancer occurrence among the comparison or control groups compared to the chloroprene groups was: 2.67 times lower in workers with chemicals unrelated to chloroprene; 6.3 times lower in workers in non-chemical industries and 17.5 times lower in workers in cultural and civic institutions.

The limitations of the above two studies[232,233] have been recently cited by IARC[220] and include: failure to distinguish prevalent from incident cases, to document completeness of case ascertainment among the exposure group, to adjust for effect of age and sex, to measure the extent of exposure, to control for the potential confounding effect of smoking and to furnish pathology information on cell type (particularly important in the study of reported skin cancer).

A study of cancer mortality among two cohorts of males engaged in the production and polymerization of chloroprene in the U.S. was recently reported by Pell[234]. One cohort consisted of 270 men first exposed between 1931-1948 and the other of 1576 men first exposed between 1942-1957. The number of lung cancer deaths in each cohort (3 in the first and 16 in the second) were about the same as was expected on the basis of U.S. or company wide rates. However, the risk of digestive cancer (19 versus 13.3) and of lymphatic and hematopoietic cancer (7 versus 4.5) were slightly elevated when contrasted with company wide experience.

Among maintenance mechanics in the study cohort of 1576, there were 8 lung cancer cases (4 living and 4 deceased) which accounted for approximately 37% of the lung cancers found in the total study cohort. In contrast, only 17% of the total cohort was composed of maintenance mechanics.

It must be noted that these mechanics whose tasks include the general maintenance in the reactor areas, installation of equipment and the replacement of leaking pipe fittings would be expected to have a potential for exposure (perhaps to high levels) to chloroprene[220]. A number of limitations to the above Pell study[234] were enumerated[220]. These include: no data were presented on any potential confounding variables such as smoking history and other occupational exposure; no specific exposure information based on chemical measurements was provided; no data presented on cell type analysis of the malignancies; and methodological shortcomings of combining of workers engaged in chloroprene monomer production and those in polymerization. There may also have been selective removal of high risk, high exposure workers from the cohort examined, in that retirees, disabled workers and former chloroprene individuals exposed in job categories not <u>actively</u> involving chloroprene were not included in the inception cohort. Additionally, the major limitations in interpreting this study were cited by IARC[220] to be: (a) the period of follow-up for the cohort is still quite incomplete for an adequate latent period and thus there may eventually be demonstration of statistically significant excesses and (b) the power of this study is limited due to the small number of person-years of exposure.

No carcinogenic effects of chloroprene have been noted to date in animal studies[220] involving oral[235] and intratrachael[235] administration to rats and dermal application[235,236] and subcutaneous injection[235,236] of chloroprene to mice and rats. A number of additional studies are in progress to investigate the carcinogenicity by oral administrations to rats and by inhalation exposure in rats and hamsters[237-239].

Exposure of S. typhimurium TA 100 and TA 1530 strain to 0.5-8% of chloroprene vapor in air in the absence of any metabolic activation system caused a linear increasing mutagenic response, (reaching 3 times the spontaneous mutations rate at a concentration of 8%)[149]. Exposure to a higher concentration (20%) caused a strong toxicity in the bacteria. This mutagenic and/or toxic effect could be caused by a direct action of chloroprene or more likely, by one of its enzymic (bacteria) or non enzymic breakdown products. Up to a 3-fold increased mutagenic response was found when a fortified 9000 g liver supernatant from either phenobarbitone-treated or untreated mice was added to such assays supporting an enzymic formation of mutagenic metabolite(s) from chloroprene[149] (probably an oxirane (epoxide) in analogy with vinyl chloride, vinylidene chloride, and trichloroethylene)[10,149]. Liver supernatant from some human biopsies also enhanced mutagenicity of chloroprene[10].

Treatment of male Drosophila for 3 days with 5.7 mM and 11.4 mM chloroprene resulted in an increase of X-linked recessive lethal mutations from 0.8 ± 0.04% in the control to 0.58 ± 0.3% and 1.0 ± 0.4% respectively[240].

Vapor of chloroprene (0.04-1.0 ppm)[241-244] induced dominant lethal mutations in sperm and chromosome aberrations in bone marrow cells of rats and of mice exposed to 1.83-3.5 mg/m^3[243]. Mixtures of chloroprene and methyl methacrylate[24,246] chloroprene, dodecylmercaptan and ammonia[245] also induced chromosomal aberrations in bone-marrow cells of rats.

An increase of chromosomal aberrations have been reported in cultured peripheral lymphocytes from workers occupationally exposed to chloroprene[243,244,247] or chloroprene and methyl methacrylate[245]. For example, Katosova[247] noted a significant rise in the number of chromosome aberrations in blood cultures of workers exposed to an average chloroprene concentration of 18 ppm for 2 to more than 10 years. In addition, decreases in motility and number of sperm were noted in exposed male workers as well as a three-fold excess of miscarriages in the wives of chloroprene workers have also bee noted[224].

Testicular atrophy and reduction in the numbers and mobility of sperm in rats with non-atrophoid testicles have been noted to an exposure level of chloroprene down to 0.06 ppm[241,244], while spermatogenesis in C57B1/6 mice was affected after 2 months exposure to 0.32 ppm chloroprene[243,248].

Sanotiskii[243] reported that the threshold for chronic effects of chloroprene on animals based upon the indicators of general systemic effect is 1.69 ± 0.087 mg/m^3, about the same as the maximum permissible concentration (MPC) in the USSR formerly adopted. The threshold concentration based on specific indicators (e.g., embryotropic, gonadotropic and mutagenic effects) was 0.15 ± 0.0059 mg/m^3 (e.g., one order of magnitude below the former MPC). Neither embryotoxic nor teratological effects have been noted after exposure of pregnant rats to 90.5 mg/m^3 (25 ppm) of chloroprene 4 hours/day from day 1 until day 12 and day 3 until day 20 of gestation[249]. The biotransformation of chloroprene has been postulated by Haley[250] to occur in an analogous fashion to vinyl chloride and vinylidene chloride, e.g., via the formation of the epoxide by the action of mixed function oxidases. The intermediate epoxide would give rise to the aldehyde or combine with glutathione and subsequently form a mercapturic acid derivative. The known oxidation of chloroprene in positions 1 and 2 as well as the decreased tissue -SH content would appear to lend some support to this postulated biotransformation.

6. Trans-1,4-dichlorobutene (1,4-dichloro-2-butene; $H\overset{Cl}{\underset{H}{C}}-C=C-\overset{H}{\underset{Cl}{C}}-H$) is employed in the U.S. mainly as an intermediate in the manufacture of hexamethylenediamine and chloroprene. Hexamethylenediamine is further used as a chemical intermediate in the production of nylon 66 and 612 polyamide resins, while chloroprene is used in the production of polychloroprene rubber. While the U.S. production of hexa-methylenediamine and polychloroprene rubber in 1975 was 340 and 143.9 million kg respectively, the percentage originally derived from trans-1,4-dichlorobutene is not known[251].

Trans-1,4-dichlorobutene has been shown to be weakly carcinogenic by sub-cutaneous and intraperitoneal administration in ICR/HA Swiss mice but not carcinogenic in mice via skin application[252].

Trans-1,4-dichlorobutene produced mutations in S. typhimurium TA 100 strains[253] with the mutagenic effect enhanced by liver microsomal fractions from mouse or humans. It has also been reported mutagenic in E. coli[254] and S. cerevisiae[255].

It has been suggested that trans-1,4-dichlorobutene-2 could conceivably be metabolized to an epoxide intermediate which is analogous in structure to open-chain β-chloroethers[252].

7. Hexachlorobutadiene (1,1,2,3,4,4-hexachlorobutadiene; HCBD; perchloro-butadiene; $Cl_2-C=\overset{Cl}{C}-\overset{Cl}{C}=CCl_2$) is normally obtained in commercial quantities as a by-product in some chlorinated hydrocarbon processes (e.g., perchloroethylene production). It is found in the tarry wastes (HEX wastes) along with hexachlorobenzene, hexachloro-ethane and other chlorinated by-products[256]. In 1974, although no HCBD was produced in the United States, 200,000 to 500,000 pounds were reported imported in the same year[256]. The production of perchloroethylene, trichloroethylene, carbon tetrachloride and chlorine in the U.S. in 1972 produced HCBD (thousands of pounds) as follows: 8,670; 3,000; 2,790; and 70[255]. The production of perchloroethylene, trichloroethylene, and carbon tetrachloride accounts for 99% of the HCBD in the United States[256].

Approximately 10 million pounds of HCBD and 5 million pounds of hexachlorobenzene (HCB) are generated annual as hex waste in the United States[257].

The largest use for HCBD in the U.S. is for the recovery of "snift" or chlorine-containing gas in chlorine plants. HCBD is also used as a chemical intermediate to produce lubricants, as a solvent and in heat transfer and hydraulic fluids.

HCBD (analogous to HCB) is highly resistant to chemical, biological and physical degradation and hence is a stable environmental pollutant.

HCBD in the ppb range has been found in water, soil, selected aquatic organisms[258] and food[257,259] (fish, eggs, milk, vegetables) samples taken along the lower Mississippi River in Louisiana.

There is a paucity of carcinogenicity and mutagenicity information on hexachloro-butadiene. In one limited study, no tumors were found in rats after 6 months administration of HCBD at levels of 2-7 mg/kg in the diet[260].

Recent chronic toxicological studies at Dow Chemical Co. in the U.S.[261,262] suggest the possibility of a threshold level for HCBD. For example, in a study where male and female Sprague-Dawley rats were maintained on diets containing 20, 2.0, 0.2 mg/kg/day of HCBD for two years, the lowest dosage caused no observed adverse effects. Ingestion of the intermediate dose level of 2.0 mg/kg/day caused some degree of toxicity, affecting primarily the kidney in which increased renal tubular epithelial hyperplasia as well as an increase of urinary excretion of coproporphyrin was noted. Ingestion of the highest dose level (20 mg/kg/day) resulted in renal tubular adenomas and adenocarcinomas, some of which metastarized to the lung[261].

A review of health data on Dow employees working in areas where HCBD has been found, revealed no abnormalities which could be attributed to the chemical. It was also noted that wastes containing the material are being recycled or incinerated in specially designed facilities[262].

HCBD has been found mutagenic (with and without activation) in Salmonella typhimurium TA 1535 and TA 100[263]. Schwetz et al[264] described the results of a reproduction study in Sprague-Dawley fed diets containing 0.2, 2.0, or 20 mg/kg/day for 90 days prior to mating, 15 days during mating, and subsequently throughout gestation and lactation. Signs of toxicity among the adult rats were observed at the two higher dose levels and included decreased weight gain and food consumption as well as alterations in kidney structure. There was no effect on pregnancy or neonatal survival and development. No toxic effects were observed among the adults at a dose level of 0.2 mg/kg/day or among the neonates at dose levels of 0.2 or 2.0 mg/kg/day[264].

8. Allyl Chloride (3-chloro-1-propene; 3-chloro-propylene; chlorallylene; $CH_2=CHCH_2Cl$) is the most important of all commercial allyl compounds. It is reactive, both as an organic halide and as an olefin. In contrast to the vinyl halides which are characteristically inert in either S_N1 or S_N2 nucleophilic substitution reactions, allyl halides are very reactive, much more than corresponding saturated compounds in both S_N1 and S_N2 reactions. The double bond facilities breaking the bond to the functional group in displacement and substitution reactions. In addition, the allyl group when introduced into other molecules, is usually reactive hence permitting many syntheses of potential commercial interest[265].

Although other reactions for the preparation of allyl chloride are known, the high-temperature substitutive chlorination of propylene is believed to be the only route used commercially at present[265]. Allyl chloride is mainly used as a monomer in the production of various plastics and resins that are used per se or incorporated into surface coatings adhesives, etc. A number of useful specialty resins are derived from allyl esters and polyesters that may be made directly from allyl chloride. Resin uses also include a number of copolymers and inter-polymers of allyl chloride with acrylonitrile, vinylidene cyanide, styrene and diallyl esters developed to provide special properties, while allyl chloride also serves as a catalyst or modifier in the production of other resins.

A number of commercially important compounds are made directly from allyl chloride, e.g., "first-generation derivatives", glycerol, epichlorohydrin and allyl alcohol. Medicinal derivatives of allyl chloride include allyl-substituted barbiturates, and mercury diuretics derived from allylamine, as well as the anesthetic, cyclopropane[265].

The production of allyl chloride in the U.S. in 1973 was estimated to be about 300 million pounds most of it produced by two manufacturers[266]. No data are available on the amounts of allyl chloride produced elsewhere.

NIOSH estimates that approximately 5000 workers are potentially exposed to allyl chloride during its manufacture or use. NIOSH has submitted to OSHA a criteria document recommending adherence to the present Federal standard of 1 ppm of allyl chloride as a time-weighted average for up to a 10-hour workday, 40-hour work week and proposed the addition of a 3 ppm ceiling concentration for any 15-minute period[266].

The MAC[267] for allyl chloride in the USSR is 3 mg/m^3.

Data are sparse concerning environmental exposure levels of allyl chloride. In one study of an allyl chloride production facility in the USSR, the concentration of allyl chloride in air of production-working areas varied between 6.4-140 mg/m^3. Workers in this facility showed early renal function impairment, higher glomerular filtration of creatinine (40%) and urea (50%) considerable hypernitremia, moderate kaliemia, higher blood Cl$^-$ (80%) and dysproteinemia[267].

There is a paucity of information regarding the chronic effects of allyl chloride. In a limited study, allyl chloride did not induce tumors in rats, guinea pigs and rabbits exposed to 3 ppm of the agent by inhalation for up to six months[268].

In the standard Salmonella mutagenicity assay in which the bacteria as well as the test agent are incorporated in the agar overlay[269], allyl chloride did not exhibit any significant mutagenic activity[270]. This initial inactivity of allyl chloride under the above conditions can be ascribed to its volatility. When steps are taken to minimize dissipation of allyl chloride vapors into the atmosphere (e.g., via impregnation of filter discs with the test material then placing them on the surface of the agar plates containing the microorganisms, and the plates then sealed in separate plastic bags) mutagenic activity of allyl chloride (0.1, 1 and 10 µl/plate) was then demonstrable for S. typhimurium TA 100 and TA 1535, but not TA 1538. Mutagenic activity was not significantly enhanced by microsome preparations derived from the livers of rats induced with Aroclor[269]. Hence these findings suggest that allyl chloride acts as a direct acting base-substitution mutagen since it affects only those strains TA 1535 and to some extent TA 100 capable of detecting such mutations[270].

Allyl chloride (at 10 µl) is mutagenic in the E. coli DNA polymerase deficient (E. coli pol A$^+$/pol A$^-$) disc test[271,272] procedure, preferentially inhibiting the growth of the pol A-, strain which is indicative of DNA modifying activity[270]. Allyl chloride (at levels of 18.4, 24.5 and 30.7 x 10^{-5}M) induces gene conversion in S. cerevisiae D4[270].

Aspects of the biotransformation of allyl halides have been described by Kaye et al[273]. Allyl mercapturic acid, its sulphoxide, and 3-hydroxypropylmercapturic acid were identified as urinary metabolites following subcutaneous administration of allyl chloride, bromide and iodide to rats[273]. S-allyl glutathione and S-allyl-2-cysteine were also detected in the bile of a rat dosed with allyl chloride. The study of Kaye et al[272] did not establish with certainty the pathway or pathways whereby allyl halides give rise to the formation of 3-hydroxypropylmercapturic acid. There are a number of pathways by which this may occur because the allyl halide can undergo reaction either at the double bond or at the site of the halogen atom. Although it is not known whether metabolic formation of a 3-halogenopropanol ($HOCH_2CH_2CH_2X$) from an allyl halide can occur, if such a reaction did take place it would probably be followed by the formation of 3-hydroxypropylmercapturic acid since the excretion of this mercapturic acid has been reported following the administration of 3-chloropropanol to rats[274]. The conversion of allyl chloride to S-allyl glutathione and the conversion of the latter compound to allyl mercapturic acid was demonstrated by the work of Kaye et al[272]. Additional potentially possible metabolic pathways for allyl chloride could involve the epoxidation of the double bond to form epichlorohydrin (e.g., $CH_2=CHCH_2Cl \rightarrow \underset{\underset{O}{\diagdown \diagup}}{CH_2\text{-}CHCH_2Cl}$) a known carcinogen and mutagen. The subsequent oxidation products of epichlorohydrin are glycidol ($\underset{\underset{O}{\diagdown \diagup}}{H_2C\text{-}CH}\text{-}CH_2OH$) and glycidaldehyde ($\underset{\underset{O}{\diagdown \diagup}}{H_2C\text{-}CHCHO}$) both are mutagenic and the latter is carcinogenic as well. The second metabolic pathway involves conversion to allyl alcohol, then acrolein, then acrylic acid, viz., $H_2C=CH_2CH_2Cl \rightarrow H_2C=CHCH_2OH \rightarrow CH_2=CHCHO \rightarrow CH_2=CHCOOH$. Acrolein (vinyl aldehyde) is mutagenic in <u>Drosophila</u> and <u>S. typhimurium</u> (strains TA 1538 and TA 98).

References

1. Environmental Protection Agency, Scientific and technical assessment report on vinyl chloride and polyvinyl chloride, Office of Research and Development, EPA, Washington, DC, June, 1975
2. IARC, Monographs on the Evaluation of Carcinogenic Risk of Chemicals to Man, Vol. 7, International Agency for Research on Cancer, Lyon (1974) 291-310
3. IARC, Monographs on the Evaluation of Carcinogenic Risk of Chemicals to Man, Vol. 17, International Agency for Reserach on Cancer, Lyon, in press
4. Sakarnis, V. F., 1,2-Dichloroethane-induced chromosome non-divergence of the X-chromosome and recessive sex-linked lethal mutation in Drosophila melanogaster Genetika, 5 (1969) 89-95
5. Rapoport, I. A., The reaction of genetic proteins with 1,2-dichloroethane, Akad. Nauk. SSR Dokl. Biol. Sci., 134 (1960) 745-747
6. McCann, J., Spingarn, N. E., Kobori, J., and Ames, B. N., Carcinogens as mutagens: Bacterial tester strains with R factor plasmids, Proc. Natl. Acad. Sci., 72 (1975) 979-983
7. McCann, J., Simon, V., Streitweiser, D., and Ames, B. N., Mutagenicity of chloroacetaldehyde, a possible metabolic product of 1,2-dichloroethane, chloroethanol, and cyclophosphamide, Proc. Natl. Acad. Sci., 72 (1975) 3190-3193
8. Brem, H., Stein, A. B., and Rosenkranz, H. S., The mutagenicity and DNA-modifying effect of haloalkanes, Cancer Res., 34 (1974) 2576-2579
9. U.S. International Trade Commission, Synthetic Organic Chemicals, U.S. Production and Sales, 1976, USITC Publication No. 833, U.S. Government Printing Office, Washington, DC (1977) pp. 183, 187, 303, 332
10. Bartsch, H., and Montesano, R., Mutagenic and carcinogenic effects of vinyl chloride, Mutation Res., 32 (1975) 93-114
11. Levinson, C., Vinyl chloride: A case study of the new occupational health hazard, ICF (Publ.) Geneva (1974)
12. IARC, Report on a working group on vinyl chloride, Lyon, IARC Internal Technical Report No. 74/005 (1974)
13. Keane, D. P., Stehaugh, R. B., and Townsend, P. L., Vinyl chloride: How, where, who-future, Hydrocarbon Processing, February (1973) 99-110
14. Bertram, C. G., Minimizing emissions from vinyl chloride plants, Environ. Sci. Technol., 11 (1977) 864-868
15. U.S. Federal Register, Part 61, National Emission Standards for Hazardous Air Pollutants, Vol. 41, No. 205, U.S. Government Printing Office, Washington, DC (1976) pp. 46560-46573
16. Gordon, S. J., and Meeks, S. A., A study of the gaseous pollutants in the Houston, Texas area, AICHE Symp Ser., 73 (1977) 84-94
17. U.S. Federal Register, National Emission Standards for Hazardous Air Pollutants, U.S. Government Printing Office, Washington, DC 42 (1977) 28154-28159
18. Cook, W. A., Giever, P. M., Dinman, B. D., and Magnuson, H. J., Occupational acroosteolysis, II. An industrial hygiene study, Arch. Env. Hlth., 22 (1971) 74-82
19. Lange, C. E., Juhe, S., Stein, G., and Veltman, G., Die sogenannte vinyl chloride krankheit, eine berufsbeding systemkierose? Int. Arc. Arbeits Med., 32 (1974) 1-32
20. Filatova, V. S., and Gronsberg, E. S., Sanitary hygienic conditions of work in the production of polychlorvinyl tar and measures of improvement, Gig. Sanit. 22 (1957) 38-42

21. Murdock, I. A., and Hammond, A. R., A practical method for the measurement of vinyl chloride monomer (VCM) in air, Ann. Occup. Hyg., 20 (1977) 55-61
22. Orusev, T., Popovski, P., Bauer, S., and Nikolova, K., Occupational risk in the production of poly(vinyl chloride), God. Zb. Med. Fak. Skopje, 22 (1976) 33-38
23. Bol'shakov, A. M., Working conditions in the production of synthetic leather, Gig. Vop. Proizvod. Primen. Polim. Mater. (1969) 47-52
24. Heath, C. W., Jr., Falk, H., and Creech, J. L., Jr., Characteristics of cases of angiosarcoma of the liver among vinyl chloride workers in the United States, Ann. NY Acad. Sci. (1974)
25. Baretta, E. D., Stewart, R. D., and Mutchler, J. E., Monitoring exposures to vinyl chloride vapor: Breath analysis and continuous air sampling, Air Ind. Hyg. Ass. J., 30 (1969) 537-544
26. Milby, T. H., Cancer Control Monograph: Vinyl Chloride (Draft Final Report), U.S. Dept. Health Education and Welfare, National Cancer Institute, Washington, DC (1977) pp. 18-22, 113
27. Anon, CIA argues case against zero VCM exposure limits, European Chem. News, May 24 (1974) p. 24
28. U.S. Federal Register, Vinyl chloride polymers in contact with food, 40, No. 171 U.S. Government Printing Office, Washington, DC, (1975) 40529-40537
29. Saggessee, M. F., Wakeman, I. B., and Owens, F. V., PVC with no VCM, Modern Packaging, Sept. (1976) pp. 19-21, 62
30. Shackelford, W. M., and Keith, L. H., Frequency of organic compounds identified in wate,r EPA 600/4/76-002, U.S. Environmental Protection Agency, Athens, GA (1976) pp. 129-130
31. Environmental Protection Agency, Preliminary Assessment of Suspected Carcinogens in Drinking Water, An Interim Report to Congress, Office of Toxic Substances, EPA Washington, DC, June (1975)
32. Lu, P. Y., Metcalf, R. L., Plummer, N., and Mandel, D., The environmental fate of three carcinogens: Benzo(a)pyrene, benzidine and vinyl chloride evaluated in laboratory model ecosystems, Arch. Environ. Contam. Toxicol., 6 (1977) 129-142
33. Roesli, M., Zimmerli, B., and Marek, B., Residues of vinyl chloride monomer in edible oils, Mitt. Geb. Lebensmittelunters. Hyg., 66 (1975) 507-511
34. Fuchs, G., Gawell, B. M., Albanus, L., and Storach, S., Vinyl chloride monomer levels in edible fats, Var Foeda, 27 (1975) 134-135
35. Anon, FDA to propose ban on use of PVC for liquor use, Food Chem. News, May 14 (1973) pp. 3,4
36. Hoffman, D., Patrianakos, C., and Brunnemann, K. D., Chromatographic determination of vinyl chloride in tobacco, Anal. Chem., 48 (1976) 47-50
37. Hedley, W. H. Cheng, J. T., McCormick, R. J., and Lewis, W. A., Sampling of automobile interiors for vinyl chloride monomer, U.S. Protection Agency, National Tech. Info. Services (NTIS), Springfield, VA (1976)
38. Maltoni, C., The value of predictive experimental bioassays in occupational and environmental carcinogenesis, Ambio, 4 (1975) 18
39. Maltoni, C., and Lefemine, G., Carcinogenicity bioassays of vinyl chloride, I. Research plans and early results, Environ. Res., 7 (1974) 387-405
40. Viola, P. L., Bigotti, A., and Caputo, A., Oncogenic response of rat skin, lungs, and bones to vinyl chloride, Cancer Res., 31 (1971) 516-519
41. Maltoni, C., La ricerca sperimentale ei suoa risultati, Sapere, 776 (1974) 42-45

42. Maltoni, C., and Lefemine, G., La potenzialita dei saggi sperimentali nella predizione dei rischi oncogeni ambientali. Un essempio: Il cloruro di vinile, Rend. Sci. Fis. Mat. Nat., 66 (1974) 1-11
43. Maltoni, C., and Lefemine, G., Carcinogenicity bioassays of vinyl chloride, current results, Ann. NY Acad. Sci., 246 (1975) 195-218
44. Keplinger, M. L., Goode, J. W., Gordon, D. E., and Calandra, J. C., Interim results of exposure of rats, hamsters, and mice to vinyl chloride, Ann. NY Acad. Sci., 246 (1975) 219-224
45. Maltoni, C., Recent findings on the carcinogenicity of chlorinated olefins, Environ. Hlth. Persp., 21 (1977) 1-5
46. Lee, C. C., Bhandari, J. C., Winston, J. M., House, W. B., Peters, P. J., Dixon, R. L., and Woods, J. S., Inhalation toxicity of vinyl chloride and vinylidene chloride, Environ. Hlth. Persp., 21 (1977) 25-32
47. Holmberg, B., Kronevi, T., and Winell, M., The pathology of vinyl chloride exposed mice, Acta. Vet. Scand., 17 (1976) 328-342
48. Caputo, A., Viola, P. L., and Bigotti, A., Oncogenicity of vinyl chloride at low concentrations in rats and rabbits, IRCS, 2 (1974) 1582
49. Creech, J. L., and Johnson, M. N., Angiosarcoma of liver in the manufacture of polyvinyl chloride, J. Occup. Med., 16 (1974) 150
50. Holmberg, B., and Molina, M. J., The industrial toxicology of vinyl chloride: A review, Work Environ. Hlth., 11 (1974) 138-144
51. Ott, M. G., Langner, R. R., and Holder, B. B., Vinyl chloride esposure in a controlled industrial environment, long-term mortality experience in 594 employees, Arch. Environ. Hlth., 30 (1975) 333-339
52. Infante, P. F., Wagoner, J. K. and Waxweiler, R. J., Carcinogenic, mutagenic and teratogenic risks associated with vinyl chloride, Mutation Res., 41 (1976) 1 131-142
53. IARC, Internal Technical Report No. 75/001, Report of a Working Group on Epidemiological Studies on Vinyl Chloride Exposed People, International Agency for Research on Cancer, Lyon (1975)
54. Monson, R. R., and Peters, J. M., Proportional mortality among vinyl chloride workers, Lancet., 1 (1974) 397-398
55. Tabershaw, I. R., and Gaffey, W. R., Mortality study of workers in the manufacture of vinyl chloride and its polymers, J. Occup. Med., 16 (1974) 509-518
56. Infante, P. F., Oncogenic and mutagenic risks in communities with polyvinyl chloride production facilities, Ann. NY Acad. Sci., 271 (1976) 49-57
57. Byren, D., Engholm, G., Englund, A., and Westerholm, P., Mortality and cancer morbidity in the group of Swedish VCM and PVC production workers, Environ. Hlth. Persp., 17 (1976) 167-170
58. Fox, A. J., and Collier, P. F., Mortality experience of workers exposed to vinyl chloride monomer in the manufacture of polyvinyl chloride in Great Britain, Brit. J. Ind. Med., 34 (1977) 1-10
59. Fox, A. J., and Collier, P. F., Low mortality rates in industrial cohort studies due to selection for work and survival in the industry, Brit. J. Prev. Soc. Med., 30 (1976) 225-230
60. Waxweiller, R. J., Stringer, W., Wagoner, J. K., Jones, J., Falk, H., and Carter, C., Neoplastic risk among workers exposed to vinyl chloride, Ann. NY Acad. Sci., 271 (1976) 40-44
61. Ducatman, V., Hirschhorn, K., and Selikoff, I. J., Vinyl chloride exposure and human chromosome aberrations, Mutation Res., 31 (1975) 163-169

62. Kilian, D. J., Picciano, D. J., and Jacobson, C. B., Industrial monitoring: A cytogenetic approach, Ann. NY Acad. Sci., 269 (1975) 4-11
63. Heath, L. W., Jr., Dumont, C. R., Gamble, J., and Waxweiler, R. J., Chromosomal damage in men occupationally exposed to vinyl chloride monomer and other chemicals, Env. Res., 14 (1977) 68-72
64. Purchase, I. F. H., Richardson, C. R., and Anderson, D., Chromosomal and dominant lethal effects of vinyl chloride, Lancet, 2 (1976) 410-411
65. Funes-Cravioto, F., Lambert, B., Lindsten, J., Ehrenberg, L., Natarahan, A. T., and Osterman-Golkar, S., Chromosome aberrations in workers exposed to vinyl chloride, Lancet, 1 (1975) 459
66. Leonard, A., Deat, G., Leonard, E. D., Lefevre, M. J., Ducuypen, L. J., and Nicaise, C. I., Cytogenetic investigations on lymphocytes from workers exposed to vinyl chloride, J. Toxicol. Environ. Hlth., 2 (1977) 1135-1142
67. Hansteen, I. L., Hillestad, L., and Thiis-Evensen, E., Chromosome studies on workers exposed to vinyl chloride, Mutation Res., 38 (1976) 112
68. Szentesi, I., Hornyak, E., Unguary, G., Czeizel, A., Bognar, Z., and Timer, M., High rate of chromosomal aberration in PVC workers, Mutation Res., 37 (1976) 313-316
69. Infante, P. F., Wagoner, J. K., McMichael, A. J., Waxweiler, R. J., and Falk, H., Genetic risks of vinyl chloride, Lancet, 1 (1976) 734-735
70. Bartsch, H., Malaveille, C., Barbin, A., Tomatis, L., and Montesano, R., Mutagenicity and metabolism of vinyl chloride and related chompounds, Environ. Hlth. Persp., 17 (1976) 193
71. Fishbein, L., Industrial mutagens and potential mutagens, I. Halogenated derivatives, Mutation Res., 32 (1976) 267-308
72. Bartsch, H., Malaveille, C., and Montesano, R., Human, rat and mouse liver-mediated mutagenicity of vinyl chloride in S. typhimurium strains, Int. J. Cancer, 15 (1975) 429-437
73. Rannug, U., Johansson, A., Ramel, C., and Wachtmeister, C. A., The mutagenicity of vinyl chloride after metabolic activaton, Ambio, 3 (1974) 194-197
74. McCann, J., Simon, V., Streitsweiser, D., and Ames, B. N., Mutagenicity of chloroacetaldehyde, A possible metabolic product of 1,2-dichloroethane, chloroethanol, and cyclophosphamide, Proc. Natl. Acad. Sci., 72 (1975) 3190-3193
75. Malaveille, C., Bartsch, H., Barbin, A., Camus, A. M., and Montesano, R., Mutagenicity of vinyl chloride, chloroethylene oxide, chloroacetaldehyde and chloroethanol, Biochem. Biophys. Res. Commun., 63 (1975) 363-370
76. Garro, A. J., Guttenplan, J. B., and Milvy, P., Vinyl chloride dependent mutagenesis: Effects of liver extracts and free radicals, Mutation Res., 38 (1976) 81-88
77. Andrews, A. W., Zawistowski, E. S., and Valentine, C. R., A comparison of the mutagenic properties of vinyl chloride and methyl chloride, Mutation Res., 40 (1976) 273-276
78. Greim, H., Bonse, G., Radwan, Z., Reichert, D., and Henschler, D., Mutagenicity in vitro and potential carcinogenicity of chlorinated ethylene as a function of metabolic oxirane formation, Biochem. Pharmacol., 24 (1975) 2013-2017
79. Loprieno, N., Barale, R., Baroncelli, S., Bauer, C., Bronzetti, G., Cammellini, A., et al., Evaluation of the genetic effects induced by vinyl chloride monomer (VCM) under mammalian metabolic activation: studies in vitro and in vivo, Mutation Res., 40 (1976) 85-96
80. Loprieno, N., Barale, R., Baroncelli, S., Bartsch, H., Bronzetti, G., et al., Induction of gene mutations and gene conversions by vinyl chloride metabolites, in yeast, Cancer Res., 36 (1977) 253-257

81. Drozdowicz, B. Z., and Huang, P. C., Lack of mutagenicity of vinyl chloride in two strains of Neurospora crassa, Mutation Res., 48 (1977) 43-50
82. Magnusson, J., and Ramel, Mutagenic effects of vinyl chloride in Drosophila melanogaster, Mutation Res., 38 (1976) 115
83. Verburgt, F. G., and Vogel, E., Vinyl chloride mutagenesis in Drosophila melanogaster, Mutation Res., 48 (1977) 327-336
84. Anderson, D., Hodge, M. C. E., and Purchase, I. F. H., Vinyl chloride: dominant lethal studies in male CD-1 mice, Mutation Res., 40 (1976) 359-370
85. Drevon, C., Kuroki, T., and Montesano, R., Microsome-mediated mutagenesis of a Chinese hamster cell line by various chemicals, In "Progress in Genetic Toxicology" (eds) Scott, D., Bridges, B. A., and Sobels, F. H., Elsevier/North Holland Biomedical Press, Amsterdam (1977)
86. Hussain, S., and Osterman-Golkar, S., Comment on the mutagenic effectiveness of vinyl chloride metabolites, Chem. Biol. Interactions, 12 (1976) 265-267
87. Huberman, E., Bartsch, H., and Sachs, L., Mutation induction in Chinese hamster V-79 cells by two vinyl chloride metabolites, chloroethylene oxide and 2-chloroacetaldehyde, Int. J. Cancer, 16 (1975) 639-644
88. Rosenkranz, S., Carr, H. S., and Rosenkranz, H. S., 2-Haloethanols: Mutagenicity and reactivity with DNA, Mutation Res., 26 (1974) 367-370
89. Elmore, J. D., Wong, J. L., Laumbach, A. D., and Streips, U. N., Vinyl chloride, mutagenicity via the metabolites chlorooxirane and chloroacetaldehyde monomer hydrate, Biochim. Biophys. Acta, 442 (1976) 405-419
90. Walling. C., and Fredericks, P. S., Positive halogen compounds, IV. Radical reactions of chlorine and t.butyl hypochlorite with some small ring compounds, J. Am. Chem. Soc., 84 (1962) 3326-3331
91. Daly, J. W., Jerina, D. M., and Witkop, B., Arene oxides and the NIH shift, The metabolism, toxicity and carcinogenicity of aromatic compounds, Experientia, 28 (1972) 1129
92. Bartsch, H., Vinyl chloride (VCM): An example for evaluating adverse biological effects in short-term tests, Mutation Res., 46 (1977) 200-201
93. Mattern, I. E., Van Der Zwaan, W. B., and Willems, M. J., Mutagenicity testing of urine from vinyl chloride (VCM) treated rats using the Salmonella test system, Mutation Res., 46 (1977) 230-231
94. Jensen, S., Jernelov, A., Lange, R., and Palmark, K. H., In FAO Technical Conference on Marine Pollution, Fir: MP/70E.88; FAO, Rome (1970)
95. Jensen, S., Lange, R., Berge, G., Polmark, K. H., and Renberg, K., On the chemistry of EDC-tar and its biological significance, Proc. R. Soc. Lond. (Biol.) 189 (1975) 333-346
96. Rannug, U., and Ramel, C., Mutagenicity of waste products from vinyl chloride industries, J. Toxicol. Env. Hlth., 2 (1977) 1019-1029
97. Antweiler, H., Studies on the metabolism of vinyl chloride, Env. Hlth. Persp., 17 (1976) 217-219
98. Bonse, G., and Henschler, D., Chemical reactivity, biotransformation and toxicity of polychlorinated aliphatic compounds, CRC Crit. Revs. Toxicol., 4 (4) (1976) 395-409
99. Hefner, R. E., Jr., Watanabe, P. G., and Gehring, P. J., Preliminary studies on the fate of inhaled vinyl chloride monomer in rats, Ann. NY Acad. Sci., 246 (1975) 133-148
100. Hefner, R. E., Jr., Watanabe, P. G., and Gehring, P. J., Preliminary studies of the fate of inhaled vinyl chloride monomer (VCM) in rats, Env. Hlth Persp. 11 (1975) 85-95

101. Green, T., and Hathaway, D. E., The chemistry and biogenesis of the S-containing metabolites of vinyl chloride in rats, Chem. Biol. Interactions, 17 (1977) 137-150
102. Green, T., and Hathaway, D. E., The biological fate of vinyl chloride in relation to its oncogenicity, Chem. Biol. Interact., 11 (1975) 545-562
103. Muller, G., and Norpoth, K., Bestimmung zweier urinmetabolite des vinyl chlorids, Naturwiss, 62 (1975) 541
104. Muller, G., Norpoth, K., and Eckard, R., Identification of two urine metabolites of vinyl chloride by GC-MS-investigations, Int. Arch. Occup. Environ. Hlth., 38 69-75
105. Watanabe, P. G., McGowan, G. R., Madrid, E. O., and Gehring, P. J. Fate of (^{14}C) vinyl chloride following inhalation exposure in rats, Toxicol. Appl. Pharmacol. 37 (1976a) 49-59
106. Watanabe, P. G., McGowan, G. R., and Gehring, P. J., Fate of (^{14}C) vinyl chloride after single oral administration in rats, Toxicol. Appl. Pharmacol., 36 (1976b) 339-35
107. Watanabe, P. G., and Gehring, P. J. (1976) Dose-dependent fate of vinyl chloride and its possible relationship to oncogenicity in rats, Env. Hlth. Persp., 17, 145-152
108. Plugge, H., and Safe, S., Vinyl chloride metabolism-A review, Chemosphere, 6 (1977) 309-325
109. Reynolds, E. S., Moslen, M. T., Szabo, S., Jaeger, R. J., and Murphys, S. D., Hepatotoxicity of vinyl chloride and 1,1-dichloroethylene, Am. J. Patho., 81 (1975a) 219-232
110. Reynolds, E. S., Moslen, M. Y., Szabo, S. and Jaeger, R. J., Vinyl chloride-induced deactivation of cytochrome P-450 and other components of the liver mixed function oxidase system: An in vivo study, Res. Comm. Chem. Pathol. Pharmacol. 12 (1975b) 685-694
111. Salmon, A. G., Cytochrome P-450 and the metabolism of vinyl chloride, Cancer Letters, 2 (1976) 109-114
112. Ivanetich, K. M., Aronson, I. & Katz, I. D., The interaction of vinyl chloride with rat hepatic microsomal cytochrome P-450 in vitro, Biochem. Biophys. Res. Commun., 74 (1977) 1411-1418
113. Fjellstedt, T. A., Allen, R. H., Duncan, B. K. & Jakoby, W. B., Enzymatic conjugation of epoxides with glutathione, J. Biol. Chem., 248 (1973)
114. Pabst, M. J., Habig, W. H., and Jakoby, W. B., Mercapturic acid formation: The several glutathione transferases of rat liver, Biochem. Biophys. Res. Commun., 52 (1973) 1123-1128
115. Barbin, A., Bresil, H., Croisy, A., Jacquignon, P., Malaveille, C., Montesano, R., and Bartsch, H., Liver-microsome-mediated formation of alkylating agents from vinyl bromide and vinyl chloride, Biochem. Biophys. Res. Commun., 67 (1975) 596-603
116. Wessling, R. A., and Edwards, F. G., Vinylidene chloride polymers In: "Encyclopedia of Polymer Science and Technology", (ed) Bikales, N. M., Vol. 14, John Wiley and Sons, New York (1971) pp. 540-579
117. Wessling, R. A., and Edwards, F. G., Poly(vinylidene chloride) In Mark, H. F., McKetta, J. J., and Othmer, D. F. (eds) Encyclopedia of Chemical Technology, 2nd ed., Vol. 21, Interscience, New York (1970) p. 275-303
118. Dow Chemical Co., Published references and literature review pertaining to toxicological properties of vinylidene chloride monomer, 1976
119. IARC, Monographs on the Evaluation of Carcinogenic Risk of Chemicals to Man, Vol.17, International Agency for Research on Cancer, Lyon (1978) in press

120. U.S. Environmental Protection Agency, Status Assessment of Toxic Chemicals. 15. Vinylidene Chloride, Industrial Environmental Research Laboratory, Environmental Protection Agency, Cincinatti, Ohio, Sept. 6 (1977)
121. Anon, Vinylidene chloride linked to cancer, Chem. Eng. News, Feb. 28 (1977) 6-7
122. Anon, Rap for film wrap, Chem. Week, Oct. 16 (1974) 20
123. ACGIH, "Threshold Limit Values for Chemical Substances in Workroom Air Adopted by ACGIH", American Conference of Governmental Industrial Hygienists, Cincinatti, Ohio (1976) p. 30
124. Ottm, M. G., Fishbeck, W. A., Townsend, J. C., and Schneider, E. J., A health study of employees exposed to vinylidene chloride, J. Occup. Med., 18 (1976) 735-738
125. Jaeger, R. J., Vinyl chloride monomer: Comments on its hepatotoxicity and interaction with 1,1-dichloroethane, Ann. NY Acad. Sci., 246 (1975) 150-151
126. Altman, P. D., and Dittmer, D. S., "Environmental Biology" Federation of American Societies for Environmental Biology, (1966)
127. Hushon, J., and Kornreich, M., Air pollution assessment of vinylidene chloride, U.S. National Technical Information Service (NTIS), Springfield, VA, Rept. No. P. B. 256,738 (1976) p. 40
128. Eurocop-Cost, "A Comprehensive List of Polluting Substances which have been Identified in Various Fresh Waters, Effluent Discharges, Aquatic Animals and Plants, and Bottom Sediments" 2nd ed., EUCO/MDU/73/76, Commission of the European Communities, Luxembourg (1976) p. 41
129. Shackelford, W. M., and Keith, L. H., "Frequency of Organic Compounds Identified in Water" EPA-600/4-16-062; U.S. Environmental Protection Agency, Athens, GA (1976) p. 130, 133, 134
130. U.S. Environmental Protection Agency "Preliminary Assessment of Suspected Carcinogens in Drinking Water", Washington, DC (1975) p. II-3
131. Kiezel, L., Liszka, M., and Rutkowski, M., Gas chromatographic determination of trace impurities in distillates of vinyl chloride monomer, Chem. Anal. (Warsaw) 20 (1975) 555-562
132. Vlasov, S. M., and Bodyagin, G. N., Gas chromatographic analysis of trichloroethylene, Tr. Khim. Khim. Tekhnol., 1 (1970) 161-162
133. Kurginyan, K. A., and Shirinyan, V. T., Identification and quantitative determination of some impurities in chloroprene, Arm. Khim. Zh., 22 (1969) 61-65
134. Birkel, T. J., Roach, J. A. G., and Sphon, J. T., Determination of vinylidene chloride in saran films by electron capture gas-solid chromatography and confirmation by mass spectrometry, J. Assoc. Offic. Anal. Chem., 60 (1977) 1210-1213
135. Lehman, A. J., Chemicals in foods: A report to the Association of Food and Drug Officials on current developments, Assoc. Food and Drug Officials, U.S. Quart. Bull., 15 (1951) 82-89
136. Anon, Food additives, packaging material for use during irradiation of pre-packaged foods, Fed. Register, 33 (1968) 4659
137. Anon, Vinylidene chloride: No trace of cancer at DOW, Chem. Eng. News, March 14 (1977) pp. 21-22
138. Haley, T. J., Vinylidene chloride: A review of the literature, Clin. Toxicol., 8 (1975) 633-643
139. U.S. Environmental Protection Agency, Health and Environmental Impacts, Task 1. Vinylidene chloride, Environmental Protection Agency, Washington, DC October (1976)
140. Maltoni, C., Recent findings on the carcinogenicity of chlorinated olefins, Environ. Hlth. Persp., 21 (1977) 1-5

141. Maltoni, C., Cotti, G., Morisi, L., and Chieco, P., Carcinogenicity bioassays of vinylidene chloride, research plans and early results, Med. Lavoro, 68 (1977) 241-262
142. Maltoni, C., Recent findings on the carcinogenicity of chlorinated olefins, NIEHS Conference on Comparative Metabolism and Toxicity of Vinyl Chloride and Related Compounds, Bethesda, MD May 2-4 (1977)
143. Viola, P. L., Carcinogenicity studies on vinylidene chloride (VDC), NIEHS Conference on Comparative Metabolism and Toxicity of Vinyl Chloride and Related Compounds, Bethesda, MD, May 2-4 (1977)
144. Rampy, L. W., Toxicity and carcinogenicity studies on vinylidene chloride, NIEHS Conference on Comparative Metabolism and Toxicity of Vinyl Chloride and Related Compounds, Bethesda, MD, May 2-4 (1977)
145. Norris, J. M., The MCA-Toxicology Program for Vinylidene Chloride, Presented at 1977 European Tappi Meeting, Hamburg, Germany, Jan. 26 (1977)
146. Rampy, L. W., Quast, J. F., Humiston, C. G., Balmer, M. F., and Schwetz, B. A., Results of two-year toxicological studies in rats of vinylidene chloride incorporated in the drinking water or administered by repeated inhalation, Abstracts of 17th Annual Meeting of Society of Toxicology, San Francisco, March 12-16 (1978) 51-53
147. Winston, J. M., Lee, C. C., Bhandari, J. C., Dixon, R. C., and Woods, J. S., A study of the carcinogenicity of inhaled vinyl chloride and vinylidene chloride (VDC) in rats and mice, Abstracts Int. Congress of Toxicology, Toronto, March 3-April 2 (1977) p. 32
148. Lee, C. C., Toxicity and carcinogenicity of vinyl chloride compared to vinylidene chloride, NIEHS Conferences on Comparative Metabolism and Toxicity of Vinyl Chloride and Related Compounds, Bethesda, MD, May 2-4 (1977)
149. Bartsch, H., Malaveille, C., Montesano, R., and Tomatis, L., Tissue-mediated mutagenicity of vinylidene chloride and 2-chlorobutadiene in Salmonella typhimurium, Nature, 255 (1975) 641-643
150. Anderson, D., Hodge, M. C. E., and Purchase, I. F. A., Dominant lethal studies with halogenated olefins vinyl chloride and vinylidene dichloride in male CD-1 mice, Environ. Hlth Persp., 21 (1977) 71-78
151. Short, R. D., Minor, J. L., Winston, J. M., and Lee, C. C., A dominant lethal study in male rats after repeated exposures to vinyl chloride or vinylidene chloride, J. Toxicol. Env. Hlth., 3 (1977) 965-968
152. Hathaway, D. E., Comparative mammalian metabolism of vinylidene and vinyl chloride in relation to outstanding oncogenic potential. NIEHS Conference on Comparative Metabolism and Toxicity of Vinyl Chloride Related Compounds, Bethesda, MD, May 2-4 (1977)
153. IARC, Monographs on the Evaluation of Carcinogenic Risk of Chemicals to Man, Vol. 11, International Agency for Research on Cancer, Lyon, (1976) 263-276
154. U.S. Environmental Protection Agency, Status Assessment of Toxic Chemicals, 13. Trichloroethylene, Industrial Environmental Research Laboratory, Environmental Protection Agency, Cincinatti, Ohio, Sept. 6 (1977)
155. U.S. International Trade Commission, Preliminary Report on U.S. Production of Selected Synthetic Organic Chemicals, August, September and Cumulative Totals, 1975, S.O.C. Series C/P-75-9, U.S. Govt. Printing Office, Washington, DC, Nov. 5 (1975) p.3
156. Anon, Solvent growth rates scrambled by strong environmental signals: hydrocarbons, ketones, Chem. Mktg. Reptr., May 8 (1972) p. 3,47
157. Anon, Trichloroethylene-report abstract, Chem. Econ. Newsletter (SRI Inst.) Nov.-Dec. (1975)

158. Lloyd, J. W., Moore, R. M., and Breslin, P., Background information on trichloroethylene, J. Occup. Med., 17 (1975) 603-605
159. Utidjian, H. M. D., Criteria for a recommended standard. Occupational exposure to trichloroethylene, I. Recommendations for a trichloroethylene standard, J. Occup. Med., 16 (1974) 192
160. Dept. of Labor, Occupational Safety and Health Administration, Trichloroethylene: Proposed occupational exposure standard (29 CFR Part 1910), Federal Register, Oct. 20 (1975) 40,49032.
161. Aviado, D. M., Simaan, J. A., Zakharis, S., and Ulsamer, A. G., Methyl chloroform and trichloroethylene in the environment, CRC Press, Cleveland, Ohio (1976) 44-89
162. Kleinfeld, M., and Tabershaw, I., Trichloroethylene toxicity, Report of five fatal cases, Arch. Ind. Hyg., 10 (1954) 134-141
163. McConnell, G., Ferguson, O. M., and Pearson, C. R., Chlorinated hydrocarbons in the environment, Endeavor, 34 (1975) 13-27
164. Garner, D. N., and Dzierlenca, P. S., Organic chemical producers data base, Vol. II, Final Report for Contract No. 68-02-1319, Task Number 15, Radian Corp., Austin, Texas, August (1976)
165. National Academy of Science "Assessing Potential Ocean Pollutants", Washington, DC (1975)
166. Radian Corporation, "Draft Report on Control Techniques for Volatile Organic Emissions from Stationary Sources" Contract No. 68-02-2608, Task 12, Austin, TX July (1977)
167. Stanford Research Institute "Chemical Economics Handbook", Menlo Park, CA November (1975)
168. Glass, W. I., A survey of trichloroethylene degreasing baths in Auckland, Occup. Hlth. Bull., December, 1961
169. Takamatsu, M., Health hazards in workers exposed to trichloroethylene vapour, Kumanoto Med. J., 15 (1962) 43-54
170. Corbett, T. H., Retention of anesthetic agents following occupational exposure, Anesth. Analg. Current Res., 52 (1973) 614-618
171. Lillian, D., Sign, H. B., Appleby, A., Lobban, L., Arnts, R., Gumpert, R., et al., Atmospheric fates of halogenated compounds, Env. Sci. Technol., 9 (1975) 1042-1048
172. Grimsrud, E. P., and Rasmussen, R. A., Survey and analysis of halocarbons in the atmosphere by gas chromatography-mass spectrometry, Atmos. Environ., 9 (1975) 1014-1017
173. Anon, Carcinogenic and other suspect substances, Toxic Materials News, 3 (17) (1976) 15
174. Murray, A. J., and Riley, J. R., Occurrence of some chlorinated aliphatic hydrocarbons in the environment, Nature, 242 (1973) 37-38
175. Environmental Protection Agency, Draft Report for Congress: Preliminary Assessment of Suspected Carcinogens in Drinking Water, Office of Toxic Substances, Washington, DC., October 17, 1975
176. Bellar, T. A., Lichtenberg, J. J., and Kroner, R. C., The occurrence of organohalides in chlorinated drinking waters, J. Am. Waterworks Assoc., 66 (1974) 703-706
177. Dowty, B. J., Carlisle, D. R., and Laseter, J. L., New Orleans drinking water sources tested by gas chromatography-mass spectrometry, Env. Sci. Technol., 9 (1975) 762-765

178. Camisa, A. G., Analysis and characteristics of trichloroethylene wastes, J. Water Pollut. Control Fed., 47 (1975) 1021-1031
179. Gracian, J., and Martel, J., Determinacion de residuos de disolventes enaceties refinados comestibles I. Grassa Aceites (Seville), 23 (1972) 1-6
180. Anon, Trichloroethylene: FDA cites cancer link in proposing ban as food additive, Chem. Reg. Reptr., 1 (1977) 993
181. Anon, FDA proposes ban on trichloroethylene, Chem. Eng. News, October 3 (1977) 16
182. National Institutes of Health, HEW News, June 14 (1976)
183. Rudali, G., A propos de l'activite oncogene de quecques hydrocarbures halogenes utilises en therapeutique, UICC Monograph, 7 (1967) 138-143
184. Henschler, D., Eder, E., Neudecker, T., and Metzler, M., Carcinogenicity of trichloroethylene: Fact or artifact, Arch. Toxicol., 37 (1977) 233-236
185. Dept. of Health, Eduation & Welfare, Memorandum on carcinogenicity of trichloroethylene, Washington, DC, March 20 (1975)
186. Starks, F. W., Stabilization of chlorinated hydrocarbons, U.S. Patent 2,818,446, October 25 (1956)
187. Shahin, M. M., and Von Borstel, R. C., Mutagenic and lethal effects of α-benzene hexachloride, dibutyl phthalate and trichloroethylene in Saccharomyces Cerevisiae, Mutation Res., 48 (1977) 173-180
188. Cerna, M., and Kypenova, Mutagenic activity of chloroethylene analyzed by screening system tests, Mutation Res., 46 (1977) 214-215
189. Euler, H. H., Tier experimentelle untersuchung einer industrie-noxe, Arch. Gynaekol., 204 (1967) 258
190. Kelley, J. M., and Brown, B. R., Jr., Biotransformation of trichloroethylene, Int. Anesthesiol. Clin., 12 (1974) 85
191. Henschler, D., Metabolism and mutagenicity of halogenated olefins-a comparison of structure and activity, Environ. Hlth. Persp., 21 (1977) 61-64
192. Bonse, G., Henschler, D., Chemical reactivity, biotransformation, and toxicity of polychlorinated aliphatic compounds, CRC Crit. Revs. Toxicology (1976) 395-409
193. Van Duuren, B. L., Chemical structure, reactivity and carcinogenicity of halohydrocarbons, Environ. Hlth. Persp., 21 (1977) 17-23
194. Dantel, T. W., Metabolism of ^{36}Cl-labeled trichloroethylene and tetrachloroethylene in the rat, Biochem. Pharmacol., 12 (1963) 795
195. Butler, T. C., Metabolic transformations of trichloroethylene J. Pharmacol. Exp. Therap., 97 (1949) 84-92
196. Powell, J. F., Trichloroethylene: Absorption, elimination and metabolism, Brit. J. Ind. Hyg., 2 (1945) 142-145
197. Byington, K. H., and Leibman, K. C., Metabolism of trichloroethylene by liver microsomes. II. Identification of the reaction product as chloral hydrate, Molec. Pharmacol., 1 (1965) 247-254
198. Scansetti, G., Rubino, G. F., and Trompeo, G., Studio sully intossicazione cronica da trielina III. Metabolismo del trichloroetilene, Med. Lavoro., 50 (1959) 743-754
199. Uehleke, H., Poplawski, S., Bonse, G., and Henschler, D., Spectral evidence for 2,2,3-trichloro-oxirane formation during microsomal trichloroethylene oxidation, Arch. Toxicol., 37 (1977) 95-105
200. Maynert, E. W., Foreman, R. L., and Watanabe, T., Epoxides as obligatory intermediates in the metabolism of olefins, J. Biol. Chem., 245 (1970) 5234

201. Banerjee, S., Van Duuren, B. L., and Goldschmidt, B. M., Microsome-dependent covalent binding of the carcinogen, trichloroethylene, to cellular macromolecules, Proc. Am. Assoc. Cancer, 18 (1977) 34
202. Van Duuren, B. L., and Banerjee, S., Covalent interaction of metabolites of the carcinogen trichloroethylene in rat hepatic microsomes, Cancer Res., 36 (1976) 2419
203. Uehleke, H., and Poplawski-Tabarelli, S., Irreversible binding of ^{14}C-labeled trichloroethylene to mice liver constituents in vivo and in vitro, Arch. Toxicol. 37 (1977) 289-294
203a. Hardie, D. W. F., Chlorocarbons and chlorohydrocarbons, tetrachloroethylene, In R. E. Kirk and D. F. Othmer (eds) Encyclopedia of Chemical Technology, 2nd ed., Vol. 5, John Wiley and Sons, New York (1964) pp. 195-203
204. Anon, Chem. Marketing Reptr. May 8 (1972) pp. 3, 47
205. Anon, NIOSH issues intelligence bulletin citing carcinogenicity of perchloroethylene, Chem. Reg. Reptr., 1 (44) (1978) 1539
206. Anon, Toxic Materials News, 3 (1976) 71, 409
207. Winell, M., An international comparison of hygienic standards for chemicals in the work environment, Ambio, 4 (1) (1975) 34-36
208. Pearson, C. R., and McConnell, G., Chlorinated C-1 and C-2 hydrocarbons in the environment, Proc. Roy. Soc. London (B) 189 (1975) 305-312
209. Loechner, F., Perchloroethylene in the environment, Umwelt, 6/7 (1976) 434-438
210. Utzinger, R., and Schlatter, C., A review of the toxicity of trace amounts of tetrachloroethylene in water, Chemosphere, 9 (1977) 517-524
211. Giger, W., and Molnar, E., Tetrachloroethylene in contaminated ground and drinking water, Bull. Env. Contam. Toxicol., 19 (1978) 475-480
212. U.S. Environmental Protection Agency, Draft Report for Congress: Preliminary Assessment of Suspected Carcinogens in Drinking Water, Office of Toxic Substances, Washington, DC, October 17 (1975)
213. Neeley, W. B., Branson, D. R., and Blau, G. E., Partition coefficient to measure bioconcentration potential of organic chemicals in fish, Environ. Sci. Technol., 8 (1974) 1113-1115
214. Weisburger, E. K., Carcinogenicity studies of halogenated hydrocarbons, NIEHS Conference on Comparative Metabolism and Toxicity of Vinyl Chloride and Related Compounds, Bethesda, MD May 2-4 (1977)
215. Anon, NCI Clearinghouse Subgroup Finds Tris, Tetrachloroethylene Carcinogenic, Toxic Materials News, 4 (1977) 60
216. Kylin, B., Sumegi, and Yllner, S., Hepatotoxicity of inhaled trichloroethylene and tetrachloroethylene, long-term exposure, Acta Pharmacol. Toxicol., 22 (1965) 379-385
217. Henschler, D., Metabolism and mutagenicity of halogenated olefins-a comparison of structure and activity, Environ. Hlth. Persp., 21 (1977) 61-64
218. Williamson, D. G., and Cvetanovic, R. J., Rates of reactions of ozone with chlorinated and conjugated olefins, J. Am. Chem. Soc., 90 (1968) 4248
219. Bauchwitz, P. S., Chloroprene In R. E. Kirk and D. F. Othmer (eds) "Encyclopedia of Chemical Technology", 2nd ed., Vol. 5, John Wiley and Sons, New York (1964) pp. 215-231
220. IARC, Monographs on the Evaluation of Carcinogenic Risk of Chemicals to Man, Vol. 7, International Agency for Research on Cancer, Lyon (1978) in press

221. Bellringer, F. J., and Hollis, C. E., Make chloroprene from butadiene, Hydrocarbon Processing, 47 (1968) 127-130
222. Lloyd, J. W., Decoufle, P., and Moore, R. M., Background information on chloroprene, J. Occup. Med., 17 (1975) 263-265
223. Van Oss, J. F., "Chemical Technology: An Encyclopedic Treatment", Vol. 5, Barnes and Noble, New York (1972) 482-483
224. Infante, P. F., and Wagoner, J. K., Chloroprene: Observations of carcinogenesis and mutagenesis, In Origins of Human Cancer, Cold Spring Harbor, New York Sept. 7-14 (1976) 77
225. Anon, Exposure limits for chloroprene, Chem. Eng. News, Aug. 22 (1977) 14
226. Sassu, G. M., Zilio-Grandi, F., and Conte, A., Gas-chromatographic determination of impurities in vinyl chloride, J. Chromtog., 34 (1968) 394-398
227. Kurosaki, M., Taima, S., Hatta, I., and Nakamura, A., Identification of high-boiling materials as by-products in vinyl chloride manufacture, Kogyo Kagaku Zasshi, 71 (1968) 488-491
228. Panina, L. A., and Fain, B. S., Chromatographic impurities in acrylonitrile, Zavod. Lab., 34 (1968) 283
229. Mnatsakagan, A. V., Pogasyan, U. G., Apoyan, K. K., Gofmekler, V. A., and Kanayan, A. S., Embryotoxic action of emissions of the chloroprene synthetic rubber industry using as materials for study the progeny forst generation of white rats, Tr. Erevan. Gas. Inst. Usoversh. Vrachei., 5 (1972) 155-158
230. Apoyan, K. K., Abeshyan, M. M., Gofmekler, V. A., Mnatsakanyan, A. V., Mutafyan, G. A., Poposyan, U. G., and Tarverdyan, A. K., Spectrophotometric method for determining chloroprene in air, Gig. Sanit., 35 (1970) 61-64
231. Buyanov, A. A., and Svishchev, G. A., Protection of the atmosphere from shoe-factory industrial emissions, Zv. Vyssh. Ucheb. Zaved. Tekhnol. Legk. Prom., 3 (1973) 68-71
232. Khachatryan, E. A., Lung cancer morbidity among people working with chloroprene, Vop. Oncol., 18 (1972) 85-86
233. Khachatryan, E. A., The role of chloroprene in the process of skin neoplasm formation, Gig. Tr. Prof. Zabol., 18 (1972) 54-55
234. Pell, S., Mortality of workers exposed to chloroprene, Presented at Conference of American Occupational Health, Boston, April 26 (1977), J. Occup. Med. (1978) in press
235. Zil'fyan, V. N., Fichidzhyan, B. S., Garibyan, D. K., and Pogosova, A. M., Experimental study of chloroprene for carcinogenicity, Vop. Onkol., 23 (1977) 61-65
236. Zil'fyan, V. N., Results of testing chloroprene for carcinogenicity, J. Exp. Clin. Med. Akad. Sci. Armen. SSR, 15 (1975) 54-57
237. Toxicology Information Programme, A long-term inhalation study of beta-chloroprene to rats, Tox-Tips, 3 (1976) 4
238. Toxicology Information Programme, Chronic inhalation of beta-chloroprene in hamsters, Tox-Tips, 13 (1977) 31
239. IARC, Information Bulletin on the Survey of Chemicals Being Tested for Carcinogenicity, International Agency for Research on Cancer, No. 16, Lyon (1976) p. 43
240. Vogel, E., Mutagenicity of carcinogens in Drosophila as a function of genotype controlled metabolism, In "In-vitro Metabolic Activation in Mutagenesis Testing", (eds) deSerres, F. J., Fouts, J. R., Bend, J. R., and Philpot, R. M., Elsevier North Holland Biomedical Press, Amsterdam (1976)
241. Davtyan, R. M., Toxicological characteristics of the action of chloroprene on the reproductive function of male rats, Toksikol. Gog. Prod. Neftekheim. Prozvod. Vses-Konf. Dokl., 2nd, 1971 (Publ. 1972) 95-97; Chem. Abstr., 80 91720A (1971)

242. Davtyan, R. M., Fumenko, V. N., and Andreeva, G. P., Effect of chloroprene on the generative function of mammals (males), Chem. Abstr., 79 (1973) 65-66
243. Sanotskii, I. V., Aspects of toxicology of chloroprene: Immediate and long-term effects, Environ. Hlth. Persp., 17 (1976) 85-93
244. Volkova, Z. A., Establishing the value of the MPEL of chloroprene in the air of a work area, Gig. Trud. Prof. Zabol., 3 (1976) 31
245. Bagramjam, S. B., and Babajan, E. A., Cytogenetic study of the mutagenic activity of chemical substances isolated from Nairit latexes MKA and LNT-1, Biol. Zh. Arm., 27 (1974) 102-103
246. Bagramjam, S. B., Pogosyan, A. S., Babajan, E. A., Ovanesjan, R. D., and Charjan, S. M., Mutagenic effect of small concentrations of volatile substances, emitted from polychloroprene latexes LNT-1 and MKH, during their combined uptake by the animals, Biol. Zh. Arm., 29 (1976) 98-99
247. Katosova, L. D., Cytogenic analysis of peripheral blood of workers engaged in the production of chloroprene, Gig. Tr. Prof. Zabol., 10 (1973) 30-33
248. Sanotskii, I. V., Correlations between the gonadotypes, embryotropic and blastomogenic types of effect of chemical compounds, Materially Nauch. Sessii Posvy. Sozdan. SSR Tbilisi., (1972) 117-118
249. Culik, R., Kelly, D. P., and Clary, J. J., β-chloroprene (2-chlorobutadiene-1,3) Embryotoxic and teratogenic studies in rats, Toxicol. Appl. Pharmacol., 37 (1976) 172
250. Haley, T. J., Chloroprene (2-chloro,1-3-butadiene). What is the evidence for its carcinogenicity, Toxicol. Annual (1977) in press
251. U.S. International Trade Commission, Synthetic Organic Chemicls United States Production And Sales of Elastomers, 1975 Preliminary, June, Washington, DC., U.S. Government Printing Office, (1976) p. 2
252. Van Duuren, B. L., Goldschmidt, B. M., and Siedman, I., Carcinogenic activity of di- and tri-functional α-chloroethers and of 1,4-dichlorobutene-2 in ICR/HA Swiss mice, Cancer Res., 35 (1975) 2553-2557
253. Bartsch, H., Malaveille, C., Barbin, A., Planche, G., and Montesano, R., Alkylating and mutagenic metabolites of halogenated oleins produced by human and animal tissues, Proc. Am. Assoc. Cancer Res., (1976) p. 17, Toronto, May 4-8, Proc. 67th Ann. Meeting of Amer. Assoc. Cancer Research
254. Mukai, F. H., and Hawryluk, I., The mutagenicity of some haloethers and haloketones, Mutation Res., 21 (1973) 228
255. Loprieno, N., Mutagenicity assays using yeasts with carcinogenic compounds, Second Meeting of Scientific Committee of the Carlo Erba Foundation, Dec. 12 (1975) pp. 129-140
256. EPA, Survey of Industrial Processing Data. Task I-hexachlorobenzene and hexachlorobutadiene pollution from chlorocarbon processes, Environmental Protection Agency, Washington, DC, June (1975)
257. Yurawecz, M. P., Dreifuss, P. A., and Kamps. L. R., Determiantion of hexachloro-1,3-butadiene in spinach, eggs, fish, and milk by electron capture gas-liquid chromatography, J. Ass. Off. Anal. Chem., 59 (1976) 552-557
258. Laska, A. C., Bartell, c. K., and Laseter, J. L., Distribution of hexachlorobenzene and hexachlorobutadiene in water, soil and selected aquatic organisms along the lower Mississippi River, Louisiana, Bull. Env. Contam. Toxicol., 15 (1976) 535-542
259. Yip, G., Survey for hexachloro-1,3-butadiene in fish, eggs, milk and vegetables, J. Ass. Off. Anal. Chem., 59 (1976) 559-561
260. Murzakaev, F. G., Action exerted by low hexachlorobutadiene doses on the activity of the central nervous system and morphological changes in animals so poisoned, Gig. Tr. Prof. Zabol., 11 (1967) 23-28; Chem. Abstr., 67 (1967) 31040A

261. Kociba, R. J., Keyes, D. G., Jersey, G. C., Ballard, J. J., et al., Results of a two-year chronic toxicity study with hexachlorobutadiene (HCBD) in rats, Abstract of 16th Annual Meeting of Society of Toxicology, Toronto, Canada, March 27-30 (1977)
262. Anon, Dow Tests of carcinogen indicate safe exposure levels, Chem. Ecology October (1976) 4
263. Simmon, V., Mutagenic halogenated hydrocarbons, Presented at Meeting of Structural Parameters Associated with Carcinogenesis, Annapolis, MD, Aug. 31- Sept. 2 (1977)
264. Schwetz, B. A., Smith, F. A., Humiston, c. G., Quast, J. F., and Kociba, R. J., Results of a reproduction study in rats fed diets containing hexachlorobutadiene, Toxicol. Appl. Pharmacol., 42 (1977) 387-398
265. Pilorz, B. H., Allyl chloride, In: Kirk-Othmer Encyclopedia of Chemical Technology, 2nd ed., Vol. 5, Wiley-Interscience, New York (1964) 205-215
266. Anon, Adherence to 1 ppm standard for allyl chloride exposure recommended, Occup. Hlth. Safety Letter, Oct. 1 (1976) 5
267. Alizade, G. A., Guselinov, F. G., Agamova, L. P., Guseinova, R. S., and Aleskerov, F. A., Functional state of the kidneys of workers in contact with allyl chloride, Azerb. Med. Zh., 53 (1976) 54-59; Chem. Abstr., 86 (1977) 194353M
268. Torkelson, T. R., Wolf, M. A., Oyen, F., and Rowe, V. K., Vapor toxicity of allyl chloride as determined in laboratory animals, Am. Ind. Hyg. Assoc. J. 20 (1959) 217-223
269. Ames, B. N., McCann, J., and Yamasaki, E., Methods for detecting carcinogens and mutagens with the Salmonella/mammalian-microsome mutagenicity test, Mutation Res., 31 (1975) 347-364
270. McCoy, E. C., Burrows, L., and Rosenkranz, H. S., Genetic activity of allyl chloride, Mutation Res., 57 (1978) 11-15
271. Rosenkranz, H. S., Gutter, B., and Speck, W. T., Mutagenicity and DNA-modifying activity: A comparison of two microbial assays, Mutation Res., 41 (1976) 61-70
272. Rosenkranz, H. S., Speck, W. T., and Gutter, B., Microbial assay procedures: Experience with two systems, In: "In Vitro Metabolic Activation in Mutagenesis Testing" (eds) deSerres, F. J., Fouts, J. R., Bend, J. R.,and Philpot, J., Elsevier/North Holland, Amsterdam (1976) 337-363
273. Kaye, C. M., Clapp, J. J., and Young, L., The metabolic formation of mercapturic acids from allyl halides, Xenobiotica, 2 (1972) 129-139
274. Kaye, C. M., Ph. D. Thesis, University of London, London, England (1971)

CHAPTER 12

HALOGENATED SATURATED HYDROCARBONS

Organohalogen compounds are probably the most ubiquitous in occurrence. They are extensively used as solvents, aerosol propellants, dry-cleaning fluids, refrigerants, flame-retardants, fumigants, degreasing agents, and intermediates in the production of other chemicals, textiles and plastics. Many of their end-use products contain halogen. A number of these, because of their use as pesticides and aerosol propellants and their high chemical stabilities, have become distributed throughout the biosphere[1-3]. A number of chlorinated hydrocarbons may occur in drinking water via contamination of water supplies by effluents from chemical industry or in part, as in the case of chloroform or other trihalomethanes, via disinfection of water supplies by chlorine[4-6].

The haloalkanes react with nucleophilic substrates, but the rates at which these reactions take place are highly dependent on the nature of the halogen and the presence of other substituent groups in the molecule. The valence electrons are most firmly held by the fluorine nucleus and least firmly held by the iodine nucleus as indicated by bond lengths and C-X bond energies (e.g., 108 and 53 k cal/mol respectively). All of these elements are more electronegative than hydrogen on the Pauling scale, and the energies of C-X bonds are less than those of C-H bonds, except for fluorine. Thus alkyl iodides are much more reactive with nucleophilic substrates than are alkyl fluorides and many of these compounds can replace hydrogen substituted on group V-A and VI-A elements in S_N-1 and S_N-2 reactions. A variety of compounds containing hetero atoms are thus capable of acting as nucleophiles toward alkyl halides, e.g., H_2O, H_2S, ROH and RSH. More reactive nucleophiles, the corresponding anions HO^-, HS^-, RO^-, RS^- can also react with a variety of alkyl halides. The nature of the halogen atoms also governs volatility which could be a factor in determining whether exposure to them is likely to occur by inhalation, dermal contact, or other routes[1].

While the chemical reactivity of halogenated aliphatic compounds is decisively determined by the halogen substitution(s), the influence is completely different in alkanes, alkenes and alkynes[7]. For example, in general chlorine substitution exerts a stabilizing effect due to steric protection by the relatively bulky substituent. This interferes with the electron-withdrawal effect of the chlorine substituent. In the case of chlorinated alkanes, the result is a destabilization of C-C and C-Cl bonds. Carbon-carbon and carbon-chlorine fissions under formation of free radicals are consequent reactions in the metabolic conversions of C_1 or C_2 compounds[7]. As was noted in the previous chapter, this behavior contrasts with the class of chlorinated ethylenes where the electron-withdrawal effect dominated over the mesomeric donator effect of the involved carbon atom, hence decreasing the electron density in the double bond resulting in a chemical stabilization against electrophilic attacks[8].

1. <u>Methyl Chloride</u> (chloromethane; monochloromethane; CH_3Cl) is produced by two principal methods: (a) the action of hydrogen chloride on methanol, with the aid of a catalyst, in either the vapor or the liquid phase and (2) the chlorination of methane[9,10]. The current use pattern (%) for methyl chloride is as follows: in the

production of silicones, 38; tetramethyl lead, 38; butyl rubber, 4; methyl cellulose, 4; herbicides, 4; quaternary amines, 3; and miscellaneous, 9. Methyl chloride is also used in making methyl mercaptan, an intermediate in the manufacture of jet fuel additives and fungicides[10]. Methyl chloride is also employed as a starting material in the manufacture of methylene chloride, chloroform, carbon tetrachloride and of various bromochloro- and chlorofluoromethanes[10]. Miscellaneous uses include: in solvent degreasing operations, in the formation of carbonated quaternized acrolein-copolymer anion exchanges[11]; quaterinization of tertiary amines[12] and as a paint remover. Originially methyl chloride was used almost entirely as a refrigerant, but this use has been largely preempted by the chlorofluoromethanes.

Information regarding the numbers of workers employed in the production of many of the saturated aliphatic hydrocarbons as well as in allied use categories is not available. The current threshold limit value-time weighted average (TLV-TWA) for a normal 8-hour workday or 40 hour workweek for methyl chloride is 210 mg/m^3 (100 ppm) and the TLV for short term exposure limit (TLV-STEL) is 260 mg/m^3 (125 ppm)[13].

The current TLV's (mg/m^3) in several countries are as follows: Federal Republic of Germany, 105; Democratic Republic of Germany, 100; Czeckoslovakia, 100. The MAC for methyl chloride in the USSR is 5 mg/m^3[14].

Methyl chloride has been found in tobacco smoke[15] suggesting that an additional portion of the population may be potentially exposed to this agent.

Methyl chloride, without metabolic activation, has been reported to be highly mutagenic in S. typhimurium strains TA 1535[16,17] and TA 100[17] which can detect mutagens causing base-pair substitutions.

2. <u>Methylene Chloride</u> (dichloromethane, CH_2Cl_2) can be produced by chlorinating either methane or methyl chloride, or by reduction of chloroform or carbon tetrachloride, manufacturing processes are based on the chlorination route. When the starting material is methane or methyl chloride, the other chloromethanes are coproducts of the methylene chloride, their relative proportions depending on the process operating conditions. Whereas conditions of chlorination cannot be devised in which methylene chloride is the exclusive product, various industrial processes for methane chlorination have been developed to produce a large preponderance of the particular chloromethane desired[9,18]. Methane chlorination plants are usually quite large with total production in excess of 160 million kg/year while most plants hydrochlorinating methanol vary in size from 7 million to 45 million kg/year[9].

The current use pattern (5) for methylene chloride in the U.S. is as follows: paint remover, 40; exports, 20; aerosol sprays, 17; chemical specialties (mostly solvent degreasing), 10; plastics processing, 6; all other, 7.

Methylene chloride is used in the textile industry as a solvent for cellulose triacetate in rayon yarn manufacture, and in the production of PVC fiber. It is also used in the extraction of naturally occurring, heat sensitive substances, such as edible fats, cocoa, butter, the beer flavoring in hops[18], and as a substitute for trichloroethylene for decaffeinating coffee[17].

Methylene chloride demand has grown in the past decade, competing partially with trichloroethylene and perchloroethylene, primarily as a paint remover and in solvent degreasing operations[9]. Increasing utility of methylene chloride as a substitute in aerosol products for the controversial chlorofluorocarbon 11 and 12 is also forecast[19]. This is predicted on the favorable properties, e.g., methylene chloride acts as a solvent, vapor pressure depressant and flame depressant. However, the problem of the compound's tendency to hydrolyze forming hydrochloric acid in small amounts which tends to corrode containers has prevented earlier, greater acceptance[19].

The estimated annual production of methylene chloride is about 500 million kg[17].

The 1976 adopted threshold limit value-time weighted average (TLV-TWA) concentration for a normal 8 hr workday or 40 hour workweek for methylene chloride in the U.S. is 720 mg/m^3 (200 ppm) and that of the TLV-short term exposure is 900 mg/m^3 (250 ppm)[13].

The 1976 OSHA environmental standard is 500 ppm, 8 hr TWA, 1000 ppm acceptable ceiling and a 2000 ppm maximum (5 minutes in 2 hours)[20]. The NIOSH recommendation for environmental exposure limit of methylene chloride is 75 ppm TWA, 500 ppm ceiling (15 minutes), and the TWA to be lowered in the presence of carbon monoxide[20].

There is a paucity of data concerning the carcinogenic potential of methylene chloride. Interim results of a two year inhalation study of Dow Chemical indicate "no evidence of cancer in test animals". The test involved nearly 2000 animals which were exposed to methylene chloride levels as high as 3500 ppm[19].

Methylene chloride recently has been shown to be mutagenic in S. typhimurium strains TA 98 and TA 100. The administration of rat-liver homogenate did not appear to be essential, though it slightly increased the number of mutations[21]. Methylene chloride when tested in desiccators by the procedure of Simmon et al[16] was found to be quite mutagenic in the Salmonella/microsome assay with TA 100, but it did not increase mitotic recombination in S. cerevisiae D3[16]. Methylene chloride induced transformed cells (cell line F 1706) which grew in agar and formed tumors when injected into immunosuppressed mice[22].

Dihalomethanes has been shown to be biotransformed in vitro and in vivo to carbon monoxide and inorganic halide[23]. The rate of metabolism of the dihalomethane was found to be dichloro > dibromo > bromochloro > dichloromethane.

3. Chloroform (trichloromethane, CHCl$_3$) is made principally via the chlorination of methane or the hydrochlorination of methanol with lesser amounts produced by the limited reduction of carbon tetrachloride[24,25].

A representative technical-quality chloroform contains not more than the following impurities (ppm): methylene chloride, 200; bromochloromethane, 550; carbon tetrachloride, 1500. Chloroform is usually stabilized by the addition of 0.6-1% ethanol (or thymol, t.butyl phenyl, or n-octyl phenol 0.0005-0.01%) to avoid photochemical transformation to phosgene and hydrogen chloride[24].

Chloroform is used in extensive quantities principally in the manufacture of chlorofluoromethane (fluorocarbon-22) (ClF$_2$HC), for use as a refrigerant and as an aerosol propellant and as a raw material for the production of fluorinated resins (e.g., Teflon, polytetrafluoroethylene, PTFE)[25]. Other uses of chloroform include: extractant and industrial solvent in the preparation of dyes, drugs, pesticides, essential oils, alkaloids, photographic processing, industrial dry cleaning, as a fumigant, as an anesthetic, in pharmaceuticals and toiletries (until recently in mouthwashes, dentrifices), hair tinting and permanent-waving formulations and in fire extinguishers (with carbon tetrachloride)[24-26]. It is also found in most chemistry laboratories.

The U.S. chloroform consumption pattern in 1974 is estimated to have been as follows: for synthesis of fluorocarbon 22 used in refrigerants and propellants, 51.3%; for synthesis of fluorocarbon 22 used in plastics, 24%; and miscellaneous uses, 24.7%. In 1975 U.S. production of fluorocarbon 22 amounted to 59.9 million kg[27].

In 1975, five U.S. companies produced a total of 119 million kg of chloroform. Annual chloroform production in Western Europe has been estimated in the 10-50 million kg range with the United Kingdom, France and the Federal Republic of Germany being the major producing and Benelux and Italy the minor producing countries. The estimated world production of chloroform in 1973 was 245,000 tons[28]. NIOSH estimates that 40,000 persons are exposed occupationally to chloroform in the U.S. The majority of these are workers where chloroform is used in relatively small amounts. These industries include those producing biological products, pharmaceutical preparations, paint and allied products, and surgical supplies, as well as hospitals, paper milling, petroleum refining and metal industries[26].

The current OSHA environmental standard ceiling value for chloroform in workplace air is 50 ppm[20,29]. NIOSH recommended in 1976 that the limit be reduced to 2 ppm (9.78 mg/m^3) as determined by a 45-liter sample taken over a period not to exceed one hour[20]. ACHIG[13] recommended in 1976 a TLV-TWA of 10 ppm (50 mg/m^3) reduction from the previous 25 ppm (120 mg/m^3).

Chloroform is widely distributed both in the atmosphere[28,30] and water[4,31] (including municipal drinking water primarily as a consequency of chlorination[4,31]. The ubiquitous occurrence of chloroform in the atmosphere (usually in the parts-per-billion range)[30,32,33] cannot be accounted for solely from direct production emission data[30]. One suggested additional source of $CHCl_3$ arises from the extensive use of chlorine in the bleaching of paper pulp[30]. A conversion efficiency as low as 6% in the bleaching process would supply a global source of $CHCl_3$ (via a haloform reaction) of magnitude of 3×10^5 tons/year[30].

Decomposition of perchloroethylene could provide an additional source of atmospheric $CHCl_3$ via the photolysis of dichloroacetylchloride[30]. Additionally chloroform, as well as carbon tetrachloride, may arise naturally by reaction between chlorine and methanol in the atmosphere[28,33].

Chloroform is removed from the atmosphere primarily by reaction with OH[30]. Chloroform slowly undergoes decomposition on prolonged exposure to sunlight in the presence or absence of air and in the dark, when air is present. The products of oxidative breakdown include phosgene, hydrogen chloride, chlorine, carbon tetrachloride and water[24].

The hydrolytic-oxidative reaction half-life of chloroform in sealed ampules is about 15 months and the evaporation rate of 90% of chloroform (1 ppm) from dilute aqueous solution is about 70 min[34].

A survey of 80 American cities by EPA found chloroform in every water system in levels ranging from < 0.3-311 µg/liter (ppb)[31,35]. Based on arbitrarily assumed water consumption of 8 liters/day, maximum human consumption would be about 25 mg/day equivalent to 0.036 mg/kg (70 kg man) at the highest chloroform level[35]. The haloform reaction was suggested as the source of $CHCl_3$ andother haloforms produced during the chlorination of municipal water supplies[5,6,34]. The yield of $CHCl_3$ in chlorination of water is about 1-3% by weight[36] and represents a lower limit to the potential conversion efficiency since the production of haloforms is limited by the availability of dissolved organic material. The main source of chloroform precursors in natural waters is humic material[5,6,37].

While chloroform residues in potable water are generally thought to be derived only from the organic matter in the raw water upon chlorination, it is important to note the formation of chloroform from potential precursors such as polyelectrolytes (PE's). These are certain high-molecular-weight polymers which are commonly used as coagulants or coagulant aids for the treatment of potable water. Recently, formation of chloroform was reported at concentrations of a few micrograms/liter from widely used polyelectrolyte coagulants and coagulant aids[38]. Most of the 10 commercial polyelectrolyte formulations tested reacted with chlorine to form chloroform under thermal conditions with the reaction strongly activated by ultraviolet irradiation. The possibility of chloroform formation from PE's was first raised in the report on the National Organics Reconnaisance Survey for Halogenated Organics in Drinking Water[36,39].

Recently, EPA proposed to set a limit of 100 ppb on chloroform and related trihalomethanes that are formed during the disinfection process at municipal water treatment plants[40]. The trihalomethane standard would apply to community water systems serving more than 75,000 people. Cities with pure drinking sources would be exempt.

Chloroform has been found in foodstuffs (ppm) in the United Kingdom in 1973 including dairy produce, 1.4-33, meat, 1-4; oils and fats, 2-10; beverages, 0.4-18; and fruits and vegetables 2-18[28].

Carcinogenic effects in laboratory animals have been reported in two studies. In a study reported in 1945, hepatomas were found in 7 of 10 female mice fed 30 doses of $CHCl_3$ at 4-day intervals of approximately 600 or 1200 mg/kg/dose over a 4-month period. The other 3 female mice died within the first week of the experiment. Male mice receiving similar doses died within the first week[41].

In the recent NCI study[26,42,43], Osborne-Mendel rats were fed $CHCl_3$ in corn oil at 90 and 180 mg/kg body weight for males and at 100 and 200 mg/kg for females for 111 weeks. A significant increase in epithelial tumors of the kidney in treated male rats was observed. Of the 13 tumors of renal tabular cell epithelium observed in 12 of the 50 high dose male rats, 10 were carcinomas and 3 adenomas, 2 of the carcinomas were found to have metastasized. Two carcinomas and 2 adenomas of renal tubular epithelium were found among the 50 low dose (half maximum tolerated dose) male rats. An increase in thyroid tumors in $CHCl_3$-treated female rats was also seen, but the NCI does not consider these to be significant findings.

A highly significant increase in hepatocellular carcinomas was observed in both sexes of B6C3F mice fed $CHCl_3$ for 92-93 weeks at 138 and 277 mg/kg doses for males and at 238 and 477 mg/kg doses for females. The incidence of hepatocellular carcinoma was 98% for males and 95% for females at the high dose and 36% for males and 80% for females at the low dose compared with 6% in both matched and colony control males. Nodular hyperplasia of the liver was observed in many low dose male mice that had not developed hepatocellular carcinoma[26,42].

No excess of neoplasms were found in female mice of the ICI/CFLP strain given toothpaste by intragastric intubation 6 days/week for 80 weeks at dose levels up to 60 mg $CHCl_3$/kg/day or in males at 17 mg $CHCl_3$/kg/day. There was no overall increase in tumors in $CHCl_3$-treated mice compared with the vehicle controls. However, ICI/CFLP males given toothpaste equivalent to 60 mg $CHCl_3$/kg/day but not males of the C57Bl or CBA strains developed more renal tumors than the control. In ICI/CFLP male mice given $CHCl_3$ in arachis oil or arachis oil alone, there were 12/52 renal tumors in the $CHCl_3$-treated group compared with 1/52 in the controls; there were 5/52 renal tumors in mice given 60 mg $CHCl_3$/kg/day in toothpaste. All the renal tumors were of the benign type except for 2 in a $CHCl_3$-treated group, neither of which showed evidence of metastasis. It was concluded that there was no evidence of a carcinogenic effect in male or female mice given toothpaste containing the equivalent of 17 mg $CHCl_3$/kg/day corresponding to approximately 100 times the estimated rate of daily ingestion during normal use of toothpaste containing 3.5% of chloroform[44].

Studies on individual differences in chloroform toxicity in mice have been reviewed recently by Hill[46]. These studies indicate that inbred strain-related differences within a species may help to point out the types of variability in response that one might encounter on a genetically heteogeneous organism, like man. Statistics like median lethal or toxic dose in randomly bred animals may, on the other hand, only serve to conceal the genetic variability that may exist, according to Hill[46]. Two genetic variations in chloroform toxicity were described. In one, animals more sensitive to $CHCl_3$-induced death, were found to be more susceptible to renal toxicity, while in the other, sensitive animals were unable to repair damaged renal tubules. An absolute sex-related difference in regard to kidney damage, but no liver damage that occurs on $CHCl_3$ administration was assessed. Androgens play on integral role in modulating the sex-related difference and perhaps the strain-related difference in sensitivity to kidney damage. It is important to note that tumors have been produced in experimental animals after long-term exposure to $CHCl_3$ in the same organs that are affected by single doses of $CHCl_3$, e.g., hepatic tumors in mice and renal tubular tumors in male rats[41,42]. The effects of long-term, low-level exposure to chloroform in man are largely unknown.

Chloroform did not induce mutations after incubation with liver or rabbit microsomes and S. typhimurium TA 1535 and TA 1538[46,47]. Simmon et al[17] were unable to convincingly demonstrate the mutagenicity of chloroform in extensive tests in the Salmonella/microsome assay, in suspension and in desiccators. Chloroform was also non-mutagenic when tested in E. coli K12 for base-pair substitution[47] and when tested at the 8-azaguanine locus on the chromsomes of Chinese hamster lung fibroblast cells in culture[48].

Inhalation of chloroform (300 ppm for 7 h/day on days 6-15 of gestation) has resulted in a significant increase in the incidence of fetal resorptions, a decrease in fetal weight and length and a reduction in conception rate (15% compared with 88% in controls).

Incidence of resorption was dose related as was the increase in indicators of retarded fetal development[49]. Inhaled chloroform while showing only slight evidence of teratogenicity was markedly embryotoxic whereas the effect of chloroform administered orally in the rat and rabbit was limited to a mild fetotoxicity[50]. No teratogenic effects were noted in rats and rabbits after oral administration of chloroform during days 6-18 of gestation, but offspring were smaller than normal[51].

Mouse strain differences suggest intermediate or multifactorial genetic control of chloroform-induced renal toxicity and death. The chloroform dose lethal to 50% of animals was 4 times higher in C57BL/6J males than in DBA/2J males while twice as much $CHCl_3$ accumulated in the kidneys of the sensitive as the resistant strains. First generation offspring were midway between parenteral strains for both parameters[52].

Chloroform and other halogenated hydrocarbons produce pathological effects by localizing in target tissues and binding covalently to cellular macromolecules[53-55].

The metabolism of chloroform has recently been reviewed by Charlesworth[56]. Chloroform undergoes biotransformation by microsomal enzymes in the cells of the liver[57-59].

^{14}C from ^{14}C-chloroform was extensively excreted in the expired air of mice, rats, and monkeys when administered orally. Most of the dose was excreted unchanged in monkeys and mice as $^{14}CO_2$ with the rat intermediate in excretion pattern. Three metabolites were detected in the urine, one of which was identified as ^{14}C-urea[60].

No evidence for the formation of methylene chloride via a $CHCl_2$ free radical was reported by Hathway[61]. Pohl et al[62] recently reported on the mechanism of the metabolic activation of chloroform. In vitro studies with $^{13}CHCl_3$ suggested that the C-H bond of $CHCl_3$ is oxidized by a cytochrome P-450 monooxygenase to produce trichloromethanol. This intermediate would spontaneously dehydrochlorinate to yield phosgene, which could then bind covalently to protein. The hepatotoxicity of $CHCl_3$ was compared with that of deuterium-labeled chloroform ($CDCl_3$) in phenobarbital pretreated rats. Only the $CHCl_3$ treated animals exhibited extensive centrolobular necrosis of the liver. This deuterium isotope effect on toxicity supports the *in vitro* mechanism of activation by establishing that *in vivo* the C-H bond is involved in its metabolic activation and hepatotoxicity.

4. <u>Carbon Tetrachloride</u> (tetrachloromethane; perchloromethane; CCl_4) is manufactured primarily via the chlorination of methane, and to a limited extent by the chlorination of carbon disulfide. A technical grade CCl_4 generally contains not more than the following impurities (ppm): bromine, 20; chloroform, 150 and water 200. Stabilizers such as alkyl cyanamide; 0.34-1% diphenylamine; thiocarbamides; hexamethylene tetramine, etc. are added to the technical grade product to prevent photo decomposition[63].

Carbon tetrachloride is produced in enormous quantities, e.g., over 1.0 billion pounds in the U.S. in 1970 and approximately 140 million pounds in Japan in 1974 and France in 1969[64]. In the U.S., the majority of the CCl_4 produced in 1970 was employed in the production of fluorocarbons, e.g., approximately 700 million pounds (69%) for dichlorofluoromethane (CF_2Cl_2) and about 260 million pounds (26%) for trichlorofluoromethane ($CFCl_3$). The remaining 5% of the production in the U.S.

(in 1970) (approximately 50 million pounds) were used as: (a) grain fumigants (alone or mixed with 2-25% ethylene bromide or chloride); (b) fire extinguishers (with 10% $CHCl_3$ or trichloroethylene); (c) solvent for oils, fats, resins, and rubber cements; (d) cleaning agent for machinery and electrical equipment; (e) degreasing metal fabricated parts; (f) in synthesis of nylon-7 and other organic chlorination processes; (g) and pharmaceutically in treatment of hookworm and liver fluke in cattle and sheep[64].

The number of workers engaged in the production and use applications of carbon tetrachloride, as well as their respective levels of exposure, is not known.

The current OSHA environmental standard for carbon tetrachloride is 10 ppm (65 mg/m^3), 8 hour TWA 25 ppm (160 mg/m^3); with a 200 maximum ceiling (5 min in 4 hrs)[20]. The NIOSH recommendation for environment exposure limit is a 2 ppm (13 mg/m^3) ceiling (60 min)[20].

Hygienic standards for permissible levels of carbon tetrachloride in the working environment in various countries were as follows (mg/m^3)[14]: Sweden, 65; Czechoslovakia, 50; Federal Republic of Germany, 65 and German Deomocratic Republic, 50; the maximum permissible concentration (MCP) in the USSR is 20 mg/m^3 [65].

Losses of carbon tetrachloride to the global environment are considerable. For example, it has been estimated by Singh et al[66] that by 1973 there had been accumulative worldwide emission of about 2.5 million metric tons and this had grown at a rate of 60,000 metric tons for the last 30 years. McConnell et al[28] estimated the 1974 loss of CCl_4 to the global environment to be in the order of 1 million tons. The occurrence of CCl_4 in the atmosphere cannot be accounted for from direct production emission data[28,33,66-68]. Mean aerial concentrations of CCl_4, e.g., 0.071 ppb, over the North and South Atlantic[33], 0.2-0.7 ppb CCl_4 over land and sea stations of the United Kingdom[69], 0.01-0.03 ppb over the Los Angeles Basin[70], and at sub ppb in rural areas and much higher concentrations in urban areas in the U.S.[67] have all been reported.

Considerable atmospheric carbon tetrachloride formation is believed to result via photo decomposition of chloroalkenes in the troposphere[67]. Another possibility is the production of CCl_4 from the irradiation of methyl chloride (presumably via a heterogeneous reaction)[71]. Methyl chloride is associated with the smoke of grass and forest fires. The concentration of carbon tetrachloride in the air is approximately one-tenth that of methyl chloride which is present at a concentration of approximately 1 ppb[71].

By virtue of its tropospheric stability and the absence of any apparent physical or biological removal mechanism, carbon tetrachloride (analogously to CCl_3F and $CCl_2F-CClF_2$) would be expected to be a precursor of stratospheric ozone-destroying chlorinations[67].

The results of computer simulation of the distribution of CCl_4 in the biosphere by Neeley[68] which consider an imput of 60,000 metric tons/year for 42 years supports the speculation that the presence of this halocarbon in the environment is due entirely to anthropogenic sources. This modeling work indicates that CCl_4 is essentially in equilibrium between the air and water and it is not necessary to consider the ocean as either a source or s sink based on the data available.

Carbon tetrachloride is found in many sample waters (rain, surface, portable and sea) in the sub-ppb range[28]. CCl_4 has been found in 10% of the U.S. drinking water supplies at levels of < 2-3 ppb in a recent EPA survey of 80 cities[13].

Assuming that all drinking water contains the maximum level of CCl_4 (3 µg/liter) the maximum human consumption would be about 24 µg/day equivalent to 0.34 µg/kg/day for a 70 kg/man, based on a water consumption of 8 liters/day.

Thirteen halogenated hydrocarbons have been identified recently in samples of New Orleans drinking water and 5 halogenated hydrocarbons were found in the pooled plasma from 8 subjects in that area. Carbon tetrachloride and tetrachloroethylene were found in both the plasma and drinking water. Considerable variation in the relative concentrations of the halogenated hydrocarbons was noted from day to day in the drinking water. In view of the lipophilic nature of CCl_4, it was suggested that a bioaccumulation mechanism may be operative, if drinking water was the only source of such materials[72].

Carbon tetrachloride has been found in foodstuffs (ppm) including: daily produce (0.2-14), meat (7-9), oils and fats (0.7-18), beverages (0.2-6) and fruit and vegetables (3-8)[28].

The use of CCl_4 as a fumigant, alone or in admixture with ethylene dichloride, ethylene dibromide, methyl bromide in the disinfestation of stored grain has resulted in residues of CCl_4 in foodstuffs, e.g., levels reaching over 5 ppm and up to 0.07 ppm in bread baked from fluor treated with CCl_4-ethylene dichloride[73].

Carbon tetrachloride has been found in human tissues as follows (ppm, wet weight): kidney, 1-3; liver, 1-5 and body fat, 4-13.6[28].

Carbon tetrachloride has produced liver tumors in the mouse, hamster, and rat following several routes of administration including inhalation and oral[64,74]. A number of cases of hepatomas appearing in men several years after carbon tetrachloride poisoning have also been described[75,76].

Carbon tetrachloride has not been found mutagenic in the Salmonella/microsome assay with tester strains TA 1535, TA 1538, TA 98 and TA 100 when tested in agar[17,46,47,77,78] or in a desiccator[17]. It also gave negative results when tested in E. coli[47], Streptomyces coelicolor[78] and Aspergillus nidulans[78].

The synergistic effect of CCl_4 on the mutagenic effectivity of cyclophosphamide in the host-mediated assay with S. typhimurium has been reported. CCl_4 did not effect the mutagenicity of cyclophosphamide when tested in vitro with S. typhimurium strains G46 and TA 1950, nor was it mutagenic when assayed in a spot-test with the above strains of S. typhimurium[79].

Carbon tetrachloride administered to pregnant Sprague-Dawley rats at 300 or 1000 ppm 7 hr/day on days 6 through 15 of gestation was not highly embryotoxic but caused some degree of retarded fetal development such as delayed ossification of sternebral. However, CCl_4 caused a greater degree of maternal toxicity (compared with 1,1-dichloroethane and methyl ethyl ketone). This included decreased weight gains and food consumption during pregnancy, increased SGPT activity, and altered absolute and relative liver weight[80].

The chemical pathology of CCl_4 liver injury is generally viewed as an example of lethal cleavage[81], e.g., the splitting of the CCl_4-Cl bond which takes place in the mixed function oxidase system of enzymes located in the hepatocellular endoplasmic reticulum. While two major views of the consequences of this cleavage have been suggested, both views take into account the high reactivity of presumptive free radical products of a homolytic cleavage of the CCl_3-Cl bond. One possibility is the direct attack (via alkylation) by toxic free radical metabolites of CCl_4 metabolism on cellular constituents, especially protein sulfhydryl groups[82]. In homolytic fission, the two odd-electron fragments formed would be trichloromethyl and monatomic chlorine free radicals (e.g., $CCl_4 + e \rightarrow \cdot CCl_3 + Cl^-$). Fowler[83] detected hexachloroethane (CCl_3CCl_3) in tissues of rabbits following CCl_4 intoxication.

An alternative view has emphasized peroxidative decomposition of lipids of the endoplasmic reticulum as a key link between the initial bond cleavage and pathological phenomena characteristic of CCl_4 liver injury. This CCl_4 binds to cytochrome P-450 apoprotein and is cleaved at that locus to yield extremely short-lived free radicals which initiate peroxidative decomposition of polyenoic lipids. The autocatalytic decomposition of the lipid spreads from the initial locus, and lipid peroxides and hydroperoxides probably also move to more distant sites where they decompose to yield new free radicals. Rapid breakdown of structure and function of the endoplasmic reticulum is due to decomposition of the lipid, and to attack on protein functional groups, especially sulhydryl groups by lipid peroxides[81].

The reduction of a number of polyhalogenated methanes (CCl_4, CBr_4, $CHCl_3$, CCl_3F, CCl_3Br, CCl_3CN, and CHI_3) under anaerobic conditions by liver microsomal cytochrome P-450 to form complexes with ferrous cytochrome P-450 has been recently reported[84]. The evolution of carbon monoxide suggested prior reductive metabolism of the polyhalogenated methanes to a carbene ligand. The question remains whether carbene complex formation is implicated in any way in the well-known hepatoxicity of the polyhalogenated methanes.

Vairns et al[85] recently reported that carbon tetrachloride at 10.3 mM/kg administered intragastrically to rats, was the most active of various polychlorinated hydrocarbons (e.g., 1,1,2,2-tetrachloroethane and pentachloroethane) in decreasing cytochrome P-450 content and the overall drug hydroxylation activities in rat liver. It was also noted that epoxide hydratase activity, and to a lesser extent UDP-glucuronosyltransferase activity, declined significantly after CCl_4 treatment.

5. __Methyl Chloroform__ (1,1,1-trichloroethane; alpha-trichloroethane; CH_3CCl_3) is produced primarily from isolated vinyl chloride and vinylidene chloride which are made from ethylene dichloride. The initial step is the hydrochlorination of vinyl chloride to 1,1-dichloroethane, then thermal chlorination of the latter to yield methyl chloroform in yields of over 95%[86,87]. In another process, methyl chloroform is produced by the continuous non-catalytic chlorination of ethane. (Ethyl chloride, vinyl chloride, vinylidene chloride, and 1,1-dichloroethane may also be produced by this process by modification of the operating conditions)[86]. Over 60% of methyl chloroform is produced from vinyl chloride, almost 30% from vinylidene chloride and the remainder from chlorination of ethane. Plant capacities in the U.S. range from 14 million to 155 million kg/year[86].

Although methyl chloroform is considerably more stable than either carbon tetrachloride or trichloroethylene, small amounts of stabilizing substances are always added to the commercial product. These can be glycol diesters, nitroaliphatic hydrocarbons, 1,4-dioxane, morpholine or a variety of alcohols[87].

Methyl chloroform is used in the U.S. principally (approximately 70%) as a cold solvent for metal degreaing and electrical and electronic equipment cleaning. It is increasingly being substituted for trichloroethylene in solvent applications, largely because it contributes less to smog formation, is relatively inert in the troposphere than trichloroethylene and because the latter is a suspect carcinogen[88]. Historically, it has been a common substitute for carbon tetrachloride[89]. Approximately 15% of methyl chloroform produced is for use in aerosol formations (both as a solvent and as a low-pressure propellant) for adhesives and polishes and miscellaneous uses. Methyl chloroform has also been used with other chlorinated solvents (e.g., methylene chloride) in combination with nitrous oxide and/or carbon dioxide in aerosol propellant combinations[89]. There are a number of consumer aerosol products containing methyl chloroform[90]. These include: cleaners (oven and spot removers containing from 25 to 70% methyl chloroform), waxes and polishes, automotive and specialty products[89,90].

In 1970 the U.S. production of methyl chloroform was 1.7×10^5 tons and by 1976 it had increased to 3.0×10^5 tons[91,92]. Worldwide capacity for methyl chloroform in 1976 was 5.3×10^5 tons with 90% utilization[93]. Additional capacity figures have been recently announced: Europe, 9×10^4 tons for the last quarter of 1976; U.S., 2.2×10^5 tons for 1978; and Japan 7.3×10^4 tons for 1979[94,95]. Hence, by 1979 the world production capacity of methyl chloroform will be 9.1×10^5 tons. NIOSH reports that about 100,000 workers are exposed to methyl chloroform.

The current OSHA environmental standard for methyl chloroform is 350 ppm (1900 mg/m^3) for an 8 hr TWA[20]. The NIOSH recommendation for an environmental exposure limit is 350 ppm ceiling (15 minutes)[20]. The current ACGIH threshold limit value for short-term exposure is 450 ppm (2,375 mg/m^3)[13].

Hygienic standards for permissible levels of methyl chloroform in the working environments in various countries are as follows (mg/m^3)[14]: Federal Republic of Germany, 1080; the German Democratic Republic, 500; Sweden, 540; Czeckoslovakia, 500; the maximum permissible concentration (MCP) in the USSR is 20 mg/m^3.

Methyl chloroform has been found in the stratosphere at an average level of 80 parts-per-trillion (ppt)[97]. Model calculations by McConnell and Schiff[93] show that about 15% of the methyl chloroform released into the atmosphere will reach the stratosphere where it is rapidly photolyzed to yield Cl and ClO, which can catalytically destroy ozone. Time scenarios based on past production figures and reasonable projections for future release rates lead to a steady-state ozone depletion due to this solvent about 20% as large as those resulting from the continuous release of chlorofluoromethanes (F-11 and F-12) at 1973 rates[93].

The potential threat of methyl chloroform depletion of ozone was further underscored recently by the National Academy of Sciences[98].

Aspects of toxicity and metabolism of methyl chloroform have recently been reviewed by Aviado et al[89].

No evidence of cancer-causing activity has been found in recent NCI animal studies of methyl chloroform[96]. However, because of shortened lifespan among the test animals, NCI did not consider the test adequate to make a positive or negative determination of cancer-causing potential. In these studies, methyl chloroform was given by gavage to groups of rats and mice of each sex at two dose levels. Starting dosages of methyl chloroform in rats proved so toxic that the experiment was stopped and a new group of rats were placed on test at dose levels of 1500 mg/kg body weight and 750 mg/kg. Averaged over the entire study, mouse dosages of methyl chloroform amounted to 5666 mg/kg for the high dose and 2833 mg/kg low dose.

In rats, various tumors occurred both in methyl chloroform-treated and untreated animals and could not be related to dosage of the chemical; liver cancer and neoplastic nodules appeared in some rats treated with carbon tetrachloride, a positive control.

More untreated mice had tumors than did methyl chloroform-treated animals, but the early death rate among high-dosed mice influenced this result. Virtually all the mice treated with carbon tetrachloride developed liver cancer and virtually all were dead before the end of the study. Although firm conclusion could not be made, the experience in mice suggested a lack of carcinogenicity with methyl chloroform.

There is a paucity of data concerning the mutagenicity of methyl chloroform. In one study methyl chloroform when tested in a desiccator with S. typhimurium TA 100 was weakly mutagenic with and without S-9 mix prepared from livers of B6C3F1 male mice[17]. No signs of maternal, embryonal or fetal toxicity were observed when pregnant rats and mice were exposed by inhalation to 4.75 mg/l methyl chloroform for 6 hrs/day on days 6 to 15 of gestation[99].

Methyl chloroform is rapidly absorbed through the lungs and the G.I. tract. It may also be absorbed in toxic quantities through the intact skin. Following inhalation of methyl chloroform for 8 hrs, rats excreted trichloroacetic acid and trichloroethanol in the urine[100].

A study of Carlson[101] suggests that a metabolite of methyl chloroform is more toxic than the parent compound as the pretreatment of rats with phenobarbital increased the hepatotoxicity of methyl chloroform as measured by serum glutomic-pyruvic and serum glutomic-oxaloacetic transaminase activities.

6. <u>1,1,2-Trichloroethane</u> ($ClCH_2CHCl_2$) is produced either directly or indirectly from acetylene, or indirectly from ethylene[87]. Direct production from acetylene is effected by reacting acetylene with a mixture of hydrogen chloride and chlorine in the presence of a catalyst (e.g., ferric, mercuric or antimony chloride). The manufacture of 1,1,2-trichloromethane by chlorination of 1,2-dichloroethane is in effect production from ethylene, viz., $CH_2=CH_2 + Cl_2 \rightarrow CH_2ClCH_2Cl + Cl_2 \rightarrow CH_2ClCHCl_2 + HCl$

The principal uses of 1,1,2-trichloroethane include: as an intermediate in the production of 1,1-dichloroethylene[87], industrial solvent to make adhesives, lacquers and coatings[101], and as a solvent for chlorinated rubber[87].

There are no available data on the number of individuals involved in the production of 1,1,2-trichloroethane, its use applications or the levels of exposure.

The current adopted TLV-TWA for 1,1,2-trichloroethane exposure (skin) is 10 ppm (45 mg/m^3) and the tentative TLV short-term exposure is 20 ppm (90 mg/m^3)[13].

1,1,2-Trichloroethane is considerably more toxic than its isomer, 1,1,1-trichloroethane (methyl chloroform)[89].

Recent NCI bioassay studies indicate that treated mice developed hepatocellular carcinomas and adrenal gland pheochromycytomas. The malignant nature of the hepatocellular carcinomas was evident based on their cellular characteristics and lung metastases. The evidence was not convincing that 1,1,2-trichloroethane was carcinogenic in the treated rats, under the conditions of the test[102,103].

There is a paucity of data concerning the mutagenicity of 1,1,2-trichloroethane. When tested in a desiccator in the Salmonella/microsome assay with TA 100 tester strain, it would found to be negative[17].

7. **1,1,2,2-Tetrachloroethane** (s-tetrachloroethane; acetylene tetrachloride; $CHCl_2CHCl_2$) is produced by the reaction of acetylene with chlorine in the presence of various catalysts such as ferric chloride, antimony pentasulfide or a mixture of sulfur dichloride and ferric chloride[87].

In the absence of air, moisture, and light, 1,1,2,2-tetrachloroethane is stable, even at high temperatures. When exposed to the air, dehydrochlorination slowly takes place with formation of trichloroethylene and traces of phosgene. Under pyrolytic conditions, tetrachloroethane splits off hydrogen chloride and chlorine, with trichloroethylene, tetrachloroethylene, penatchloroethane and hexachloroethane being formed[87].

1,1,2,2-Tetrachloroethane is a starting point of chlorinated solvent manufacture in many producing countries. It is also used as a solvent and to a minor extent as an insecticidal fumigant[17].

The numbers of workers involved in the production and use application of 1,1,2,2-tetrachloroethane or their levels of exposure are not known.

The current OSHA environmental standard for 1,1,2,2-tetrachloroethane is 5 ppm (35 mg/m^3) (skin) for an 8-hr TWA[20]. The tentative ACGIH tolerated limit value-short term exposure limit (skin) is 10 ppm (70 mg/m^3)[13]. The NIOSH recommendation for environmental exposure limit is 1 ppm (time weighted average)[20].

Hygienic standards for permissible levels of 1,1,2,2-tetrachloroethane in the working environment in various countries are as follows (mg/m^3)[14]: Federal Republic of Germany, 7; the Democratic Republic of Germany, 10 and the MPC in the USSR is 5 mg/m^3.

Orally administered technical grade 1,1,2,2-tetrachloroethane was found to be a liver carcinogen in both sexes of B6C3F1 mice, but did not provide evidence of carcinogenicity when tested in Osborne-Mendel rats[104,105]. Ninety-percent (44/49) of high-dose (108 mg/kg/day) and 13/50 (20%) of low-dose (62 mg/kg/day) male mice treated for 78 weeks developed hepatocellular carcinomas. Liver cancer was also observed in female mice treated with 76 and 43 mg/kg/day for 78 weeks[105].

1,1,2,2-Tetrachloroethane is mutagenic towards E. coli pol A+/pol A- and in S. typhimurium TA 1530 and TA 1535 but not in TA 1538[106].

8. Hexachloroethane (perchloroethane, CCl_3CCl_3) is produced when any lower aliphatic straight-chain hydrocarbon or chlorinated hydrocarbon is reacted with excess chlorine in the presence of chlorination catalysts at temperatures above 200°C. The industrial processes[87] include: (a) the chlorination of tetrachloroethylene, in the presence of ferric chloride at 100-140°C; (2) the photochemical reaction of tetrachloroethylene under pressure with liquid chlorine at less than 60°C; (3) via the chlorination of chlorobutadiene, 1,2-dichloroethane, chloropentanes or chloroparaffins.

At temperatures above 71°C, decomposition of hexachloroethane can occur with the formation of tetrachloroethylene and carbon tetrachloride.

Hexachloroethane has been used as a plasticizer for cellulose esters, as an accelerator in rubber, a retardant in fermentation processes, as a constituent of various fungicidal and insecticidal formulations, as a moth-repellant (either alone or in admixture with para-dichlorobenzene, as an additive to fire-extinguishing fluids, in the formulation of extreme pressure lubricants and in the production of smoke (fog) candles and grenades[8]

Data are not available concerning the production quantities, numbers of workers engaged in the production of hexachloroethane or in its use applications or the respective exposure levels.

The current ACGIH threshold limit value-time weighted average for exposure to hexachloroethane (skin) is 1 ppm (10 mg/m^3) and the tentative threshold limit value for short-term exposure is 3 ppm (30 mg/m^3)[13].

Hexachloroethane has been recently reported by NCI to be carcinogenic in B6C3F1 mice inducing hepatocellular carcinoma in both sexes following administration in corn oil by gavage at 1179 and 590 mg/kg/day for male and female mice for 5 days/week for 78 weeks[107].

No evidence was provided for the carcinogenicity of hexachloroethane in male and female Osborne-Mendel rats administered the compound in corn oil by gavage at levels of 423 and 212 mg/kg/day for male and female rats for 5 days/week for 78 weeks[107]. This failure to observe a carcinogenic effect may have been due to their early death, as evidence by the association between increased dosage and accelerated mortality[103].

There are no data concerning the mutagenicity of hexachloroethane.

9. Miscellaneous Chloro, Bromo, Iodo Derivatives

Although the emphasis above has been on saturated chlorinated hydrocarbons, selected primarily on their production volumes, reported carcinogenicity and/or mutagenicity as well as populations at potential risk, it should be noted that other related chlorinated, as well as brominated and iodinated derivatives based on á-priori structural considerations are also potentially hazardous compounds worthy of consideration. The numbers of individuals engaged in the production or use application or the levels of exposure of these compounds are generally not known.

a. **Ethyl chloride** (chloroethane; monochloroethane; CH_3CH_2Cl) is produced by the free radical chlorination of ethane; by the addition of hydrogen chloride to ethylene or by the action of chlorine on ethylene in the presence of the chlorides of copper or iron.

More than 80% of the total 1973 U.S. production of ethyl chloride (approximately 680 million pounds) was used for the production of tetraethyl lead, used as a gasoline antiknock agent. The remainder was used for a variety of applications including: (a) manufacture of ethyl cellulose plastics, dyes, and pharmaceuticals; (b) use as a solvent; (c) use as a refrigerant; and (d) to a limited extent as a topical anesthetic[108].

The U.S. production of ethyl chloride in the years 1974 to 1976 was 662, 575 and 642 million pounds respectively[108]. The future level of production of ethyl chloride is largely dependent on the consumption of tetraethyl lead as a gasoline additive. In the event the EPA's program to reduce the lead content of gasoline to 0.5 gram/gallon by late 1979 is implemented, large decreases in the consumption of lead alkyls would result. It should be noted that U.S. consumption of lead alkyls, mostly tetraethyl lead, decreased (on a 100% lead alkyl basis) from 660 million pounds in 1973 to 548 million pounds in 1975.

Combined U.S. production of ethyl cellulose and ethyl hydroxyethyl cellulose, both of which are made with ethyl chloride amounted to approximately 13 million pounds in 1973. Of the 10 million pounds domestically consumed, 7 million pounds involved surface cotaings and the remainder consumed was in plastics, adhesives, and inks[108].

The current ACGIH adopted threshold limit value-time weighted average for ethyl chloride is 1000 ppm (2,600 mg/m^3) and the tentative TLV short term exposure is 1,250 ppm (3,250 mg/m^3)[13].

Hygienic standards for permissible levels of carbon tetrachloride in the working environment in various countries are as follows (mg/m^3)[14]: Federal Republic of Germany, 2600; Democratic Republic of Germany, 2000. The MPC for ethyl chloride in the USSR is 50 mg/m^3.

There are no available data as to the carcinogenicity and/or mutagenicity of ethyl chloride.

b. **Methyl iodide** (iodomethane, CH_3I) is produced from methanol, iodine and red phosphorus or from the reaction of potassium iodide and methyl sulfate. Methyl iodide is used primarily as a methylating agent in the preparation of pharmaceuticals and in organic synthesis[109]. It also is used to a limited extent in microscopy due to its high refractive index, and as a reagent in testing for pyridine.

The production of methyl iodide in the U.S. in 1973 amounted to 8.9×10^3 lbs[108]. The current TLV-time weighted average in the U.S. for methyl iodide (skin) is 5 ppm (28 mg/m^3). The tentative TLV-short term exposure level is 10 ppm (56 mg/m^3).

It should also be noted that methyl iodide can be formed in nuclear reactor environments[110]. An ambient concentration of 80×10^{-12} by volume (0.08 ppb) of methyl iodide has been reported in the air of New Brunswick, NJ[111].

Methyl iodide is carcinogenic in BD strain rats by subcutaneous administration inducing local sarcomas after single or repeated injections[112,113]. In a limited study in A/He mice, methyl iodide caused an increased incidence of lung tumors after intraperitoneal injection[114].

Methyl iodide has been reported to be mutagenic in S. typhimurium TA 100 when plates were exposed to its vapors or tested in desiccators[17]. The mutagenic activity was slightly enhanced in the presence of rat liver microsomal fraction (9000 x g)[17,77]. Methyl iodide did not increase the back mutation frequency in Aspergillus nidulans to methionine in dependence at concentrations of 0.01-0.1 M for 5-15 minutes[115], but did increase mitotic recombination in S. cerevisiae D3[17].

c. <u>Other alkyl bromides and iodides</u>. A number of lower chain (C_1-C_4) alkyl bromides and iodides could also be considered as potentially hazardous substances[116] since a high percentage of those tested have been shown to be carcinogenic and/or mutagenic. For example, of 15 alkyl halides evaluated by Poirier et al[114], in studies where male and female A/Heston (A/He) strain mice were injected i.p. with 3 doses of alkyl halides 3 times/week for a total of 24 doses, nine were found to cause lung tumors at a confidence level of $p < 0.05$ and one at a level of $p < 0.01$ (Table 1). The four most active carcinogens were the alkyl iodides, e.g., methyl-, n-butyl-, isopropyl- and n-propyl, which would be expected because of the high reactivities of these compounds with electron rich centers[117]. However, it should be noted that ethyl iodide was not carcinogenic, and sec.butyl iodide ranked seventh in activity. Of the remaining active compounds, three were bromides (e.g., sec.butyl-, isobutyl-, and tert.butyl-) and two were chlorides (e.g., tert.butyl- and sec.butyl-). Poirier et al[114] concluded that chemicals with primary structures are less active than those with secondary or tertiary structures, although there were exceptions to this (e.g., methyl iodide which was most active, and tert.butyl iodide which was negative.

Table 1. Alkyl Halides Causing Pulmonary Adenomas in Mice with Confidence Level of < 0.05[a]

Rank	Compound	Lowest Effective Dose (mmol/kg)
1	Methyl iodide	0.31
2	n-Butyl iodide	2.6
3	Isopropyl iodide	7.0
4	n-Propyl iodide	17.6
5	sec-Butyl bromide	21.8[b]
5	Isobutyl bromide	21.8
5	tert-Butyl bromide	21.8
6	tert-Butyl chloride	32.4
7	sec-Butyl iodide	32.6
8	sec-Butyl chloride	35.0

[a] Poirier et al., 1975[114]
[b] Significant at $p < 0.01$

When 12 alkyl halides were tested for mutagenicity to histidine-independent revertants of E. coli WP2 (hcr-) by 8 hour exposures to the chemicals, all ethyl, propyl and butyl halides studied were mutagenic to one or more strains[118]. Chain branching generally increased mutagenic activity, these compounds being 5 to 20 times more active than the corresponding n-isomers. In general, increasing the chain lengths of the alkyl halides diminished their mutagenic activities. This does not correlate well with the in vitro reactivities of these compounds with nucleophilic substrates, as increased chain length has only minor effects on reactivity for compounds containing two or more carbon atoms[117].

The alkyl chlorides were generally less mutagenic than the bromides and iodides[118] which correlates well with the bond energies of these compounds[117].

Three of the four butyl bromide isomers (n-, sec-, and tert.butyl) were mutagenic in assay in desiccators when tested with S. typhimurium TA 100[17].

In the methyl halide series, the order of mutagenic activity was bromide > chloride > iodide when tested in desiccators in S. typhimurium TA 100[17].

It should also be noted that the trihalomethanes, bromodichloromethane, chlorodibromomethane and bromoform were mutagenic only when tested in desiccators but not when incorporated into agar when tested with S. typhimurium TA 100[17].

It is estimated that the total exposure of the alkyl bromide group as a whole is high (e.g., the total production or exposure is estimated to be at least 10^{10} grams)[116].

One commercially important alkyl bromide is methyl bromide (bromomethane, CH_3Br) which is an intermediate in organic synthesis, as a grain fumigant and degreasing wool. Methyl bromide is prepared industrially by the action of hydrobromic acid on methanol.

The current TLV ceiling value for methyl chloride in the U.S. is 20 ppm (80 mg/m^3)[14]. The ACGIH adopted TLV-time weighted average (skin) is 15 ppm (60 mg/m^3) and the tentative TLV-short term exposure limit is 15 ppm[13].

Hygienic standards for permissible levels of methyl bromide in the working environments in various countries as tabulated by Winell[14] are as follows (mg/m^3): Federal Republic of Germany, 80; the Democratic Republic of Germany, 50; the MPC in the USSR is 1 mg/m^3.

10. Fluorocarbons

Increasing interest and anxiety has focused on the fluorocarbons (chloroformmethanes) because of the significant emissions, their potential environmental impact, including destruction of the protective ozone layer[119-122] with subsequent projected increases in deaths due to skin cancer[122-125].

Since their introduction 40 years ago as refrigerants and later as propellants for self-contained aerosol products, the fluorocarbons have generally been considered to have an extremely low order of biological activity[126] and as a consequence, their use patterns and production have increased enormously. Between 1960 and the present, the world production was estimated to have grown exponentially with a doubling time of 3.5 years[122,127]. These chemicals (mostly known as "freons"), unlike CCl_4, have no natural sources or sinks in the troposphere, possess relative chemical inertness and high volatility[119,120]. The major fluorocarbons of environmental as well as toxicological concern are trichlorofluoromethane ($CFCl_3$; F-11, P-11) and dichlorodifluoromethane (CF_2Cl_2; F-12, P-12) which are now used almost exclusively as aerosol propellant gases (approximately 50-60%); refrigerants (approximately 20-30%) and for a number of other purposes including blowing agents for the production of foam, as solvents, and fire extinguishers[129]. Other fluorocarbons (e.g., F-22) are used as feedstocks for fluorocarbon resins.

Fluorocarbons are produced worldwide by 6 companies in the U.S. and by an estimated 48 companies in more than 23 countries. World production of $CFCl_3$ and CF_2Cl_2 (excluding the USSR and Eastern European countries) was 1.7 billion pounds in 1973, which represents an 11% growth over the 1972 production[124]. Approximately half of that world production and use occurs in U.S. The total U.S. production of fluorocarbons has been doubling every 5 to 7 years since the early 1950's. U.S. production of $CFCl_3$ and CF_2Cl_2 in 1974 is estimated to have been 859 million pounds and it is believed that the market will continue to grow by about 5 to 6% a year[127]. Approximately 13.8 billion pounds of $CFCl_3$ and CF_2Cl_2 have been produced to date in the world[124].

Approximately 3 billion aerosol cans (containing $CFCl_3$ and CF_2Cl_2 as propellants) are sold yearly in the U.S. for use as follows (in millions of units, for 1972): personal products (e.g., shaving creams, cosmetics, perfumes, deodorants, anti-perspirants, hair sprays) 1490; household products (e.g., window cleaners, air freshners, oven cleaners, furniture polishes) 699; coatings and finishing products (e.g., paint sprays) 270; insect sprays, 135; lubricant and degreasers, 100; and automotive products, 76. Miscellaneous uses have included: mold releases, silicone sprays and in some foods[124].

The most commonly used procedure for the production of fluorocarbons is the successive replacement of the chlorine atoms of chlorocarbon by fluorine using anhydrous HF with partially fluorinated antimony pentachloride as a catalyst[124].

Approximately 4000 people are directly engaged in fluorocarbon production, sales and research by 6 companies at 15 plants in the U.S.[124]. No estimate of the numbers of individuals exposed to the myriad number of use applications (e.g., primarily aerosol) of the fluorocarbons is available. The TLV time weighted average for $CFCl_3$ in the US is 5600 mg/m^3 (1000 ppm)[13]. The tentative threshold limit value-short term exposure limit is 7,000 mg/m^3 (1,250 ppm).

The major source for the release of fluorocarbons into the atmosphere results from their use as propellants, the annual loss estimated to be in excess of 650 million pounds[129]. Lesser amounts are released from uses as foaming agents, refrigerants, fire extinguishers and solvents. Approximately 1% of fluorocarbon is lost during production and 1% during transportation and storage amounting to approximately 10 million pounds respectively in the U.S.[124]. Atmospheric emission rates for $CFCl_3$ alone have increased from an average of 0.14 billion pounds per year between 1961 and 1965 to 0.51 billion pounds per year between 1971 and 1972 with the U.S. and Canada accounting for 44% of the world emissions[130]. The yearly world wide emission of CF_2Cl_2 into the troposphere in 1974 is estimated to be nearly 1 billion pounds[67].

Both $CFCl_3$ and CF_2Cl_2 are not appreciably decomposed in the lower atmosphere and are virtually inert chemically in the troposphere. Since they have very low solubility in water, they are not washed out of the atmosphere by precipitation. These fluorocarbons are found in the atmosphere in concentrations that appear to be consistent with the total release to date[124].

$CFCl_3$ and CF_2Cl_2 are ubiquitous and have been monitored at background levels in the 100-500 parts-per-trillion (ppt) range. In 1974, the Northern Hemisphere (the primary source of $CFCl_3$) contained about 100 ppt by volume in clean air, compared to about 60 ppt in the Southern Hemisphere. Local fluorocarbon sources such as urban areas produced high local concentrations and unrepresentative ratios of CF_2Cl_2 to $CFCl_3$ that differ from world wide averages. In clean air, there appears to be a seasonal variation in $CFCl_3$ that may be due to seasonal differences in releases of fluorocarbons to the atmosphere and/or seasonal differences in the intensity of vertical mixing in the troposphere[124]. Mean levels of CF_2Cl_2 and $CFCl_3$ ranging from 0.32-6.27 and 0.13-1.44 ppb respectively have been found at 8 locations in the U.S.[67]. During extended periods of inversion, the ambient concentrations of the ubiquitous fluorocarbons such as CF_2Cl_2 and $CFCl_3$ may obtain values as high as 100-500 times their minimum concentrations[67,131]. Fluorocarbon use patterns suggest increasing concentrations going from the background environment to urban areas to human dwellings. Concentrations in homes may vary in the 200-500,000 ppt range, depending on the sporadic use of aerosols and leaks from refrigerant applications[132]. It is expected that highly populated centers may have average concentrations 10-15 times the global concentrations of $CFCl_3$. An average level of 650 ppt (0.65 ppb) and a lower concentration of 110 ppt of $CFCl_3$ has been reported in the Los Angeles Basin[133].

Fluorocarbon levels in homes and in the immediate vicinity of factories where they are used may be as high as 0.5 ppm which is several thousand times as high as background levels. However, these levels are still a factor of more than a thousand below the threshold limit value of part per thousand[132]. Concentrations of Freon-12 in a beauty shop have been reported to range from 370 ppb[132] to 3000 ppb[134].

The most important sink for atmospheric $CFCl_3$ and CF_2Cl_2 appears to be stratospheric photolytic dissociation to $CFCl_2$ + Cl and to CF_2Cl + Cl, respectively at altitudes of 20-40 km. Each of the reactions creates two odd-electron species - one Cl atom and one free radical[119]. Thus, in addition to the NO_x cycle of ozone destruction, photodissociated atomic chlorine is a free radical that can catalytically cause the removal of stratospheric ozone as follows: (1) $Cl + O_3 \rightarrow ClO + O_2$; (2) $ClO + O \rightarrow Cl + O_2$. The net result is $O + O_3 \rightarrow 2O_2$. It is estimated that the ClO_x cycle may be 3 times more efficient in destroying ozone than the NO_x cycle on a molecule for molecular basis[119]. Stratospheric photolysis occurs between 185 and 227 nanometers. Fluorocarbons are estimated to take about 10 years after their release at ground level to reach stratospheric heights at which they are photolyzed. Fluorocarbons may remain in the atmosphere for 40-150 years, and concentrations can be expected to reach 10 to 30 times present levels[119,121].

The best current estimates are that fluorocarbon production and release to the environment to date may have resulted in a current reduction in average ozone concentration, most likely between 0.5 and 1% and possibly as large as 2% and eventually may result in as much as 1.3 to 3% reduction in the equilibrium ozone concentration[124].

Current model calculations predict that if release of fluorocarbons were to continue at the 1972 rate, a maximum reduction of about 7% in the equilibrium ozone concentration would be expected after several decades[124].

Calculations based on observed changes in incidence of skin cancer with variations with latitude for each percent ozone reduction range from 2100 to 15,000 (6000 medium) additional cases of non-melanoma skin cancer per year in light-skinned individuals in the U.S. at steady state[124].

An important aspect of the theoretical reduction of ozone by $CFCl_3$ and CF_2Cl_2 is its projected decrease for about 10 years after cessation of releases to the atmosphere and the very slow subsequent recovery to normal values taking place over many tens of years[124].

It is also important to note other model calculations concerning the destruction of ozone by fluorocarbons. It has been suggested[135] that if the fluorocarbon's tropospheric lifetime is as short as 10 years, representing about 80% destruction in the troposphere, current long-term predictions of ozone depletion would be reduced by about 80%. Hence for the same fluorocarbon usage, the model shows that there would be a 1.2% ozone depletion by the year 2100 rather than the approximately 10-15% depletion predicted by current theoretical calculations[136] which assume photolysis to be the only removal mechanism.

Long-term photolysis studies in simulated sunlight of both $CFCl_3$ and CF_2Cl_2 in ambient air samples and in air samples with 1 ppm of hydrocarbon and 1 ppm of nitrogen oxides indicate that the two fluorocarbons are photochemically stable even when photolyzed for several weeks[132]. The stratospheric half-life of 1 ppb of $CFCl_3$ has been calculated to be 69 hr based on UV photolysis experiments with zero air and irradiation equivalent to 400 times the solar irradiance at the top of the atmosphere[67]. Residence times in the atmosphere of 10 years for CF_2Cl_2[127] and 30 years for $CFCl_3$ have been reported[137] while Wofsy et al[136] estimated the lifetimes of the above fluorocarbons to be 45 and 68 years respectively.

Although earlier reports considered the fluorocarbons to be of an extremely low order of biological activity[126] reports during the 1960's of fatalities attributed to the use and/or abuse of aerosols[138], particularly of those used in the treatment of bronchial asthma have made it necessary to re-examine the toxicity of the propellants.

Propellants used in aerosols have the potential to produce bronchoconstriction, reduce pulmonary compliance, depress respiratory minute volume, reduce mean blood pressure and accelerate heart rate in dogs[139]. $CFCl_3$ (F-11) the most widely used low-pressure propellant was found to be the most toxic propellant, it exerted all of the above undesired effects except bronchoconstriction and a reduction of pulmonary complicance. Cardiac arrhythmia was the most serious sign of toxicity to acute inhalation of $CFCl_3$ in the mouse, rat, dog and monkey[140] and may account for the sudden deaths associated with the use and abuse of aerosols[141].

There is a paucity of data concerning both the carcinogenicity and mutagenicity of the fluorocarbons. Epstein et al[142] reported that in mice, the combined administration of $CFCl_3$ (F-11) and the insecticidal synergist piperonyl butoxide, increased the incidence of malignant hepatoma and postulated that piperonyl butoxide may so alter the in vivo dehalogenation of the propellant that carcinogenic compounds are formed.

Mutation studies with <u>Drosophila melanogaster</u> exposed to four flourinated hydrocarbon gases were described by Foltz and Fuerst[143]. The compounds studied were: Freon C-318 (octafluorocyclobutane; $CF_2CF_2CF_2CF_2$); Genetron-23 (Fluoroform; CHF_3); Genetron-152A (1,1-difluoroethane; CH_3CHF_2) and perfluorobutene-2 (Octafluoro-2-butene; $CF_3CF=CFCF_3$). Two special genetic techniques were employed. The Basc technique, with Muller-5 females and wild-type males[144] was used for scoring sex-linked recessive lethal mutations that arise in the germ line of the treated paternal male. All four fluorinated hydrocarbons were found to significantly increase mutation rates in progeny over control levels with Genetron-23 being the most mutagenic of the gases.

Freons 12, 21, and 22 (CF_2Cl_2, $CHFCl_2$ and CHF_2Cl respectively) have been reported to be non-mutagenic when tested without metabolic activation in <u>S. typhimurium</u> TA 1538 strain in contrast to methyl chloride and vinyl chloride which were mutagenic when tested similarly[145].

Trichlorofluoromethane (F-11) did not induce mutation after incubation with liver microsomes and <u>S. typhimurium</u> TA 1535 and TA 1538[46].

A number of fluorocarbons used as refrigerants and aerosols (including F-11 and F-12) in the presence of S-9 homogenate from Aroclor-1254 induced rats, induced base-pair substitution mutations as evidenced by their genetic activity with <u>S. typhimurium</u> strains G-46, TA 1535 and TA 100, no frame shift mutations in TA 1537, TA 1538 and TA 98 could be detected. While positive results were obtained with most of the fluorocarbons in a time exposure method and showed a time-related increase in the revertant frequency, the results from exposure to various volumes of the gas in a closed chamber yielded positive results but did not show evidence of a dose response[146].

The biotransformation and elimination of $(^{14}C)CFCl_3$ and $(^{14}C)CF_2Cl_2$ has been studied in beagles[147]. Less than 1% of inhaled fluorocarbons was biodegraded to metabolites after a short (6-20 min) period of inhalation. Limitations posed by the presence of radioactive impurities (e.g., $(^{14}C)CCl_4$ and 1.4% $(^{14}C)CHCl_3$ and $(^{14}C)CClF_3$ and/or $(^{14}C)CF_4$ in CF_2Cl_4) prevent excluding the possibility that any metabolic conversion occurred. However, it was suggested that $CFCl_3$ and CF_2Cl_2 (analogous with fluorocarbon anesthetics) are relatively refractory to biotransformation after inhalation. It was cautioned that extrapolation of these conclusions to other routes of administration (e.g., oral) or to longer exposure and/or to other fluorocarbons should be guarded[147].

Very low activity was associated with soluble protein or RNA added to incubation mixtures of $^{14}CCl_3F$ with liver microsomes suggesting extremely low biological activity[46]. Under anaerobic reducing conditions, CCl_3F forms complexes with ferrous cytochrome P450. Carbon monoxide was detected as a metabolic product of the interaction[84]. Other reactions occur as well in this system as evidence by the formation of $CHCl_2F$ from CCl_3F under similar incubation conditions (analogous to $CHCl_3$ formed from CCl_4)[148,149]. It should also be noted that active metabolites of both of these substrates bond irreversibly to microsomal proteins and lipids[150].

References

1. Burchfield, H. P., and Storrs, E. E., Organohalogen Carcinogens, In "Advances in Modern Toxicology" (eds) Kraybill, H. F., and Mehlman, M. A., Vol. 3, Hemisphere Press, Washington, DC (1977) pp. 319-351
2. Fishbein, L., Industrial mutagens and potential mutagens, I. Halogenated aliphatic derivatives, Mutation Res., 32 (1976) 267-308
3. Van Duuren, B. L., Chemical structure, reactivity, and carcinogenicity of halohydrocarbons, Environ. Hlth. Persp., 21 (1977) 17-23
4. U.S. Environmental Protection Agency, A Report: Assessment of health risks from organics in drinking water, Office of Research and Development, Washington, DC, April 30 (1975)
5. Morris, J. C., Formation of halogenated organics by chlorination of water supplies, EPA Rept. 600/1-75-002, National Technical Information Service (NTIS), Springfield, VA, March (1975)
6. Rook, J. J., Formation of haloforms during chlorination of natural water, Water Treatment Exam, 23 (1974) 234
7. Henschler, D., Metabolism and mutagenicity of halogenated olefins-a comparison of structure and activity, Environ. Hlth. Persp., 21 (1977) in press
8. Bonse, G., and Henschler, D., Chemical reactivity, biotransformation and toxicity of polychlorinated aliphatic compounds, CRC Crit. Rev. Toxicol., 4 (1976) 395
9. Cownheim, F. A.,and Moran, M. K., Methyl chloride-methylene chloride, In Faith, Keyes and Clark's Industrial Chemicals, 4th ed., John Wiley, New York (1975) 530-538
10. Hardie, D. W. F., Methyl Chloride, In: Kirk-Othmer's Encyclopedia of Chemical Technology, 2nd ed., Vol. 5, Interscience, New York (1964) 100-111
11. Clemens, D. H., and Lange, R. J., Ion-exchange resins from acrolein copolymers, U.S. Patent 3,813,353, May 28 (1974) Chem. Abstr., 81 (1974) P121747E
12. Mitsuyasu, T., and Tsuji, J., Quaternary ammonium salts, Japan Kokai, 7,426,209 Chem. Abstr., 81 (1974) P33374
13. ACHIG, Threshold Limit Values for Chemical Substances in Workroom Air Adopted by ACGIH for 1976, American Conference of Government Industrial Hygienists, Cincinatti, Ohio (1976)
14. Winell, M., An international comparison of hygienic standard for chemicals in the work environment, Ambio, 4 (1975) 34-37
15. Chopra, N. M., and Sherman, L. R., Systematic studies on the breakdown of p,p'-DDT in tobacco smokes. Investigation of methyl chloride, dichloromethane, and chloroform in tobacco smoke, Anal. Chem., 44 (1972) 1036
16. Andrews, A. W., Zawistowski, E. S., and Valentine, C. R., A comparison of the mutagenic properties of vinyl chloride and methyl chloride, Mutation Res., 40 (1976) 273-276
17. Simmon, V. F., Kauhanen, K., and Tardiff, R. G., Mutagenic activity of chemicals identified in drinking water, In "Progress in Genetic Toxicology", (eds) Scott, D., Bridges, B. A., and Sobels, F. H., Elsevier/North Holland Biomedical Press, Amsterdam (1977) 249-286
18. Hardie, D. W. F., Methylene Chloride, In "Kirk-Othmer's Encyclopedia of Chemical Technology" 2nd ed., Vol. 5, Interscience, New York (1964) 111-119
19. Anon, Methylene chloride passes early tests, Chem. Eng. News, May 9 (1977) 6
20. NIOSH, Summary of NIOSH recommendations for occupational health standards, October, 1977, Chem. Reg. Reptr., 1 (40) (1977) 1376-1391

21. Jongen, W. M. F., Alink, G. M., and Koeman, J. H., Mutagenic effect of dichloromethane on Salmonella typhimurium, Mutation Res., 56 (1978) 245-248
22. Price, P., Personal communication in reference 16
23. Stevens, J. L., Andersn, M. W., Metabolism of dihalomethanes to carbon monoxide (CO) by isolated rat hepatocytes (IRH), Pharmacologist, 18 (1976) 246
24. Hardie, D. W. F., In "Kirk-Othmer's Encyclopedia of Chemical Technology", 2nd ed., Vol. 5, Interscience, New York (1964) 119-127
25. IARC, Monographs on the Evaluation of Carcinogenic Risk of Chemicals to Man, Vol. 1, International Agency for Research on Cancer, Lyon (1972) 61-65
26. NIOSH, Current Intelligence Bulletin: Chloroform, National Institute for Occupational Safety and Health, Rockville, MD, March 15 (1976)
27. U.S. International Trade Commission, Synthetic Organic Chemicals, United States Production and Sales of Miscellaneous Chemicals, 1975, Preliminary, U.S. Govt. Printing Office, Washington, DC (1976) p. 10
28. McConnell, G., Ferguson, O. M., and Pearson, C. R., Chlorinated hydrocarbons in the environment, Endeavor, 34 (1975) 13-27
29. Federal Register, Vol. 39, June 27 (1974) No. 125 Part II, page 23451
30. Yung, Y. L., McElroy, M. B., and Wofsy, Atmospheric halocarbons: A discussion with emphasis on chloroform, Geophys. Res. Letters, 2 (1975) 397-399
31. U.S. Environmental Protection Agency, Draft Report for Congress: Preliminary Assessment of Suspected Carcinogens and Drinking Water, Office of Toxic Substances, Washington, DC, October 17 (1975)
32. Lovelock, J. E., Atmospheric halocarbons and stratospheric ozone, Nature, 252 (1974) 292
33. Lovelock, J. E., Maggs, R. J., and Wade, R. J., Halogenated hydrocarbons in and over the Atlantic, Nature, 241, (1973) 194
34. Bellar, T. A., Lichtenberg, J. J., and Kroner, R. C., The occurrence of organohalides in chlorinated drinking waters, J. Am. Waterworks Assoc., 66 (1974) 703-706
35. Murphy, S. P., A Report: Assessment of health risk from organics in drinking water. Ad-Hoc Study Group fo the EPA SCience Advisory Board-Hazardous Materials Advisory Committee, April 30 (1975)
36. Symons, J. M., Suspect carcinogens in water supplies, Interim Report to Congress Water Supply Research Laboratory, National Environmental Research Center, Cincinatti, Ohio (1975)
37. Anon, Chloroform precursors found in natural waters, Chem. Eng. News, June 6 (1977) 6-7
38. Kaiser, K. L. E., and Lawrence, J., Polyelectrolytes: Potential chloroform precursors, Science, 196 (1977) 1205-1206
39. Symons, J. M., and Roebeck, G. G., Treatment processes for coping with variation in raw-water quality, J. Am. Waterworks Assoc., 67 (1975) 142-145
40. Anon, Government moves to control drinking water contaminants, Chem. Ecology, Feb. (1978) pp. 1,3
41. Eschenbrenner, A. B., Induction of hepatomas in mice by repeat oral administration of chloroform, with observations on sex differences, J. Natl. Cancer Inst., 5 (1945) 251-255
42. National Cancer Institute, Report on Carcinogenesis Bioassay of Chloroform, Carcinogenesis Program National Institutes of Health, Bethesda, MD, March 1 (1976)
43. Powers, M. B., and Welker, M. W., Evaluation of the oncogenic potential of chloroform long-term oral administration in rodents, 15th Annual Meeting Society of Toxicology, Atlanta, GA, March 14-18 (1976)

44. Roe, F. J. C., Palmer, A. K., Worden, A. N., and Van Abbe, N. J., Safety evaluation of toothpaste containing chloroform. 1. Long term studies in mice, International Congress on Toxicology, Toronto, March 30-April (1977) p. 26
45. Hill, R. N., Differential toxicity of chlorfrom in the mouse, Ann. NY Acad. Sci., 298 (1977) 170-175
46. Uehleke, H., Werner, T., Greim, H., and Kramer, M., Metabolic activation of haloalkanes and tests in vitro for mutagenicity, Xenobiotica, 7 (1977) 393-400
47. Uehleke, H., Greim, H., Kramer, M., and Werner, T., Covalent binding of haloalkanes to liver constituents, but absence of mutagenicity on bacteria in a metabolizing test system, Mutation Res., 38 (1976) 114
48. Sturrock, J., Lack of mutagneic effect of halothane or chloroform on cultured cells using the azaguanine test system, Brit. J. Anaest., 49 (1977) 207-210
49. Schwetz, B. A., Leong, B. K. J., and Gehring, P. J., Embryo and fetotoxicity of inhaled chloroform in rats, Toxicol. Appl. Pharmacol., 29 (1974) 348-357
50. Thompson, D. J., Warner, S. D., and Robinson, V. B., Teratology studies on orally adminsitered chloroform in the rat and rabbit, Toxicol. Appl. Pharmacol., 29 (1974) 348-357
51. Thompson, D. J., Warner, S. D., and Robinson, V. B., Teratology studies on orally administered chloroform in the rat and rabbit, Toxicol. Appl. Pharmacol., 29 (1974) 348-357
52. Hill, R. N., Clemens, T. L., Lin, D. K., and Vesell, E. S., Genetic control of chloroform toxicity in mice, Science, 190 (1975) 159-161
53. Ilett, K. F., Reid, W. D., Sipes, I. G., and Krishna, G., Chloroform toxicity in mice; correlations of renal and hepatic necrosis with covalent binding of metabolites to tissue macromolecules, Exp. Mol. Pathol., 19 (1973) 215
54. Reid, W. D., and Krishna, G., Centrobular heptatic necrosis related to covalent binding of metabolites of halogenated aromatic hydrocarbons, Exp. Mol. Pathol., 18 (1973) 80-95
55. Brodie, B. B., Reid, W. D., Cho, A. K., Sipes, G., Krishna, G., and Gillette, J. R., Possible mechanism of liver necrosis caused by aromatic organic compounds, Proc. Natl. Acad. Sci. (USA) 68 (1971) 160
56. Charlesworth, F. A., Patterns of chloroform metabolizing, Food Cosmet. Toxicol. 14 (1976) 59-60
57. Scholler, K. L., Modification of the effects of chloroform on the rat liver, Brit. J. Anaest., 42 (1960) 603
58. Cohen, E. N., Metabolism of the valatile anesthetics, Anesthesiology, 35 (1971) 193
59. Van Dyke, R. A., Biotransformation of volatile anesthetics with special emphasis on the role of metabolism in the toxicity of anesthetics, Can Anaesth. Soc. J. 20 (1973) 2
60. Brown, D. M., Langley, P. F., Smith, D., and Taylor, D. C., Metabolism of chloroform. I. Metabolism of ^{14}C-chloroform by different speices, Xenobiotica 4 (1974) 151-163
61. Hathway, D. E., Chemical, biochemical and toxicological differences between carbon tetrachloride and chloroform, Arznei. Forsch., 24 (1974) 173-176
62. Pohl, L. R., Bhooshan, B., and Krishna, G., Mechanism of the metabolic activation of chloroform, Abstract of 17th Annual Meeting of Society of Toxicology, San Francisco, CA, March 12-16 (1978) 39
63. Hardie, D. W. F., Carbon tetrachloride, In "Kirk-Othmer's Encyclopedia of Chemical Technology", 2nd ed., Vol. 5, Interscience, New York (1964) pp. 128-139
64. IARC, Monographs on the Evaluation of Carcinogenic Risk of Chemicals to Man, Vol. 1, International Agency for Research on Cancer, Lyon (1972) pp. 53-60

65. Roschin, A. V., and Timofeevskaya, C. A., Chemical substances in the work environment: Some comparative aspects of USSR and U.S. hygienic standards, Ambio, 4 (1975) 30-33
66. Singh, H. B., Fowler, D. P., and Peyton, T.O., Atmospheric carbon tetrachloride: Another man-made pollutant, Science, 192 (1976) 1231
67. Lillian, D., Singh, H. B., Appleby, L. Lobban, R., Arnts, R., Gumpert, R., Hague, R., Toomey, J., Kazazis, J., Antell, M., Hansen, D., and SCott, B., Atmospheric fates of halogenated compounds, Env. Sci. Technol., 9 (1975) 1042-1048
68. Neeley, S. B., Material balance analysis of trichlorofluoromethane and carbon tetrachloride in the atmosphere, Sci. Total Environ., 8 (1977) 267-274
69. Murray, A. J., and Riley, J. P., Occurrence of some chlorinated aliphatic hydrocarbons in the environment, Nature, 242 (1973) 37-38
70. Simmonds, P. G., Kerrin, S. L., Lovelock, J. E., and Shair, F. H., Distribution of atmospheric halocarbons in the air over the Los Angeles basin, Atmospheric Environment, 8 (1974) 209-216
71. Lovelock, J., Halogenated hydrocarbons in the atmosphere, International Symposium on Chemical Effect of Environmental Quality, Munchen/Neuherberg, Sept. 9-10 (1975)
72. Dowty, B., Carlosle, D., Laseter, J. L., and Storer, J., Halogenated hydrocarbons in New Orleans Drinking water and blood plasma, Science, 187 (1975) 75-77
73. Wit, S. L., Besemer, A. F. H., Das, H. A., Goedkoop, W., Loostes, F. E., Rept. No. 36/39 Toxicology, National Institute of Public Health, Bilthaven, Netherlands (1969)
74. Warwick, G. P., In "Liver Cancer" International Agency for Research on Cancer, Lyon (1971) pp. 121-157
75. Tracey, J. P., and Sherlock, P., Hepatoma following carbon tetrachloride poisoning, New York St. J. Med., 68 (1968) 2202
76. Rubin, E., and Popper, H., The evolution of human cirrhosis deduced from observations in experimental animals, chlorofluorocarbons in the atmosphere, Medicine, 46 (1967) 163
77. McCann, J., Choi, E., Yamasaki, E., and Ames, B. N., Detection of carcinogens as mutagens in the Salmonella/microsome test: Assay of 300 chemicals, Proc. Natl. Acad. Sci., 72 (1975) 5135-5139
78. Bignami, M., Cardamone, G., Carere, A., Comba, P., Dogliotti, E., Morpurgo, G., and Ortali, V. A., Mutagenicity of chemicals of industrial and agricultural relevance in Salmonella, Streptomyces and Aspergillus, Cancer Research, (1978) in press
79. Braun, R., and Schoneich, J., The influence of ethanol and carbon tetrachloride in the host-mediated assay with Salmonella typhimurium, Mutation Res., 31 (1975) 191-194
80. Schwetz, B. A., Leong, B. K. J., and Gehring, P. J., Embrytotoxicity and fetotoxicity of inhaled carbon tetrachloride, 1,1-dichloroethane and methyl ethyl ketone in rats, Toxicol. Appl. Pharmcol., 28 (1974) 852-864
81. Rechnagel, R. O., and Glende, E. A., Jr., Carbon tetrachloride: An example of lethal cleavage, CRC Crit. Revs. In Toxicology, (1973) 263-297
82. Butler, T. C., Reduction of carbon tetrachloride in vivo and reduction of carbon tetrachloride and chloroform in vitro by tissues and tissue constituents, J. Pharmcol. Exp. Therap., 134 (1961) 311
83. Fowler, S. S. L., Carbon tetrachloride metabolism in the rabbit, Brit. J. Pharmacol., 37 (1969) 773

84. Wolf, C. R., Mansuy, D., Nastainczyk, W., Deutschmann, G., and Ullrich, V., The reduction of polyhalogenated methanes by liver microsomal cytochrome P-450, Mol. Pharmacol., 13 (1977) 698-705
85. Vainio, H., Parkki, M. G., and Marniemi, J., Effects of aliphatic chlorohydrocarbons on drug-metabolizing enzymes in rat liver in vivo, Xenobiotica, 6 (1976) 599-604
86. Lowenheim, F. A., and Moran, M. K., 1,1,1-Trichloroethane, In, Faith, Keyes and Clark's Industrial Chemicals, 4th ed., John Wiley, New York (1975) 836-842
87. Hardie, D. W. F., Other Chloroethanes, In Kirk-Othmer Encyclopedia of Chemical Technology, 2nd ed., Vol. 5, Interscience, New York (1964) 154-170
88. Anon, TSCA Regulation of Methyl Chloroform Suggested by EPA Air Office, Pesticide Toxic Chem. News, 6 (2) (1977) 24
89. Aviado, D. M., Simaan, J. A., Zakhari, S., and Ulsamer, A. G., "Methyl Chloroform and Teichloroethylene in the Environment", CRC Press, Cleveland (1976)
90. Gleason, M. N., Gosselin, R. E., Hodge, H. C., and Smith, R. P., "Clinical Toxicology of Commercial Products", 3rd ed., Williams and Wilkins, Baltimore (1969)
91. Blackford, J. L., In "Chemical Economics Handbook" Stanford Research Institute Menlo Park, CA (1972) p. 697.3031A
92. U.S. International Trade Commission Reports, "Synthetic Organic Chemicals", Series C/P-76-8, Government Printing Office, Washington, DC, October 20 (1976)
93. McConnell, J. C., and Schiff, H. I., Methyl chloroform: Impact on stratospheric ozone, Science, 199 (1978) 174-177
94. Anon, Chem. Marketing Reptr., 211 (1977) 9
95. Anon, Chem. Week., 119 (1976) 32
96. National Cancer Institute, Carcinogenesis Bioassay of 1,1,1-trichloroethane, Office of Cancer Communications, NCI, Bethesda, MD (1977)
97. Anon, Draft MRI Report on Solvents Hits Methyl Chloroform for Ozone Depletion, Pesticide Toxic Chem. News, 5 (43) (1977) 24-25
98. Anon, Bromine, methyl chloroform pose depletion threats, New NAS Report, Pesticide Toxic Chem. News, 6 (7) (1978) 12-13
99. Schwetz, B. A., Leong, B. K. J., and Gehring, P. J., The effect of maternally inhaled trichloroethylene, perchloroethylene, methyl chloroform and methylene chloride on embryonal and fetal development in mice and rats, Toxicol. Appl. Pharmacol., 32 (1975) 84
100. Ikeda, M., and Ohtsuji, H., A comparative study of the excretion of Fujiwara reaction-positive substances in urine of humans and rodents given trichloro- or tetrachloro derivatives of ethane and ethylene, Brit. J. Ind. Med., 29 (1972) 99
101. Carlson, G. P., Effect of phenobarbital and 3-methyl cholanthrene pretreatment on the hepatotoxicity of 1,1,1-trichloroethane and 1,1,2-trichloroethane, Life Sci., 13 (1973) 67-73
102. Anon, 1,1,2-Trichloroethane and hexachloroethane appear to pose human cancer risks, Pesticide Toxic Chem. News, 6 (9) (1978) 30-31
103. National Institutes of Health, Seventh Meeting of the Data Evaluation/Risk Assessment Subgroup of the Clearinghouse on Environmental Carcinogens, NIH, Bethesda, MD, January 18 (1978)
104. Anon, Reports given Clearinghouse Subgroup by NCI Bioassay Program Staff, Chem. Reg. Reptr., 1 (1977) 990

105. National Cancer Institute, Bioassay of 1,1,2,2-Tetrachloroethane for Possible Carcinogenicity, Office of Cancer Communications, NCI, Bethesda, MD (1977)
106. Rosenkranz, H. S., Mutagenicity of halogenated olefins and their derivatives, Environ. Hlth. Persp., 21 (1977) in press
107. National Cancer Institute, Bioassay of hexachloroethane for possible carcinogenicity, DHEW Publication No. (NIH) 78-1318, National Institutes of Health, Bethesda, MD January 4 (1978)
108. Stanford Research Institute, A Study of Industrial Data on Candidate Chemicals for Testing, Prepared for Office of Toxic Substances, U.S. Environmental Protection Agency, EPA-560/5-77-066, Washington, DC, August (1977)
109. Hart, A. W., Gergel, M. G., and Clarke, J., Iodine, In Kirk, R. E., and Othmer, D. F. (eds) Encyclopedia of Chemical Technology, 2nd ed., Vol. 11, John Wiley and Sons, New York (1966) pp. 862-863
110. Barnes, R. H., Kircher, J. F., and Townley, C. W., Chemical-equilibrium studies or organic-iodide formation under nuclear-reactor-accident conditions, AEC Accession No. 43166 Rept. No. BMI-1781, Chem. Abstr., 66 (1966) 71587V
111. Lillian, D., and Birsingh, H., Absolute determination of atmospheric halocarbons by gas phase coulometry, Anal. Chem., 46 (1974) 1060-1063
112. Preussman, R., Direct alkylating agents as carcinogens, Food Cosmet. Toxicol. 6 (1968) 576-577
113. Druckrey, H., Kruse, H., Preussmann, R., Invankovic, S., and Lanschütz, C., Carcerogene alkylierende substanzen. III. Alkyl-halogenide, -sulfate, sulfonate, und ringespannte heterocyclen, Z. Krebsforsch., 74 (1970) 241-270
114. Poirier, L. A., Stoner, G. D., and Shimkin, M. B., Bioassay of alkyl halides and nucleotide base analogs by pulmonary tumor response in strain A mice, Cancer Res., 35 (1975) 1411-1415
115. Moura Duarte, F. A., Efeitos mutagenicos de alguns esteres de acidos inorganicos em aspergillus nidulans (EIDAM) winter, Ciencia e Cultura, 24 (1971) 42-52
116. DHEW Clearinghouse on Environmental Carcinogens, Organohalides, Chemical Selection Group, Bethesda, MD, Dec. 19 (1977)
117. Burchfield, H. P., and Storrs, E. E., Organohalogen Carcinogens, In "Advances in Modern Toxicology" Vol. 3, (eds) Kraybill, H. F., and Mehlman, M. A., Hemisphere Press, Washington, DC (1977) 329
118. Simmon, V. F., Personal communication, Structural correlates of carcinogenesis and mutagenesis meeting, 2nd Office of Science Summer Symposium, U.S. Food and Drug Administration, Annapolis, MD, August 31-Sept. 2 (1977)
119. Molina, M. J., and Rowland, F. S., Stratospheric sink for chlorofluoromethanes: chlorine catalyzed destruction of ozone, Nature, 249 (1974) 810-812
120. Cicerone, R. J., Stolarski, R. S., and Walters, S., Stratospheric ozone destruction by man-made chlorofluoromethanes, Science, 185 (1974) 1165-1167
121. Rowland, M. S., and Molina, M. J., Chlorofluoromethanes in the environment, Atomic Energy Commission Report No. 1974-1, Univ. of California, Irving, 1974
122. Crutzen, P. J., Estimates of possible variations in total ozone due to natural causes and human activities, Ambio, 3 (1974) 201-210
123. Hammond, A. C., Ozone destruction: problems' scope grows, its urgency recedes, Science, 186 (1975) 335-338
124. Council on Environmental Quality, Fluorocarbons and the Environment, Council on Environmental Quality, Federal Council for Science and Technology, Washington, DC, June (1975)
125. National Resources Defense Council, Petition of Concern to the Consumer Safety Commission, Chem. Tech. (1975) 22-27

126. Clayton, J. W., Fluorocarbon toxicity and biological action, Fluor. Chem. Revs. 1 (1967) 197-252
127. Lovelock, J. E., Maggs, R. J., and Wade, R. J., Halogenated hydrocarbons in and over the Atlantic, Nature, 241 (1973) 194
128. Kaiser Aluminum and Chem. Co., At Issue Fluorocarbons, Kaiser Aluminum and Chemical Corp., Oakland, Cal., June 1975
129. McCarthy, R. L., Fluorocarbons in the environment, presented at Am. Geophysical Union Meeting, San Francisco, CA (1974)
130. Hoffman, C. S., personal communication in ref. 67.
131. Lovelock, J. E., Atmospheric halocarbons and stratospheric ozone, Nature, 252 (1974) 292
132. Hester, N. E., Stephens, E. R., and Tayler, O. C., Fluorocarbons in the Los Angeles basin, J. Air. Pollut. Control. Assoc., 24 (1974) 591-595
133. Simmonds, P. G., Kerrin, S. L., Lovelock, J. E., and Shair, F. H., Distribution of atmospheric halocarbons in the air over Los Angeles bain, Atmos. Environ., 8 (1974) 209-216
134. Bridbord, K., Brubaker, P. E., Gay, B., Jr., and French, J. G., Exposure to halogenated hydrocarbons in the indoor environment, Env. Hlth. Persp., 11 (1975) 215-220
135. Anon, Fluorocarbon effect on atmosphere may be over estimated, a new MCA study says, Pesticide Chem. News, 4 (1976) 7
136. Wofsy, S. C., McElroy, M. B., and Sze, N. D., Freon consumption: implications for atmospheric ozone, Science, 187 (1974) 535-537
137. Su, C. W. and Goldberg, E. D., Chlorofluorocarbons in the atmosphere, Nature, 245 (1973) 27
138. Reinhardt, C. F., Azar, A., Maxfield, M. E., Smith, P. F., and Mullin, L. S., Cardia arrhthmias and aerosol, "Sniffing", Arch. Env. Hlth., 22 (1971) 265
139. Belej, M. A., and Aviado, D. M., Cardiopulmonary toxicity of propellants for aerosols, J. Clin. Pharmacol., 15 (1975) 105-115
140. Aviado, D. M., Toxicity of aerosol propellants in the respiratory and circulatory systems, IX. Summary of the most toxic: Trichlorofluoromethane (F-11), Toxicology, 3 (1975) 311-319
141. Aviado, D. M., Toxicity of aerosols, J. Clin. Pharmacol., 15 (1975) 86-104
142. Epstein, S. S., Joshi, S., Andrea, J., Clapp, P., Falk, H., and Mantel, N., Synergistic toxicity and carcinogenicity of "Freons" and piperonyl butoxide, Nature, 214 (1967) 526
143. Foltz, V. C., and Fuerst, R., Mutation studies with Drosophila melanogaster exposed to four fluorinated hydrocarbon gases, Env. Res., 7 (1974) 275-285
144. Spencer, W. P., and Stern, C., Experiments to test the validity of the linear R-dose/mutation frequency relation in Drosophila at low dosage, Gentics, 33 (1948) 43
145. Andrews, A. W., Silverman, S. J., and Valentine, C. E., The identification of endogenous and exogenous mutagenic compounds, 4th Carcinogenesis Bioassay Program, Orlando, Fla., February, 1976
146. Jagannath, D. R., Goode, S., and Brusick, D. J., Mutagenicity of fluorocarbons, 9th Annual Meeting of Environmental Mutagen Society, San Francisco, CA, March 9-13 (1978) 67
147. Blake, D. A., and Mergner, G. W., Inhalation studies on the biotransformation and elimination of (^{14}C)-trichlorofluoromethane and (^{14}C)-dichlorodifluoromethane in beagles, Toxicol. Appl. Pharmacol., 30 (1974) 396-407

148. Uehleke, H., Hellmer, K. H., and Tabaretti, S., Binding of ^{14}C-carbon tetrachloride to microsomal proteins in vitro and formation of chloroform by reduced liver microsomes, Xenobiotica, 3 (1973) 1-11
149. Wolf, C. R., King, L. J., and Parke, D. V., Anaerobic dechlorination of trichlorofluoromethane by liver microsomal preparations in vivo, Biochem. Soc. Trans., 3 (1975) 175-176
150. Uehleke, H., A comparative study of the binding of carbon tetrachloride and chloroform to microsomal proteins, Excerpta Med. Int. Congr. Ser., 311 (1973) 119-129

CHAPTER 13

ALKANE HALIDES, HALOGENATED ALKANOLS AND HALOGENATED ETHERS

A. Alkane Halides

Compounds containing halogen atoms on adjacent carbon atoms (vicinal substitution) are generally more reactive than those containing a single halogen.

1. Ethylene dichloride (1,2-dichloroethane; $ClCH_2CH_2Cl$) is produced industrially by the oxychlorination of ethylene. The estimated U.S. domestic consumption pattern for ethylene dichloride in 1976 was 86% as an intermediate for vinyl chloride; approximately 3% each as an intermediate for 1,1,1-trichloroethane (methyl chloroform and ethylene amines); 2% each as intermediate for tetrachloroethylene (perchloroethylene) 1,1-dichloroethane, and trichloroethylene and as a lead scavenger for motor fuels[1]. Ethylene chloride is also used in various solvent applications, and as a component of fumigants (with ethylene dibromide) for grain, upholstery and carpets[2].

U.S. production of ethylene chloride in recent years has been as follows (in billions of pounds): 1974 (9.17); 1975 (7.98); and 1976 (7.92). U.S. imports of ethylene dichloride in 1974 were estimated to be 75 million pounds and that exported in 1974 and 1975 totaled 369 and 130 million pounds respectively. In 1975, there were 11 ethylene dichloride producing companies in the U.S. with a total capacity of 13.8 billion pounds/year. Approximately 20×10^3 tons of ethylene chloride was produced world wide in 1973[3]. The current U.S. consumption of ethylene chloride is about 10 billion pounds[4].

A total of 2 million workers in the U.S. may receive some exposure to ethylene dichloride, with perhaps 200,000 receiving a substantial exposure primarily during its use as a solvent in textile cleaning and metal degreasing, in certain adhesives, and as a component in fumigants[4]. At least 29 U.S. firms use ethylene dichloride as an ingredient in at least 45 fumigant-insecticide products[5]. Low level exposure of gas station attendants can also occur from its use as a gasoline additive[4]. According to NIOSH, 36 U.S. firms put ethylene chloride into mixes with ethylene dibromide for this use[5].

Both the current OSHA and ACGIH environmental standards is 50 ppm, for 8 hr TWA, 100 ppm acceptable ceiling, 20 ppm maximum ceiling (5 min in 3 hrs)[6,7]. The NIOSH recommended in 1976 for a reduction from 50 to 5 ppm TWA and a 15 ppm ceiling (15 min)[7].

In a recent EPA National Organics Reconnaisance Survey in the U.S., ethylene chloride was found in 11 raw water locations at levels of < 0.2-31 µg/l and 26 finished water locations (32.9% of total) at levels of 0.2-6 µg/l[8].

In 1977, NCI reported preliminary results from studies indicating that male and female Osborne-Mendal rats administered ethylene dichloride in corn oil by gastric intubation for 78 weeks had significant excesses of site-specific malignant and non-malignant tumors. Specifically, of 50 male rats exposed daily to an average of 50-75 mg/kg ethylene dichloride, 20 developed malignant tumors; 6 of the animals had hemangiosarcomas of the spleen and 3 had squamous cell carcinomas of the fore stomach. Of 50 male rats exposed daily to 100-150 mg/kg, 20 developed tumors, 9 of which were hemangiosarcomas of unnamed internal organs. Female rats receiving 100-150 mg/kg appeared to exhibit a statistically significant excess of adenocarcinomas of the mammary glands[5,9].

B6C3F1 mice were administered ethylene dichloride in corn oil by gastric intubation for 78 weeks. The high and low time-weighted average doses for the male mice were 195 and 97 mg/kg/day respectively, and 299 and 149 mg/kg/day respectively for the female mice. The compound was found to be carcinogenic causing mammary adenocarcinoma and endometrial tumors in female mice and alveolar/bronchrolar adenomas in mice of both sexes[9]. The male mice also appeared to have a statistically significant excess of carcinomas of the stomach[5].

In December, 1977, the Manufacturing Chemists Association (MCA) informed EPA of an ongoing inhalation study on ethylene dichloride conducted by Dr. C. Maltoni of the Bologna Tumor Center[10]. According to an interim report on the study, no evidence of any exceptional tumor in rats and mice has been found. In this study, Swiss mice and Sprague-Dawley rats were exposed to ethylene dichloride at concentrations of 5, 10, 50 and 150 ppm for 100 weeks of a two-year study[10].

Ethylene dichloride, without activation, is a weak mutagen in S. typhimurium TA 1530, TA 1535 and TA 100 tester strains[11-14] (when tested in agar[11,12,14] or in a desiccator[13]) and negative when tested at 125 mg/plate with S-9 activation in TA 1538 and TA 98[14]. It should be noted that in a comparison of metabolites of ethylene dichloride, chloroacetaldehyde was hundreds of times more active on a molar basis than chloroethanol and chloracetic acid in reverting TA 100[11].

Ethylene dichloride is also mutagenic in E. coli (DNA polymer and deficient pol A-strain[15]) and increased the frequency of recessive lethals and induced chromosome disjunction in Drosophila.

Ethylene is not mutagenic in S. coelicolor or A. nidulans[14].

2. Ethylene dibromide (1,2-dibromoethane; sym-dibromoethane; ethylene bromide; DBE; EDB; $BrCH_2CH_2Br$) is prepared commercially by the reaction of ethylene and bromine. In 1974, five U.S. companies reported production of 150 million kg[19], and imports were reported to be 1770 kg[20]. The reported production of ethylene dibromide in the U.S. in 1975 was 125 million kg. Ethylene dibromide is currently produced in the United Kingdom, Benelux, France, Spain, Italy and Switzerland with production estimated to be about 3-30 million kg/year. Another estimate of West European production for 1974 was 40 million kg[18]. In Japan in 1975, 1.4 million kg were produced by two companies, and 667 thousand kg were imported[21].

The major areas of utility of ethylene dibromide include: (1) as a lead scavenger in tetralkyl lead gasoline and antiknock preparations; (2) as a soil and grain fumigant; (3) as an intermediate in the synthesis of dyes and pharmaceuticals and (4) as a solvent for resins, gums and waxes.

Approximately 97 million kg of ethylene dibromide were formulated into tetraalkyl lead antiknock mixes in the U.S. in 1974. The concentration of ethylene dibromide is variable. Motorfuel anti-knock mixes contain approximately 18% by weight of ethylene dibromide (2.8 g/liter), while aviation gasoline anti-knock mixes contain about 36% by weight[18].

Its use in the U.S. has been declining in these mixes since the early 1970's. In late 1976, one source estimated that U.S. consumption of ethylene dibromide would drop at an annual rate of 10% through 1980 as a result of a continued drop in usage of lead-based anti-knock motor mixes[1].

In the U.S., ethylene dibromide is registered for use as a soil fumigant on a variety of vegetable, fruit and grain crops[22]. An estimated 5 million pounds were used in the U.S. in 1975 as a fumigant (mostly on agricultural crops).

NIOSH estimates that some 9000 employees are potentially exposed to ethylene dibromide although this figure would increase to 660,000 if gasoline station attendants are included[23].

The current OSHA environmental standard for ethylene dibromide is 20 ppm (150 mg/m^3) for an 8 hr time-weighted exposure in a 40 hr work week; 30 ppm (230 mg/m^3) acceptable ceiling; and a 50 ppm (380 mg/m^3) maximum peak (5 min)[7,24]. NIOSH has recommended a 1 mg/m^3 ceiling (15 min) environmental exposure limit[7,23].

The chief sources of ethylene dibromide and dichloride emissions are from automotive sources via evaporation from the fuel tank and carburetor of cars operated on leaded fuel. Emissions from these sources hae been estimated to range from 2 to 25 mg/day for 1972 through 1974 model-year cars in the U.S.[25].

Very limited and preliminary air monitoring data for ethylene dibromide show air concentration values of 0.07-0.11 μg/m^3 (about 0.01 ppb) in the vicinity of gasoline stations along traffic arteries in 3 major U.S. cities; 0.2-1.7 μg/m^3 (about 0.1 ppb) at an oil refinery and 90-115 μg/m^3 (10-15 ppb) at ethylene dibromide manufacturing sites in the U.S., suggesting that it is present in ambient air at very low concentrations[25].

It should be noted that the increased use of unleaded gasoline should result in lower ambient air levels of ethylene dibromide from its major sources of emissions[25,26].

Ethylene dibromide has also been found in concentrations of 96 μg/m^3, up to a mile away from a U.S. Dept. of Agriculture's fumigation center[26].

Air concentrations of ethylene dibromide have been measured in and around citrus fumigation centers in Florida. Average air levels to which site personnel were exposed ranged from 370 to 3100 μg/m^3 inside the facilities and from about 0.1 to 29 μg/m^3 outside[27].

The Criteria and Evaluation Division of EPA has recently made a preliminary estimate concerning the exposure of professional applicators involved with exposure of professional applicators involved with ethylene dibromide soil fumigation applications. These individuals applying ethylene dibromide for 30-40 days per year would receive a total annual inhalation dose of 3-40 mg/kg and farmer-applicators applying ethylene metabolized to chloroacetic acid[87]. 2-Chloroethanol is also considered a likely metabolic product of vinyl chloride which is carcinogenic and mutagenic[90-92] as well as a precursor of chloroacetaldehyde from the metabolism of 1,2-dichloroethane[12].

2-Chloroethanol when tested for carcinogenicity in rats at 10 mg/kg and lower doses by sub-cutaneous administration twice/week for 1 year did not increase the tumor incidence comparable with those of controls[87].

2-Chloroethanol is mutagenic in S. typhimurium TA 1530[93], TA 1535[94] and TA 100[12], with and without activation. In the presence of rat-liver microsomes, 2-chloroethanol was activated to a form which was active on S. typhimurium TA 100 and very slightly active on TA 1535[72]. Similar results were obtained for TA 100 using human liver extracts. 2-Chloroethanol at higher concentrations (2.1 µg x 10^4) was weakly mutagenic directly (without microsomes) for TA 100 and showed a trace of activity for TA 1535. The enhanced reversion of TA 100 relative to TA 1535 after metabolic activation could suggest that chloroacetaldehyde which has similar specificity in reversion of TA 100 is the active metabolite of 2-chloroethanol[12].

2-Chloroethanol (1 M aqueous solution) caused base pair substitutions directly in S. typhimurium TA 1535 when reacted for 1 hour at 25°C (pH 7.4). 2-Chloroethanol caused a significant increase of the mutation frequency, twice the control value, only at the highest concentration tested (1M)[94].

The order of reactivity of 2-haloethanols in E. coli pol A$^+$/pol A$^-$ (to cause mutations of the base substitution type) was 2-iodo- > 2-bromo- > 2-chloroethanol[52]. In the Salmonella assay with TA 1530 and 1535 the order was 2-bromo- > 2-iodo- > 2-chloroethanol[52,95]. The mutagenicity order of activity when tested in Klebsiella pneumoniae was iodoethanol > bromoethanol > chloroethanol. This order of activity correlated with the decrease of bond dissociation energies between the halogens and carbon atoms[96].

2-Chloroethanol has been found to effect a significant reduction in activities of drug-metabolizing enzymes (e.g., aminopyrine, N-demethylase, coumarin-3-hydroxylase) and a marked decrease in glucose-6-phosphatase in both sexes of Albino-Wistar rats given dose levels of 20 mg/kg s.c. daily for 7 days[97]. Although there is no information on the interaction between 2-chloroethanol and its metabolites and liver microsomal enzymes in other species, the data on rats in the above study suggested that care should be exercised in the determination of the tolerance limits in various products containing residues of 2-chloroethanol or the precursor, ethylene oxide. As cited earlier, 2-chloroethanol is metabolically converted to 2-chloroacetaldehyde by hepatic enzymes localized in the microsomal fraction[12] with its in vitro formation documented[98,99]. 2-Chloroacetaldehyde has been shown to be a chemically reactive compound[12,100,101] although its tumor production potential has not yet been demonstrated.

2. <u>1-Chloro-2-propanol</u> (propylene chlorohydrin; Cl-CH$_2$CH$_2$CH$_3$) and <u>2-chloro-1-propanol</u> (propylene chlorohydrin; HO-CH$_2$-CH(Cl)-CH$_3$) are prepared in mixture by the chlorohydrination of propylene. The U.S. production of this mixture is estimated to have been greater than 1.78 billion pounds in 1976[1].

Propylene chlorohydrins apparently have no commercially significant uses other than as an intermediate for propylene oxide production[1]. Approximately 60% of the propylene oxide produced in the U.S. is made by the reaction of propylene with hypochlorous acid, followed by treatment of the resulting propylene chlorohydrins with slaked lime or caustic soda. The other 40% is made by the catalytic peroxidation of propylene[1].

U.S. consumption of propylene oxide is expected to grow at an annual rate of 9-10% through 1980, largely as a result of growth in propylene oxide-derived polyols used in polyurethanes and propylene glycols used in polyester resins[1].

The propylene chlorohydrins can also arise from the use of propylene oxide for the sterilization of a broad spectrum of foodstuffs including powdered and flaked foods, cereals, dried fruits, spices, cocoa, flour and egg powder[85]. It has been reported that conditions for effective sterilization also resulted in the formation of non-volatile residual chloropropanols[86,102]. In some foodstuffs the level of 1-chloro-2-propanol may reach 47 ppm which suggests that human intake of this chemical may be substantial. For example, it was calculated that the consumption of 1 lb of fumigated food may result in the intake of as much as 25 mg of 1-chloro-2-propanol[103].

Information as to the carcinogenic and/or mutagenic potential of the chloropropanols is extremely scant. The mutagenicity of a preparation of 1-chloro-2-propanol containing 25% 2-chloro-1-propanol (at levels of 2.2 to 22 mg/plate) was demonstrated in S. typhimurium TA 1530, but not TA 1538 (indicating base-substitution type mutations)[103].

3. **2,3-Dibromo-1-propanol** ($BrCH_2$-$CHBrCH_2OH$) has been produced in the U.S. in 1976 in quantities greater than 10 million pounds[1]. It is used primarily as an intermediate in the preparation of several flame retardants including tris(2,3-dibromopropyl phosphate) (Tris) (in which it is present as an impurity at 0.2%)[72]. It is also used as a reactive flame retardant and as a chemical intermediate in the preparation of insecticides and pharmaceuticals[104,105].

The total U.S. production of 2,3-dibromo-1-propanol will undoubtedly decrease as a result of concern about the mutagenic and carcinogenic properties of Tris (see Chapter 7) (The Consumer Product Safety Commission banned the sale of Tris-treated sleepwear in April, 1977).

Information is lacking concerning the numbers of workers involved in the production of 2,3-dibromo-1-propanol, its use applications and exposure levels. No occupational standard for 2,3-dibromo-1-propanol has yet been established by OSHA.

2,3-Dibromo-1-propanol has been detected by EPA in industrial effluent discharges (e.g., 0.5×10^{-3} g/l)[106,107]. No other data are available on the environmental occurrence or fate of 2,3-dibromo-1-propanol. It is also known to be a hydrolysis product of the flame retardant Tris[72].

No literature is available concerning animal carcinogenicity tests with 2,3-dibromo-1-propanol.

2,3-Dibromo-1-propanol has been shown to be mutagenic in S. typhimurium TA 1535 and TA 100 both without metabolic activation and with metabolic activation from the liver S-9 fraction of Aroclor pretreated rats[72]. In TA 100, it was more potent than ethylene dibromide and less potent than tris (2,3-dibromopropyl) phosphate[73]. 2,3-Dibromo-1-propanol is not active in S. typhimurium TA 1538 either with or without metabolic activation[72].

While no information is available on the metabolism of 2,3-dibromo-1-propanol, it is believed to be probably oxidized to carboxylic acid and conjugated with glucuronic acid as are other primary alcohols and chlorinated primary alcohols[108]. 2,3-Dibromo-1-propanol has been detected in the urine following application of Tris to the skin of a rat, and in vitro, rat liver hydrolyzed Tris to 2,3-dibromo-1-propanol. However it has not been found in human urine following exposure to the fabrics for up to nine days.

C. Haloethers

Haloethers, primarily alphachloromethyl ethers, represent a category of alkylating agents of increasing concern[110-127] due to the establishment of a casual relationship between occupational exposure to two agents of this class and lung cancer in the U.S. Germany and Japan[112-127]. These haloethers are bis(chloromethyl)ether and chloromethyl methyl ether.

1. Bis(chloromethyl)ether (BCME; $ClCH_2OCH_2Cl$) and chloromethyl methyl ether (methyl chloromethyl ether, CMME, $ClCH_2OCH_3$) are agents which have been widely used in industry as chloromethylation agents in organic synthesis for (1) preparation of anion-exchange resins; (2) formation of water repellants and other textile-treating agents; (3) manufacture of polymers and (4) as solvents for polymerization reactions. Anion-exchange resins (modified polystyrene resins which are chloromethylated and then treated with a tertiary amine or with a polyamine) have been produced in the U.S., France, Federal Republic of Germany, German Democratic Republic, Italy, The Netherlands, United Kingdom, USSR and Japan. (No data are available on the quantities produced)[112,113].

Other chloromethylated compounds of commercial significance include: chloromethyl diphenyloxide, 1-chloromethyl-naphthalene, dodecyl benzyl chloride, and di(chloromethyl)toluene. These compounds are largely used as intermediates (captively) and no data are available on their total production.

BCME can be produced by saturating a solution on paraformaldehyde in cold sulphuric acid with hydrogen chloride. BCME has been primarily manufactured in the U.S. as a chemical intermediate. No information is available on the volume produced in the U.S. as well as in other countries[112].

CMME can be produced by the reaction of methanol, formaldehyde and anhydrous hydrogen chloride. (It should be noted that commercial grades of CMME can be contaminated with 1% to 8% BCME)[115,116]. The number of U.S. producers of CMME has dropped from three in 1969 to one in 1973. No information is available on world production of CMME.

The potential for BCME formation increases with available formaldehyde and chloride[128-130] (in both gaseous and liquid phases), viz., $2Cl^- + 2HCHO + 2H^+ \rightarrow ClCH_2\text{-}OCH_2Cl + H_2O$.

The reaction is believed to be an equilibrium much in favor of the reactants. The extent of hazard from the combination of formaldehyde and HCl to form BCME is unknown at present, and to date, the results appear scanty and disparate[119].

The hydrolytic reactions of BCME and CMME can be depicted as follows:

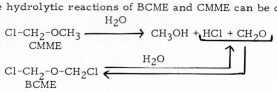

Potential sources of human exposure to BCME appear to exist primarily in areas including (1) its use in chloromethylating (cross-linking) reaction mixtures in anion-exchange resin production[128]; (2) segments of the textile industry using formaldehyde containing reactants and resins in the finishing of fabric and as adhesives in the laminating and flocking of fabrics[130] and (3) in non-woven industry which uses the binders, thermosetting acrylic emulsion polymers comprising methylol acrylamide, since a finite amount of formaldehyde is liberated on the drying and curing of these bonding agents[130].

NIOSH has confirmed the spontaneous formation of BCME from the reaction of formaldehyde and hydrochloric acid in some textile plants and is now investigating the extent of possible worker exposure to the carcinogen[131]. However, this finding has recently been disputed by industrial tests in which BCME was not formed in air by the reaction of textile systems employing hydrochloric acid and formaldehyde[132].

Although mixtures of HCl and formaldehyde in moist air did not form BCME at a detection limit of 0.1 ppb, even with the reactants at their TLV's (e.g, 100 ppm each) low parts per billion levels of BCME were formed when higher concentrations of these reactants (e.g., 300-500 ppm) were present together in moist air[129].

However, it has also been reported that BCME was not observed in the aqueous or gas phase (with detection limits of 9 ppb and 1 ppb, respectively) when aqueous HCl and formaldehyde were reacted together at concentrations up to 2000 ppb at ambient temperatures for 18 h[133]. A recent kinetic study of the hydrolysis of BCME and CMME in aqueous solution showed that they were both hydrolyzed very fast with half-lives in the order of 10-20 sec and < 1 sec (extrapolated to pure water) respectively[134]. Studies involving the stabilities of BCME and CMME in humid air as determined by the kinetics of their hydrolysis rates showed dependence on the construction material of ther eactor and showed that BCME was much more stable than CMME (the upper limits of the rates of hydrolysis taking place in gas phase were 0.00047 min^{-1} ($t_{\frac{1}{2}}$, 25h) and 0.0018 min^{-1} ($t_{\frac{1}{2}}$, 6.5h) respectively[135].

Collier[136] reported that BCME was stable in air at 10 and 100 ppm levels in 70% relative humidity for at least 18 h.

Alvarez and Rosen[137] reported that BCME forms rapidly in a chloromethylating medium (HCl-HCHO-ZnCl$_2$) at low levels (100 ppm) and attains a fairly steady concentration of 300-500 ppm. Although BCME is considered reactive, it is not readily decomposed by water or aqueous base because of its low solubility. In homogeneous media such as methanol-water, the rate of hydrolysis is much faster, while the reaction of BCME with aqueous ammonia is very fast and a reaction path to hexamethylene tetramine is suggested[137].

Information as to the number of employees engaged in the production of BCME and CMME as well as in their use applications is not available.

Both BCME and CMME together with 12 other chemicals are subject to an Emergency Temporary Standard on certain carcinogens under an order made by the Occupational Safety and Health Administration, Dept. of Labor on 26 April, 1973[138].

Regulations published recently by OSHA in U.S., specifically list both BCME and CMME as human carcinogens[114]. A high incidence of predominantly oat-cell carcinoma in a small population of workers exposed to BCME strongly suggests that exposure to this compound constitutes a serious human lung cancer hazard[112,119-124]. Epidemiological studies on an industry-wide basis in the United States have disclosed some 30 cases of lung cancer in association with BCME and CMME[119,120]. Five cases of lung cancer among 32 individuals exposed to BCME over a 15 year period in Japan have been reported[121]. A retrospective study in Germany (1954-62) disclosed 6 cases of lung cancer among 18 men employed in a testing laboratory[122]. Two additional cases of lung cancer were found among a group of 50 production workers. The exposure period ranged from 6-9 years, and the latent period from first exposure to diagnosis was from 8 to 16 years.

One study based on 4 cases of oat cell lung cancer observed among 111 workers in the U.S. exposed to CMME (and its associated BCME) followed for 5 years, suggests as increased risk of lung cancer[113]. Evidence for the existence of a specific lung cancer risk was further supported by the retrospective identification of a total of 14 cases, aged 35-55 years, all of whom had been employed in the production of CMME[116,139]. Albert et al[115] recently examined the lung cancer mortality experience with respect to intensity and duration of exposure in 6 of the 7 chemical companies that account for virtually all of the CMME used in the United States. The study included about 1800 workers who were exposed in the period 1948-72 and about 8000 workers not exposed to CMME from the same plants who served as controls. The age adjusted death rate for respiratory cancer in the CMME exposed group as a whole was 2.5 times that in the control group. There was also a gradation of lung cancer risk according to intensity and duration of exposure and the time elapsed since the onset of exposure. The data demonstrated that BCME and/or CMME are human respiratory carcinogens within 25 years of initial exposure, if the exposure was very intense. It should be noted that when CMME hydrolyizes among the products of hydrolysis are HCl and formaldehyde. Thus while the actual substance or substances involved may not be exactly defined, it appears clear that in operational terms, commercial grade CMME must be considered a carcinogen, although of lower order of activity than BCME[140].

The carcinogenicity of BCME and CMME by skin application to mice and by subcutaneous administration to mice and rats[110,119], the induction of lung adenomas by intrapentoneal injection of BCME in newborn mice[141] and by inhalation of CMME and

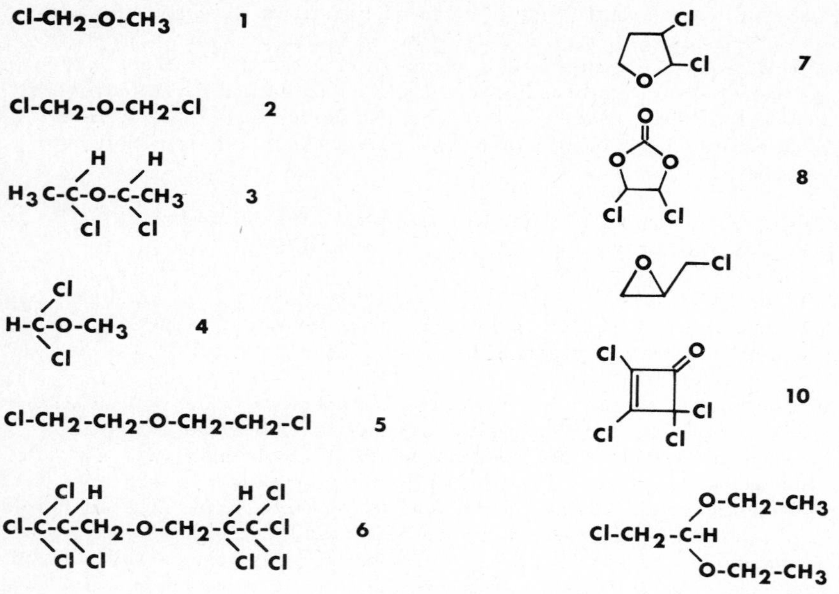

Cl-CH$_2$-O-CH$_3$ 1

Cl-CH$_2$-O-CH$_2$-Cl 2

H$_3$C-CH(Cl)-O-CH(Cl)-CH$_3$ 3

H-C(Cl)$_2$-O-CH$_3$ 4

Cl-CH$_2$-CH$_2$-O-CH$_2$-CH$_2$-Cl 5

Cl$_2$(Cl)C-C(Cl)-CH$_2$-O-CH$_2$-C(Cl)-C(Cl)$_2$Cl 6

2,3-Dichlorotetrahydrofuran 7

1,2-Dichloroethylene carbonate 8

Epichlorohydrin 9

Perchlorocyclobut-2-enone 10

Cl-CH$_2$-CH(O-CH$_2$-CH$_3$)$_2$ 11

Figure 1. Structure of 11 chloroethers tested for carcinogenic activity.[2]

TABLE 1
CARCINOGENICITY OF HALOGENATED COMPOUNDS*

Compound	Carcinogenicity of Halogenated Compounds Mouse Skin		Subcutaneous Injection in Mice: Sarcomas at Injection Site/ Group Size	Subcutaneous Injection in Rats: Sarcomas at Injection Site/ Group Size
	Whole Carcinogen Mice with Papillomas/ Total Mice/Group	Initiating Agent		
CMME, 1	0	12/40 (5)	10/30	1/20
BCME, 2	13/20 (12)	5/20 (2)	-	7/20
Bis(a-chloroethyl)ether, 3	-	7/20 (0)	4/30	-
a, a-Dichloromethyl methyl ether, 4	0/20 (0)	3/20 (1)	-	-
BIS(β-chloroethyl)ether, 5	-	3/20 (0)	2/30	-
Octachloro-di-n-propyl ether, 6	0/20 (0)	3/20 (1)	-	-
2,3-Dichlorotetrahydrofuran, 7	-	5/20 (1)	1/30	-
1,2-Dichloroethylene carbonate, 8	-	3/20 (0)	2/30	-
Epichlorohydrin, 9	-	0/20 (0)	2/50	-
Perchlorocyclobut-2-enone, 10	-	-	1/30	-
Monochloroacetaldehyde diethyl acetal, 11	0/20	1/20 (0)	-	-

*From Reference 111
Number of mice with carcinomas given in parentheses

BCME[142], and the induction of squamous carcinomas of the lung and esthesioneuro-epitheliomas in rats by inhalation exposure[125,143] of 0.1 ppm BCME 5 hr/day, 5 days/week through their lifetime as well as in groups of rats given 10, 20, 40, 60, 80 and 100 exposures to 0.1 ppm BCME and then held until death, have all been reported[125,143].

Van Duuren et al[110,111,144] suggested that the α-haloethers be classifed with the biologically active alkylating agents (e.g., nitrogen mustards, epoxides, β-lactones, etc.). The high chemical reactivity of the α-haloethers is attributed to the reactivity of the halogen atom in displacement reactions. In comparing the carcinogenicity of 11 chloroethers[111,119] (Figure 1, Table 1), in general, bifunctional α-chloroethers are more active than their monofunctional analogs. As the chain length increases, activity decreases, and as chlorine moves further away from the ether oxygen, carcinogenic activity also decreases. It was also noted that in a general way, the more carcinogenically active compounds are the most labile; as stability increases, carcinogenicity also decreases[119].

A comparison of the carcinogenic and mutagenic activity (in E. coli and S. typhimurium microbial systems) of a number of haloethers has been described (Table 2)[119,145,146]. The agents shown are direct-acting and do not require metabolism for mutagenic activity. BCME had been previously listed as a mutagen in the assay of 300 chemicals in the Salmonella/microsome test[147].

The potential genetic risk of occupational exposures in Czeckoslovakia to BCME and CMME was recently reported[148]. Scoring 200 cells per person of 12 workers exposed for 2 years, 6.7% of aberrant cells in peripheral lymphocytes was detected with 9 persons having more than 5% of aberrant cells, compared to a value of only 2% on control non-exposed workers. Job-rotation as well as periodic absence from exposure, lowered the percentage of aberrant cells in BCME and CMME workers.

TABLE 2

COMPARISON OF CARCINOGENIC AND MUTAGENIC (IN MICROBIAL SYSTEMS) ACTIVITY*

	Biological Activity of Halogenated Compounds		
No.	Compound	Mutagenic Activity	Carcinogenic Activity
1	Chloromethyl methyl ether	+	+
2	Bis(chloromethyl) ether	+	+
3	Bis(a-chloroethyl)ether	+	+
4	a,a-dichloromethyl ether	+	+
5	Bis-(B-chloroethyl)ether	-	+
6	Octachloro-di-n-propyl ether	not tested	+
7	2,3-dichlorotetrahydrofuran	-	+
8	1,2-dichloroethylene carbonate	-	-
9	Epichlorohydrin	+	+
10	Perchlorocyclobutenenone	+	-
11	Chloroacetaldehyde diethyl acetal	+	-
12	Dimethyl carbamyl chloride	+	+

* From Reference 111

2. Miscellaneous Haloethers

While BCME and CMME have received the most attention of the haloethers because of their human carcinogenic activity, it is also important to note that other haloethers have industrial utility or have been found as by-products in the environment. For example, bis(2-chloroethyl)ether (BCE) (structure 5, Figure 1) has been extensively used in the paint and varnish industry as a solvent for many resins, including glyceryl phthalate resins, ester gums, paraffin, gum camphor, castor, linseed and other fatty oils, turpentine, polyvinyl acetate and ethyl cellulose. It is also used as a solvent for rubber or cellulose esters, but only in the presence of 10 to 30% alcohol. In the textile industry it is used for grease-spotting and the removal of paint and tar brand marks from raw wool and is incorporated in scouring and fueling soaps. Its other uses include its utility as an extractive for lubricating oils in the petroleum industry and its application as a soil insecticide (fumigant).

Bis(2-chloroisopropyl)ether (dichloroisopropyl ether; bis(2-chloro-1-methylethyl)-
$$\text{ether; } CH_3\text{-}\underset{\underset{CH_3}{|}}{\overset{\overset{Cl}{|}}{C}}\text{-O-}\underset{\underset{CH_3}{|}}{\overset{\overset{Cl}{|}}{C}}\text{-}CH_3)$$
has found use primarily as a solvent for fats, waxes and greases, as an extractant, in paint and varnish removers, in spotting and cleaning solutions, and in textile processing. Its use as a chemical intermediate appears to have been very limited.

The 1975 U.S. production of bis(2-chloroisopropyl)ether is estimated to have been greater than 30 million pounds[1]. Bis(2-chloroisopropyl)ether and BCE are by-products of the chlorohydrin processes for making propylene oxide and ethylene oxide respectively. Although the last U.S. plant to use the chlorohydrin process for ethylene oxide stopped production in 1973, propylene glycol is still produced by the chlorohydrin process. Production losses, principally to wastewaters, are estimated to be about a million pounds per year[49]. Both these haloethers are water soluble and fairly stable in aqueous media. Bis(2-chlorethyl)ether and bis(2-chloroisopropyl)ether have been found in U.S. rivers as a result of industrial outfall from ethylene and propylene glycol production[150,151] in amounts ranging from 0.2 to 5 µg/l in the case of bis(2-chloroisopropyl)ether[150]. BCE has also been identified in waterways in the Netherlands[15]

Both bis(2-chloroethyl)- and bis(2-chloroisopropyl)ethers have been found in the municipal drinking water in the U.S.[150,153]. In the case of bis(2-chloroethyl)-ether the amounts are in the range of 0.02-0.12 µg/l (ppb) with low distribution in the 80 cities screened[152]. The possibility exists that haloethers could be formed as a consequence of chlorination water procedures[152]. However, it should be noted that since both BCME and CMME hydrolyze readily, they would not be expected to remain as such for prolonged periods in waste streams from plants where they are produced or used or if they were produced as a consequence of chlorination water treatment.

The carcinogenicity of BCE is equivocal. While hepatomas have been reported in mice fed BCE[154], its action as an initiating agent and when administered s.c. in mice appears marginal (Table 1)[119].

Bis(2-chloroethyl)ether has also been reported to be mutagenic in Drosophila[155]. However, while it was not mutagenic in the standard Salmonella/microsome assay or towards E. coli WP2 using a similar procedure[156], it was mutagenic when tested with

S. typhimurium strains TA 1535 and TA 100 and was weakly mutagenic in strains TA 1538 and TA 98 and E. coli WP2 when tested in desiccators to contain the volatile fumes[13,157]. In suspension assays, BCE was mutagenic when tested with S. typhimurium TA 1535 and TA 100 and with S. cerevisiae D_3[13,156].

While both BCE and bis(2-chloroisopropyl)ether were mutagenic in S. typhimurium TA 1535 and TA 100 when tested in suspension or in desiccators and did not require S-9 mix for mutagenic activity, the mutagenic activity of bis(2-chloroisopropyl)ether was significantly enhanced by S-9 mix (including and S-9 mix prepared from human liver)[13]. BCE was not mutagenic in host-mediated assays when given as a single oral dose or when administered for 2 weeks prior to the injection of the S. typhimurium into the peritoneal cavity[13,156]. BCE as well as bis(2-chloroisopropyl)ether did not induce heritable translocations when tested in mice[157].

The fate of BCE in rats after acute oral administration of 40 mg/kg was recently reported[158]. One major urinary metabolite was thiodiglycolic acid (TDGA), a lesser metabolite was 2-chloroethanol-β-D-glucuronide. The presence of these two metabolites demonstrates that cleavage of the ether linkage is a major step in biotransformation of bis(2-chloroethyl)ether.

References

1. Stanford Research Institute, A Study of Industrial Data on Candidate Chemicals for Testing, Prepared for U.S. Environmental Protection Agency, EPA-560/5-77-06, Menlo Park, CA, August (1977)
2. Stanford Research Institute, Chemical Economics Handbook, Stanford Reserach Institute, Menlo Park, CA (1972)
3. McConnell, G., Ferguson, O. M., and Pearson, C. R., Chlorinated hydrocarbons in the environment, Endeavor, 34 (1975) 13-27
4. Baier, E. J., Statement on ethylene dichloride before the Subcommittee on Oversight and Investigations House Committee on Interstate and Foreign Commerce, Washington, DC, January 23 (1978)
5. Anon, Ethylene chloride appears to be positive in Cancer Institute Bioassay, Pesticide Toxic Chem. News, 5 (50) (1977) 26-27
6. ACHIG, Threshold Limit Values for Chemical Substances in Workroom Air Adopted by ACGIH for 1976, American Conference of Government Industrial Hygienists, Cincinatti, Ohio (1976)
7. NIOSH, Summary of NIOSH recommendations for occupational health standards, October, 1977, Chem. Reg. Reptr., 1 (40) (1977) 1376-1391
8. Environmentl Protection Agency, Draft Report for Congress: Preliminary Assessment of Suspected Carcinogens in Drinking Water, Office of Toxic Substances Washington, DC, October 17 (1975)
9. National Cancer Institute, Bioassay of 1,2-dichloroethane for possible carcinogenicity, Carcinogenesis Testing Program, DHEW Publ. No. (NIH)78-1305, National Institutes of Health, Bethesda, MD, January 10 (1978)
10. Anon, MCA informs EPA of ongoing inhalation test on ethylene dichloride, Pest. Toxic Chem. News, 6 (7) (1978) 18
11. McCann, J., Spingarn, N. E., Kobori, J., and Ames, B. N., Detection of carcinogens as mutagens: Bacterial tester strains with G factor plasmids, Proc. Natl. Acad. Sci. (USA) 72 (1975) 979-983
12. McCann, J., Simmon, V., Streitwiser, D., and Ames, B. N., Mutagenicity of chloroacetaldehyde, a possible metabolic product of 1,2-dichloroethane (ethylene dichloride), chloroethanol (ethylene chlorohydrin), vinyl chloride and cyclophosphamide, Proc. Natl. Acad. Sci. (USA) 72 (1975) 3190-3193
13. Simmon, V. F., Kauhanen, K., and Tardiff, R. G., Mutagenic activity of chemicals identified in drinking water, In: Progress in Genetic Toxicology (eds) Scott, D., Bridges, B. A., and Sobels, F. H., Elsevier/North-Holland Biomedical Press, Amsterdam (1977) pp. 249-256
14. Bignami, M., Cardamone, G., Carere, A., Comba, P., Dogliotti, E., Morpurgo, G., and Ortali, V. A., Mutagenicity of chemicals on industrial and agricultural relevance in Salmonella, Streptomyces and Aspergillus, Cancer Research (1978) in press
15. Brem, H., Stein, A. B., and Rosenkranz, H. S., The mutagenicity and DNA-modifying effect of haloalkanes, Cancer Res., 34 (1974) 2576-2579
16. Shakarnis, V. F., 1,2-Dichloroethane-induced chromosome non-disjunction and recessive sex-linked lethal mutation in Drosophila melanogaster, Genetika, 5 (12) (1969) 89-95
17. Rapoport, I. A., Reaction of gene proteins with ethylene chloride, Akad. Nauk. SSSR Dokl. Biol. Sci., 134 (1960) 745-747
18. IARC, Monographs on the Evaluation of the Carcinogenic Risk of Chemicals to Man, Vol. 15, International Agency for Research on Cancer, Lyon (1977) 195-209

19. U.S. International Trade Commission, "Synthetic Organic Chemicals", U. S. Production and Sales, 1974, ITC Publ. 776, U.S. Government Printing Office, Washington, DC, (1976) pp. 203,209
20. U.S. Department of Commerce, U.S. Imports for Consumption and General Imports, FT 246/Annual, 1974, Bureau of the Census, U.S. Government Printing Office, Washington, DC (1975) p. 206
21. Muto, T., Noyaku Yoran, Tokyo, Nippon Plant Boeki Association (1976) p. 35
22. U.S. Environmental Protection Agency, EPA Compendium of Registered Pesticides, U.S. Government Printing Office, Washington, DC (1970) pp. III-E-0.1-III-E-9.5
23. Anon, Ethylene dibromide: Institute recommends ceiling limit one milligram, engineering controls, Chem. Reg. Reptr., $\underline{1}$ (36) (1977)
24. U.S. Occupational Safety and Health Administration, Air Contaminants, U.S. Code of Federal Regulations, Title 29, (1976) part 1910.1000 (e) p. 30
25. Environmental Protection Agency, Sampling and Analysis of Selected Toxic Substances, Task II-Ethylene Dibromide, Final Report, Office of Toxic Substances, Environmental Protection Agency, Washington, DC (1975) Sept.
26. Anon, Ethylene dibromide "ubiquitous" in air, EPA report says, Toxic Materials News, $\underline{3}$ (1976) 12
27. Anon, EDB presents close to 100% cancer risk for citrus fumigators, EPA finds, Pesticide Chem. News, $\underline{5}$ (46) (1977) 3-4
28. Environmental Protection Agency, EPA notice of rebuttable presumption against registration and continued registration of pesticide products containing ethylene dibromide, Federal Register, $\underline{42}$, December 14 (1977) 63134; Chem. Reg. Reptr. $\underline{1}$ (40) (1977) 1436-1448
29. Berck, B., Fumigant residues of carbon tetrachloride, ethylene dichloride, and ethylene dibromide in wheat, flour, bran, middlings and bread, J. Agr. Food Chem., $\underline{22}$ (1974) 977-984
30. Wit, S. L., Besemer, A. F. H., Das, H. A., Goedkoop, W., Loostes, F. E., and Meppelink, Rept. No. 36/39, Toxicology (1969) National Institute of Public Health, Bilthaven, Netherlands
31. Fishbein, L., Potential hazards of fumigant residues, Env. Hlth. Persp., $\underline{14}$ (1976) 39-45
32. Olomucki, E., and Bondi, A., Ethylene dibromide fumigation of cereals, I. Sorption of ethylene dibromide by grain, J. Sci. Food Agr., $\underline{6}$ (1955) 592
33. Dumas, T., Inorganic and organic bromide residues in foodstuffs fumigated with methyl bromide and ethylene dibromide at low temperatures, J. Agr. Food Chem., $\underline{21}$ (1973) 433-436
34. Olson, W. A., Haberman, R. T., Weisburger, E. K., Ward, J. M., and Weisburger, J. H., Induction of stomach cancer in rats and mice by halogenated aliphatic fumigants, J. Natl. Cancer Inst., $\underline{51}$ (1973) 1993-1995
35. Powers, M. B., Voelker, R. W., Page, N. P., Weisburger, E. K., and Kraybill, H. F., Carcinogenicity of ethylene dibromide (EDB) and 1,2-dibromo-3-chloropropane (DBCP) after oral administration in rats and mice, Toxicol. Appl. Pharmacol. $\underline{33}$ (1975) 171-172
36. Ames, B. N., The detection of chemical mutagens with enteric bacteria, In: "Chemical Mutagens: Principles and Methods for the Detection", Vol. 1 (ed) Hollaender, A., Plenum Press, New York (1971) 267-282
37. Alper, M. D., and Ames, B. N., Positive selection of mutants with deletions of the gal-chl region of the Salmonella chromosome as a screening procedure for mutagens that cause deletions, J. Bacteriol., $\underline{121}$ (1975) 295-296

38. Buselmaier, V. W., Rohrborn, G., and Propping, P., Mutagenitäts-untersuchungen mit pestiziden im host-mediated assay und mit dem dominanten-letla test an der maus, Biol. Zbl., 91 (1972) 311-325
39. Buselmaier, W., Rohrborn, G., and Proppin, P., Comparative investigations on the mutagenicity of pesticides in mammalian test systems, Mutation Res., 21 (1973) 25-26
40. DeSerres, F. J., and Malling, H. V., Genetic analysis of Ad-3 mutants of Neurospora crassa produced by ethylene dibromide, a commonly used pesticide, EMS Newsletter, 3 (1970) 36-37
41. Malling, H. V., Ethylene dibromide: a potent pesticide with high mutagenic activity, Genetics, 61 (1969) 539
42. Sparrow, A. H., Schairer, L. A., Mutagenic response of Tradescantia to treatment with X-rays, EMS, DBE, ozone, SO_2, N_2O and several insecticides, Mutation Res. 26 (1974) 445
43. Sparrow, A. H., Schairer, L. A., Interaction of exposure time and gaseous mutagen concentration on somatic mutation frequency in Tradescantia, Mutation Res., 31 (1975) 318
44. Nauman, C. H., Villalobos-Pietrini, R., and Sautkulis, R. C., Response of a mutable close of Tradescantia to gaseous chemical mutagens and to ionizing radiation, Mutation Res., 26 (1974) 444
45. Sparrow, A. H., Schairer, L. A., and Villalobos-Pietrini, R., Comparison of somatic mutation rates induced in tradescantia by chemical and physical mutagens, Mutation Res., 26 (1974) 265-276
46. Vogel, E., and Chandler, J. L. R., Mutagenicity testing of cyclamate and some pesticides in Drosophila melanogaster, Experientia, 30 (1974) 621-623
47. Epstein, S. S., Arnold, E., Andrea, J., Bass, W., and Bishop, Y., Detection of chemical mutagens by the dominant lethal assay in the mouse, Toxicol. Appl. Pharmacol., 23 (1972) 288-325
48. Fahrig, R., Comparative Mutagenicity Studies with Pesticides, In: "Chemical Carcinogenesis Assay" IARC Sci. Series No. 10, International Agency for Research on Cancer, Lyon (1974)
49. Meneghini, R., Repair replication of opossum lymphocyte DNA: Effect of compounds that bind to DNA, Chem. Biol. Interactions, 8 (1974) 113-126
50. Clive, D., Flamm, W. G., Machesko, M. R., and Bernheim, N. J., A mutational assay system using the thymidine kinase locus in mouse lymphoma cells, Mutation Res., 16 (1972) 77-87
51. Kristoffersson, U., Genetic effects of some gasoline additives, Hereditas, 78 (1974) 319
52. Rosenkranz, H. S., Mutagenicity of halogenated olefins and their derivatives, Environ. Hlth. Persp., 21 (1977) in press
53. Ames, B. N., McCann, J., and Yamasaki, F., Methods for the detection of carcinogens and mutagens with the Salmonella/mammalian-microsome mutagenicity test, Mutation Res., 31 (1975) 347
54. Nauman, C. H., Sparrow, A. H., and Schairer, L. A., Comparative effects of ionizing radiation and two gaseous chemical mutagens on somatic mutation induction in one mutable and two non-mutable clones of tradescantia, Mutation Res., 38 (1976) 53-70
55. Ehrenberg, L., Osterman-Holkar, S., Singh, D., Lundquist, U., On the reaction kinetics and mutagenic activity of methylating and β-halogenoethylating gasoline derivatives, Radiat. Biol., 15 (1974) 185-194

56. Edwards, K., Jackson, H., and Jones, A. R., Studies with alkylating esters, II. A chemical interpretation through metabolic studies of the infertility effects of ethylene dimethanesulfonate and ethylene dibromide, Biochem. Pharmacol., 19 (1970) 1783-1789
57. Amir, D., The sites of the spermicidal action of ethylene dibromide in bulls, J. Reprod. Fert., 35 (1973) 519-525
58. Alumot, E., The mechanism of action of ethylene dibromide on laying hens, Res. Rev., 41 (1972) 1-11
59. Jones, A. R., and Edwards, K., The comparative metabolism of ethylene dimethanesulfonate and ethylene dibromide, Experientia, 24 (1968) 1100-1101
60. Plotnick, H. B., and Conner, W. L., Tissue distribution of ^{14}C-labeled ethylene dibromide on the guinea pig, Res. Commun. Chem. Path. Pharmacol., 13 (1976) 251-258
61. Nachtomi, E., The metabolism of ethylene dibromide in the rat. The enzymic reaction of glutathione in vitro and in vivo, Biochem. Pharmacol., 19 (1970) 2853-2860
62. Winell, M., An international comparison of hygienic standards for chemicals in the work environment, Ambio, 4 (1975) 34-36
63. DeLorenzo, F., Degl'innocenti, S., Ruocco, A. Silengo, L., and Cortese, R., Mutagenicity of pesticides containing 1,3-dichloropropane, Cancer Res., 37 (1977) 1915-1917
64. IARC, Monographs on the Evaluation of the Carcinogenic Risk of Chemicals to Man, Vol. 15, International Agency for Research on Cancer, Lyon (1977) 139-147
65. Anon, Dibromochloropropane: Sterility said to be permanent for some as hearings on OSHA proposed rule begin, Chem. Reg. Reptr., 1 (40) (1977) 1423
66. OSHA, OSHA Emergency Temporary Standard for Occupational Exposure to Dibromochloropropane, Federal Register, 42 Sept. 9 (1977) 45536; Chem. Reg. Reptr. 1 (26) (1977) 924-937
67. Johnson, D. E., and Lear, 1,2-Dibromo-3-chloropropane: Recovery from soil and analysis by GLC, J. Chromatog. Soc., 7 (1969) 384-385
68. U.S. Environmental Protection Agency, EPA Compendium of Registered Pesticides, Vol. II, Fungicides, Nematicides, Part II, U.S. Government Printing Office, Washington, DC (1974) pp. D-25-00.01-D-25-00.13
69. U.S. Department of Agriculture, Farmers Use of Pesticides in 1971, Quantities, Agricultural Economic Rept. No. 252, Washington, DC (1971) p. 54
70. California Department of Food and Agriculture, Pesticide Use Report 1975, Sacramento (1976) pp. 51-53
71. Environmental Protection Agency, EPA Notice of Intent to Suspend/Conditionally Suspend Registration of Pesticide Products Containing Dibrmochloropropane, Federal Register, 42 Sept. 26 (1977) 48915; Chem. Reg. Reptr., 1 (1977) 1002-1008
72. Prival, M. J., McCoy, E. C., Gutter, B., and Rosenkranz, H. S., Tris(2,3-dibromopropyl)phosphate: Mutageicity of a widely used flame retardant, Science 195 (1977) 76-78
73. Blum, A., and Ames, B. N., Flame-retardant additives as possible cancer hazards, Science, 195 (1977) 17-23
74. Rosenkranz, H. S., Genetic activity of 1,2-dibromo-3-chloropropane, a widely used fumigant, Bull. Environ. Contam. Toxicol., 14 (1975) 8-12
75. Wolff, S., Personal communication, 9th Annual Meeting Environmental Mutagen Society, San Francisco, March 9-13 (1978)
76. Biles, R. W., Connor, T. H., Trieff, N. M., Ramanutam, V. M. S., and Legator, M. S., The absence of mutagenicity in unadulterated DBCP, 9th Annual Meeting of Environmental Mutagen Society, San Francisco, CA, March 9-13 (1978) 36-37

77. Faydysh, E. V, Rakhmatullaev, N. N., and Varshavskii, V. A., The cytoxic action of nemagon in a subacute experiment, Med. Zh. Uzbekistana, 1 (1970) 65-65
78. Torkelson, T. R., Sadek, S. E., Rowe, V. K., Kodama, J. K., Anderson, H. H., Loquvam, G. S., and Hine, C. H., Toxicologic investigations of 1,2-dibromo-3-chloropropane, Toxicol. Appl. Pharmacol., 3 (1961) 545-549
79. Reznik, Y. B., and Sprinchan, G. K., Experimental data on the gonadotoxic effect of Nemagon, Gig. Sanit, 6 (1975) 101-102
80. Rakmatullaev, N. N., Hygienic characteristics of the nematocide Nemagon in relation to water pollution control, Hyg. Sanit, 36 (1971) 344-348
81. Anon, Senate Pesticide Worker Safety Hearings Open with DBCP Inquiry, Pesticide Toxic Chem. News, 6 (3) (1977) 33-35
82. Anon, Sperm count rise in DBCP-exposed workers, Chem. Eng. News, Dec. 19 (1977) 7
83. Anon, Dibromochloropropane: Dow study of Michigan workers finds "normal" distribution of sperm counts, Chem. Reg. Reptr.. 1 (36) (1977) 1298
84. Anon, More tests link DBCP to worker sterility, Chem. Eng. News, Sept. 5 (1977) 5-6
85. Fishbein, L., Degradation and residues of alkylating agents, Ann. NY Acad. Sci., 163 (1969) 869-894
86. Wesley, F., Rouke, B., and Darbishire, O., The formation of persistent toxic chlorohydrins in foodstuffs by fumigation with ethylene oxide and propylene oxide, J. Food Sci., 30 (1965) 1037-1042
87. Balazs, T., Toxicity of ethylene oxide and chloroethanol, FDA By-Lines, 3 (1976) 150-155
88. Johnson, M. K., Metabolism of chloroethanol in the rat, Biochem. Pharmacol. 16 (1967) 185-199
89. Blair, A. H., and Vallee, B. L., Some catalytic properties of human liver dehydrogenase, Biochemistry, 5 (1966) 2026-2034
90. Bartsch, H., and Montesano, R., Mutagenic and carcinogenic effects of vinyl chloride, Mutation Res., 32 (1975) 93-114
91. Watanabe, R. E., Hefner, R. E., Jr., and Gehring, P. G., Preliminary studies of the fate of inhaled vinyl chloride monomer in rats, Ann. NY Acad. Sci., 246 (1975) 135-148
92. Loprieno, N., Barale, R., Baroncelli, S., Boveri, C., Baroncelli, etal., Evaluation of the genetic effects of vinyl chloride monomer (VCM) under the influence of liver microsomes, Mutation Res., 40 (1976) 85-96
93. Malaveille, C., Bartsch, H., Barbin, C., Camus, A. M., and Montesano, R., Mutagenicity of vinyl chloride, chloroethylene oxide, chloroacetaldehyde and chloroethanol, Biochem. Biophys. Res. Commun., 63 (1975) 363-370
94. Rannug, U., Gothe, R., and Wachtmeister, C. A., The mutagenicity of chloroethylene oxide, chloroacetaldehyde, 2-chloroethanol and chloroacetic acid, conceivable metabolites of vinyl chloride, Chem. Biol. Interactions, 12 (1976) 251-263
95. Rosenkranz, S., Carr, H. S., and Rosenkranz, H. S., 2-Haloethanols: Mutagenicity and reactivity with DNA, Mutation Res., 26 (1974) 367-370
96. Voogd, C. E., and Van Der Vet, P., Mutagenic action of ethylene halogenhydrins, Experientia, 25 (1969) 85-86
97. Fuer, G., Balazs, T., Farkus, R., and Ilyas, M. S., Effect of 2-chloroethanol on hepatic microsomal enzymes in the rat, J. Toxicol. Env. Hlth., 3 (1977) 569-576
98. Bartsch, H., Malaveille, C., and Montesano, R., Human, rat and mouse liver-mediated mutagenicity of vinyl chloride in S. typhimurium strains, Int. J. Cancer, 15 (1975) 429-437

99. Gothe, R., Callerman, C. J., Ehrenberg, L., and Wachmeister, C. A., Trapping with 3,4-dichloro-benzenethiol of reactive metabolites found in vitro from the carcinogen vinyl chloride, Ambio, 3 (1974) 234-236
100. Huberman, E., Bartsch, H., and Sachs, L., Mutation induction in Chinese hamster V79 cells by two vinyl chloride metabolites, chloroethylene oxide and 2-chloroacetaldehyde, Int. J. Cancer, 16 (1975) 639-644
101. Van Duuren, B. L., Goldschmidt, B. M., Katz, C., Siedman, I., and Paul, J. S., Carcinogenic activity of alkylating agents, J. Natl. Cancer Inst., 53 (1974) 695-700
102. Ragelis, E. P., Fisher, B. S., and Klimeck, A. B., Isolation and determination of chlorohydrins in foods fumigated with ethylene oxide or with propylene oxide, J. Ass. Offic. Anal. Chem., 51 (1968) 707-715
103. Rosenkranz, H. S., Wlodkowski, T. J., and Bodine, S. R., Chloropropanol, a mutagenic residue resulting from propylene oxide sterilization, Mutation Res., 30 (1975) 303-304
104. Hawley, G. G. (ed) The Condensed Chemical Dictionary, Van Nostrand Reinhold Company, New York (1971) p. 278
105. DHEW, Clearinghouse on Environmental Carcinogens, Chemical Selection Subgroup Bethesda, MD Dec. 19 (1977)
106. Webb, R. C., Environmental Protection Technology Series, EPA-R2-73-277, Washington, DC (1973)
107. Commission of the European Communities, European Cooperation and Coordination in the Field of Scientific and Technical Research, COST-Project 64b, "A Comprehensive List of Polluting Substances Which Have Been Identified in Various Fresh Waters, Effluent Discharges, Aquatic Animals and Plants, and Bottom Sediments", Second Ed. (1976) p. 39
108. Williams, R. T., Detoxification Mechanisms, 2nd Edition, John Wiley & Sons, New York (1969) p. 59
109. St. John, L. E., Eldefrawi, M. E., and Lisk, D. J., Studies of possible absorption of a flame retardant from treated fabrics worn by rats and humans, Bull. Environ. Contam. Toxicol., 15 (1976) 192-197
110. Van Duuren, B. L., Goldschmidt, B. M., Katz, C., Langseth, L., Mercado, C., and Sivak, A., Alpha-haloethers: A new type of alkylating carcinogen, Arch. Env. Hlth., 16 (1968) 472-476
111. Van Duuren, B. L., Katz, c., Goldschmidt, B. M., Frenkel, K., and Sivak, A., Carcinogenicity of haloethers. II. Structure-activity relationships of analogs of bis(chloromethyl)ether, J. Natl. Cancer Inst., 48 (1972) 1431-1439
112. IARC, Bis(chloromethyl)ether, In Monograph No. 4., Lyon (1974) pp. 231-238
113. IARC, Chloromethylmethylether, In Monograph No. 4, Lyon (1974) pp. 239-245
114. OSHA, Occupational Safety & Health Standards: Carcinogens, Fed. Reg., 39 (20) (1974) 3768-3773; 3773-3776
115. Albert, R. E., Pasternack, B. S., Shore, R. E., Lippmann, M., Nelson, N., and Ferris, B., Mortality patterns among workers exposed to chloromethyl ethers, Env. Hlth. Persp., 11 (1975) 209-214
116. Figueroa, W. G., Raszkowski, R., and Weiss, W., Lung cancer in chloromethyl methyl ether workers, New Engl. J. Med., 288 (1973) 1094-1096
117. Weiss, W., and Figueroa, W. G., The characteristics of lung cancer due to chloromethyl ethers, J. Occup. Med., 18 (1976) 623-627
118. Weiss, W., Chloromethyl ethers, cigarettes, cough and cancer, J. Occup. Med., 18 (1976) 194-199
119. Nelson, N., The chloroethers-occupational carcinogens: A summary of laboratory and epidemiology studies, Ann. NY Acad. Sci., 271 (1976) 81-90

120. Nelson, N., The carcinogenicity of chloroethers and related compounds, Meeting on Origins of Human Cancer, Cold Springs Harbor Laboratory, New York Sept. 7-14 (1976) p. 8
121. Sakabe, H., Lung cancer due to exposure to bis(chloromethyl)ether, Ind. Hlth., 11 (1973) 145
122. Thiess, A. M., Hey, W., and Zeller, H., Zur toxikologie von dichlorodimethyl aether-verdacht auf kanzerogene wirkung auch beim menschen, Zbl. Arbeits Med., 23 (1973) 97
123. Reznik, G., Wagner, H. H., and Atay, Z., Lung cancer following exposure to bis(chloromethyl)ether: A case report, J. Environ. Path. Toxicol., 1 (1977) 105-111
124. Bettendorf, U., Gewerblich induzierte lungenkarzinome nach inhalation alkylierender verbindungen, Verh. Dtsch. Ges. Path., 60 (1976) 457
125. Laskin, S., Kuschner, J., Drew, R. T., Cappiello, V. P., and Nelson, N., Tumors of the respiratory tract induced by inhalation of bis(chloromethyl)ether, Arch. Environ. Hlth., 23 (1971) 135-136
126. Defonso, L. R., and Kelton, S. C., Jr., Lung cancer following exposure to chloromethyl methyl ether, Arch. Environ. Hlth., 31 (1976) 125-130
127. Pasternack, B. S., Shore, R. E., and Albert, R. E., Occupational exposure to chloromethyl ethers, J. Occup. Med., 19 (1977) 741-745
128. Rohm & Hass Co., News release: Reaction of formaldehyde and HCl forms bis-CME Rohm & Hass, Philadelphia, PA, Dec. 27 (1972)
129. Kallos, G. J., and Solomon, R. A., Investigation of the formation of bis(chloromethyl)ether in simulated hydrogen chloride-formaldehyde atmospheric environments, Amer. Ind. Hyg. Assoc. T., 34 (1973) 469-473
130. Hurwitz, M. D., Assessing the hazard from BCME in formaldehyde-containing acrylic emulsions, Amer. Dyestuff Reptr., 63 (1974) 62-64, 77
131. Anon, Industry's problems with cancer aired, Chem. Eng. News, 53 (1975) 4
132. Anon, Dow says bis(chloromethyl)ether does not form during textile operations, Toxic Materials News, 3 (20) (1976) 157
133. Tou, J. C., and Kallos, G. J., Study of aqueous HCl and formaldehyde mixtures for formation of bis(chloromethyl)ether, Am Ind. Hyg. Assoc. J., 35 (1974) 419-42?
134. Tou, J. C., Westover, L. B., Sonnabend, L. F., Kinetic studies of bis(chloromethyl ether hydrolysis by mass spectrometry, J. Phys. Chem., 78 (1974) 1096
135. Tou, J. C. and Kallos, G. T., Kinetic study of the stabilities of chloromethyl ether and bis(chloromethyl)ether in humid air, Anal. Chem., 46 (1974) 1866-1869
136. Collier, L., Determination of BH-chloromethylether at the PPB level in air samples by H164 resolution mass spectroscopy, Env. Sci. Technol., 6 (1972) 930
137. Alvarez, M., and Rosen, R. T., Formation and decomposition of bis(chloroethyl) ether in aqueous solution, Int. J. Environ. Anal. Chem., 4 (1976) 241-246
138. Department of Labor, Occupational Safety and Health Standards, Federal Register 38 (85) (1973) 10929
139. Nelson, N., Carcinogenicity of haloethers, New Engl. J. Med., 288 (1973) 1123
140. Laskin, S., Drew, R. T., Capiello, V., Kuschner, M., and Nelson, N., Inhalation carcinogenicity studies of alpha haloethers. II. Chronic inhalation studies with chloromethyl methyl ethers, Arch. Env. Hlth., 30 (1975) 70-72
141. Gargus, J. L., Reese, W. H., Jr., and Rutter, H. A., Induction of lung adenomas in new born mice by bis(chloromethyl)ether, Toxicol. Appl. Pharmacol., 15 (1969) 92-96

142. Leong, B. K., MacFarland, H. N., and Reese, W. H., Jr., Induction of lung adenomas by chronic inhalation of bis(chloromethyl)ethers, Arch. Env. Hlth., 22 (1971) 663-666
143. Kuschner, M., Laskins, S., Drew, R. T., Cappiello, V., and Nelson, N., Inhalation carcinogenicity of alpha-haloethers, III. Lifetime and limited period inhalation studies with BCME at 0.1 ppm, Arch. Env. Hlth., 30 (1975) 73-77
144. Van Duuren, B. L., Carcinogenic epoxides, lactones and halo-ethers and their mode of action, Ann. NY Acad. Sci., 163 (1969) 633-651
145. Mukai, F., and Troll, W., The mutagenicity and initiating activity of some aromatic amine metabolites, Ann. NY Acad. Sci., 163 (1969) 828
146. Mukai, F., and Hawryluk, I., The mutagenicity of some haloethers and haloketones, Mutation Res., 21 (1973) 228
147. McCann, J., Choi, E., Yamasaki, E., and Ames, B. N., Detection of carcinogens as mutagens in the Salmonella/microsome test: Assay of 300 chemicals, Proc. Natl. Acad. Sci., 72 (1975) 5135-5139
148. Zudova, Z., and Landa, K., Genetic risk of occupational exposures to haloethers, Mutation Res., 46 (1977) 242-243
149. DeSerres, F., National Institute of Environmental Health Sciences, Personal communication, Dec. (1975)
150. Kleupfer, R. D., and Fairless, B. J., Characterization of organic components in municipal water supply, Env. Sci. Technol., 6 (1972) 1062-1063
151. Rosen, A. A., Skeel, R. T., and Ettinger, M. B., Relationship of river water to specific organic ontaminants, J. Water Pollut. Control Fed., 35 (1963) 777-782
152. Zoetman, B. C. J., Rijks Instituut Voor Drink Water Voorziening, Personal communication cited in reference 30.
153. Environmental Protection Agency, Draft Report for Congress: Preliminary Assessment of Suspected Carcinogens in Drinking Water, Office of Toxic Substances, Washington, DC, October 17 (1975)
154. Innes, J. R., Ulland, B. M., Valerio, M. C., Petrucelli, C., Fishbein, L., Hart, E. R., Pallotta, A. J., Bates, R. R., Falk, H. L., Gart, J. T., Klein, M., Mitchell, I., and Peters, J., Bioassay of pesticides and industrial chemicals for tumorigenicity in mice: a preliminary note, J. Natl. Cancer Inst., 42 (1969) 1101
155. Auerbach, C., Robson, J. M., and Carr, J. G., The chemical production of mutations, Science, 105 (1947) 243
156. Simmon, V. F., Kauhanen, K., and Tardiff, R. G., Mutagenicity assays with bis(2-chloroethyl)ether, Abstracts of Meeting of International Congress of Toxicology, Toronto, March 3-April 2 (1977) p. 31
157. Jorgenson, T. A., Rushbrook, C. J., Newell, G. W., and Tardiff, R. G., Study of the mutagenic potential of bis(2-chloroethyl) and bis(2-chloroisopropyl) ethers in mice by heritable translocation test, Abstracts of 8th Annual Meeting Environmental Mutagen Society, Colorado Springs, Colorado, Feb. 13-17 (1977) p. 76
158. Lingg, R. D., Kaylor, W. H., Glass, J. W., Pyle, S. M., and Tardiff, R. G., Fate of bis(2-chloroethyl)ether in rats after acute oral administration, Abstracts 17th Annual Meeting Society of Toxicology, San Francisco, CA, March 12-16 (1978) pp. 59, 61.

CHAPTER 14

HALOGENATED ARYL DERIVATIVES

A. Chlorinated Benzenes

Chlorinated benzenes are formed progressively and simultaneously when chlorine reacts with benzene at elevated temperatures in the presence of chlorination catalysts[1]. Once substitution occurs, the chlorine constituents exert an orienting effect on further chlorine atoms entering the ring resulting in certain isomers being preferentially formed[1]. Most aryl halides are usually much less reactive than alkyl or allyl halides toward nucleophilic reagents in either S_N1- or S_N2-type reactions[2].

Chlorinated benzenes are composed of twelve chemical species: one mono-, three di-, three tri-, three tetra-, one penta-, and one hexachlorobenzene. Most of these are not only important intermediates for the production of many kinds of chemicals but they are also extensively employed singly or in combination in the home, industry, and agriculture for a variety of applications including: solvents for pesticides, heat transfer agents, insect repellants, pesticides, deodorants and as additives for rubber products.

1. Chlorobenzene (monochlorobenzene; ⌬-Cl) is used principally as illustrated in the U.S. consumption pattern in 1974[3]: for pesticides and degreasing operations, 49%; as intermediate for production of chloronitrobenzenes (in dye and pesticide intermediates), 30%; intermediate for diphenyloxide synthesis, 8%; intermediate for DDT production, 7%; other uses, 6%. U.S. production of chlorobenzene in recent years has been as follows (millions of pounds): 1974 (379); 1975 (306) and 1976 (329). U.S. imports totaled 1.49 million pounds in 1974 and 8.37 million pounds in 1975[3].

It is estimated that the consumption of chlorobenzene will grow at an average annual rate of 1-2% in the U.S. provided governmental regulations do not significantly restrict its use as a solvent for pesticides[3].

2. ortho-Dichlorobenzene (1,2-dichlorobenzene; o-dichlorobenzol; DCB;) has been produced commercially in the U.S. for over 50 years, primarily by procedures involving the chlorination of benzene or monochlorobenzene. Mixtures of the dichlorobenzenes can be produced by chlorination of chlorobenzene at 150-190°C in the presence of ferric chloride[1,4]. The two principal dichlorobenzenes in the residue or in the mixture can be separated by fractional distillation or by crystallization of the para-isomer. When chlorinated in the presence of aluminum analogous chlorobenzene yields a mixture of o- and p-dichlorobenzenes plus a very small amount of the meta isomer. A technical grade of ortho-dichlorobenzene available in the U.S. typically contains 98.7% by weight of the ortho-isomer and 1.3% of the meta- and para-isomers combined. Other grades of ortho-dichlorobenzenes are also available, e.g., one typically contains 83% of the ortho-isomer, and 17% of the meta- and para-isomers combined.

Table 1. Industries Identified as Possible HCB Sources and Potential Origin of HCB Wastes.

Industry/Type	Potential Origin of HCB Wastes
Basic HCB production/distribution	HCB production operation
Chlorinated solvents production	Reaction side-product in the production of chlorinated solvents, mainly, carbon tetrachloride, perchloroethylene, trichloroethylene, and dichloroethylene
Pesticide production	Reaction side-product in the production of Dacthal, simazine, mirex, atrazine, propazine, and pentachloronitrobenzene (PCNB)
Pesticide formulation/distribution	Formulation, packaging and disbution of HCB-containing pesticides
Electrolytic chlorine production	Chlorine attack on the graphite anode or its hydrocarbon coating
Ordnance and pyrotechnics production	Use of HCB in the manufacture of pyrotechnics, and trace bullets and other ordnance items
Sodium chlorate production	Similar to electrolytic chlorine production where graphite anodes are used
Aluminum manufacture	Use of HCB as a fluxing agent in smelting
Seed treatment industry	Use of HCB in seed protectant formulations
Pentachlorophenol (PCP) production	Reaction by-product of PCB production by chlorination of phenol
Wood preservatives industry	Use of HCB as a wood preserving agent
Electrode manufacture	Use of HCB as a porosity control in the manufacture of graphite anodes
Vinyl chloride monomer production	By-product in the manufacture of vinyl chloride monomer
Synthetic rubber production	Use of HCB as a peptizing agent in the production of nitroso and styrene rubbers for tires

Table 2. Estimated Total Quantity of Hexachlorobenzene Contained in U.S. Industrial Wastes, By-products, and Products in 1972[42]

Product	U.S. Production in 1972 (1000 lb)	Estimated HCB Produced (1000 lb) High	Low
Perchloroethylene	734,800	3,500	1,750
Trichloroethylene	427,000	450	230
Carbon Tetrachloride	997,000	400	200
Chlorine	19,076,000	390	160
Dacthal	2,000	100	80
Vinyl Chloride	4,494,000	27	0
Atrazine, Propazine, Simazine	112,000	9	5
Pentachloronitrobenzene	3,000	6	3
Mirex	1,000	2	1
Total	25,846,800	4,884	2,429

HCB can be produced industrially from some isomers of BHC by one of the following processes:

(A) C$_6$H$_6$Cl$_6$ $\xrightarrow{300-400°\text{C} \atop -3\text{HCl}}$ C$_6$H$_3$Cl$_3$ $\xrightarrow{3\text{O}_2}$ C$_6$Cl$_6$ + 3HCl

(B) C$_6$H$_6$Cl$_6$ $\xrightarrow{\text{atm O}_2}$ C$_6$Cl$_6$O + 3H$_2$O

It is possible that with a process similar to (B), among the products of environmental degradation of BHC isomers, HCB may also be formed[45].

While hexachlorobenzene is used as a fungicide (primarily to control bunt of wheat) it can be formed as an impurity during the synthesis of the widely used herbicide DCPA (dimethyltetrachloroterephthalate, Dacthal) which can contain 10-14% HCB[46]. Pentachlorobenzene (PCNB, quintozene) a pesticide, contains 1-6% HCB. HCB is also an intermediate in the production of pentachlorophenol (PCP) a widely used herbicide and fungicide. Technical pentachlorophenol can contain up to 13% impurities, a part of which could consist of residual HCB[47]. The estimated total quantity of HCB waste generated in the pesticide industry is 1,655 tons per year. The HCB is present mainly in tars and still bottoms from the manufacturing operations[43].

Of the total HCB generated by the chlorinated solvents industry, 92% is discharged in the waste streams which are disposed of on land or are incinerated while only 8% (210 tons) of HCB are recovered for sale[43].

The prevalence of various disposal methods is shown in Table 3 in terms of HCB (and HCB-containing wastes) handled in the U.S., and the number of facilities (on-site and off-site) which utilize the disposal methods. The data in Table 3 indicate that based on the total quantity of waste handled land disposal is currently the most prevalent method for ultimate disposal of HCB waste. Among land disposal methods, the use of

TABLE 3. Prevalence of Methods Used for Ultimate Disposal of HCB wastes.

Disposal Method	Industry Type	Plant Sites No.	Plant Sites % of total	HCB Wastes Quantity (tons/year)	HCB Wastes % of total	HCB-Containing Wastes Quantity (tons/year)	HCB-Containing Wastes % of total
Land disposal							
Sanitary landfill	Chlorinated solvents	1	5.3	208	7.7	281	1.1
Industrial landfill	Chlorinated solvents	3	15.7	1,050	38.9	7,000	28.7
	Pesticide industry	1	5.3	50	1.9	67	0.3
	Electrolytic chlorine	1	5.3	—	—	156	0.6
Deep well disposal	Chlorinated solvents	2	10.5	225	8.3	11,660	47.7
(Subtotal)		8	42.1	1,533	56.8	19,164	78.4
Incineration							
Without by-product recovery	Chlorinated solvents	4	21.0	213*	7.9	3,991*	16.4
	Pesticide industry	2	10.5	200	7.4	266	10.0
	Electrolytic chlorine	1	5.3	not available	—	not available	—
With by-product recovery	Chlorinated solvents	1	5.3	750**	27.8	1,000**	4.1
(Subtotal)		8	42.1	1,163	43.1	5,257	21.5
Resource recovery (excluding incineration)	Chlorinated solvents	1	5.3	not available	—	not available	—
Discharge to waste treatment plants	Pesticide formulation/distribution	1	5.3	small; data on exact quantity not available	—	not available	—

(Continued on p. 275)

TABLE 3. (continued)

Disposal Method	Industry Type	Plant Sites No.	% of total	HCB Wastes Quantity (tons/year)	% of total	HCB-Containing Wastes Quantity (tons/year)	% of total
Emission to atmosphere	Pesticide Industry	1	5.3	not available	--	not available	--
TOTAL		19	100%	2,696***	100%	24,421***	100%

*Includes a very small quantity of HCB wastes (400 lb per year) from three plant sites engaged in chlorinated solvents manufacture. These wastes are extremely dilute (10 to 40 ppm HCB) and were not included in the total waste quantities in order to avoid gross distortion of "HCB-Containing Waste Quantities" handled by incineration.
**Waste quantities are based on 1970-71 data supplied by the off-site disposal contractor then handling the waste. Waste is assumed to contain 75 percent HCB based on data for other plants.
***Does not include 1,400 tons per year of HCB waste (1,750 tons of HCB-containing waste) temporarily stored under cover at one pesticide production site. Also does not include 210 tons of HCB which is recovered for sale from 284 tons of HCB-containing wastes at one chlorinated solvents plant.

landfills is the most prevalent method, accounting for the disposal of 56.8% of all HCB wastes. Ranked next to land disposal is incineration which accounted for the destruction of a minimum of 1,163 tons of HCB per year contained in a waste mixture in excess of 5,257 tons per year. Compared to land disposal and incineration, the quantities of waste discharged to sewage treatment plants and to the atmosphere appears to be very small[43].

Mismanagement of HCB wastes and use of products have resulted in several serious incidents of environmental contamination[48].

HCB has been found recently (with hexachlorobutadiene) in water, soil and selected aquatic organisms along the lower Mississippi River in Louisiana. Highest levels were found downstream from heavily industrialized areas[49]. Air immediately adjacent to production facilities has been shown to have concentrations from 1.0 to 23.6 µg/m^3. HCB has also been identified in the municipal drinking water supply of Evansville, Indiana[50].

HCB has also been identified in surface water and industrial effluents in USSR, Germany, and Italy[45]. The average pollution levels differ greatly, e.g., from 2.5 ppt for Italian surface waters to 130 ppt for the waters of the River Rhine in Holland, which suggests both fungicidal and industrial as original sources of the contamination[45].

HCB has been found in the intake, effluent, and activated sludge in several sewage plants in Tokyo at average levels of 0.0013, 0.0011 and 38 ppb respectively[28].

A major HCB contamination episode occurred in southern Louisiana (US) in 1972 where HCB levels were detected in beef cattle far in excess of the tolerance level of 0.3 ppm in beef fat then in effect[51,52]. The source of HCB in these animals originated primarily from the industrial plants in the area which were engaged in the production of perchloroethylene, carbon tetrachloride, synthetic rubber, and agricultural chemicals (atrazine, simazine and propazine herbicides). HCB waste from at least two of these plants had been dumped in off-site landfills.

The hazardous characteristics of HCB have been well documented and include its potential for bioaccumulation in the food chain[13,53,54], environmental persistence due to physical, chemical and biological stability[13,41,42] and from its toxicity[13,41,42].

HCB has been found in fish from a number of river systems in Canada[55] and in the U.S.[56-58], in juvenile seals at levels of 0.067 ppm (wet wgt) collected in coastal waters off the Netherlands[59], in low levels in Japanese fish and shellfish[13], in wild birds in the Netherlands[60], and in the eggs of ferns collected in Canada and in the Netherlands[57,61].

Studies by Laseter et al have indicated that HCB is bioaccumulative in bass to levels 44,000 times the concentration of HCB in the surrounding aquatic environment[54].

HCB is very stable and lipophilic. The monitoring of human adipose tissues collected from across the U.S. has shown that approximately 95% of the populace has trace residues of HCB[62]. HCB residues have also been found in adipose tissue and blood samples in Australian, Japan, Turkey and United Kingdom[62-68].

In a study conducted in Louisiana, in 86 persons who resided in an HCB-contaminated area, the mean HCB concentration in plasma samples was 0.0036 ppm[66]. A group of farm workers in the U.S. occupationally exposed to HCB-contaminated Dacthal had blood HCB concentrations averaging 0.040 ppm, but there was no evidence of porphyria in this group[69].

An Australian study of occupationally exposed individuals showed levels ranging from 0 to 0.095 ppm[65]. Another Australian study showed that all perrenal fat samples examined contained HCB at an average concentration of 1.25 ppm with a range from a trace to 8.2 ppm[63].

Human adipose tissues from Japan contained HCB levels ranging from 0.38 to 1.48 ppm in one study[62] and 0.10 to 0.42 ppm with a mean of 0.21 ppm in a later study[70].

Significant residues of HCB have been reported in human milk from the Netherlands[71], Australia[72], Germany[64] and Sweden[75]. For example, levels of HCB ranging from 0.07 to 0.22 ppm (on a fat content basis) have been found in breast milk from samples in Sweden[73].

In Turkey, HCB was the causative agent in a severe outbreak of porphyria cutanea tarda symptomatica involving several thousand people in 1955 and 1959[67,74,75]. The doses of HCB ingested were estimated to have been 1 to 4 mg/kg for several months to two years[67].

The carcinogenic activity of HCB in Syrian golden hamsters was recently demonstrated by Cabral et al[76]. Doses ranging from 50 to 200 ppm of HCB administered in the diet for life resulted in a significant induction of hepatomas, haemangio-endotheliomas and thyroid adenomas. Further, an increased number of tumor-bearing animals and of tumors per animal was noted, as were a shortened lifespan and a reduced latency period for onset of liver tumors. These effects may indicate that HCB behaves like certain carcinogens shown to have multipotential activity[76]. Cabral et al[76] in a purely quantitative approach indicated that an intake of 4-16 mg/kg/day of HCB in their hamster studies above was within the range of the estimated quantities accidently consumed by the Turkish people for several months at a time resulting in severe toxic porphyria[67].

The carcinogenicity of HCB has also been demonstrated recently in Swiss mice fed for life a diet containing 50, 100 and 200 ppm. The hepatoma incidence was 10% in both males and females treated with the median dose of HCB, and 29.7 and 14.0% respectively, in females and males treated with the highest dose of HCB. None of the hepatomas metastasized or occurred in the control group[77].

HCB has been found mutagenic in Saccharomyces cerevisiae (Ceppo 632/4 strain) using reversion from histidine and methionine as a measure of the induced mutation[78].

HCB at doses of 70 or 221 mg/kg did not induce dominant lethal mutations in the rat[79]. HCB-pretreatment of rats led to an increase in liver microsomal 2,4-diaminoanisole activation to a mutagen[80] (when tested by the Ames assay in Salmonella)[81] after a dose of 10 mg/kg i.p. and to an increase in ethylmorphine N-demethylase after a dose of 50 mg/kg i.p.

Recently evidence was reported[82-85] for the urinary excretion of the metabolites pentachlorophenol, 2,4,5-trichlorophenol, tetrahydrohydroquinone and pentachlorothiophenol in the rat following administration of HCB. (However, no urinary phenolic metabolites of HCB have been found in rabbits[86]). Mehendale et al[83] also identified pentachlorobenzene and tetrachlorobenzene as HCB metabolites in the rat and found that the reductive dechlorination of HCB was catalyzed by an enzyme located in the microsomal fraction of liver, lung, kidney and intestine. It was also reported that 70% of the total dose was found remaining 7 days after administration[83]. Following a single dose of HCB to male Wistar rats, the biological half-life was estimated to be about 60 days[87].

Aspects of the nature of the trace contaminants in commercial HCB bears particular emphasis. For example, Villaneuva et al[88] reported hepta- and octadichlorodibenzofurans, octachlorodibenzo-p-dioxin, octa-, and nona- and decachlorobiphenyls, 1-pentachlorophenol, 2,2-dichloroethylene, hexacyclopentadiene, pentachloroiodobenzene and heptachlorotropilium in trace amounts in various samples of commercial HCB. The toxicities of some of the chlorinated dibenzo-p-dioxins and chlorodibenzofurans have been reported[89-93].

6. <u>Brominated Benzenes</u>. The extent of industrial utilization of brominated (as well as iodinated) benzene derivatives are relatively small compared to their chlorinated analogs. In general, very much less is known of their environmental and health effects.

a. <u>Bromobenzene</u> (phenylbromide; ⟨O⟩-Br) is prepared industrially by the action of bromine or benzene in the presence of iron powder. It is used as a chemical intermediate (e.g., for the preparation of Grignard reagents such as phenyl magnesium bromide); as an additive for motor oil, and as a solvent for crystallization on a large scale where a heavy liquid is required. The production of bromobenzene in the years 1972-1975 was estimated to be greater than 500 kg per year[94].

No occupational standard for bromobenzene has been established by OSHA.

Bromobenzene has been detected by EPA in tap water from New Orleans, although no quantitative measurement was reported[95].

No animal carcinogenicity studies have been reported for bromobenzene. When tested at doses ranging from 10-750 µg/plate, it was inactive in <u>S. typhimurium</u> TA 100, TA 98, TA 1535 and TA 1537 with metabolic activation by liver S-9 fractions from Aroclor-pretreated rats[96]. Bromobenzene was also inactive when tested <u>in vitro</u> with S. cerevisiae D3[97] but mutagenic when tested in the host-mediated assay with S. cerevisiae D3[98].

Bromobenzene is inactive in the transformation of Golden Syrian hamster embryo cells[99].

Bromobenzene appears to be rapidly absorbed and metabolized in animals. The urinary metabolites which have been detected in animals include: 2-bromophenol, 3-bromophenol, 4-bromophenol, bromocatechol, bromophenyldihydrodiol and p-bromophenylmercapturic acid[94,100,101]. Glucuronide and sulfate conjugates of the bromophenols, bromocatechol and dihydrodiol are also formed; the proportion of metabolites excreted is species dependent[101].

Bromobenzene has been demonstrated in vitro and in vivo to bind with liver proteins in rats when liver glutathione levels have been extensively depleted. This suggests that bromobenzene is initially activated to an electrophile (bromobenzene-3,4-oxide) and conjugated with glutathione in the liver[100], or can alkylate cellular macromolecules in the liver[94]. This epoxide may also be excreted in the bile and be reabsorbed through the enterohepatic circulation[102].

b. para-Dibromobenzene (1,4-dibromobenzene; Br–⌬–Br) is prepared by the catalytic bromination of benzene. It is used principally as a chemical intermediate in the production of drugs, dyestuffs and other organic compounds.

Production of para-dichlorobenzene is believed to be significantly greater than 500 kg/year[94]. No occupational standard for para-dibromobenzene has been established by OSHA or have there been TLV's or MAC's adopted elsewhere.

An unspecified isomer of dibromobenzene has been detected in finished drinking water at three locations and in one sample of river water as reported in recent EPA surveys[95].

No information appears to be available on the carcinogenicity and mutagenicity of para-dibromobenzene.

Data on the biotransformation of para-dichlorobenzene suggests that it may be metabolically activated via an arene oxide to a potential electrophile. Two major metabolites, 2,4- and 2,5-dibromophenol were identified in the urine and to a lesser extent in the feces of rabbits given single i.p. injections of 50 mg/kg of para-dibromobenzene. It was suggested that para-dibromobenzene is metabolized to the phenols via formation of 1,2-arene oxide, which is then cleaved to yield the 1- or 2-carbonium ion intermediate. A 1,2-H or 1,2-Br shift then yields the dienone, which then enolizes to yield 2,4- or 2,5-dibromophenol[101,103] It should also be noted that 2,4-dibromophenol has been found to be a tumor promoter in mouse skin[102].

7. Benzyl Halides. Most aryl halides are usually much less reactive than alkyl or allyl halides toward nucleophilic reagents in either S_N1- or S_N2-type reactions. However, in contrast to phenyl halides, benzyl halides are quite reactive, are analogous in reactivity to allyl halides and are hence readily attacked by nucleophilic reactants in both S_N1- and S_N2-displacement reactions. This reactivity is related to the stability of the benzylcation, the positive charge of which is expected to be extensively delocalized[104].

a. Benzyl Chloride (⌬ –CH_2Cl; chloromethylbenzene; α-chlorotoluene) is made by the chlorination of toluene used principally (65-70%) in the U.S. as an intermediate in the manufacture of butylbenzylphthalate, a vinyl resin plasticizer, while the remaining 30-35% is employed as an intermediate in the production of benzyl alcohol, quaternary ammonium chlorides and benzyl derivatives such as benzyl acetate, cyanide, salicylate and cinnamate, which are used in perfumes and pharmaceutical products[105].

Suggested uses of benzyl chloride include: in the vulcanization of fluororubbers[106] and in the benzylation of phenol and its derivatives for the production of possible disinfectants[107]. Annual production of benzyl chloride in the U.S. in recent years is estimated to be: (millions of pounds) 1974 (99); 1975 (70); and 1976 (90)[3].

The current OSHA environmental standard for benzyl chloride is 1 ppm (5 mg/m^3) for an 8 hr time-weighted exposure in a 40 hr work week. The hygienic standards for permissible levels of benzyl chloride in the working environment in various countries were as follows (mg/m^3)[15]: Federal Republic of Germany, 5; Democratic Republic of Germany, 5; the MAC in the USSR is 0.5 mg/m^3.

Benzyl chloride has been shown to induce local sarcomas in rats treated by subcutaneous injection[108,109].

Benzyl chloride was reported to be weakly mutagenic in S. typhimurium TA 100 strain[110]. It is active at doses of 0.1, 1.0 and 10.0 µg/ml in the transformation of Golden Syrian hamster embryo cells[99].

References

1. Hardie, D. W. F., Chlorocarbons and chlorohydrocarbons: Chlorinated benzenes, dichlorobenzenes, In: Kirk-Othmer Encyopedia of Chemical Technolgoy, 2nd ed., Vol. 5, John Wiley & Sons, New York (1964) pp. 253-267
2. Roberts, J. D., and Caserio, M. C., "Modern Organic Chemistry" W. A. Benjamin Inc., New York (1967) p. 571
3. Stanford Research Institute, A study of industrial data on candidate chemicals for testing, EPA-560/5-77-006; Menlo Park, CA, August (1977) pp. 4-150-4-158
4. IARC, Monographs for the Assessment of Carcinogenic Risk to Man, Vol. 7, International Agency for Research on Cancer, Lyon (1974) pp. 231-244
5. PPG Industries, ortho-Dichlorobenzene Bulletin 30B, Pittsburgh, PPG Industries, Inc (1970)
6. Skelly, J. K., Evans, D. G., and Broadbent, B., Dyeing of natural and synthetic fibers, Ger. Offen. 2,421,507, 21 Nov., 1974; Chem. Abstr., 82 (1975) 87588N
7. Matsuda, S., and Takagi, K., Dyeing polyolefin products, Japan Patent 74,24190 20 June, 1974, Chem. Abstr., 82 (1975) 113109D
8. Shioya, K., and Katahira, S., Phenol resin adhesive, Japan Patent 74 38,094, 15 October, 1974, Chem. Abstr., 82 (1975) 126258B
9. Marchitto, M. J., and Landriscina, V. J., Jr., Solid stock lubricant for tapping holes, Chem. Abstr., 82 (1975) 114026H, U.S. Patent 3,847,824, 12 November 1974
10. Tilhaud, R., Composition for dissolving petroleum sediments, Fr. Demande 2,211,547, 19 July 1974, Chem. Abstr., 82 (1975) 158526D
11. Shimosaka, Y., and Ozaki, M., Transparency improver for acrylonitrile polymers, Japan Patent 74 10,105; 15 May, 1974; Chem. Abstr., 82 (1975) 87201F
12. Chemical Information Services, Directory of West European Chemical Producers, Oceanside, New York (1973)
13. Morita, M., Chlorinated benzenes in the environment, Ecotoxicol. Environ. Safety, 1 (1977) 1-6
14. ACGIH, Threshold Limit Values for Chemical Substances in Workroom Air Adopted by ACGIH for 1976, American Conference of Industrial Hygienists, Cincinatti, Ohio, (1976)
15. Winell, M., An international comparison of hygienic standards for chemicals in the work environment, Ambio, 4 (1975) 34-36
16. Jacobs, A., Blangetti, M., Hellmund, E., and Koelle, W., Accumulation of organic compounds, identified as harmful substances in Rhine water, in the fatty tissues of rats, Krebs Forschungszentrum Karlsruhe, KFK 1969UF (1974) 6-7; Chem. Abstr., 82 (1975) 39285Q
17. Grob, K., and Grob, G., Organic substances in potable water and its precursor, Part II, Applications in the area of Zurich, J. Chromatog., 90 (1974) 303-313
18. Parsons, D. L., On early tumor formation in pure-line mice treated with carcinogenic compounds and the associated blood and tissue changes, J. Path. Bact., 54 (1942) 321-330
19. Hollingsworth, R. L., Rowe, V. K., Oyen, F., Hoyle, H. R., and Spencer, H. C., Toxicity of para-dichlorobenzene. Determinations on experimental animals and human subjects, Arch. Ind. Hlth., 14 (1956) 138-147
20. Hollingsworth, R. L., Rowe, V. K., Oyen, F., Torkelson, R. R., Adams, E. M., Toxicity of o-dichlorobenzene, Studies on animals and industrial experience, Arch. Ind. Hlth., 17 (1958) 180-187
21. Azouz, W. M., Parke, D. V., and Williams, R. T., The metabolism of halogenbenzenes ortho- and para-dichlorobenzenes, Biochem. J., 59 (1955) 410-415

22. Reid, W. D., and Krishna, G., Centrolobular hepatic necrosis related to covalent binding of metabolites of halogenated aromatic hydrocarbons, Exp. Mol. Path. 18 (1973) 80-99
23. PPG Industries, para-Dichlorobenzene Bulletin 30C, Pittsburgh, PPG Industries Inc. (1972)
24. Nysted, L. N., Polyethylene-paraffin-metal salt hydrate supports for perfumes and fumigants, Fr. Demande 2,223,443, 25 Oct. 1974; Chem. Abstr., 82 (1975) P157400W
25. Ellis, W. R., Perfumed lavoratory cleansing compositions, Brit. Patent 1,364,460; 21 August, 1974, Chem. Abstr., 82 (1975) 5578X
26. PPG Industries, Paradichlorobenzene, Bulletin 30C, Pittsburgh Plate & Glass Industries, Inc. Pittsburgh, PA (1972)
27. Campbell, R. W., Arylenesulfide polymers, US Patent 3,870,686; 11 March, 1975; Chem. Abstr., 82 (1975) 171718N
28. Morita, M., and Ohi, G., Para-dichlorobenzene in human tissue and atmosphere in Tokyo metropolitan area, Environ. Pollut., 8 (1975) 269-273
29. Girard, R., Tolot, F., Martin, P., and Bourret, J., Hemopathies graves et exposition a des derives chlores du benzene (a propos de 7 cas), J. Med. Lyon 50 (1969) 771-773
30. Hallowell, M., Acute haemolytic anaemia following the ingestion of para-dichlorobenzene, Arch. Dis. Children, 34 (1959) 74-75
31. Pagnotto, L. D., and Walkley, J. E. Urinary dichlorophenol as an index of para-dichlorobenzene exposure, Amer. Ind. Hyg. Assoc. J.. 26 (1965) 137-142
32. Winholz, M. (ed) The Merck Index, 9th ed., Merck & Co., Inc. Rahway, NJ (1976) p. 9318
33. Coate, W. B., Schoenfisch, W. H., Lewis, T. R., and Busey, W. M., Chronic inhalation exposure of rats, rabbits and monkeys to 1,2,4-trichlorobenzene, Arch. Environ. Hlth., 32 (1977) 249-255
34. Jondorff, W. F., Parke, D. V., and Williams, R. T., Studies in detoxification.66. The metabolism of halogenbenzenes, 1,2,4- and 1,3,5-trichlorobenzene, Biochem. J., 61 (1955) 512-521
35. Gillette, J. R., Mitchell, J. R., and Brodie, B. D., Biochemical mechanisms of drug toxicity, Ann. Rev. Pharmacol.. 14 (1974) 271-288
36. Daly, J. W., Jerina, D. M., and Witkop, B., Arene oxides and NIH shift. Metabolism, toxicity and carcinogenicity of aromatic compounds, Experientia 28 (1972) 1129-1149
37. Azouz, W. M., Parke, D. V., and Williams, R. T., Studies in detoxification. LI. Determinations of catechols in urine and the formation of catechols in rabbits receiving halobenzenes and other compounds. Dihydroxylation in vivo. Biochem. J., 55 (1) (1953) 146-151
38. Knight, R. H., and Young, L., Biochemical studies of toxic agents. IX. Occurrence of premercapturic acids, Biochem. J., 70 (1958) 111-119
39. Gillette, J. R., A perspective on the role of chemically reactive metabolites of foreign compounds in toxicity. I. Correlation of changes in covalent binding of reactivity metabolites with changes in the incidence and severity of toxicity. Biochem. Pharmaco., 23 (1974) 2785-2794
40. Cameron, G. R., Thomas, J. C., Ashmore, A. S., Buchan, J. L., Warren, E. H., and Huges, A. W. M., The toxicity of certain chlorine derivatives of benzene with special reference to o-dichlorobenzene, J. Pathol. Bacteriol., 44 (1937) 281-296

41. U.S. Environmental Protection Agency, Status Assessment of Toxic Chemicals, No. 7-Hexachlorobenzene, Industrial Environmental Research Laboratory, Cincinatti, Ohio Sept. 6 (1977)
42. Mumma, C. E., Lawless, E. W., Survey of Industrial Processing Data. Task I-Hexachlorobenzene and Hexachlorobutadiene Pollution From Chlorocarbon Processes, Contract No. 68-01-2105, U.S. Environmental Protection Agency, Washington, DC, June (1976)
43. Quinlivan, S. C., Ghassemi, M., and Leshendok, T. V., Sources, characteristics and treatment and disposal of industrial wastes containing hexachlorobenzene, J. Hazardous Mater., 1 (1975/77) 343-359
44. Winteringham, F. P. W., Comparative ecotoxicology of halogenated hydrocarbons residues, Ecotox. Environ. Safety, 1 (1977) 407-425
45. Leoni, V., D'Arca, S. U., Experimental data and critical review of the occurrence of hexachlorobenzene in the Italian environment, Sci. Total Env., 5 (1973) 253-272
46. Wapensky, L. A., Collaborative study of gas chromatographic and infrared methods for dacthal formulations, J. Assoc. Off. Anal. Chem., 52 (1969) 1284-1290
47. Melnikov, N. N., Halogen derivatives of aromatic hydrocarbons, Residue Revs. 36 (1971) 67-82
48. U.S. Environmental Protection Agency, Environmental Contamination from Hexachlorobenzene, EPA 560/6-76-014, Washington, DC, April 1976
49. Laska, A. L., Bartell, C. K., and Laseter, J. L., Distribution of hexachlorobenzene and hexachlorobutadiene in water, soil and selected aquatic organisms among the lower Mississippi River, Louisiana, Bull. Env. Cont. Toxicol., 15 (1976) 535-542
50. Kleopfer, R. D., and Fairless, B. J., Characterization of organic compounds in a municipal water supply, Env. Sci. Technol., 6 (1972) 1036-1037
51. U.S. Environmental Protection Agency, HCB Contamination of Cattle in Louisiana, In: Hazardous Waste Disposal Damage Report, Document No. 3, S.W.-151.3 Washington, DC (May 1976)
52. Lazar, E. C., Damage incidents from improper land disposal, J. Hazardous Mater., 1 (1975/76) 157
53. Avrahami, M., and Steel, R. T., Hexachlorobenzene: I. Accumulation and elimination of HCB in sheep after oral dosing, New Zealand J. Agr. Res., 15 (1975) 476
54. Laseter, J. L., Bartell, C. K., Laska, A. L., Holmquist, D. G., and Condie, D. B., An ecological study of hexachlorobenzene, EPA 560/6-76-009, U.S. Environmental Protection Agency, Washington, DC, 1976
55. Zitko, V., Polychlorinated biphenyls and organochlorine pesticides in some freshwater and marine fishes, Bull. Env. Contam. Toxicol., 6 (1971) 464-470
56. Holden, A. V., Pesticide Monit. J., 4 (1970) 117
57. Holden, A. V., International cooperative study of organochlorine and mercury residues in wildlife, 1969-1971, Pesticide Monit. J., 7 (1973) 37-52
58. Johnson, J. L., Stalling, D. L., and Hogon, J. W., Hexachlorobenzene (HCB) residues in fish, Bull. Environ. Contam. Toxicol., 6 (1971) 464
59. Koeman, J. H., Tennoever de Brau, M. C., DeVos, R. H., Chlorinated diphenyls in fish, mussels and birds from the River Rhine and the Netherlands coastal area, Nature, 221 (1969) 1126-1128
60. Vos, J. G., Breeman, H. A., Benshop, H., and Rijksfac, M., Occurrence of the fungicide HCB in wild birds and its toxicological importance. A preliminary communication, Landbouwwetensch. Gent., 33 (1968) 1263-1268

61. Gilbertson, M., and Reynolds, L. M., Hexachlorobenzene (HCB) in the eggs of common terns in Hamilton Harbour, Ontario, Bull. Environ. Contam. Toxicol., 7 (1972) 371-373
62. Curley, A., Burse, V. W., Jennings, R. W., Villaneuva, E. C., Tomatis, L., and Akazaki, K., Chlorinated hydrocarbon pesticides and related compounds in adipose tissue from people of Japan, Nature, 242 (1973) 338-340
63. Brady, M. N., and Siyali, D. S., Hexachlorobenzene in human body fat, Med. J. Aust., 1 (1972) 158-161
64. Acker, L., and Schulte, E., Uber des vorkommen von chlorierten biphenylen und hexachlorobenzoil neben chlorierten insektiziden in huamn milch und menschlichen fettgewebe, Naturwiss, 57 (1970) 497
65. Siyali, D. S., Hexachlorobenzene and other organochlorine pesticides in human blood, Med. J. Aust., 2 (1972) 1063-1066
66. Burns, J. E., and Miller, F. M., Hexachlorobenzene contamination: Its effects in a Louisiana population, Arch. Env. Hlth., 30 (1975) 44-48
67. Cam. C., and Nigogosyan, G., Acquired toxic porphyria cutanea tarda due to hexachlorobenzene, J. Amer. Med. Assoc., 183 (1963) 88-91
68. Abbott, D. C., Collins, G. B., and Goulding, R., Organochlorine pesticide residue in human fat in the United Kingdom, 1969-1971, Brit. Med. J., 2 (1972) 553-556
69. Burns, J. E., Miller, F. M., Gomes, E. D., and Albert, R. A., Hexachlorobenzene exposure from contaminated DCPA in vegetable spraymen, Arch. Environ. Hlth., 29 (1974) 192-194
70. Morita, M., Mimura, S., Ohi, G., Yagyu, H., and Nishizawa, T., A systematic determination of chlorinated benzenes in human adipose tissue, Environ. Pollut., 9 (1975) 175-179
71. Tuinstra, L. G. M. T., Organochlorine residues in human milk in the Leiden Region, Ned. Melk-Zuiveltijdschr., 25 (1971) 24-32
72. Siyau, D. S., Polychlorinated biphenyls, hexachlorobenzene, and other organochlorine pesticides in human milk, Med. J. Aust., 2 (1973) 815-818
73. Westöö, G., and Noren, K., Organochlorine contaminants in human milk, Stockholm 1967-1977, Ambio, 7 (1978) 62-64
74. Schmid, R., Cutaneous porphyria in Turkey, New Engl. J. Med., 263 (1960) 397-398
75. Ockner, R. K., and Schmid, R., Acquired porphyria in man and rat due to hexachlorobenzene intoxication, Nature, 189 (1961) 499
76. Cabral, J. R. P., Shubik, P., Millner, T., and Raitano, F., Carcinogenic activity of hexachlorobenzene in hamsters, Nature, 269 (1977) 510-511
77. Cabral, J. R. P., Mollner, T., Raitano, T., and Shubik, P., Carcinogenesis study in mice with hexachlorobenzene, Abstracts of 17th Annual Meeting of Society of Toxicology, San Francisco, CA, March 12-16 (1978) 209
78. Guerzoni, M. E., DelCupolo, L., and Ponti, L., Attivita mutagenica degli antiparassitari, Riv. Sci. Tecn. Alim. Nutr. Um., 6 (1976) 161-165
79. Simon, G. S., Kipps, B. R., Tardiff, R. G., and Borzelleca, J. F., Failure of kepone and hexachlorobenzene to induce dominant lethal mutations in the rat, Abstracts of 17th Annual Meeting Society of Toxicology, San Frnacisco, CA, March 12-16 (1978) 223, 225
80. Dybing, E., and Aune, T., Hexachlorobenzene induction of 2,4-diamincanisole mutagenicity in vitro, Acta Pharmacol. Toxicol., 40 (1977) 575-583
81. Ames, B. N., Durston, W. E., Yamasaki, E., and Lee, F. D., Carcinogens are mutagens: A simple test system combining liver homogenates for activation and bacteria for detection, Proc. Natl. Acad. Sci. (US), 70 (1973) 2281-2285

82. Lui, H., and Sweeney, G. D., Hepatic metabolism of hexachlorobenzene in rats, FEBS Letter, 51 (1975) 225-226
83. Mehendale, H. M., Fields, M., and Matthews, H. B., Metabolism and effects of hexachlorobenzene on hepatic microsomal enzymes in the rat, J. Agr. Food Chem., 23 (1975) 261-265
84. Koss, G., Koransky, W., and Steinbach, Studies on the toxicology of hexachlorobenzene II Identification and determination of metabolites, Arch. Toxicol., 35 (1976) 107-114
85. Renner, G., and Schuster, K. P., 2,4,5-Trichlorophenol, a new urinary metabolite of hexachlorobenzene, Toxicol. Appl. Pharmacol., 39 (1977) 355-356
86. Kohli, J., Jones, D., and Safe, S., The metabolism of higher chlorinated benzene isomers, Can. J. Biochem., 54 (1976) 203-208
87. Morita, M., and Oishi, S., Clearance and tissue distribution of hexachlorobenzene in rats, Bull. Environ. Contam. Toxicol., 14 (1975) 313-318
88. Villaneuva, E. C., Jennings, R. W., Burse, V. W., and Kimbrough, R. D., Evidence of chlorodibenzo-p-dioxin and chlorodibenzofuran in hexachlorobenzene, J. Agr. Food Chem., 22 (1974) 916-917
89. Williams, D. T., Cunningham, H. M., and Blanchfield, B. J., Distribution and excretion studies of octachlorodibenzo-p-dioxin in the rat, Bull. Env. Contam. Toxicol., 7 (1972) 57-62
90. Higginbotham, G. R., Huang, A., Firestone, D., Verrett, J., Ress, J., and Campbell, A. D., Chemical and toxicological evaluations of isolated and synthetic chloro derivatives of dibenzo-p-dioxin, Nature, 220 (1968) 702-703
91. Huff, H. E., and Wassom, J. S., Health hazards from chemical impurities: Chlorinated dibenzodioxins and chlorinated dibenzofurans, Intern. J. Env. Studies, 6 (1974) 13-17
92. King, M. E., Shefner, A. M., and Bates, R. R., Carcinogenesis bioassay of chlorinated dibenzodioxins and related chemicals, Env. Hlth. Persp., 5 (1973) 163-170
93. Schwetz, B. A., Norris, J. M., Sparschu, G. L., Rowe, V. K., Gehring, P. J., Emerson, J. L., and Herbig, C. G., Toxicology of chlorinated dibenzo-p-dioxins, Env. Hlth. Persp., 5 (1973) 87-109
94. National Cancer Institute, Clearinghouse on Environmental Carcinogens, Chemical Selection Sub-Group: Bromobenzene and p-dibromobenzene, Bethesda, MD Dec. 19 (1977)
95. Shackelford, W. M., and Keith, L. H., Frequency of Organic Compounds Identified in Water, EPA/600/4-76/062; Environmental Protection Agency, Athens, GA (1976)
96. McCann, J. E., Choi, E. S., Yamasaki, E., and Ames, B. N., Detection of carcinogens in the Salmonella/microsome test: Assay of 300 Chemicals, Proc. Natl. Acad. Sci., 72 (1975) 5135-5139
97. Simmon, V. F., Kauhanen, K., and Tardiff, R. G., Mutagenic activity of chemicals identified in drinking water, In: Progress in Genetic Toxicology, (eds) Scott, D., Bridges, B. A., and Sobels, F. H., Elsevier/North Holland Biomedical Press (1977) 249-258
98. Simmon, V. F., In vitro mutagenicity assays with Saccharmyces cerevisiae D3, J. Natl. Cancer Inst. (1978) in press
99. Dunkel, V. C., Wolff, J. S., and Pienta, R. J., In vitro transformation as a presumptive test for detecting chemical carcinogens, Cancer Bulletin, 29 (1977) 167-174

100. Searle, C. E. (ed) Chemical Carcinogens, American Chemical Society Monograph No. 173, Washington, DC (1976) p. 745
101. Kasperek, G. J., and Bruice, T. C., The mechanism of the aromatization of arene oxides, J. Am. Chem. Soc., 94 (1972) 198
102. Boutwell, R. K., and Bosch, D. K., The tumor-promoting action of phenol and related compounds for mouse skin, Cancer Res., 19 (1959) 413-424
103. Ruzo, L. O., Safe, S., and Hutzinger, O., Metabolism of bromo derivatives in the rabbit, J. Agr. Food Chem., 24 (1976) 291-293
104. Roberts, J. D., Caserio, M. C., "Modern Organic Chemistry", W. A. Benjamın, Inc., New York (1967) p. 571
105. IARC, Monographs on the Evaluation of Carcinogenic Risk of Chemicals to Man, Vol. 11, International Agency for Research on Cancer, Lyon (1976) 217-223
106. Okada, S., and Iwa, R., Fluorolefin elastomer stocks, Japan Kokai 7,414,540, Feb. 8 (1974) Chem. Abstr., 81 (1974) P50806K
107. Janata, V., Simek, A., and Nemeck, O., Benzyl phenols, Czeck Patent 152,190 Feb. 15 (1974) Chem. Abstr., 81 (1974) P25347D
108. Preussmann, R., Direct alkylating agents as carcinogens, Food Cosmet. Toxicol., 6 (1968) 576-577
109. Druckrey, H., Kruse, H., Preussmann, R., Ivankovic, S., and Lanschütz, C., Cancerogene alkylierende substanzen. III. Alkyl-halogenide, -sulfate, sulfonate und ringespannte heterocyclen, Z. Krebsforsch., 74 (1970) 241-270
110. McCann, J., Springarn, N. E., Kobori, J., and Ames, B. N., Carcinogens as mutagens: Bacterial tester strains with R factor plasmids, Proc. Natl. Acad. Sci., 72 (1975) 979-983

CHAPTER 15

HALOGENATED POLYAROMATIC DERIVATIVES

1. Polychlorinated Biphenyls

Halogenated biphenyls, as is typical of aryl halides, generally are quite stable to chemical alteration. Polychlorinated biphenyls (PCB's) (first introduced into commercial use more than 45 years ago) are one member of a class of chlorinated aromatic organic compounds which are of increasing concern because of their apparent ubiquitous dispersal, persistence in the environment, and tendency to accumulate in food chains, with possible adverse effects on animals at the top of food webs, including man[1-7].

Polychlorinated biphenyls are prepared by the chlorination of biphenyl and hence are complex mixtures containing isomers of chlorobiphenyls with different chlorine contents[8]. It should be noted that there are 209 possible compounds obtainable by substituting chlorine for hydrogen or from one to ten different positions on the biphenyl ring system. An estimated 40-70 different chlorinated biphenyl compounds can be present in each of the higher chlorinated commercial mixture[9,10]. For example, Arochlor 1254 contains 69 different molecules, which differ in the number and position of chlorine atoms[10].

It should also be noted that certain PCB commercial mixtures produced in the U.S. and elsewhere (e.g., France, Germany, and Japan) have been shown to contain other classes of chlorinated derivatives, e.g., chlorinated naphthalenes and chlorinated dibenzofurans[7,11-14]. The possibility that naphthalene and dibenzofuran contaminate the technical biphenyl feedstock used in the preparation of the commercial PCB mixtures cannot be excluded. Table 1 illustrates the structures of the chlorinated biphenyls; chlorinated naphthalenes; chlorinated dibenzofurans and lists the extent of chlorination as well as the number of chlorinated derivatives possible.

The PCBs have been employed in a broad spectrum of applications because of their chemical stability, low volatility, high dielectric constants, nonflammability, and general compatability with chlorinated hydrocarbons. The major areas of utility have included: heat exchangers and dielectric fluids, in transformers and capacitors, hydraulic and lubricating fluids, diffusion pump oils, plasticizers for plastics and coatings, ingredients of caulking compounds, printing inks, paints, adhesives and carbonless duplicating, flame retardants, extender for pesticides, and electrical circuitry and component.

The rates and routes of transport of the PCBs in the environment[1,2,5,6] and their accumulation in ecosystems[1,2,15-20], and toxicity[1,3,5,7,21-23] have been reviewed.

It is generally acknowledged that the toxicological assessment of commercially available PCBs has been complicated by the heterogeneity of the isomeric chlorobiphenyls and by marked differences in physical and chemical properties that influence the rates of absorption, distribution, biotransformation and excretion[1,7,21-29].

TABLE 1
STRUCTURES, EXTENT OF POSSIBLE CHLORINATION AND NUMBER OF CHLORINATED DERIVATIVES OF CHLORINATED BIPHENYLS, NAPHTHALENES AND DIBENZO-p-DIOXINS

Name	Structure	Extent of Chlorination Possible	Number of Chlorinated Derivatives Possible
Chlorinated biphenyls		x=1-10	209
Chlorinated naphthalenes		m=1-8	75
Chlorinated dibenzofurans		m=1-8	135
Chlorinated dibenzo-p-dioxins		m=1-8	75

The lower chlorine homologs of PCBs (either examined individually per se or in Aroclor mixtures) are reported to be more rapidly metabolized in the rat then the higher homologs[30-34]. Sex-linked differences were also disclosed (e.g., the biological half-life of Aroclor 1254 in adipose tissue of rats fed 500 ppm was 8 and 12 weeks in males and females respectively)[32].

The lowest PCB homolog found from Aroclor 1254 in human fat was pentachlorobiphenyl[35].

The metabolism of many PCB isomers have consistently shown the formation of various hydroxylated urinary excretion products. For example, the metabolism of 4,4'-dichlorobiphenyl in the rat yielded four monohydroxy-, four dehydroxy- and two trihydroxy metabolites (Scheme 1)[36]. The structure of the major metabolites in the rat are consistent with epoxidation of the biphenyl nucleus followed by epoxide ring opening accompanied by a 1,2-chlorine shift (NIH shift). The formation of minor rat metabolites, 4-chloro-3'-biphenylol appeared to occur via reductive dechlorination[36].

Urinary metabolites of 2,5,2',5'-tetrachlorobiphenyl (TCB) in the non-human primate included: TCB; monohydroxy-TCB; dehydroxy-TCB; hydroxy-3,4-dihydro-3,4-dehydroxy-TCB; and trans-3,4'-dihydro-3,4-dihydro-TCB[37].

In studies involving the metabolism of 2,2',4,4',5,5'-hexachlorobiphenyl by rabbits rats and mice[38,39], it was shown that the rabbit excreted hexachlorobiphenylol-, pentachlorobiphenylol-, and methoxypentachlorobiphenylol compounds[39,40] (Scheme 2)[40], while rats and mice excreted only a hexachlorobiphenylol[38].

Scheme I

Rat metabolism of 4,4'-dichlorobiphenyl

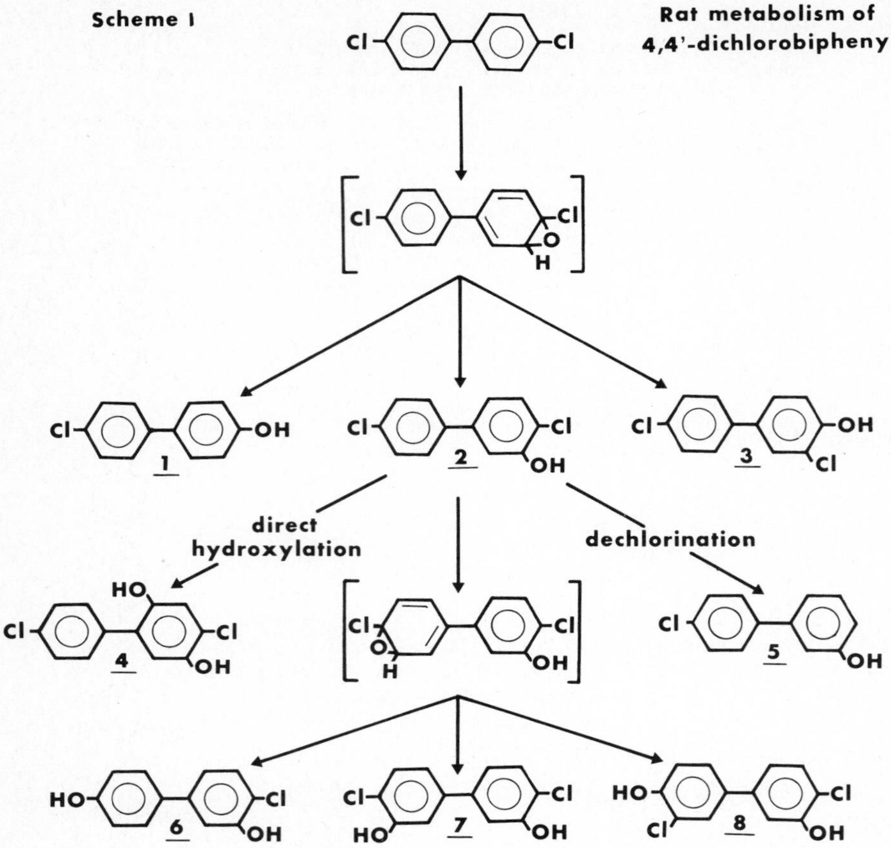

These results and previous studies by Gardner and co-workers[41] as well as in vitro metabolism studies with 4-chlorobiphenyl[42] indicate that PCBs are metabolized via metabolically activated arene oxide intermediates.

It is also of potential importance to note the presence of methyl sulfone metabolites of PCB (as well as DDE) recently found in seal blubber[43]. The toxicological significance of these metabolites have not been elucidated to date.

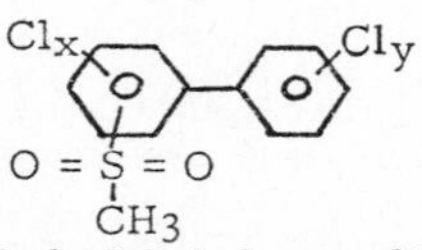

Methyl Sulphone of PCB

$(X + Y = 3 - 7)$

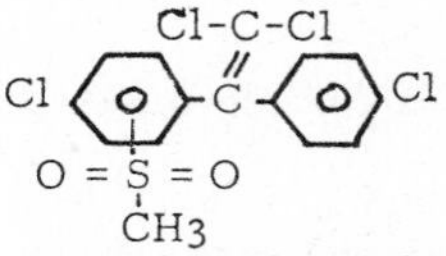

Methyl Sulfone of DDE

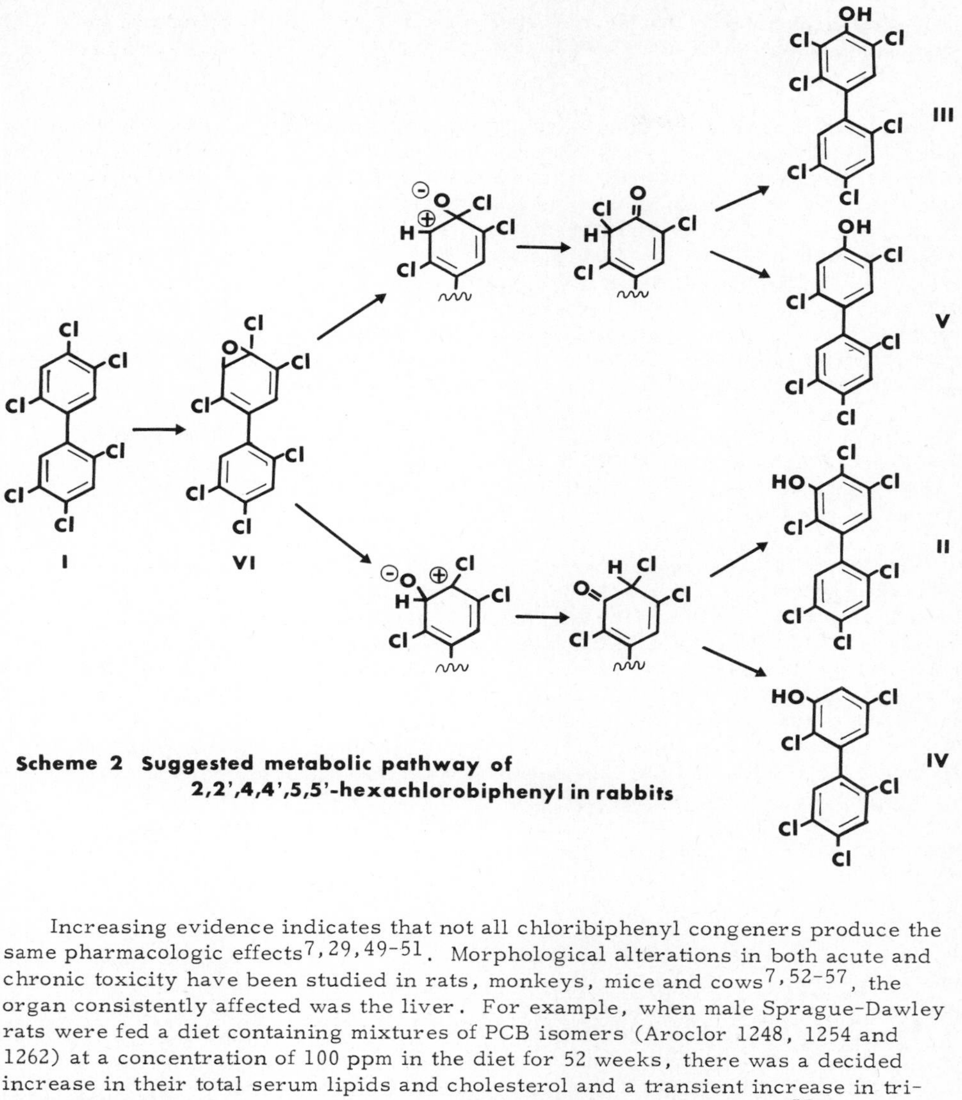

Scheme 2 Suggested metabolic pathway of
2,2',4,4',5,5'-hexachlorobiphenyl in rabbits

Increasing evidence indicates that not all chloribiphenyl congeners produce the same pharmacologic effects[7,29,49-51]. Morphological alterations in both acute and chronic toxicity have been studied in rats, monkeys, mice and cows[7,52-57], the organ consistently affected was the liver. For example, when male Sprague-Dawley rats were fed a diet containing mixtures of PCB isomers (Aroclor 1248, 1254 and 1262) at a concentration of 100 ppm in the diet for 52 weeks, there was a decided increase in their total serum lipids and cholesterol and a transient increase in triglycerides accompanied by distinct morphological changes in the liver[57]. Generalized liver hypertrophy and focal areas of hepatocellular degeneration were followed by a wide spectrum of repair processes. The tissue levels of PCB were greater in the animals receiving the high chlorine mixtures and high levels persisted in these tissues even after the PCB treatment have been discontinued.

Indirect effects of PCB exposure are related to increased microsomal enzyme activity and include alateration in metabolism of drugs, hormones, and pesticides[7,44,58-60]. A large portion of the human population has detectable levels of PCB in adipose tissue[2,63] and recent preliminary reports have revealed that PCBs have been found in 48 of 50 samples of mothers' milk in 10 states[61-63] (the average levels in the 48 samples was 2.1 ppm).

A recent final report of the EPA survey of PCBs in mothers milk revealed that 29.7% of the 1,038 samples were greater than or equal to 0.05 ppm (whole milk basis); while 6.7% were greater[62]. In Canada, the mean for a 1970 survey of PCBs in whole mother's milk was 0.006 ppm and in 1976 it was 0.012 ppm. The range was the same for urban and rural areas in both 1970 and 1975. With respect to the 1975 Canadian survey, the maximum value reported was 0.068 ppm and 98 of the 100 samples were greater than or equal to 0.001 ppm. PCBs on a whole mother's milk basis in Japan were generally 0.03 ppm in samples surveyed between 1971 and 1976[62]. The highest mean value reported was 0.05 ppm where as the highest individual value was 0.31 ppm. For most countries in Europe, Australian and African (Ghana) the samples of whole mother's milk had levels of 0.02 ppm or less or non-detectable[62].

The 1971-74 adipose levels of PCB in the U.S. showed that of 6500 samples examined, 77% contained PCB (e.g., 26% contained <1 ppm; 44% contained 1-2 ppm and 7% contained > 2 ppm[63]. There were no sex differences and the levels of PCB increased with age[63].

The 1971-1974 PCB ambient water levels were as follows: of 4472 water samples, 130 were in the range of 0.1 to 4.0 ppb (detection limit: 0.1-1.0 ppb)[63]; of 1544 sediment samples, 1157 were in the range of 0.1 to 13,000 ppb (limit of detection 0.1-1.0 ppb)[63]. In a recent accidental plant discharge episode, Hudson River sediments near Fort Edward, New York were found to contain 540-2980 ppm PCBs[64]. Soil levels of PCBs in the 1971-1974 period ranged from 0.001-3.33 ppm (average 0.02 ppm) in 1,434 samples taken from 12 of 19 metropolitan areas[63]. In limited surveys in ambient air, an average of 100 ng/cm of PCB was found for each of 3-24 hr samples from Miami, Florida, Jacksonville, Florida and Fort Collins, Colorado[63]. Where PCBs have been found in food, it has generally been in fish samples from the Great Lakes area[64] with contamination arising mainly through the environment. Whereas in the past, milk and dairy products, eggs, poultry, animal feeds, infant foods as well as paper food packaging received PCBs principally from agricultural and industrial applications[64].

Although there has been a sharp curtailment of PCB production and dispersive use applications from a record high of 70 million lbs in 1969, it is believed that it will take several years for ecosystems such as Lake Michigan to cleanse itself of the compounds even if no new input is made[65]. Due to its inertness and high adsorption coefficient, the PCBs have accumulated in the bottom sediments. The final sink for PCB is predicted to be degradation in the atmosphere, with some fraction being buried in underlying sediments of lakes[65]. Figure 3 illustrates the possible routes of loss of PCBs into the environment as determined for 1970.

It is important to note that even with the ceasation of PCB production per se, other environmental sources of PCB may exist. For example, it has been reported that some PCBs are products of DDT photolysis[4,66,67] (Figure 4). Uyeta et al[68] recently reported the photoformation of PCBs from the sunlight irradiation of mono-, di-, tri-, tetra-, and hexachlorobenzenes.

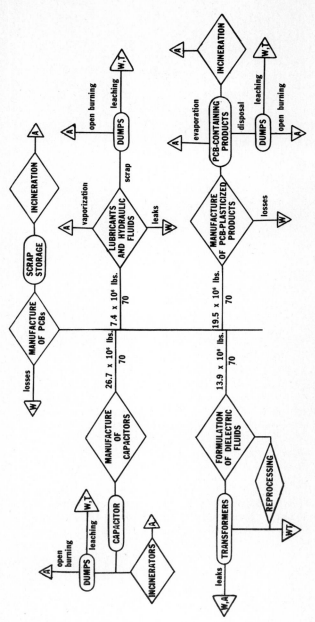

Fig. 3 POSSIBLE ROUTES OF TRANSPORT OF PCBs INTO THE ENVIRONMENT.

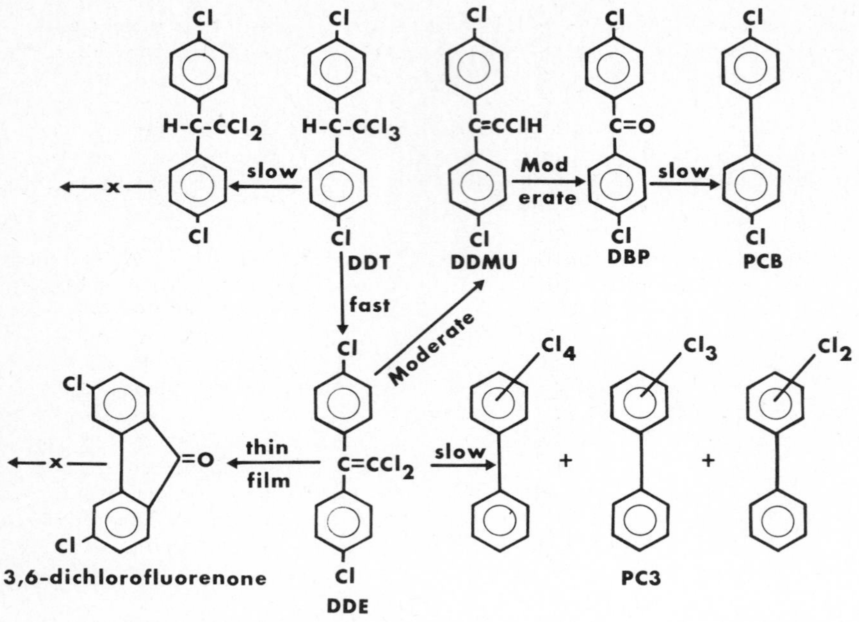

Fig. 4 Proposed scheme for the degradation of DDT vapor in sunlight

Gaffney[68a] recently reported the formation of various mono-, di-, and trichlorobiphenyls resulting from the final chlorination of municipal wastes containing biphenyl. Laboratory chlorination of influent and effluent from a municipal waste treatment facility also resulted in the formation of these and other chloroorganic substances such as di- and trichloro-benzenes.

Previous clinical aspects of human poisoning in Japan ("Yusho" disease) involving at least 1000 people consuming rice bran oil contaminated with Kanechlor 400 (a PCB containing 48% chlorine with 2,4,3',4'-, 2,5,3',4'-, 2,3,5,4'- and 3,4,3,4'-tetrachlorobiphenyl, and 2,3,5,3',4'-pentachlorobiphenyl)[69] are well documented[1,7,70,71]. It has also been claimed that there are an estimated 15,000 victims of "Yusho" disease although only 1081 persons have been officially diagnosed as such[72].

It has been very recently reported by Hirayama[73] that five of the Yusho victims died of liver cancer within 5 years after consuming the contaminated cooking oil.

Recent reports of high cancer rates among Mobil Oil employees at its Paulsboro, N.J. refinery exposed to PCBs (Aroclor 1254) have suggested a possible link between PCB exposure and skin (melanoma) or pancreatic cancer[74-76]. The Mobil study indicated that 8 cancers developed between 1957 and 1975 among 92 research and development and refinery workers exposed for 5 or 6 years in the late 1940's and early 1950's to varying levels of Aroclor 1254. Of the 8 cancers, 3 were malignant melanomas and two were cancers of the pancreas. NIOSH said "this is significantly more skin cancer (melanoma) and pancreatic cancer than would be expected in a population of this size, based on the Third National Cancer Survey"[75].

It should be noted that Monsanto Co. could find no casual relationship between cancer and PCB exposure at its plant in Sauget, Ill. The Monsanto study was based on a review of the records of more than 300 current and former employees at the Illinois plant which had been engaged in PCB production since 1936[77].

Earlier indications of the carcinogenicity of PCBs were reported in 1972 by Nagasaki et al[78] who cited the hepatocarcinogenicity of Kaneclor-500 in male dd mice fed 500 ppm of the PCB. The hepatomas appeared similar to those induced by the gamma-isomer of benzene hexachloride (BHC)[79,80], whereas Kaneclor-400 and Kaneclor-300 had no carcinogenicity activity in the liver of mice. The Kaneclor-500 sample contained 55.0% pentachlorobiphenyl, 25.5% tetrachlorobiphenyl, 12.8% hexachlorobiphenyl and 5.0% trichlorobiphenyl. Later studies by Ito et al[81] also demonstrated that Kaneclor-500 not only induced hepatic neoplasms in mice when fed at levels of 500 ppm in the diet for 32 weeks but also promoted the induction of tumors by alpha-PHC and beta-BHC. Kimbrough et al[82] reported the induction of liver tumors in Sherman strain female rats fed 100 ppm of Aroclor 1260 in their diet for approximately 21 months. Recent studies also suggest that PCBs exert a potent promoting action in experimental azo dye hepato carcinogenesis[83].

Conflicting evidence to date exists concerning mutagenic effects of mixtures of PCBs[63,42,84-89]. Tests on *Drosophila* with PCB of mixed degrees of chlorination did not indicate any chromosome-breaking effects[85]. However, it was suggested by Ramel[84], that PCBs may have an indirect bearing on mutagenicity and carcinogenicity since they induce enzymatic detoxification enzymes in liver microsomes.

No chromosomal aberrations have been observed in human lymphocyte cultures exposed to Aroclor 1254 at 100 ppm levels[86]. Keplinger et al[87] employing a dominant lethal assay, reported no evidence of mutagenic effects of Aroclors. Green et al[88,89] reported a lack of mutagenic activity as measured by dominant lethal test for male Osborne-Mendel rats subjected to 4 different regimens of Aroclor 1242 and 1254. While the above studies of Green et al[88,89] were negative in regard to chromosomal mutations, they do not entirely rule out the possibility that PCBs may induce point mutations. However, to date there are no known reports in the literature concerning the induction of point mutations by PCBs in laboratory model systems.

A recent comparison of the mutagenic activity of Aroclor 1254 (average chlorine content, 4.96% Cl/molecule), 2,2',5,5'-tetrachlorobiphenyl (4 Cl/molecule), Aroclor 1268 (average chlorine content, 9.7 Cl/molecule), Aroclor 1221 (average chlorine content, 1.15 Cl/molecule) and 4-chlorobiphenyl showed that as the degree of chlorination decreased the mutagenicity to *Salmonella* typhimurium TA 1538 strain (in the presence of liver homogenate[90]) increased[42]. This strain is sensitive to frameshift mutagens[90]. The influence of the degree of chlorination on the mutagenicity to the mutant strain TA 1538 also complements the observations that as the chlorine content of the PCB substrate increases, the metabolic rate decreases[31,91].

It is also important to note the recent report that the in vitro metabolism of 4-chlorobiphenyl proceeds via an arene oxide intermediate and is accompanied by binding to the endogenous microsomal RNA and protein[42]. Preliminary results also suggest binding of PCB to exogenous DNA[42] which confirms an earlier report of Allen and Norback[92]. Covalent binding of the 2,5,2',5'-tetrachlorobiphenyl metabolites (e.g., <u>trans</u> dihydrodihydroxy) to cellular macromolecules was suggested by Allen and Norback[93] to be a possible pathway for the carcinogenic action of the PCBs.

Teratogenic studies appear to be thus far nondefinitive[7]. However, while the PCBs have not exhibited known or clearly defined teratogenic effects in mammals, their easy passage across the placenta suggests the potential for some form of fetal toxicity[7,72,94]. Placental transport of PCBs have been reported for the rabbit, rat[95], mouse and cow as well as observed among "Yusho" patients[7,96,97].

No account of the toxicity of the polychlorinated biphenyls can be complete without stressing the possible role of trace contaminants[7,11-14], e.g., the chlorinated dibenzofurans. For example, embryotoxicity of the PCBs Clophen A069 and Phenoclor DP-6 has been attributed chlorinated dibenzofurans present as trace contaminants in the commercial preparations[11,98]. Subsequently, tetra-, penta-, and hexachloro-dibenzofurans were detected in a number of American preparations of PCBs (e.g., Aroclor 1248, 1254, 1260), concentrations of the individual chloridbenzofurans were in the order of 0.1 µg/kg of the PCB. Chlorinated dibenzofurans have been considered as possible causes of embryonic mortality and birth-defects observed in PCB-feeding experiments in birds[99,100]. The chlorinated dibenzofurans are structurally related to the chlorinated dibenzo-p-dioxin (Table 1) some of which are both highly toxic and teratogenic[101].

A number of possibilities exist to account for the presence of chlorodibenzofurans in commercial PCB mixtures. One explanation considers the presence of the parent compound (dibenzofuran) in the technical grade biphenyl subjected to the chlorination process. It is also conceivable that chlorinated dibenzofurans may be produced from PCBs in the environment. Two possible mechanisms for such a transformation are illustrated in Figure 5, both of which involve hydroxy derivatives.

As cited earlier, hydroxylation is a route of metabolism of the PCBs. Polar oxygenated compounds have also been found as photolytic products of the PCBs. It should be stressed that the transformation of only 0.002% of a major constituent of an Aroclor mixture to the corresponding chlorinated dibenzofurans would produce concentrations in the mixture corresponding to the values reported by Vos et al as toxicologically significant[102,103].

To date, there have been no published findings of chlorinated dibenzofurans in aquatic samples or in foods. The extremely low levels of these trace contaminants in the original organic chemicals and/or complex mixtures (e.g., PCBs, chlorinated phenols) would stress the requirement for analytical procedures permitting the sampling, concentration and detection in the parts-per-billion to parts-per-trillion range.

2. <u>Polybrominated Biphenyls</u>

Although the overwhelming stress thus far in a consideration of the halogenated polyaromatics has focused on the polychlorinated biphenyls, it must be noted that structurally related derivatives such as the polybrominated biphenyls (PBBs) have been increasingly employed, primarily as fire retardants[105,106]. For example, the PBBs are incorporated into thermoplastics at a concentration of about oedema[122]. It has been stressed that in studies on the toxicity of such compounds, that the posisble role of small amounts of highly toxic contaminants such as the chlorinated dibenzodioxins and dibenzofurans be noted.

Hydroxylation and/or hydroxylation-dechlorination have been found to be common metabolic pathways for a number of the isomers of PCNs[126-130]. The major urinary metabolites of chloronaphthalenes (PCNs) in rabbits[126] and in pigs[127] are chloronaphthols. For example, 4-chloronaphthol and 3-chloro-2-naphthol have been identified as the major pig urinary metabolites of 1- and 2-chloronaphthalene[127].

Ruzo et al[128] reported 2,4-dihalonaphthols as the major pig urinary metabolic products following administration of 1,4-dichloro- and 1,4-dibromopaphthalene retrocarotidly. These findings are consistent with the intermediacy of arene oxides, decomposition of these intermediates is accompanied by a 1,2-H (or Cl) migration[132]. Phenolic metabolites of 1,2-dichloronaphthalenes and 1,2,3,4-tetrachloronaphthalene were also identified whereas the higher chlorinated 1,2,3,4,5,6-hexachloronaphthalene did not yield any urinary metabolites[128]. This observation is comparable with the resistance of higher chlorinated biphenyls to metabolic degradation. Higher chlorinated naphthalenes have been implicated in a number of farm animal diseases and the stability of these compounds may be a factor in their biological activity.

In uptake and distribution studies of PCNs and their metabolites in pigs it was shown that while 1- and 2-chloronaphthalene were distributed in various organ and tissue samples (e.g., brain kidney, liver, lung, heart, skeletal muscle and fat with the highest concentration in the brain and kidney, the metabolites (4-chloronaphthol and 3-chloro-2-naphthol) were concentrated in the urine, bile, kidney and liver[129].

Recent studies of Chu et al[128] also demonstrated in the rat that in addition to hydroxylation, PCNs can also undergo hydroxylation-dechlorination, following their oral administration. Thus, 1,2-dichloronaphthalene was biotransformed to the glucuronide conjugate of 5,6-dichloro-1,2-dihydroxy-1,2-dihydronaphthalene. 2,7-Dichloronaphthalene was metabolized to free and conjugated 7-chloro-2-naphthol; and 2,6-dichloronaphthalene gave rise to free and conjugated 6-chloro-2-naphthol and 2,6-dichloronaphthol[130]. These results are also conconant with arene oxides[132] as intermediate in PCN metabolism.

There are no apparent data available on the carcinogenicity or mutagenicity of either mixtures of the PCNs or individual isomers.

References

1. Panel on Hazardous Substances, Polychlorinated Biphenyls-Environmental Impact, Environ. Res., (1972) 249
2. Hammond, A. L., Chemical Pollution: Polychlorinated Biphenyls, Science, 175 (1972) 155
3. National Swedish Environment Protection Board, PCB Conference, Wenner-Gren Center, Stockholm, Sweden, Sept. 29 (1970)
4. Peakall, D., and Lincer, J., Polychlorinated Biphenyls: Another Long-Life Widespread Chemical in the Environment, Bioscience, 20 (1970) 958
5. Ahmed, A. H., PCBs in the Environment, Environment, 18 (1976) 6
6. Nisbet, I. C. T., and Sarofim, A. F., Rates and Routes of Transport of PCBs in the Environment, Env. Hlth. Persp., 1 (1972) 21
7. Dept. HEW, Final Report of the Subcommittee on the Health Effects of Polychlorinated Biphenyls and Polybrominated Biphenyls, Washington, D.C. July (1976).
8. Hubbard, H. L., Chlorinated Biphenyl and Related Compounds, In Kirk-Othmer's "Encyclopedia of Chemical Technology", 2nd ed. 5 (1964) 289
9. Hutzinger, O., Safe, S., and Zitko, V., "The Chemistry of PCBs", CRC Press, Cleveland, Ohio (1974)
10. Zitko, V., Hutzinger, O., and Safe, S., Retention Times and Electron-Capture Responses of Some Individual Chlorobiphenyls, Bull. Env. Contam. Toxicol., 6 (1971) 160
11. Vos, J. G., Koeman, J. J., Vandermaas, H. L., TenNoever de Brauw, M. C., and DeVos, Identification and Toxicological Evaluation of Chlorinated Dibenzofuran and Chlorinated Naphthalene in Two Commercial Polychlorinated Biphenyls, Food Cosmet. Toxicol., 8 (1970) 625
12. Bowes, G. W., Identification of Chlorinated Dibenzofurans in American Polychlorinated Biphenyls, Nature, 256 (1975) 305
13. Roach, J. A. G., and Pomerantz, I. H., The Findings of Chlorinated Dibenzofurans in Aroclor PCB's of Recent Manufacture, Paper No. 53, Presented at 88th Annual Meeting of Assn. of Official Analytical Chemists, Washington, D.C. October 14-17 (1974)
15. Risebrough, R. W., and deLappe, R. W., Polychlorinated Biphenyls In Ecosystems, Env. Hlth. Persp., 1 (1972) 39
15a. WHO, Polychlorinated biphenyls and terphenyls, environmental health criteria? World Health Organization, Geneva (1976)
16. Gustafson, C. G., PCBs-Prevalent and Persistent, Env. Sci. Technol., 4 (1970) 814
17. Ruopp, D. J., and deCarlo, V. J., Environmental Levels of PCBs, Int. Conf. on Environ. Sensing and Assessment, Las Vegas, Nev., Sept. 14-19 (1975)
18. Maugh, T. H., Chemical Pollutants: Polychlorinated Biphenyls Still a Threat, Science, 190 (1975) 1189
19. Kolbye, A. C., Food Exposures to Polychlorinated Biphenyls, Environ. Hlth. Persp., 1 (1972) 85-88
20. Anon, The Rising Clamor About PCBs, Env. Sci. Technol., 10 (1976) 122
21. Fishbein, L., Toxicity of Chlorinated Biphenyls, Ann. Rev. Pharmacol., 14 (1974) 139
22. Kimbrough, R. D., The Toxicity of Polychlorinated Polycyclic Compounds and Related Compounds, Crit. Revs. Toxicol., 2 (1974) 445
23. Allen, J. R., Response of the Non Human Primate to Polychlorinated Biphenyl Exposure, Fed. Proc., 34 (1975) 1675

24. Grant, D. L., Phillips, W. E. J., and Villeneuve, D., Metabolism of Polychlorinated Biphenyl (Aroclor 1254) Mixture in the Rat, Bull. Env. Contam. Toxicol., 6 (1971) 102
25. Bush, B., Tumasonis, C. F., and Baker, F. D., Toxicity and Persistence of PCB Homologs and Isomers in the Avian System, Arch. Env. Con. Tox., 2 (1974) 195
26. Sissons, D., and Welti, D., Structural Identification of Polychlorinated Biphenyls in Commercial Mixtures by Gas-Liquid Chromatography, Nuclear Magnetic Resonance and Mass Spectrometry, J. Chromatog., 60 (1971) 15
27. Webb, R. G., and McCall, A. C., Identities of Polychlorinated Biphenyl Isomers in Aroclors, J. Assoc. Off. Anal. Chem., 55 (1972) 746
28. Ecobichon, D. J., and Comeau, A. M., Comparative Effects of Commercial Aroclors on Rat Liver Enzyme Activities, Chem. Biol. Interactions, 9 (1974) 314
29. Ecobichon, D. J., and Comeau, A. M., Isomerically Pure Chlorobiphenyl Congeners and Hepatic Function in the Rat: Influence of Position and Degree of Chlorination, Toxicol. Appl. Pharmacol., 33 (1975) 94
30. Matthews, H. B., and Anderson, M. W., The Distribution and Excretion of 2,4,5,2',5'-Pentachlorobiphenyl in the Rat, Drug Metab. Dispos., 3 (1975) 211-219
31. Matthews, H. B., and Anderson, M. W., Effect of Chlorination on the Distribution and Excretion of Polychlorinated Biphenyls, Drug Metab. Dispos., 3 (1975) 371-380
32. Braunberg, R. C., Dailey, R. E., Brouwer, E. A., Kasza, L., and Blaschka, A. M., Acute, Subacute, and Residual Effects of Polychlorinated Biphenyl (PCB) in Rats. Biological Half-life in Adipose Tissue, J. Toxicol. Env. Hlth., 1 (1976) 683-688
33. Mehendale, H. M., Uptake and Disposition of Chlorinated Biphenyls by Isolated Perfused Rat Liver, Drug Metab. Dispos., 4 (1976) 124-132
34. Peterson, R. E., Seymour, J. L., and Allen, J. R., Distribution and Biliary Excretion of Polychlorinated Biphenyls in Rats, Tox. Appl. Pharmacol., 38 (1976) 609-619
35. Biros, F. J., Walker, A. C., and Medbery, A., Polychlorinated Biphenyls in Human Adipose Tissue, Bull. Env. Contam. Toxicol., 5 (1970) 317-323
36. Tulp, M. T. M., Sundström, G., and Hutzinger, O., The Metabolism of 4,4'-Dichlorobiphenyl in Rats and Frogs, Chemosphere, 6 (1976) 425-432
37. Hsu, I. C., Van Miller, J. P., Seymour, J. L., and Allen, J. R., Urinary Metabolites of 2,5,2',5'-Tetrachlorobiphenyl in the Nonhuman Primate, Proc. Soc. Exp. Biol. Med., 150 (1975) 185-188
38. Jensen, S., and Sundstrom, G., Metabolic hydroxylation of a chlorobiphenyl containing only isolated unsubstituted positions-2,2',4,4',5,5'-hexachlorobiphenyl, Nature, 251 (1974) 219
39. Hutzinger, O., Jamieson, W. D., Safe, S., Paulmann, L., and Ammon, R., Identification of metabolic dechlorination of highly chlorinated biphenyl in rabbit, Nature, 252 (1974) 698
40. Sundström, G., Hutzinger, O., and Safe, S., The Metabolism of 2,2',4,4',5,5'-Hexachlorobiphenyl by Rabbits, Rats and Mice, Chemosphere, 4 (1976) 249-253
41. Gardner, A. M., Chen, J. T., Roach, J. A. G., and Ragelis, E. P., Polychlorinated biphenyls: Hydroxylated urinary metabolites of 2,5,2',5'-tetrachlorobiphenyl identified in rabbits, Biochem. Biophys. Res. Comm., 55 (1973) 1377-1384
42. Wyndham, C., Devenish, J., and Safe, S., The In Vitro Metabolism, Macromolecular Binding and Bacterial Mutagenicity of 4-Chlorobiphenyl, A Model PCB Substrate, Res. Communs. Chem. Pathol. Pharmacol., 15 (1976) 563-570

43. Jensen, S., and Jansson, B., Methyl Sulfone Metabolites of PCB and DDE, Ambio, 5-6 (1976) 257-260
44. Vos, J. G., and Notenboom-Ram, E., Comparative Toxicity Study of 2,4,5,2',4',5'-Hexachlorobiphenyl and a Polychlroinated Biphenyl Mixture in Rabbits, Toxicol. Appl. Pharmacol., 23 (1972) 563
45. Figita, S., Tsuji, Kato, K., Saeki, S., and Tsukamoto, H., Effects of Biphenyl Chlorides on Rat Liver Microsomes, Fukuoka Acta. Med., 62 (1971) 30
46. Ecobichon, D. J., and MacKenzie, D. O., The Uterotrophic Activity of Commercial and Isomerically Pure Chlorobiphenyls in the Rat, Res. Commun. Chem. Pathol. Pharmcol., 9 (1974) 85
47. Chen, P. R., Mehendale, H. M., and Fishbein, L., Effect of Two Isomeric Tetrachlorobiphenyls on Rat and Their Hepatic Microsomes, Arch. Env. Contam. Tox., 1 (1973) 36
48. Johnstone, G. J., Ecobichon, D. J., and Hutzinger, O., The Influence of Pure Polychlorinated Biphenyl Compounds on Hepatic Function in the Rat, Toxicol. Appl. Pharmacol., 28 (1974) 66
49. Jao, L. T., Hass, J. R., and Mathews, H. B., In Vivo Metabolism of Radioactive 4-Chloro- and 4,4'-Dichlorobiphenyl in Rats, Toxicol. Appl. Pharmacol., 37 (1976) 147
50. Mathews, H. B., and Tuey, D. B., The Distribution and Excretion of 3,5,3',5'-Tetrachlorobiphenyl in the Male Rat, Toxicol. Appl. Pharmacol., 37 (1976) 148
51. Seymour, J. L., Peterson, R. E., and Allen, J. R., Tissue Distribution and Biliary Excretion of 2,5,2',5'-Tetrachlorobiphenyl (TCB) and 2,4,5,2',4',5'-Hexachlorobiphenyl (HCB) in Male and Female Rats, Toxicol. Appl. Pharmacol., 37 (1976) 171
52. Kimbrough, R. D., Linder, R. E., and Gaines, T. B., Morphological Changes in Livers of Rats Fed Polychlorinated Biphenyls: Light Microscopy and Ultrastructure, Arch. Env. Hlth., 25 (1972) 354
53. Nagasaki, H., Tomii, S., Mega, T., Marugami, M., and Ito, N., Hepatocarcinogenicity of Polychlorinated Biphenyls in Mice, Gann, 63 (1972) 805
54. Nishizumi, M., Light and Electron Microscope Study of Chlorobiphenyl Poisoning in Mouse and Monkey Liver, Arch. Env. Hlth., 21 (1970) 620
55. Platonow, N. S., and Chen, N. Y., Transplacental Transfer of Polychlorinated Biphenyls (Aroclor 1254) in Cow, Vet. Rec., 92 (1973) 69
56. Kasza, L., Weinberger, M. A., Carter, C., Hinton, D. E., Trump, B. F., and Brouwer, E. A., Acute, Subacute and Residual Effects of Polychlorinated Biphenyl (PCB) in Rats. II. Pathology and Electron Microscopy of LIver and Serum Enzyme Study, J. Toxicol. Env. Hlth., 1 (1976) 689
57. Allen, J. R., Carstens, L. A., and Abrahamson, L. J., Responses of Rats Exposed to Polychlorinated Biphenyls for 52 Weeks, 1. Comparison of Tissue Levels of PCB and Biological Changes, Arch. Env. Contam., 4 (1976) 404-419
58. Grant, D. L., Moodie, C. A., and Phillips, W. E. J., Toxicodyanamics of Aroclor 1254 in the Male Rat, Env. Physiol. Biochem., 4 (1974) 214
59. Fuhremann, T. W., and Lichtenstein, E. P., Increase in the Toxicity of Organophosphorus Insecticides to House Flies Due to Polychlorinated Biphenyl Compounds, Toxicol. Appl. Pharmacol., 22 (1972) 628
60. Villeneuve, D. C., Grant, D. L., and Phillips, W. E. J., Modification of Pentobarbital Sleeping Tunes in Rats Following Chronic PCB Ingestion, Bull. Env. Contam. Toxicol., 7 (1972) 264
61. Anon, Worrisome PCB Levels Found in Mothers' Milk in 10 States, Toxic Mat. News, 3 (1976) 141

62. U.S. Dept. Health, Education & Welfare, Meeting to Coordinate Toxicology and Related Programs: PCBs in Mothers Molk, Bethesda, MD, Jan. 31 (1978)
63. DHEW/EPA, Meeting to Review and Evaluate EPA's Sampling of PCBs in the Fat of Mothers' Milk, DHEW Committee to Coordinate Toxicology and Related Programs in Cooperation with Environmental Protection Agency, Bethesda, MD, Sept. 23, 1976
64. Anon, The Rising Clamor About PCBs, Env. Sci. Technol., 10 (1976) 122-123
65. Neeley, W. B., A Material Balance Study of Polychlorinated Biphenyls in Lake Michigan, Sci. Total Environ., 7 (1977) 117-129
66. Plimmer, J. R., and Kligebiel, U. L., PCB Formation, Science, 181 (1973) 994-995
67. Moilanen, K. W., and Crosby, Amer. Chem. Soc. Meeting, 165th, Dallas, TX April (1973)
68. Uyeta, M., Taue, S., Chikasawa, K., and Mazaki, M., Photoformation of Polychlorinated Biphenyls from Chlorinated Benzenes, Nature, 264 (1976) 583-584
68a. Gaffney, P. E., Chlorobiphenyls and PCBs: Formation during chlorination, J. Water Pollut. Control, 49 (1977) 401-404
69. Saeki, S., Tsutsui, A., Oguri, K., Yoshimura, H., and Hamana, M., Isolation and Structure Elucidation of the Amin Component of KC-400 (Chlorobiphenyls), Fukuoka Acta Med., 62 (1971) 20
70. Kuratsune, M., et al., Morikawa, Y., Hirohata, T., Nishizumi, M., Kochi, S., An Epidemiologic Study on "Yusho" or Chlorobiphenyls Poisoning, Fukuka Acta Med., 60 (1969) 513-532
71. Kuratsune, M., Yoshimura, T., Matsuzaka, J., et al., Yusho, A Poisoning Caused by Rice Oil Contaminated with Chlorobiphenyls, Env. Hlth. Persp., 1 (1972) 119
72. Umeda, G., Kanemi Yusho: PCB Poisoning in Japan, Ambio (1972) 132-134
73. Hirayma, T., Meeting on Origins of Human Cancer, Cold Spring Harbor, NY Sept. 7-14 (1976)
74. Bahn, A. K., Rosenwaike, I., Herrmann, N., et al., Melanoma After Exposure to PCBs, New Engl. J. Med., 295 (1976) 450
75. Anon, Mobil Study Suggests Possible Link Between PCBs and Human Cancer, Occup. Hlth. Safety Letter, 6 (1976) 4
76. Anon, High Cancer Rates Among Workers Exposed to PCBs, Chem. Eng. News, 54 (1976) 26
77. Anon, Monsanto Finds No Link Between Cancer and PCBs in Workers, Occup. Hlth. Safety Letter, 5 (1976) pp. 5-6
78. Nagasaki, H., Tomii, S., and Mega, T., Hepatocarcinogenicity of Polychlorinated Biphenyls in Mice, Gann, 63 (1972) 805
79. Nagasaki, H., Tomii, S., Mega, T., Marugami, M., and Ito, N., Development of Hepatomas in Mice Treated with Benzene Hexachloride, Gann, 64 (1971) 431
80. Kimura, N. T., Baba, T., Neoplastic Changes in the Rat Liver Induced by Polychlorinated Biphenyl, Gann, 64 (1973) 105
81. Ito, N., Nagasaki, H., Arai, M., Makiura, S., Sugihara, S., and Hirao, K., Histopathologic Studies on Liver Tumorigenesis Induced by Mice by Technical Polychlorinated Biphenyls and Its Promoting Effect on Liver Tumors Induced by Benzene Hexachloride, J. Natl. Cancer Inst., 51 (1973) 1637
82. Kimbrough, R. D., Squire, R. A., Linder, R. E., Strandberg, J. D., Montali, R. J., and Burse, V. W., Induction of Liver Tumors in Sherman Female Rats by Polychlorinated Biphenyl Aroclor 1260, J. Natl. Cancer Inst., 55 (1975) 1453-1459

83. Kimura, N. T., Kanematsu, I., and Baba, T., Polychlorinated Biphenyl(s) as a Promotor in Experimental Hepatocarcinogenesis in Rats, Z. Krebsforsch. Klin. Onkol., 87 (1976) 257-266
84. Ramel, C., Mutagenicity Research and Testing in Sweden, Mutation Res., 33, (1975) 79-86
85. Nilsson, B., and Ramel, C., Genetic Tests on Drosophila Melanogaster with Polychlorinated Biphenyls (PCB), Hereditas, 77 (1974) 319
86. Hoopingarner, R., Samule, A., and Krause, D., Polychlorinated Biphenyl Interactions with Tissue Culture Cells, Env. Hlth. Persp., 1 (1972) 155
87. Keplinger, M. L., et al., Toxicological Studies with Polyclorinated Biphenyls PCB Conference, Quail. Roost. Conf. Center, Rougemont, NC, Dec. (1971)
88. Green, S., Sauro, F. M., and Friemann, L., Lack of Dominant Lethality in Rats Treated with Polychlorinated Biphenyls, Food Cosmet. Toxicol., 13 (1975) 507
89. Green, S., Carr, J. V., Palmer, K. A., and Oswald, E. J., Lack of Cytogenetic Effects in Bone Marrow Spermatogonial Cells in Rats Treated with Polychlorinated Biphenyls, Bull. Env. Contam. Toxicol., 13 (1975) 14
90. Ames, B. N., Darston, W. E., Yamasaki, E., and Lee, F. D., Carcinogens are Mutagens: A Simple Test for Combining Liver Homogenates for Activation and Bacteria for Detection, Proc. Natl. Acad. Sci., 70 (1973) 2281-2285
91. Ghiasuddin, S. M., Nenzer, R. E., and Nelson, J. O., Metabolism of 2,5,2'-Trichloro-, 2,5,2',5'-Tetrachloro- and 2,4,5,2',5'-Pentachlorobiphenyl in Rat Hepatic Microsomal Systems, Toxicol. Appl. Pharmacol., 36 (1976) 187-194
92. Allen, J. R. and Norback, D. H., Pathobiological Responses of Primates to Polychlorinated Biphenyl Exposure, In National Conference on Polychlorinated Biphenyls, Proceedings, 43-49 (1975)
93. Allen, J. R. and Norback, D. H., Carcinogenic Potential of the Polychlorinated Biphenyls, In Origins of Human Cancer Meeting, Cold Spring Harbor, New York, Sept. 7-14 (1976) Abstract p. 75
94. Grant, D. L., Villeneuve, D. C., McCully, K. A., and Phillips, W. E. J., Placental Transfer of Polychlorinated Biphenyls in the Rabbit, Env. Physiol., 1 (1971) 61
95. Takagi, Y., Otake, T., Kataoka, M., Murata, Y., Aburada, S., Akasaka, S., Hashimoto, K., Uda, H., and Kitaura, T., Studies on the Transfer and Distribution of ^{14}C-Polychlorinated Biphenyls from Maternal to Fetal and Suckling Rats, Toxicol. Appl. Pharmacol., 38 (1976) 549-558
96. Kojima, T., Fukumoto, H., Makisumi, S., Chloro-Biphenyl Poisonings, Gas-Chromatographic Detection of Chloro-Biphenyls in the Rice Oil and Biological Materials, Jap. J. Legal Med., 23 (1969) 415
97. Inagami, K., Koga, T., Tomita, Y., Poisoning of Chlorobiphenyls, Shokuhin. Eisei Gaku Zasshi, 10 (1969) 415
98. Vos, J. G., and Koeman, J. H., Comparative Toxicologic Study with Polychlorinated Biphenyls in Chickens with Special Reference to Porphyria, Edema Formation, Liver Necrosis, and Tissue Residues, Toxicol. Appl. Pharm., 17 (1970) 565
99. Tumasonis, C. F., Bush, B., and Baker, F. D., PCB Levels in Egg Yolks Associated with Embryonic Mortality and Deformity of Hatched Chicks, Arch. Env. Contam. Toxicol., 1 (1973) 312
100. Gilbertson, M., and Hale, R., Early Embryonic Mortality in a Herring Gull Colony in Lake Ontario, Canad. Field-Naturalist, 88 (1974) 354

101. Schwetz, B. A., Norris, J. M., Sparschu, G. L., Rowe, V. K., Gehring, P. J., Emerson, J. L., and Gerbig, C. G., Toxicology of Chlorinated Dibenzo-p-dioxins, Env. Hlth. Persp., 5 (1973) 87
102. Vos, J. G., Koeman, J. H., Vandermaas, H. S., Debraun, M. C. and DeVos, H., Identification and Toxicological Evaluation of Chlorinated Dibenzofuran and Chlorinated Naphthalene in Two Commercial Polychlorinated Biphenyls, Food Cosmet. Toxicol., 8 (1970) 625-633
103. Vos, J. G., and Koeman, J. H., Dermal toxicity studies of technical polychlorinated biphenyls and fractions thereof in rabbit, Toxicol. Appl. Pharmacol., 19 (1971) 617
104. Sundstrom, G., Hutzinger, O., and Safe, S., Identification of 2,2',4,4',5,5'-Hexabromobiphenyl as the Major Component of Flame Retardant, Firemaster BP-6, Chemosphere, 1 (1976) 11
105. Carter, L. J., Michigan's PBB Incident: Chemical Mix-up Leads to Disaster, Science, 192 (1976) 240
106. Dent, J. G., Netter, K. J., and Gibson, J. E., Characterization of Various Parameters of Rat Hepatic Microsomal Mixed Function Oxidase after Induction by Polybrominated Biphenyls, Toxicol. Appl. Pharmacol., 37 (1976) 147
107. U.S. Environmental Protection Agency; Polybrominated Biphenyls, Washington, DC (1977)
108. Ficsor, G., and Wertz, G. F., Polybrominated Biphenyl Nonteratogenic C-mitosis Synergist in the Rat, Presented at 7th Annual Meeting of Environmental Mutagen Society, Atlanta (1976)
109. Stratton, C. L., and Sosebee, J. B., Jr., PCB and PCT contamination of the environment near sites of manufacture and use, Environ. Sci. Technol., 10 (1976) 1229-1236
110. Freudenthal, J., and Greve, P. A., Polychlorinated terphenyls in the environment, Bull. Environ. Contam. Toxicol., 10 (1973) 108-111
111. Zitko, V., Hutzinger, O., Jamieson, W. D., and Choi, P. M. K., Polychlorinated terphenyls in the environment, Bull. Env. Contam. Toxicol., 7 (1972) 200-201
112. Addison, R. F., Fletcher, G. L., Ray, S., and Doane, J., Analysis of a chlorinated terphenyl (Aroclor 5460) and its decomposition in tissues of cod (Gadus morphua), Bull. Environ. Contam. Toxicol., 8 (1972) 52
113. Doguchi, M., Polychlorinated terphenyls as an environmental pollutant in Japan, Ecotoxicol. Environ. Safety, 1 (1977) 239-248
114. Hardie, D. W. F., Chlorinated naphthalenes In: Kirk-Othmer Encyclopedia of Chemical Technology, 2nd ed., Vol. 5, Interscience, New York (1964) p. 297
115. Brinkman, U. A. T., and Reymer, H. G. M., Polychlorinated naphthalenes J. Chromatog., 127 (1976) 203-243
116. Koppers Co., Halowax®, Chlorinated naphthalene oils and waxlike solids, Technical Bulletin, Pittsburgh, PA
117. Bayer, Nibrenwachse, Technical Bulletin, Leverkusen, W. Germany (1970)
118. Beland, F. A., and Geer, R. D., Identification of chlorinated naphthalenes in halowaxes 1031, 1000, 1001, J. Chromatog., 84 (1973) 59
119. Hutzinger, O., Safe, S., and Zitkov, V., "The Chemistry of Polychlorinated Biphenyls", CRC Press, Cleveland, Ohio (1974)
120. Vos, J. G., Koeman, J. H., Van der Maas, H. L., Tennoever de Brauw, M. C., and deVos, R. H., Identification and toxicological evaluation of chlorinated dibenzofuran and chlorinated naphthalene in two commercial polychlorinated biphenyls, Food Cosmet. Toxicol., 8 (1970) 625-630

121. Stalling, D. L., and Huckins, J. N., Reverse phase thin-layer chromatography of some aroclors, halowaxes and pesticides, J. Assoc. Offic. Anal. Chem., 56 (1973) 367
122. Kimbrough, R. D., Toxicity of chlorinated hydrocarbons and related compounds, Arch. Environ. Hlth., 25 (1972) 125
123. Kover, F. D., Environmental Hazard Assessment Report-Chlorinated Naphthalenes, EPA 560/8-75-001, U.S. Environmental Protection Agency, Washington, DC (1975)
124. Kleinfeld, M., Messite, J., Swenciki, R., Clinical effects of chlorinated naphthalene exposure, J. Occup. Med., 14 (1972) 377
125. Sikes, D., and Bridge, M. E., Production of X-disease in cattle with a chlorinated naphthalene, Science, 114 (1952) 506
126. Cornish, H. H., and Block, W. D., Metabolism of chlorinated naphthalenes, J. Biol. Chem., 231 (1958) 583
127. Ruzo, L. O., Safe, S., Hutzinger, I., Platonow, N., and Jones, D., Hydroxylated metabolites of chloronaphthalenes (halowax 1031) in pig urine, Chemosphere, 4 (1975) 123
128. Ruzo, L. O., Jones, D., Safe, S., and Hutzinger, O., Metabolism of chlorinated naphthalenes, J. Agr. Food Chem., 24 (1976) 581-583
129. Ruzo, L. O., Safe, S., Jones, D., and Platonow, N., Uptake and distribution of chloronaphthalenes and their metabolites in pigs, Bull. Environ. Contam. Toxicol., 16 (1976) 233-239
130. Chu, I., Villeneuve, D. C., Secours, V., and Viau, A., Metabolism of chloronaphthalenes, J. Agr. Food Chem., 25 (1977) 881-883
131. Chu, I., Secours, V., and Viau, A., Metabolites of chloronaphthalene, Chemosphere 6 (1976) 439-444
132. Daly, J. W., Jerina, D. M., and Witkop, W. T., Arene oxides and the NIH shift. The metabolism, toxicity and carcinogenicity of aromatic compounds, Experientia 28 (1972) 1129-1264

Table 2
Tumorigenic hydrazine compounds

Compound	Species	Organ	Treatment	References
1-Acetyl-2-isonicotinoylhydrazine	Mice	Lungs	p.o.	39
$CH_3-CH_2-CH_2-CH_2-CH_2-NH-NH_2 \cdot HCl$ N-Amylhydrazine HCl	Mice	Lungs, blood vessels	p.o.	33
Benzoylhydrazine	Mice	Lungs, lymphoreticular tissue	p.o.	2, 34
$NH_2-NH-CH_2-CH_2-CH_2-CH_3 \cdot HCl$ N Butylhydrazine HCl	Mice	Lungs	p.o.	44
$NH_2-NH-CO-NH_2 \cdot HCl$ Carbamylhydrazine · HCl	Mice	Lungs, blood vessels	p.o.	43
1-Carbamyl-2-phenylhydrazine	Mice	Lungs	p.o.	41
$CH_3-CH_2-NH-NH-CH_2-CH_3 \cdot 2HCl$ 1,2-Diethylhydrazine · 2HCl	Rats	Lymphoreticular and nerve tissues, liver, ethmoturbinal	s.c.	5
$(CH_3)_2 N-NH_2$ 1,1-Dimethylhydrazine	Mice	Lungs, blood vessels, kidney, liver	p.o.	28, 37
$CH_3-NH-NH-CH_3 \cdot 2HCl$ 1,2-Dimethylhydrazine · 2HCl	Mice Hamsters Rats	Colon, lungs, blood vessels Liver, stomach, intestine, blood vessels Intestine	s.c., p.o. i.m., p.o. s.c., p.o.	46, 52 27, 36 6
$NH_2-NH-CH_2-CH_3 \cdot HCl$ Ethylhydrazine · HCl	Mice	Lungs, blood vessels	p.o.	42
$NH_2-NH_2 \cdot H_2SO_4$ Hydrazine sulfate	Mice Rats	Lungs, liver Liver, lungs	p.o. p.o.	1 32
$NH_2-NH-CH_2-CH_2OH$ 2-Hydroxyethylhydrazine	Mice	Liver	p.o.	11
1-Isonicotinoyl-2-isopropylhydrazine	Mice	Lungs	p.o.	2
N-Isopropyl-α-(2-methylhydrazino)-p-toluamide HCl	Mice Rats	Lungs, lymphoreticular tissue, kidney Breast, lungs, blood vessels	p.o., i.p. p.o., i.p.	13 14
o-Methoxybenzoylhydrazine	Mice	Lungs	p.o.	2
p-Methoxybenzoylhydrazine	Mice	Lungs	p.o.	2
$CH_3-NH-NH_2$ Methylhydrazine	Mice Hamsters	Lungs Kupffer cells, cecum	p.o. p.o.	35 40

Table 2—Continued

Compound	Species	Organ	Treatment	References
CH₃-NH-NH-CH₂-⟨phenyl⟩ 1-Methyl-2-benzylhydrazine	Rats	Central and peripheral nervous systems, bulbus olfactorius	s.c., p.o.	4
CH₃—NH—NH—CH₂—CH₂—CH₂—CH₃ · 2HCl 1—Methyl—2—butylhydrazine 2 HCl	Rats	Large intestine, bulbus olfactorius	s.c., p.o.	4
⟨phenyl⟩-NH-NH₂ · HCl Phenylhydrazine HCl	Mice	Lungs	p.o.	2

The oxidation-reduction capabilities of hydroxylamine make it useful in many applications, e.g., as a reducing agent for many metal ions, and for the termination of peroxide-catalyzed polymerizations.

Hydroxylamine (NH_2OH) and its hydrochloride or sulfate salts are used in applications including: prevention of discoloration of rayon and cellulose products[59], vinylidene chloride polymers[60] and paper-pulp[61]; as bleaching agents for phenol resin fibers[62]; modification of acrylic fibers[63]; fire-proofing of acrylic fibers[64]; in multicolor dyeing of acrylic fibers[65]; as catalysts for polymerization of acrylamide[66]; conjugated diolefins[67]; in electroplating[68]; in soldering fluxes for radiators[69]; in photographic color developers[70] and emulsions[71]; in the stabilization of water solutions of fertilizers[72]; as antishining agents in paints; and complexing agents for metals and as a laboratory reagent for the preparation and determination of oximes.

Hydroxylamine and certain hydroxylamine derivatives (e.g., CH_3NHOH and NH_2OCH_3) (as well as the closely related hydrazine) are mutagenic to different degrees in bacteria and to transforming DNA[72]. Each has in common the ability to interact specifically with pyrimidines under specific conditions including pH, concentration of reagent, and oxygen tension. Their mutagenicity also depends markedly upon the above conditions, although, in general, hydroxylamine and its analogs are appreciably more mutagenic than hydrazine.

At high concentrations and high pH (approximately pH 9), hydroxylamine (0.1 M to 1.0 M) reacts exclusively with the uracil moieties of nucleic acids.

At low pH (approximately pH 6) and high concentration (0.1 M to 1.0 M), hydroxylamine reacts exclusively with cytosine moieties of DNA, aminating only the C-4 atom[73,74]. Paradoxically, at lower concentrations of hydroxylamine, the reaction involves all four bases[74,75] analogous to that reported for higher pH's (see above). Furthermore, it is well known that hydroxylamine is highly toxic at low concentrations[76], where it is only weakly mutagenic[77], and yet, at the higher concentrations at which it is mutagenic, little or no cytotoxicity is observed. The explanation may relate to the degradation products of hydroxylamine (e.g., hyponitrous acid) rather than the compound per se[78].

The reaction of hydroxylamine with cytosine and related compounds[79] its effects and induction of mutations in transforming DNA[80-83], bacteriophage

TABLE 3

CHROMOSOMAL ABERRATIONS AND ANOMALIES PRODUCED BY HYDRAZINE DERIVATIVES

Compound	Test object	Effect
Methylhydrazine	Human leukocytes in vitro from treated patients	Chromatid aberrations
1,2-Dimethylhydrazine	Saccharomyces cerevisiae	Mitotic crossing-over
1-[(N'-Methylhydrazino)methyl]-N-isopropyl benzamide (sometimes called "methylhydrazine")	Ehrlich ascites cells	Chromatid aberrations
	Ehrlich ascites cells	Karyotype changes, mitotic inhibition
	Hematopoietic cells of rat	Chromatid aberrations
2-Benzyl-1-methylhydrazine	Ehrlich ascites cells	Chromatid aberrations, mitotic inhibition
	Ascites tumor cells in vitro	Translocations
	Ehrlich ascites and HeLa cells in vitro	No aberrations
	Bone marrow, spleen, narcissus root tips in vivo	No aberrations
	Human lymphocytes and mouse spleen cells in vitro	No aberrations
Succinic acid, mono(2,2'-dimethylhydrazide)	Root tips from soaked barley seeds, pollen mother cells from plants germinated from soaked seeds	Various chromosome abnormalities
N-(Methylhydrazinomethyl) nicotinamide	Lepidium sativum	Mitotic inhibition
2-Methylhydrazide-5-nitroquinoline-4-carboxylic acid	Lepidium sativum	Mitotic inhibition
3-Thiosemicarbazide	Meiotic cells in Vicia faba plants sprayed with compound	Various abnormalities
Isonicotinic hydrazide (isoniazid)	Human leukocytes in vitro	Chromatid aberrations, achromatic lesions
	Rat bone marrow	Gaps, breaks, deletions, fragments
Isonicotinic 2-isopropylhydrazide (iproniazide)	Vicia faba root tips	Normal karyotype, mitotic inhibition

(S13 and ΦX174[84], T4[85,86]), Neurospora[87], Saccharomyces pombe[88], E. coli[89], and induction of chromosome aberrations in human chromosomes[90], cultured Chinese hamster cells[91], mouse embryo cells[92,93] and in Vicia faba[94] have been described.

N- and O-derivatives of hydroxylamine, e.g., N-methyl- and O-methyl-hydroxylamine, have also been found to be mutagenic in transforming DNA of B. subtillis[80] and Neurospora[95], and to induce chromosome aberrations in Chinese hamster cells[92]. The selective reaction of O-methylhydroxylamine with the cytidine nucleus has been reported by Kochetov et al[96].

Among the known chemical mutagens, hydroxylamine as well as its O-methyl and N-methyl analogs are of particular interest because of their apparent specificity and ability to induce point mutations. The mutagenic activity is due largely to its reactions with cytosine residues in DNA, or RNA[97-100], optimal in slightly acid medium. Under these conditions uracil reacts to a minor extent with subsequent ring opening, so that in RNA this reaction is inactivating rather than mutagenic[101]. Adenine also reacts to a small extent with hydroxylamine. The mechanism of reaction of hydroxylamine with 1-substituted cytosine residues is illustrated in scheme 1[101]. The adduct I is a presumed intermediate, the instability of which has prevented its detection or isolation. The final products are compounds II and/or III, which are interconvertible under the conditions of the reaction[102] as shown in scheme 1.

Scheme 1

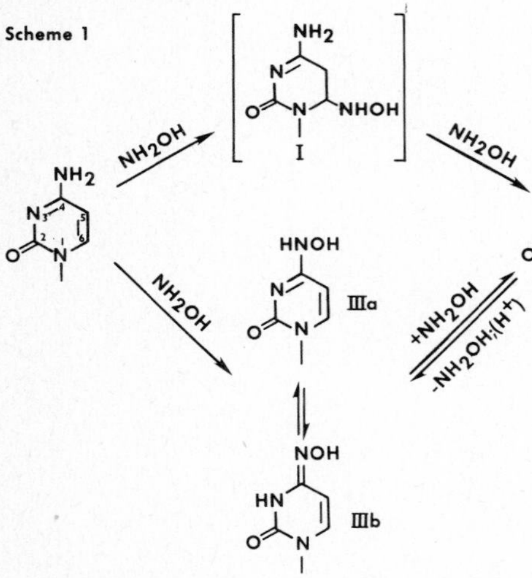

Evidence for compound III as the product responsible for hydroxylamine mutagenesis is based largely on the observation that hydroxylamine is highly mutagenic against the T-even bacteriophages[85,103,104], the DNA of which contains, in place of cytosine, free and/or glucosylated 5-hydroxymethylcytosine, and the demonstration that such 5-substituted cytosine residues react with hydroxylamine by only one pathway to give uniquely compound IV, the 5-substituted analogue of compound III (scheme 2)[105,106]. In addition, N^4-hydroxycytidine (i.e., compound III) has been shown to be highly mutagenic in two selected bacterial systems[107,108].

G46 strain. At concentrations of 100 and 1000 ppm, ETU had a relative mutagenicity activity of 2.51 and 2.44 (p<0.002) respectively[213]. ETU was also mutagenic in the repair-deficient strain TA 1530 but could not induce revertants in the frameshift mutants TA 1531, TA 1532 and TA 1964. Hence, ETU induces mutations of the base-pair substitution type with a large fraction of the induced lesions eliminated by excision repair[214].

Ethylene thiourea was non-mutagenic when tested in the rec-assay utilizing Bacillus subtilis strains H17 Rec$^+$ and M45 Rec$^-$[215]. The reversion plate assay was weakly positive with four to five times increase in the numbers of revertants when high-dose of ETU was tested against S. typhimurium TA 1535 strain. The metabolic activation test employing S-9 mix was negative[215] Negative results were obtained in the host-mediated assays using S. typhimurium G46 strain in rats[215] and mice[214,215]. However, simultaneous oral administration of ETU and sodium nitrite gave a positive result in the host-mediated assay in mice. ETU was non-cytogenetic when studied both in vitro using a Chinese hamster cell line (Don) and in vivo employing the bone-marrow cells of rats[215]. ETU was also non-mutagenic in a dominant lethal study in mice[214,215].

In the micronucleus test in which Swiss albino mice were treated twice for 24 hours with 25, 700, 1850 and 6000 mg/kg ETU, no increase in the number of erythrocytes containing micronuclei was found in treated animals[214]. ETU was also non-mutagenic when tested in Drosophila at concentrations up to 2.5%[216].

Ethylene thiourea has been found teratogenic in rats at doses that produced no apparent maternal toxicity or fetal deaths, suggesting that placental transfer of the chemical can occur[217].

In metabolic studies of 2-C^{14}-ETU and 4,5-C^{14}-ETU in pregnant rats it was shown that ETU was metabolized to CO_2 via the fragmentation of imidazoline ring and the decarboxylation step of 4 and/or 5 carbon atoms of ETU were involved[218]. A teratogenic dose of ETU was easily absorbed from rat G.I. tract, translocated into the whole body tissues, including the fetus (homogeneously except the thyroid gland) and eliminated rapidly mostly into the urine[218].

References

1. IARC, Vol., 4, International Agency for Research on Cancer, Lyon, France, pp. 127-136; 137-143
2. Biancifiori, C., and Severi, L., The Relation of Isoniazid (INH) and Allied Compounds to Carcinogenesis in Some Species of Small Laboratory Animals: A Review, Brit. J. Cancer, 20 (1966) 528
3. Kelly, M. G., O'Gara, R. W., Yancey, S. T., Gadekar, K., Botkin, C., and Oliverio, V. T., Comparative Carcinogenicity of N-isopropyl-α-(2-methylhydrazino)-p-toluamide·HCl(Procarbazine Hydrochloride), its Degradation Products, Other Hydrazines, and Isonicotinic Acid Hydrazide, J. Natl. Cancer Inst., 42 (1969) 337
4. Toth, B., Investigation on the Relationship Between Chemical Structure and Carcinogenic Activity of Substituted Hydrazines, Proc. Am. Ass. Cancer
5. Toth, B., Lung Tumor Induction and Inhibition of Breast Adenocarcinomas by Hydrazine Sulfate in Mice, J. Natl. Cancer Inst., 42 (1969) 469
6. Severi, L., and Biancifiori, C., Hepatic Carcinogenesis in CBA/Cb/Se Mice and Cb/Se Rats by Isonicotini Acid Hydrazide and Hydrazine Sulfate, J. Natl. Cancer Inst., 41 (1968) 331
    Chandra, S. V. S. G., and Reddy, G. M., Specific Locus Mutations in Maize by Chemical Mutagens, Curr. Sci., 40 (1971) 136-137
8. Lingens, F., Mutagene Wirkung Von Hydrazin Aug E. Coli-Zellen, Naturwiss, 48 (1961) 480
9. Lingens, F., Erzeugung Biochemischer Mangel Mutanten Von E. Coli Mit Hilfe Von Hydrazin Und Hydrazin Derivaten, Z. Naturforsch., 19B (1964) 151-156
10. Kimball, R. F., and Hirsch, B. F., Tests for the Mutagenic Action of a Number of Chemicals on Haemophilus Influenzae with Special Emphasis on Hydrazine, Mutation Res., 30 (1975) 9-20
11. Kimball, R. F., Reversions of Proline-Requiring Auxotrophs of Haemophilus Influenzae by N-methyl-N'-nitro-nitrosoguanidine and Hydrazine, Mutation Res., 36 (1976) 29-38
12. Jain, H. K., Raut, R. N., and Khamanker, Y. G., Base-Specific Chemicals and Mutation Analysis in Lycopersicon, Heredity, 23 (1968) 247
13. Chu, B. C. F., Brown, D. M., and Burdon, M. G., Effect of Nitrogen and Catalase on Hydroxylamine and Hydrazine Mutagenesis, Mutation Res., 20 (1973) 265-270
14. Shukla, P. T., Analysis of Mutagen Specificity in Drosophila Melanogaster, Mutation Res., 16 (1972) 363-371
15. Jain, H. K., and Shukla, P. T., Locus Specificity of Mutagens in Drosophila, Mutation Res., 14 (1972) 440-442
16. Epstein, S. S., Arnold, E., Andrea, J., Bass, W., and Bishop, Y., Detection of Chemical Mutagens by the Dominant Lethal Assay in the Mouse, Toxicol. Appl. Pharmacol., 23 (1972) 288
17. Epstein, S. S., and Shafner, H., Chemical Mutagens in the Human Environment, Nature, 219 (1968) 385
18. Röhrborn, G., Propping, P., and Buselmaier, W., Mutagenic Activity of Isoniazid and Hydrazine in Mammalian Test Systems, Mutation Res., 16 (1972) 189
19. Magee, P. D., and Barnes, Carcinogenic nitrosocompounds, Adv. Cancer Res., 10 (1967) 163

20. Druckrey, H., Preussmann, R., and Schmähl, D., Carcinogenicity and chemical structure of nitrosamines, Acta. Unio. Intern. Contra. Cancrum., 19 (1963) 510
21. Fishbein, L., Flamm, W. G., and Falk, H. L., "Chemical Mutagens", Academic Press, New York, (1970) pp. 165-167
22. Tsuji, I., Azuma, K., Kato, H., Tachimichi, H., Motegi, A., and Suzuki, O., Cross-linking accelerators for anaerobic adhesives, Japan Kokai., 74,120,889, Nov. 19 (1974), Chem. Absr., 82 (1975) P126144M
23. Azuma, K., Tsuji, I., Kato, H., Tatemichi, H., Motegi, A., Suzuki, O., and Kondo, K., Anaerobic-hardenable adhesives, Ger. Offen., 2,402,427, August 8 (1974); Chem. Abstr., 82 (1975) 59020C
23a. Schmeltz, I., Abidi, S., and Hoffmann, D., Tumorigenic agents in unburned processed tobacco-N-nitrosodiethanolamine and 1,1-dimethylhydrazine, Cancer Letters, 2 (1977) 125-132
24. Wyrobek, A. J., and London, S. A., Effect of hydrazines on mouse sperm cells, Proc. Ann. Conf. Environ. Toxicol. 4th, (1973), AD-781031, pp. 417-432, Chem. Abstr., 82 (1975) 150111U
25. Toth, B., Comparative studies with hydrazine derivatives. Carcinogenicity of 1,1-dimethylhydrazine, unsymmetrical (1,1-DMH) in the blood vessels, lung, kidneys, and liver of Swiss mice, Proc. Am. Assoc. Cancer Res., 13, (1972) 34
26. Toth, B., 1,1-Dimethylhydrazine (unsymmetrical) carcinogenesis in mice. Light microscopik and ultrastructural studies on neoplastic blood vessels, J. Natl. Cancer Inst., 50 (1973) 181
26a. Toth, B., The large bowel carcinogenic effects of hydrazines and related compounds occurring in nature and in the environment, Cancer, 40 (1977) 2427-2431
27. IARC, 1,2-Dimethylhydrazine, In Vol. 4, International Agency for Research on Cancer, Lyon, France, pp. 145-152 (1974)
28. Druckrey, H., Production of Colonic Carcinomas by 1,2-Dialkylhydrazines and Azoxyalkanes, In "Carcinoma of the Colon and Antecedent Epithelium", ed. W. J. Burdette, Thomas Publ., Springfield, Ill. (1970) pp. 267-279
29. Fiala, E. S., Kulakis, C., Bobotas, G., and Weisburger, J. H., Detection and Estimation of Azomethane in Expired Air of 1,2-Dimethylhydrazine-Treated Rats, J. Natl. Cancer Inst., 56 (1976) 1271
30. United States Rubber Co., Foamed, Vulcanized Polysulfide Rubber, Ger. Patent 1,229,717, Dec. 1, (1966), Chem. Abstr., 67 (1967) P12366V
31. Kehr, L. L., Carbazate cross-linking agent and thermosetting resins, Chem. Abstr., 73 (1970) P26293E
32. Zharkova, M. A., Kudryavtsev, G. I., Khudoshev, I. F., and Romanova, T. A., Khim. Volokna., 2 (1969) 49
33. Meincke, E. R., Color-stabilized Copolymers of Ethylene and Vinylacetate, Ger. Offen., 1,953,693, June 18 (1970) Chem. Abstr., 73 (1970) PJ6821T
34. Roechling, H., Hartz, P., and Hoerlein, G., Plant-Growth Regulators, Ger. Offen., 2,332,000, Jan. 16 (1975), Chem. Abstr., 82 (1975) 156326Q
35. Samborskii, I. V., Vakulenko, V. A., Chetverikov, A. F., Pedikova, L. N., and Nekrasova, L. G., Anion-Exchangers, USSR Patent 398,570, Sept. 27 (1973) Chem. Abstr., 81 (1974) P106565A
36. Isojima, T., Phosphor, Japan Kokai, 73 102,780, Dec. 24 (1973), Chem. Abstr., 81 (1974) P19153W
37. Bodit, F., Stoll, R., and Maraud, R., Effects of Hydroxyurea, Semicarbazide and Related Compounds on Development of Chick Embryo, C.R. Soc. Biol., 160 (1960) 960

38. Bhattacharya, A. K., Chromosome Damage Induced by Semicarbazide in Spermatocytes of a Grasshopper, Mutation Res., 40 (1976) 237
39. Mitra, A. B., Effect of pH on Hydroxylamine (HA) Phenyl Hydrazine (PH) and Semicarbazide (SC) Induced Chromosome Aberration Frequency in Mice, J. Cytol. Genet., 6 (1971) 123-127
40. Rieger, R., and Michaelis, A., Die Auslösung Von Chromosomen Aberrations Bei Vicia Faba Durch Chemische Agenzien, Die Kulturpflante, 10 (1962) 212
41. Mitchell, J. R., Long, M. W., Thorgeirsson, U. P., and Jallow, D. J., Acetylation rats and monthly liver function tests during one year of isoniazid preventive therapy, Chest, 68 (1975) 181
42. Mitchell, J. R., Thorgeirsson, U. P., Black, M., Timbrell, J. A., et al, Increased incidence of isoniazid hepatitis in rapid acetylators: Possible relation to hydrazine, Clin. Pharmacol. Ther., 18 (1975) 70
43. Mitchell, J. R., and Jollow, D. J., and Gillette, J., Relationship between metabolism of foreign compounds and liver injury, Israel J. Med. Sci., 10 (1974) 339
44. Snodgrass, W. R., Potter, W. Z., Timbrell, J. A., and Mitchell, J. R., Possible mechanism of isoniazid-related hepatic injury, Clin. Res., 22 (1974) 323A
45. Nelson, S. D., Mitchell, J. R., Timbrell, J. A., Snodgrass, W. R., and Corcoran, G. B., Isonazid and Ipronazid: Activation of Metabolites to Toxic Intermediates in Man and Rat, Science, 193 (1976)
46. Magee, P. N., and Barnes, S. M., Carcinogenic nitroso compounds, Adv. Cancer Res., 10 (1967) 163
47. Druckrey, H., Specific carcinogenic and teratogenic effects of "indirect" alkylating methyl and ethyl compounds and their dependency on stages of ontogenic developments Xenobiotica, 3 (1973) 271
48. Freese, E., Bautz, E., and Freese, E. B., The chemical and mutagenic specificity of hydroxylamine, Proc. Natl. Acad. Sci. U.S., 47 (1961) 845
49. Jain, H. K., and Raut, R. N., Differential response of some tomato genes to base specific mutagens, Nature, 211 (1966) 652
50. Lingens, F., Mutagenic effect of hydrazine on E. coli, Z. Naturforsch., 19b, (1964) 151
51. Back, K. C., and Thomas, A. A., Aerospace problems in pharmacology and toxicology, Annual Rev. Pharmacol., 10 (1970) 395
52. Juchau, M. R., and Herita, A., Metabolism of hydrazine derivatives of interest, Drug Metab. Rev., 1 (1972) 71
53. Toth, B., Synthetic and Naturally Occurring Hydrazines as Possible Causative Agents, Cancer Res., 35 (1975) 3693
54. Gowing, D. P., and Leeper, R. W., Induction of Flowering in Pineapple by Beta-Hydroxyethyl Hydrazine, Science, 122 (1955) 1267
55. Mathe, G., Schweisguth, O., Schneider, M., Amiel, J. L., Berumen, L., Brule, G., Cattan, A., and Schwarzenberg, L., Methylhydrazine in Treatment of Hodgkin's Disease, Lancet, 2 (1963) 1077-1080
56. Kimball, R. F., The mutagenicity of hydrazine and some of its derivatives, Mutation Res., 39 (1977) 111-126
57. Colvin, L. B., Metabolic fate of hydrazines and hydrazides, J. Pharm Sci., 58 (1969) 1433-1443
58. Juchau, M. R., and Horita, A., Metabolism of hydrazine derivatives of pharmacologic interest, Drug Metabolism Reviews, 1 (1973) 71-100
59. Smith, F. R., and Scharppel, J. W., Hydroxylamine-Treated Hemicellulose-Containing Regenerated Cellulose Product, U. S. Patent 3,832,277. August 27 (1974) Chem. Abstr., 81 (1974) p. 154499D

60. Moore, C., Inhibiting Discoloration of Vinylidene Chloride Polymers, U. S. Patent 3,817,895, June 18 (1974), Chem. Abstr., 81 (1974) P121930J
61. Andrews, D. H., Unbleached and semi-bleached sufite pulp, Canadian Patent, 611,510 (1960), Chem. Abstr., 55 (1961) 10888H
62. Yano, M., Mori, F., Ohno, K., Bleaching of Phenolic Resin Fibers, Japan Kokai 74 36, 972, April 5 (1974), Chem. Abstr., 81 (1974) P92968V
63. Mikhailova, L. P., Antonov, A. N., Pichkhadze, S. V., and Soshina, S. M., Carbon Fibers, U.S.S.R. Patent, 389,184, July 5 (1973), Chem. Abstr., 81 (1974) P50908V
64. Ono, M., Sahara, H., and Akasaka, M., Continuous Nonflammable Treatment of Acrylic Fibers, Japan Kokai, 74 54, 631, May 28 (1974), Chem. Abstr., 81 (1974) 154443F
65. Muroya, K., Ono, M., Sahara, H., and Akasaka, M., Acrylic Fiber Goods Having Multicolor Effects, Japan 73 28, 990, Sept. 6 (1973), Chem. Abstr., 81 (1974) P27100S
66. Das, S., Kar, K. K., Palit, S. R., Catalysts for Polymerization of Acrylamide, J. Indian. Chem. Soc., 51 (1974) 3931
67. Rakhmankulov, D. L., Makismova, N. E., Melikyan, V. R., and Isagulyants, V. I., Conjugated Diolefin, USSR Patent 434,075, June 30 (1974), Chem. Abstr., 81 (1974) P77447F
68. Rosenberg, W. E., Aqueous Acid Electroplating Baths, U. S. Patent 3,808,110, April 30 (1974), Chem. Abstr., 81 (1974) P32620R
69. Zobkiv, B. A., Kovalysko, Y. M., Egorov, G. Y. A., and Ishchenko, V. G., Flux for Soldering Radiators, USSR Patent 399,330, October 3 (1973), Chem. Abstr., 81 (1974) P531974
70. Fisch, R. S., Photographic Developer Replenisher Concentrates, U. S. Patent 3,785,824, Jan. 15 (1974), Chem. Abstr., 81 (1974) P56593K
71. Mason, L. F. A., Color developer, German Patent 1,057,875 (1959) Chem. Abstr. 55 (1961) 8137C
72. Nabiev, M. N., Amirova, A. M., Badalova, E. K., and Saibova, M., Fertilizers with Manganese Trace Element, USSR Patent 242,850, April 25 (1974), Chem. Abstr., 81 (1974) P103952P
73. Fishbein, L., Flamm, W. G., and Falk, H. L., "Chemical Mutagens" Academic Press, New York (1970) pp. 27-28
74. Freese, E., Bautz, E., and Freese, E. B., The chemical and mutagenic specificity of hydroxylamine, Proc. Natl. Acad. Sci., 47 (1961) 844
75. Tessman, I., Ishiwa, H., and Kumar, S., Mutagenic effects of hydroxylamine in vivo, Science, 148 (1965) 507
76. Gray, J. D. A., Lambert, R. A., Bacteriostatic action of oximes, Nature, 162 (1948) 733
77. Schuster, H., The reaction of tobacco mosaic virus ribonucleic acid with hydroxylamine, J. Mol. Biol., 3 (1961) 447
78. Bendich, A., Borenfreund, E., Korgold, G., Kirm, M., and Bolis, M., in "Acidi Nucleic e. Loro Fungione Biologica", pg. 214, Fondazione Basselli, Instituto Lombardo Pavia, Italy (1964)
79. Brown, D. M., and Schell, P., The reaction of hydroxylamine with cytosine and related compounds, J. Mol. Biol., 3 (1961) 709
80. Freese, E. B., and Freese, E., Two seperaple effects of hydroxylamine on transforming DNA, Proc. Natl. Acad. Sci. US, 52 (1964) 1289
81. Freese, E., and Strack, H. B., Induction of mutations in transforming DNA by hydroxylamine, Proc. Natl. Acad. Sci., 48 (1962) 1796

82. Kapadia, R. T., and Srogl, M., Inactivation of B. subtilis transforming DNA by mutagenic agents, Folio Microbiol. (Prague), 14 (1968) 51
83. Freese, E., and Freese, E. B., Mutagenic and inactivating DNA alterations, Radiation Res. Suppl. 6 (1966) 97-140
84. Tessman, I., Isiwa, H., and Kumar, S., Mutagenic effects of hydroxylamine in vivo, Science, 148 (1965) 507
85. Freese, E., Freese, E. B., and Bautz, E., Hydroxylamine as a mutagenic and inactivating reagent, J. Mol. Biol., 3 (1961) 133
86. Dhillio, E. K. S., and Dhillio, T. S., N-methyl-N'-nitro-N-nitrosoguanidine and hydroxylamine induced mutants of the rII-region of phage T4, Mutation Res., 22 (1974) 223-233
87. Malling, H. V., Hydroxylamine as a Mutagenic Agent for Neurospora Crassa, Mutation Res., 3 (1966) 470
88. Loprieno, N., Guglielminetti, R., Bonatil, S., and Abbondanoalo, A., Evaluation of the Genetic Alterations Induced by Chemical Mutagens in S. Pombe, Mutation Res., 8 (1969) 65
89. Androsov, V. V., Molecular Mechanisms of Mutations in E. Coli K-12 Under the Action of N-Nitroso-N-Methyl Urea, Dokl. Akad. Nauk. SSSR, 215 (1974) 1481
90. Engel, W., Krone, W., and Wold, U., Die Wirkung Von Thioguanin, Hydroxyl-aminunds-Bromodesoxyuridin auf Menschliche Chromosomen In Vitro, Mutation Res., 4 (1967) 353
91. Somers, C. F., and Hsu, T. C., Chromosome Damage Induced by Hydroxylamine in Mammalian Cells, Proc. Natl. Acad. Sci., 48 (1962) 937
92. Borenfreund, E., Krim, M., and Bendich, A., Chromosomal Aberrations Induced by Hyponitrite and Hydroxylamine Derivatives, J. Natl. Cancer Inst., 32 (1964) 667
93. Bendich, A., Borenfreund, E., Korngold, G. C., and Krim, M., Action of Hydroxylamine on DNA and Chromosomes, Federation Proc., 22 (1963) 582
94. Natarajan, A. T., and Upadhya, M. D., Localized Chromosome Breakage Induced by Ethyl Methane Sulfonate and Hydroxylamine in Vicia Faba, Chromosoma, 15 (1964) 156-169
95. Malling, H. V., Mutagenicity of Methylhydroxylamines in Neurospora Crassa, Mutation Res., 4 (1967) 559
96. Kochetkov, N. K., Budowsky, E. J., and Shibaeva, R. P., The Selective Reaction of O-Methylhydroxylamine with the Cytidine Nucleus, Biochim. Biophys. Acta. 68 (1963) 493-496
97. Kochetkov, N. K., and Budowsky, E. J., The Chemical Modification of Nucleic Acids Prog. Nucleic Acid Res. Mol. Biol., 9 (1969) 403-438
98. Singer, B., and Fraenkel-Conrat, H., The Role of Confirmation in Chemical Mutagenesis, Prog. Nucleic Acid Res. Mol. Biol., 9 (1969) 1-29
99. Phillips, J. H., and Brown, D. M., The Mutagenic Action of Hydroxylamine, Prog. Nucleic Acid Res. Mol. Biol., 7 (1967) 349-368
100. Banks, G. R., Mutagenesis: A Review of Some Molecular Aspects, Sci. Prog., 59 (1971) 475-503
101. Shugar, D., Huber, C. P., and Birnbaum, G. I., Mechanism of Hydroxylamine Mutagenesis. Crystal Structure and Confirmation of 1,5-Dimethyl-N4-hydroxy-cytosine, Biochim. Biophys. Acta., 447 (1976) 274-284
102. Janion, C., and Shugar, D., Preparation and Properties of Some 4-Substituted Analogs of Cytosine and Dihydrocytosine, Acta. Biochim. Pol., 15 (1968) 261-272
103. Champe, S. P., and Benzer, S., Reversal of Mutant Phenotypes by 5-Fluorouracil: An Approach to Nucleotide Sequences in Messenger-RNA, Proc. Natl. Acad. Sci. U.S., 48 (1962) 532-546

104. Schuster, H., and Vielmetter, W., Studies on the inactivating and mutagenic effects of nitrous acid and hydroxylamines on viruses, J. Chim. Phys., 58 (1961) 1005-1010
105. Janion, C., and Shugar, D., Reaction of Hydroxylamine with 5-Substituted Cytosines Biochem. Biophys. Res. Commun., 18 (1965) 617-622
106. Janion, C., and Shugar, D., Mutagenicity of Hydroxylamine: Reaction with Analogs of Cytosine, 5(6)Substituted Cytosines and Some 2-Keto-4-Ethoxy Pyrimidines, Acta. Biochim. Pol., 12 (1965) 337-355
107. Salganik, R. I., Vasjunina, E. A., Poslovina, A. S., and Andreeva, I. S., Mutagenic Action of N_4-Hydroxycytidine on Escherichia Coli B cyt$^-$, Mutation Res., 20 (1973) 1-5
108. Popowska, E., and Janion, C., N_4-Hydroxycytidine-A New Mutagen of a Base Analog Type, Biochem. Biophys. Res. Commun., 56 (1974) 459-466
109. George, D. K., Moore, D. H., Brian, W. P., and Garman, J. A., Relative Herbicidal and Growth Modifying Activity of Several Esters of N-Phenylcarbamic Acid, J. Agr. Food Chem., 2 (1954) 353
110. Ferguson, G. R., and Alexander, C. C., Heterocyclic Carbamates Having Systemic Insecticidal Action, J. Agr. Food Chem., 1 (1953) 888
111. Mowry, D. T., and Piesbergew, N. R., U. S. Patent 2,537,690 (1951), Chem. Abstr., 45 (1951) 2142d
112. Henshaw, P. S., Minimal Number of Anesthetic Treatments with Urethane Required to Induce Pulmonary Tumors, J. Natl. Cancer Inst., 4 (1943) 523
113. Cowen, P. N., Some Studies on the Action of Urethane On Mice, Brit. J. Cancer, 1, (1947) 401
114. Orr, J. W., The Induction of Pulmonary Adenomata in Mice by Urethane, Brit. J. Cancer, 1 (1947) 311
115. Jaffe, W. G., Carcinogenic Action of Ethyl Urethan on Rats, Cancer Res., 7 (1947) 107
116. Sinclair, J. G., A specific transplacental effect of urethan in mice, Texas Rept. Biol. Med. Med., 8 (1950) 623-632
117. Ferm, V. H., Severe Developmental Malformations: Malformations Induced by Urethane and Hydroxyurea in the Hamster, Arch. Pathol., 81 (1966) 174
118. Battle, H. I., and Hisaoka, K. K., Effects of Ethyl Carbamate (Urethan) on the Early Development of the Teleost (Brachydanio Rerio), Cancer Res., 12 (1952) 334
119. McMillan, D. B., and Battle, H. I., Effects of Ethyl Carbamate (Urethan) and Related Compounds on Early Developmental Processes of the Leopard Frog, Rana Pipiens, Cancer Res., 14 (1954) 319
120. Bateman, A. J., The mutagenic action of urethane, Mutation Res., 39 (1976) 75-96
121. Oehlkers, F., Chromosome breaks induced by chemicals, Heredity Suppl., 6 (1953) 95-105
122. Oehlkers, F., Die auslosung von chromosomen mutationen in der meiosis durch ein wirkung von chemikalein, Z. Vererbungslehre, 81 (1943) 313
123. Bryson, V., Carbamate induced phage resistant mutants of E. coli, Proc. 8th Int. Congress Genetics, (1948) 545
124. Latarget, R., Buu-Hoi, N. P., and Elias, C. A., Induction of a specific mutation in a bacterium by a water-soluble cancerigen, Pubbl. Staz. Zool. Napoli., 22 Suppl., (1950) 78-83
125. Bryson, V., Carbamate-induced phage-resistant mutants of E. coli, Hereditas Suppl. (1949) 545

126. Demerec, M., Bertani, G., and Flint, J., A survey of chemicals for mutagenic action on E. coli, Am. Naturalist, 85 (1951) 119-136
127. Vogt, M., Mutations aüslosung bei Drosophila durch athylurethan, Experientia, 4 (1948) 68
128. Vogt, M., Urethan induced mutations in Drosophila, Pubbl. Staz. Zool. Napoli, 22, Suppl. (1950) 114
129. Oster, I. I., The induction of mutations in Drosophila melanogaster by orally administered ethyl carbamate (urethane), Genetics, 40 (1955) 588-589
130. Freese, E. B., The effects of urethan and hydroxyurethan on transforming DNA, Genetics, 51 (1965) 953-960
131. Jensen, K. A., Kirk, I.. Kølmark, G., and Westergaard, M., Chemically induced mutations in Neurospora, Cold Spring Harbor Symp. Quant. Biol., 16 (1951) 245-261
132. Gugleimonetti, R., Bonatti, S., and Loprieno, The mutagenic activity of N-nitro-N-methyl urethane and N-nitroso-N-ethyl urethane in S. pombe, Mutation Res., 3 (1966) 152-157
133. Bateman, A., A failure to detect any mutagenic action of urethane in the mouse, Mutation Res., 4 (1967) 710-712
134. Jackson, H., Fox, B. W., and Craig, A. W., The effect of alkylating agents on male rat fertility, Brit. J. Pharmacol. Chemotherapy, 14 (1959) 149-157
135. Kennedy, G. L., Jr., Arnold, D. W., and Keplinger, M. L., Mutagenic response of known carcinogens, Mutation Res., 21 (1973) 224-225
136. Yumkawa, K., Lack of effect of urethane on the induction of dominant lethal mutations in male mice, Nat. Inst. Genet. Mishima. Ann. Rep., 19 (1968) 69-70
137. Boyland, E., and Koehler, P. C., Effects of urethane on mitosis in the Walker rat carcinoma, Brit. J. Cancer, 8 (1964) 677-684
138. Platonova, G. M., Comparative study of the mutagenic action of urethane and N-hydroxyurethane on the chromosomes of embryonic lung cells of A/SN and C57 black mice, Sov. Genet., 5 (1969) 262-263 (Engl. transl.)
139. Pogosyants, E. E., Platonova, G. M., Tolkacheva, E. N., and Ganzenko, L. F., Effect of urethane on mammalian chromosomes in vitro, Sov. Genet., 4 (1968) 902-911 (Engl. transl.)
140. Tolkacheva, E. N., Platonova, G. M., and Vishenkova, N. S., Mechanism of the mutagenic action of urethane in mammalian cells in vitro, Sov. Genet., 6 (1970) 1077-1082 (Engl. Transl.)
141. Rapoport, I. A., Derivatives of carbamic acid and mutations, Bull. Exp. Biol. Med., 23 (1974) 198-201
142. Dean, B. J., Chemically induced chromosome damage, Lab. Animal, 3 (1969) 157-174
143. Boyland, E., and Nery, R., The metabolism of urethane and related compounds, Biochem. J., 94 (1965) 198-208
144. Boyland, E., Nery, R., and Peggie, K. S., The induction of chromosome aberrations in Vicia Faba root meristems by N-hydroxyurethan and related compounds, Brit. J. Cancer, 19 (1965) 878
145. Borenfreund, E., Krim, M., and Bendich, A., Chromosomal aberrations induced by hyponitrite and hydroxylamine derivatives, J. Natl. Cancer Inst., 32 (1964) 667-679
146. Bendich, A., Borenfreund, E., Korngold, G. C., and Krim, M., Action of hydroxylamine on DNA and chromosomes, Federation Proc., 22 (1963) 582
147. Freese, E. B., Gerson, J., Taber, H., Rhaese, H. J., and Freese, E., Inactivating DNA alterations induced by peroxides and peroxide-producing agents, Mut. Res. 4 (1967) 517

148. Freese, E., Sklarow, S., and Freese, E. B., DNA damage caused by antidepressant hydrazines and related drugs, Mutation Res., 5 (1968) 343
149. Nery, R., Acylation of cytosine by ethyl N-hydroxycarbamate and its acyl derivatives and the binding of these agents to nucleic acids and proteins, J. Chem. Soc. (C), (1969) 1860-1865
150. Boyland, E., and Williams, K., Reaction of urethane with nucleic acids in vivo, Biochem. J., 111 (1969) 121-127
151. Colnaghi, M. I., Della Porta, G. D., Parmiani, G., and Caprio, G., Chromosomal changes associated with urethan leukemogenesis in mice, Int. J. Cancer, 4 (1969) 327-333
152. Dahl, G. A., Miller, E. C., and Miller, J. A., Vinyl carbamate, A potent carcinogen and a possible urethan metabolite in the mouse, Proc. Am. Ass. Cancer, 18 (1977) 6
153. Roberts, J. D., Caserio, M. C. "Modern Organic Chemistry", W. A. Benjamin, Inc., New York (1967) pp. 469-470
154. Lurie, A. P., Acetamide, In "Kirk-Othmer Encyclopedia of Chemical Technology", 2nd ed., Interscience, New York (1963) pp. 142-145
155. Dessau, F. I., and Jackson, B., Acetamide-induced liver-cell alterations in rats, Lab. Invest., 4 (1955) 387-397
156. IARC, Acetamide, In Vol. 7, International Agency for Research on Cancer, Lyon (1974) pp. 197-202
157. Jackson, B., and Dessau, F. I., Liver tumors in rats fed acetamide, Lab. Invest. 10 (1961) 909-923
158. IARC, Thioacetamide, In Vol. 7, International Agency for Research on Cancer, Lyon (1974) pp. 78-83
159. Hueper, W. C., and Conway, W. D., "Chemical Carcinogenesis and Cancers, Thomas, Springfield, IL (1964) p. 37
160. Gothoskar, S. V., Talwalker, G. V., and Bhide, S. V., Tumorigenic effect of thioacetamide in Swiss strain mice, Brit. J. Cancer, 24 (1970) 498-503
161. Gupta, D. N., Nodular cirrhosis and metastasizing tumours produced in the liver of rats by prolonged feeding with thioacetamide, J. Path. Bact., 72 (1956) 415-426
162. Production of cancer of the bile ducts with thioacetamide, Gupta, D. N., Nature, 175 (1955) 257
163. Terracini, B., and Della Porta, G., Feeding with aminoazo dyes, thioacetamide, and ethionine, Arch. Path., 71 (1961) 566-575
164. IARC, Thiourea, In Some Anti-thyroid and Related Substances, Nitrofurans and Industrial Chemicals, Vol. 7, International Agency for Research on Cancer, Lyon (1974) pp. 95-109
165. Jones, R. G., Thyroid and Antithyroid Preparations, In Kirk-Othmer Encyclopedia of Chemical Technology, Vol. 20, 2nd ed., Wiley and Sons, New York (1969) 271
166. Inukai, C., and Terama, E. K., Fireproofing of nylon fabrics, Japan Kokai 75, 160,599, 25 Dec. 1975; Chem. Abstr., 84 (1976) P152164D
167. Tsuzuki, R., Hashizume, A., and Mizushima, H., Fireproofing of polyester-cotton fabrics, Japan Kokai 75 88,399; 16 July 1975, Chem. Abstr., 84 (1976) P32530Y
168. Inukai, C., and Teramae, K., Finishing of natural and synthetic polyamide fiber textiles; Japan Kokai 76 01, 799, 8 Jan., 1976; Chem. Abstr., 84 (1976) P152173F
169. Agarwal, S. C., and Somlo, T., Solid, coldwater-soluble dye and fluorescent whitener preparations, Ger. Offen., 2,529,568; 16 Jan. 1976; Chem. Abstr., 84 (1976) P123324A
170. Throckmorton, M. C., Process of polymerization of conjugated diolefins using iron catalysts and sulfur ligands, U. S. Patent, 3,936,432; 3 Feb. 1976; Chem. Abstr., 84 (1976) P135955G

171. Gorton, A. D. T., Vulcanization characteristics of natural latex concentrate, Nr. Technol., 6 (1975) 52-64; Chem. Abstr., 84 (1976) 32303B
172. Fischer, A. A., Adhesive compositions; U. S. Patent 3,919,153; 11 Nov. 1975; Chem. Abstr., 84 (1976) P60572G
173. Park, S. H., and Yun, K. S., Electrolytic refining for copper, Kumsok Hakhoe Chi., 13 (1975) 444-453; Chem. Abstr., 84 (1976) 167777E
174. Sankov, V. M., Latyshov, A. N., and Evgrafov, V. A., Electrolyte for plating with copper-lead alloy, USSR Patent 503,941; 25 Feb. 1976; Chem. Abstr., 84 (1976) P128070Z
175. Frisch, B., and Thiele, W. R., Use of inhibitors for hydrochloric acid-pickling, Arch. Eisenhuettenwes., 46 (1975) 575-580
176. Higaki, Y., and Kataoka, J., Corrosion inhibition of metals in acid solutions, Japan Kokai, 75 51,437; 8 May, 1975; Chem. Abstr., 84 (1976) 63707R
177. Sakurai, J., Ohida, F., Kurashige, N., Koyama, S., and Takehana, T., Drying of acrylamide polymers in presence of thiourea and ethylene urea, Japan Kokai, 75 149,738; 1 Dec. 1975; Chem. Abstr., 84 (1976) 90975H
178. Conger, R. P., and Palmer, L. B., Chemical embossing of PVC foams, Polym.-Plast. Technol. Eng., 6 (1976) 185-202; Chem. Abstr., 84 (1976) P181225E
179. Ono, M., Printing of acrylic fabrics for stereopattern, Japan Kokai 7607,229; 21 Jan. 1976; Chem. Abstr., 84 (1976) P152125S
180. Merck & Co., "The Merck Index", 8th ed., Merck & Co., Rahway, NJ (1968) P. 1046
181. Purves, H. D., and Griesbach, W. E., Studies on experimental goitre. VIII. Thyroid tumours in rats treated with thiourea, Brit. J. Exp. Pathol., 28 (1947) 46-53
182. Purves, H. D., and Griesbach, W. E., Studies on experimental goitre. VII. Thyroid carcinomata in rats treated with thiourea, Brit. J. Exp. Pathol., 27 (1946) 294-297
183. Rosin, A., and Ungar, H., Malignant tumors in the eyelids and the auricular region of thiourea-treated rats, Cancer Res., 17 (1957) 302-305
184. Ungar, H., and Rosin, A., The histogenesis of thiourea-induced carcinoma of the auditory duct sebaceous (Zymbal's) glands in rats, Arch. De Vecchi Anat. Pat. 31 (1960) 419-430
185. Fitzhugh, O. G., and Nelson, A. A., Liver tumors in rats fed thiourea or thioacetamide, Science, 108 (1948) 626-628
186. Dalton, A. J., Morris, H. P., and Dubnik, C. S., Morphologic changes in the organs of female C3H mice after long-term ingestion of thiourea and thiouracil, J. Natl. Cancer Inst., 9 (1948) 201-223
187. Vazquez-Lopez, E., The effects of thiourea on the development of spontaneous tumors on mice, Brit. J. Cancer, 3 (1949) 401-414
188. McCann, J., Choi, E., Yamasaki, E., and Ames, B. N., Detection of carcinogens as mutagens in the Salmonella/microsome test: Assay of 300 chemicals, Proc. Natl. Acad. Sci., 72 (1975) 5135-5139
189. IARC, EThylene thiourea, In Some Anti-Thyroid and Related Substances, Nitrofurans and Industrial Chemicals, Vol. 7, International Agency for Research on Cancer, Lyon, pp. 45-52
190. Johnson, T. B., and Edens, C. O., Complex formations between iodine and mercaptodihydroglyoxalines, J. Amer. Chem. Soc., 64 (1942) 2706-2708
191. Fishbein, L., Environmental health aspects of fungicides. I. Dithiocarbamates, J. Toxicol. Env. Hlth., 1 (1976) 713-735
192. Bontoyan, W. R., Looker, J. B., Kaiser, T. E., Giang, P., and Olive, B. M., Survey of ethylene thiourea in commercial ethylenebisdithiocarbamate formulation, J. Assoc. Off. Anal. Chem., 55 (1972) 923-925

193. Engst, R., and Schnaak, W., Investigations of the metabolism of fungicidal ethylene bis-dithiocarbamates maneb and zineb, Z. Lebensm. Unters. Forsch., 134 (1967) 216-221
194. Engst, R., Schnaak, W., and Lewerenz, H. J., Investigation about the metabolism of the fungicides: Maneb, zineb, and nabam, V. Toxicity of degradation products, Z. Labensm. Unters. Forsch., 146 (1971) 91-94
195. Seidler, H., Haertig, M., Schnaak, W., and Engst, R., Metabolism of certain insecticides and fungicides in the rat. II. Distribution and degradation of carbon-14-labeled maneb, Nahrung, 14 (1970) 363-373
196. Truhaut, R., Fugita, M., Lich, N. P., and Chaigneau, M., C.R. Hebd. Seanc. Acad. Sci. (Paris) Ser. D. 276 (1973) 229-233
197. Vonk, J. W., Ethylene thiourea, a systemic decomposition product of nabam, Med. Landbouwhenesch. Genet., 36 (1971) 109-114
198. Blazquez, C. H., Residue determination of ethylene thiourea (2-imidazolidine) from tomato foliage, soil and water, J. Agric. Food Chem., 21 (1973) 330-332
199. Engst, R., and Schnaak, W., Metabolism of the fungicidal ethylene bis dithiocarbamates maneb and zineb. III. Pathways of degradation, Z. Lebensm. Unters. Forsch., 43 (1970) 99-103
200. Engst, R., and Schnaak, W., Residues of dithiocarbamate fungicides and their metabolites on plant foods, Residue Rev., 52 (1974) 45-67
201. Fishbein, L., and Fawkes, J., Thin-layer chromatography of metallic derivatives of ethylene bis (dithiocarmamic acid) and their degradation products. J. Chroma. 19 (1965) 364-369
202. Hylin, J. W., Oxidative decomposition of ethylene bis-dithiocarbamates, Bull. Environ. Contam. Toxicol., 10 (1973) 227-233
203. Klöpping, J. L., and Vanderkerk, G. J. M., Investigation on organic fungicides: V. Chemical constitution and fungistatic activity of aliphatic bis dithiocarbamates and isothiocyanates, Rev. Trav. Chim., 70 (1951) 949-954
204. Lopatecki, L. E., and Newton, W., The decomposition of dithiocarbamate fungicides with special reference to the volatile products, Can. J. Bot., 30 (1952) 131-138
205. Ludwig, R. A., Thorn, G. D., and Miller, D. M., Studies on the mechanism of fungicidal action of disodium ethylene bis-dithiocarbamate (nabam), Can. J. Bot. 32 (1954) 48-54
206. Newsome, W. H., Determination of ethylene thiourea residues in apples, J. Agric. Food Chem., 20 (1972) 967-969
207. Newsome, W. H., and Laver, G. W., Effect of boiling on the formation of ethylene thiourea in zineb treated food, Bull. Environ. Contam. Toxicol., 10 (1973) 151-154
208. Thorn, G. D., and Ludwig, R. A., The aeration products of disodium ethylene bis dithiocarbamates, Can. J. Chem., 32 (1954) 872-879
209. Watts, R. R., Stroherr, R. W., and Onley, J. H., Effects of cooking on ethylene bis dithiocarbamate degradation to ethylene thiorea, Bull. Contam. Toxicol., 12 (1974) 224-228
210. Yip, G., Onley, J. H., and Howard, S. F., Residues of maneb and ethylene thiourea on field-sprayed lettuce and kale, J. Assoc. Off. Anal. Chem., 54 (1971) 1373-1375
211. Ulland, B. M., Weisburger, J. H., Weisburger, E. K., Rice, J. M., and Cypher, R., Thyroid cancer in rats from ethylene thiourea intake, J. Natl. Cancer Inst., 49 (1972) 583-584
212. Innes, J. R. M., Valerio, M., Ulland, B. M., Pallotta, A. J., Petrucelli, L., Fishbein, L., Hart, E. R., Falk, H. L., Klein, M., and Peters, A. J., Bioassay of pesticides and industrial chemicals for tumorigenicity in mice, J. Natl. Cancer Inst., 42 (1969) 1101-1109

213. Seiler, J. P., Ethylene thiourea (ETU), a carcinogenic and mutagenic metabolite of ethylene bis dithiocarbamate, Mutation Res., 26 (1974) 189-194
214. Schupach, M., and Hummler, H., Evaluation of the mutagenicity of ethylene thiourea, using bacterial and mammalian test systems, Mutation Res., 38 (1976) 122
215. Shirasu, Y., Moriya, M., Tezuka, H., and Teramoto, S., Mutagenicity screening on pesticides, Origins of Human Cancer Meeting, Coldspring Harbor, NY, Sept. 7-14 (1976) p. 80
216. Mollet, P., Toxicity and mutagenicity of ethylene thiourea (ETU) in Drosophila, Mutation Res., 29 (1975) 254
217. Khera, K. S., Ethylenethiourea: Theratogenicity study in rats and rabbits, Teratology, 7 (1973) 243-252
218. Kato, Y., Odanaka, Y., Teramoto, S., and Matano, O., Metabolic fate of ethylene thiourea in pregnant rats, Bull. Env. Contam. Toxicol., 16 (1976) 546-555

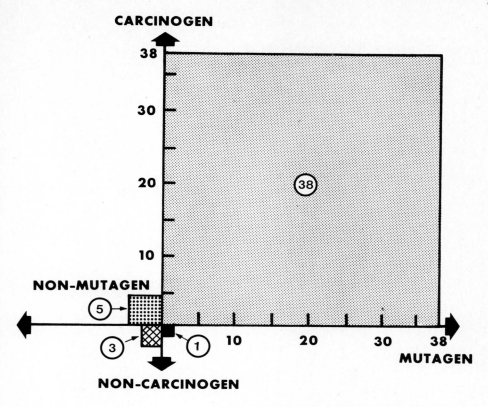

Fig. 2 Correlation between mutagenic and carcinogenic effects of *N*-nitroso compounds.

TABLE J

CARCINOGENICITY AND MUTAGENICITY OF N-NITROSO COMPOUNDS

Compound	Genetic indicator	Genetic changes	Mutagenicity assays [b]					
			A	B	C	D	E	F
Carcinogenicity [a]: species and principal target organs								
(1) N-Nitrosodimethylamine								
Rat: liver, kidney, nasal cavities	*S. typhimurium*							
Mouse: Lung, liver, kidney	human, mouse, rat, hamster, liver MS [c]	Reverse mut.	−	+				
S.G. Hamster: liver, nasal cavities	rat, mouse, hamster lung MS	Reverse mut.		−				
European hamster: liver, kidney	mouse and rat	Reverse mut.			+			
Rabbit: mastomys; guinea pig; trout; newt; aquarium fish; mink: liver	*E. coli*							
	rat liver MS	Reverse or forward mut. and preferential growth inhibition	−	+				
	rat kidney MS	Reverse mut.		−				
	mouse	Reverse mut.		+				
	B. subtilis							
	mouse liver MS	Reverse mut.	−	+				
	Saccharomyces carevisiae							
	Udenfriend hydroxylation system, mouse liver MS	Back mut., gene recombination and conversion, petite mut, canavanine resistant mut.	−	+				
	mouse	Gene recombination and conversion			+			
	Serratia marcescens	ω Mut.						
	mouse	Reverse mut.	−	−	+			
	Neurospora crassa	Reverse mut.						
	Udenfriend hydroxylation system, mouse liver MS	Forward mut.	−	−	+			
	mouse	Forward mut.			+			

Drosophila melanogaster	Recessive lethal mut.	+			
Chinese hamster cells V-79 rat liver MS	Thioguanine-resistant mut.	−	+		
Chinese hamster cells CHO-K1	Nutritional auxotropic mutant	+			
Murine leukaemic cells L5178Y/Asn− mouse	Asparagine independent mut.	−			
Mouse	Dominant lethal mut.	−	+		

(2) *N*-Nitrosodiethylamine

Rat: liver, oesophagus, nasal cavities, kidney
Mouse: liver, lung, forestomach, oesophagus, nasal cavities
S.G. hamster: trachea, larynx, nasal cavities, lung, liver
Chinese hamster: oesophagus, forestomach, liver
European hamster: nasal cavities, trachea, bronchi, larynx
Guinea-pig; rabbit; dog; pig; trout; *Brachydanio rerio*; grass parakeet; monkey: liver

S. typhimurium rat, mouse and hamster MS	Reverse mut.	−	+		
rat, mouse and hamster lung	Reverse mut.	+	−		
mouse, rat	Reverse mut.				
E. coli rat liver MS	Reverse mut.	−	+	+	
rat kidney MS	Reverse mut.	−			
mouse	Forward mut.				
Saccharomyces cerevisiae Udenfriend hydroxylation system, mouse liver MS	Back and reverse mut., gene conversion Forward and petite mut., gene recombination	−	+		
mouse	Gene recombination and conversion	+	−		
Serratia marcescens mouse	ω Mut. Reverse mut.	− −	− −		

TABLE I (continued)

Compound	Genetic indicator	Genetic changes	Mutagenicity assays [b]					
			A	B	C	D	E	F
Carcinogenicity [a]: species and principal target organs								
	Neurospora crassa	Back mut.	−	−				
	Udenfriend hydroxylation system	Forward mut.	−	+				
	mouse	Forward mut.			+			
	Drosophila melanogaster	Recessive lethal mut.						+
	Mouse	Dominant lethal				−		
	Rat liver cells in vivo	Chromosome aberrations					+	+
(3) *N*-Nitrosomethylvinylamine								
Rat: Oesophagus, pharynx, tongue, nasal cavities	*Drosophila melanogaster*	Recessive lethal mut.						
(4) *N*-Nitrosodi-*n*-propylamine								
Rat: liver, oesophagus, tongue S.G. hamster: nasal cavities, trachea [202]	*S. typhimurium* rat and hamster liver, hamster lung MS	Reverse mut.	−	+				
	rat lung, hamster and rat kidney MS	Reverse mut.		−				
	E. coli							
	rat liver MS	Reverse mut.	−	+				
	rat kidney MS	Reverse mut.		−				
	Saccharomyces cerevisiae Udenfriend hydroxylation system	Gene recombination	−	+				

(5) *N*-nitroso-*N*-(2-hydroxy-*n*-propyl)-*n*-propylamine

			S. typhimurium	
Rat: nasal cavities, lung, oesophagus liver			rat and hamster liver MS	Reverse mut. — +
S.G. hamster: nasal cavities, trachea, lung, liver [206]			hamster lung MS	Reverse mut. +
			rat lung, rat and hamster kidney MS	Reverse mut. —

(6) *N*-Nitrosodi-(2-hydroxy-*n*-propyl)amine

			S. typhimurium	
S.G. hamster: respiratory tract, liver, kidney, [201]			hamster liver MS	Reverse mut. — +
			rat liver, rat and hamster lung and kidney MS	Reverse mut. —

(7) *N*-Nitrosodi-(2-acetoxy-*n*-propyl)amine

			S. typhimurium	
			rat and hamster liver MS	Reverse mut. — +
			hamster lung and kidney MS	Reverse mut. — +
			rat lung and kidney MS	Reverse mut. —

(8) *N*-Nitroso-*N*-*n*-propyl-2-oxopropylamine

			S. typhimurium	
S.G. hamster: nasal cavities, respiratory tract, liver, kidney [199]			rat and hamster liver MS	Reverse mut. — +
			rat and hamster lung	Reverse mut. —

(9) *N*-Nitroso-*N*-di-(2-oxo-propyl)amine

			S. typhimurium	
S.G. hamster: pancreas [200]			rat and hamster liver MS	Reverse mut. — +
			rat and hamster lung and kidney, hamster pancreas MS	Reverse mut. —

(10) *N*-Nitrosomethyl-*n*-propylamine

			S. typhimurium	
Rat: nasal cavities, liver, oesophagus [206]			hamster liver MS	Reverse mut. — +
S.G. hamster: nasal cavities, trachea, lung, liver			rat liver, lung and kidney, hamster lung and kidney MS	Reverse mut. —

TABLE I (continued)

Compound Carcinogenicity [a]: species and principal target organs	Genetic indicator	Genetic changes	Mutagenicity assays [b]					
			A	B	C	D	E	F
(11) N-Nitrosodi-n-butylamine								
Rat: bladder, oesophagus, liver Mouse: oesophagus, bladder, liver, forestomach, tongue, lung S.G. hamster: bladder, trachea, lung, forestomach Chinese hamster: forestomach, bladder	*S. typhimurium* rat liver MS	Reverse mut.	−	+				
	E. coli rat liver MS rat kidney MS	Reverse mut. Reverse mut.	−	± −				
Guinea-pig: liver, bladder	*Saccharomyces cerevisiae* Udenfriend hydroxylation system	Gene recombination	−	±				
(12) N-Nitrosomethyl-n-butylamine								
Rat: oesophagus, nasal cavities, liver Mouse: eyelid, nasal cavities S.G. hamster: trachea, lung	*E. coli* rat liver MS	Reverse mut.	−	+				
(13) N-Nitrosoethyl-tert-butylamine								
Rat: negative	*E. coli* rat liver MS	Reverse mut.	−	−				
	Drosophila melanogaster	Recessive lethal mut.						−
(14) N-Nitrosodi-n-pentylamine								
Rat: liver, lung	*S. typhimurium* rat liver MS	Reverse mut.	−	+				
(15) N-Nitrosodiphenylamine								
Rat: negative	*S. typhimurium* rat liver MS	Reverse mut.	−	−				

	E. coli rat liver MS	Reverse mut.	—
	Neurospora crassa	Reverse mut.	—
(16) *N*-Nitroso-*N*-methylphenylamine			
Rat: oesophagus, pharynx, forestomach	*E. coli* rat liver MS	Reverse mut.	—
Mouse: lung	*Neurospora crassa*	Back mut.	—
(17) *N*-Nitroso-*N*-methylbenzylamine			
Rat: oesophagus	*Drosophila melanogaster*	Recessive lethal	+
Mouse: oesophagus, forestomach			
(18) *N*-Nitrosodibenzylamine			
Rat: negative	*S. typhimurium* rat liver MS	Reverse mut.	—
(19) Di-*N*,*N'*-*bis*-nitroso-*N*,*N'*-*bis*-(1-hydroxymethylpropyl)-ethylenediame	*S. typhimurium*		
Mouse: lung [19]	rat liver and lung MS	Reverse mut.	—
(20) Di-*N*,*N'*-*bis*-nitroso-*N*,*N'*-*bis*-(1-acetoxymethylpropyl)-ethylenediamine	*S. typhimurium* rat liver and lung MS	Reverse mut.	—
(21) *N*-Nitrosomorpholine	*S. typhimurium*		
Rat: liver, nasal cavities, kidney, oesophagus, ovary	human and rat liver MS	Reverse mut.	+
Mouse: liver, lung	mouse	Reverse mut.	+

TABLE I (continued)

Compound	Genetic indicator	Genetic changes	A	B	C	D	E	F
Carcinogenicity [a]: species and principal target organs								
S.G. hamster: trachea, larynx, bronchi	E. coli	Biochemical mut.	−					
	rat liver MS	Reverse mut.	−	+				
	kidney MS	Reverse mut.		−				
	Mouse	Dominant lethal mut.			−			
	Drosophila melanogaster	Recessive lethal mut. and translocation						+
(22) N-Nitrosopyrrolidine								
Rat: liver, nasal cavities, testis	S. typhimurium	Reverse mut.	−	+				
Mouse: lung	human and rat liver MS							
S.G. hamster: trachea, lung								
(23) N-Nitrosopiperidine								
Rat: oesophagus, liver, nasal cavities, larynx, trachea	S. typhimurium human and rat liver MS	Reverse mut.		+				
Mouse: forestomach, liver, lung, oesophagus	E. coli rat liver MS	Biochemical mut. Reverse mut.	−	+				
S.G. hamster: trachea, lung, larynx	rat kidney MS	Reverse mut.	−	−				
Monkey: liver								
(24) N-Nitrosopiperazine								
Rat: nasal cavities	S. typhimurium mouse	Reverse mut.			+			
(25) N-Nitroso-N'-methylpiperazine								
Rat: nasal cavities	S. typhimurium human and rat liver MS mouse	Reverse mut. Reverse mut.	−	+	+			

	+		+					+			
				+						+	
											+
				+		+					
				+	+	+	—	+			
				+							
				—							
				+							
				+							
				+							
				+							
				+							

Site / species	Organism	Mutation type
	Drosophila melanogaster	Recessive lethal mut.
(26) *N,N'*-Dinitrosopiperazine		
Rat: oesophagus, liver, nasal cavities, forestomach	*S. typhimurium* mouse	Reverse mut.
Mouse: lung, liver	*E. coli*	Biochemical mut.
	Drosophila melanogaster	Recessive lethal mut.
(27) *N*-Nitroso-*N*-methylurea		
Rat: central and peripheral nervous system, intestine, kidney, forestomach, glandular stomach, skin and annexes, jaw, bladder, uterus, vagina	*S. typhimurium* mouse	Reverse mut.
	E. coli	Reverse mut., preferential growth inhibition
	mouse	Forward mut.
Mouse: lung, haemapoietic system, forestomach, kidney, skin, liver (only new-born), central nervous system	*Saccharomyces cerevisiae*	Gene conversion and reverse mut.
S.G. hamster: intestine, pharynx, oesophagus, trachea, bronchi, oral cavity, skin and annexes and s.c. site of injection	*Serratia marcescens* mouse	Reverse mut.
European hamster: s.c. site of injection	*Drosophila melanogaster*	Recessive lethal mut.
Guinea-pig: stomach, pancreas, ear duct	Chinese hamster cells V79	8-azaguanine-resistant mut.
Rabbit: central nervous system, intestine, skin	Chinese hamster cells CHO-K1	Nutritional auxotrophic mut.
Dog: central and peripheral nervous system	Mouse	Dominant lethal mut.
	coliphage T$_2$	r-mut
	Aspergillus nidulans	Forward and reverse mut.
	B. subtilis (transforming DNA)	Fluorescent *ind*$^-$ mut., inactivation

TABLE I (continued)

Compound Carcinogenicity [a]: species and principal target organs	Genetic indicator	Genetic changes	A	B	C	D	E	F
(28) N-Nitroso-1,3,dimethylurea								
Rat: central and peripheral nervous systems, kidney	B. subtilis (tranforming DNA)	Fluorescent ind$^-$ mut., inactivation	+					
Mouse: haemapoietic system	Saccharomyces cerevisiae	Reverse mut	−					
(29) N-Nitrosotrimethylurea								
Rat: central and peripheral nervous systems, kidney, skin	Saccharomyces cerevisiae	Reverse mut.	−					
(30) N-Nitroso-N-ethylurea								
Rat: central and peripheral nervous systems, kidney, haemapoietic system, skin, intestine, ovary, uterus	E. coli	Reverse mut., preferential growth inhibition	+					
	coliphage T$_2$	r-mut	+					
Mouse: haemapoietic system, lung, central and peripheral nervous system, kidney								
(31) N-Nitroso-N-n-propylurea	Saccharomyces cerevisiae	Gene conversion	+					
(32) N-Nitroso-N-n-butylurea								
Rat: haemapoietic system, mammary glands, central nervous system, oesophagus, forestomach, ear ducts	S. typhimurium mouse	Reverse mut.	+				+	
	Saccharomyces cerevisiae	Gene conversion	+					
Mouse: haemapoietic system	E. coli							

S.G. hamster: peripheral nervous system	mouse	Forward mut.	+
	Serratia marcescens	Reverse mut.	+
	mouse	Dominant lethal mut.	−
(33) *N*-Nitroso-*N*-allylurea	*Saccharomyces cerevisiae*	Gene conversion	+
(34) *N*-Nitroso-*N*-methylurethane			
Rat: forestomach, lung, oesophagus, intestine, kidney, ovary	*E. coli*	Preferential growth inhibition; forward and reverse mut.	+
Mouse: lung, forestomach			
S.G. hamster: oesophagus, forestomach	*B. subtilis*	Preferential growth inhibition	+
Guinea-pig: pancreas, s.c. site of injection	*Saccharomyces cerevisiae*	Gene conversion, back mut., respiratory deficient mut.	+
	Serratia marcescens	Back mut.	+
	Drosophila melanogaster	Recessive lethal mut., translocation	+
	Chinese hamster cells (CHO-KI)	Nutritional auxotropic mut.	+
	Schizosaccharomyces pombe	Reverse and forward mut.	+
	Collectotrichum cocodes	Nutritional auxotrophic mut.	+
	Haemophilus influenzae	Novobiocin-resistant mut.	+
	S. typhimurium	Reverse mut.	+
	Neurospora crassa	Reverse mut.	+

TABLE I (continued)

Compound	Genetic indicator	Genetic changes	Mutagenicity assays [b]					
Carcinogenicity [a]: species and principal target organs			A	B	C	D	E	F
(35) N-nitroso-N-ethylurethane								
Rat: forestomach, intestine	E. coli	Preferential growth inhibition	+					
	Saccharomyces cerevisiae	Back and reverse mut., gene conversion	+					
	Neurospora crassa	Reverse mut.	+					
	Schizosaccharomyces pombe	Reverse and forward mut.	+					
	Collectotrichum cocodes	Nutritional auxotrophic mut.	+					
	Drosophila melanogaster	Recessive lethal mut.						+
(36) N-Nitroso-N'-D-glucosyl-2-methylurea (Streptozotocin)								
Rat: kidney	S. typhimurium mouse	Reverse mut.	+		+			
Chinese hamster: liver	Mouse	Dominant lethal mut.				+		
(37) N-Nitroso-N-methylacetamide								
Rat: forestomach	E. coli	Forward mut.	+					
	Saccharomyces cerevisiae	Reverse mut.	+					
	Drosophila melanogaster	Recessive lethal mut., translocation						+

(38) N-Nitroso-N-methylcapronamide

    Saccharomyces cerevisiae   Gene conversion   +

(39) N-Nitrosoimidazolidone

    Saccharomyces cerevisiae   Gene conversion; recombination; respiratory deficient mut., reverse mut.   +

Rat: s.c. site of injection

(40) N,N'-Dinitroso-N,N-dimethyloxamide

    Saccharomyces cerevisiae   Reverse mut.   +

Rat: liver

(41) N-Nitrosobenzothiazuram[N-nitroso-N-methyl-N'-(2-benzothiazolyl)urea]

    Saccharomyces cerevisiae   Gene conversion   +

Rat: forestomach, kidney [246]

(42) N-Nitrosopropoxur(N-nitroso-iso-propoxyphenyl-N-methylcarbamate)

    Saccharomyces cerevisiae   Gene conversion   +

(43) N-Nitrosocarbaryl(N-nitroso-1-naphthyl-N-methylcarbamate)

    Saccharomyces cerevisiae   Gene conversion   +
    Haemophilus influenzae   Temperature-sensitive mut., novobiocin-resistant mut.   +

Rat: s.c. site of injection, forestomach [54]

(44) p-Tolylsulphonylmethylnitrosamide

    Saccharomyces cerevisiae   Reverse mut.   —

Rat: **negative**
Mouse: **lung**

(45) N-Nitroso-N-methylsulpholan

    Saccharomyces cerevisiae   Reverse mut.   —

Rat: oesophagus

TABLE 1 (continued)

Compound	Genetic indicator	Genetic changes	A	B	C	D	E	F
Carcinogenicity [a]: species and principal target organs								
(46) N,N'-Dinitroso-N',N-dimethylphthalamide	Saccharomyces cerevisiae	Reverse mut.	+					
Rat: negative								
(47) N-Nitroso-N-methyl-N'-nitroguanidine	S. typhimurium	Reverse mut.	+					
Rat: glandular stomach, forestomach, intestine, s.c. site of injection	mouse	Reverse mut.			+			
Mouse: intestine, forestomach, skin (site of injection)	E. coli	Reverse and forward mut	+					
S.G. hamster: glandular stomach, intestine	mouse	Forward mut.			+			
Rabbit: lung	Saccharomyces cerevisiae	Forward and reverse mut.	+					
Dog: stomach, intestine	mouse	Gene conversion	+		+			
	Neurospora crassa	Forward, recessive lethal mut, deletions	+					
	Chinese hamster cells (CHO-KI)	Nutritional auxotrophic mut.	+				+	
	Chinese hamster cells V79	8-Azaguanine-resistant mut.	+					
	Murine leukaemia cells L5178Y/Asn	Asparagine-independent mut.	+					
	Haemophilus influenzae	Novobiocin-resistant mut.	+					
	Aspergillus nidulans	Forward and reverse mut.	+					

23. Low, H., Nitroso Compounds-Safety and Public Health, Arch. Env. Hlth., 29 (1974) 256
24. Sander, J., Kann Nitrit in der Menschlichen Nahrung Ursache Einer Krebsentstehung durch Nitrosaminebildung sein? Arch. Hyg. Bakt., 151 (1967) 22-24
25. Sander, J., Seif, F., Bakterielle Reduktion van Nitrat im Magen des Menschen als Ursache einer Nitrosamin-Bildung, Arzneim Forsch, 19 (1969) 1091-1093
26. Hawksworth, G., Hill, M. J., The Formation of Nitrosamines by Human Intestinal Bacteria, Biochem. J., 122 (1971) 28P-29P
27. Magee, P. N., Toxicity of Nitrosamines: Their Possible Human Health Hazards, Food Cosmet. Toxicol., 9 (1971) 207-218
28. Wogan, G. N., and Tannenbaum, S. R., Environmental N-nitroso compounds: Implications for public health, Toxicol. Appl. Pharmacol., 31 (1975) 373
29. Scanlan, R. A., In: "Critical Reviews in Food Technology" Cleveland, CRC Press, (1975) Vol. 5, No. 4
30. Sen, N. P., In: "Toxic Constituents of Animal Foodstuffs", edited by I. E. Liener, New York, Academic, (1974)
31. Shank, R. C., Toxicology of N-Nitrosocompounds, Toxicol. Appl. Pharmacol. 31 (1975) 361
32. Mirvish, S. S., Formation of N-Nitroso Compounds: Chemistry, Kinetics and In Vitro Occurrence, Toxicol. Appl. Pharmacol., 31 (1975) 325
33. Fiddler, W., The Occurrence and Determination of N-Nitroso Compounds, Toxicol. Appl. Pharmacol., 31 (1975) 352
34. Anon, Dupont to Limit Nitrosamine Emissions: EPA Studies Nitrosamine Problem, Toxic Materials News, 3 (1976) 111
35. Anon, EPA Advisory Unit Recommends No Immediate Regulatory Action on Nitrosamines, Toxic Materials News, 3 (1975) 139
36. Anon, Science Advisory Board Endorses Closer Look at Nitrosamines, Env. Hlth. Letter, 15 (1976) 1
37. Horvitz, D., and Cerwonka, E., U. S. Patent 2,916,426 (1959), Chem. Abstr. 54, (1960) 6370c
38. National Distillers and Chem. Corp., British Patent 817,523 (1959), Chem. Abstr. 54 (1960) 10601c
39. Maitlen, E. G., U. S. Patent 2,970,939 (1961), Chem. Abstr., 56 (1961) 11752f
40. Goring, C. A. I., U.S. Patent 3,256,083 (1966), Chem. Abstr., 65 (1966) 6253d
41. Lytton, M. R., Wielicki, E. A., and Lewis, E., U. S. Patent 2,776,946 (1957); Chem. Abstr., 51 (1957) 5466e
42. Elliot, W. E., Huff, J. R., Adler, R. W., and Towle, W. L., Proc. Ann. Power Sources Conf., 20 (1966) 67-70; Chem. Abstr., 66 (1967) 100955w
43. Middleton, W. J., U. S. Patent 3,240,765 (1966), Chem. Abstr., 64 (1966) 19826f
44. Lel'Chuck, Sh. L., and Sedlis, V. I., Zh. Prikl. Khim., 31 (1958) 128; Chem. Abstr., 52 (1958) 17787g
45. Klager, K., U. S. Patent 3,192,707 (1965).
46. Boyland, E., Carter, R. L., Gorrod, J. W., and Roe, F. J. C., Carcinogenic Properties of Certain Rubber Additives, Europ. J. Cancer, 4 (1968) 233
47. Anon, Nitrosamines Warning, Am. Ind. Hyg. Assoc. J., 37 (1976) A-9
48. Anon, Nitrosamines Reported in Industrial Cutting Oils, Occup. Hlth. Safety Letter, 5 (1976) 3-4
49. Anon, Nitrosamines Found in Cutting Oils, Toxic Materials News, 3 (1976) 156-157
50. Anon, NIOSH Issues Nitrosamine Alert, Toxic Materials News, 3 (1967) 167

51. Fine, D. H., Roundbehler, D. P., Sawicki, E., Krost, K., and DeMarrais, G. A., N-Nitroso Compounds in the Ambient Community Air of Baltimore, Maryland, Anal. Letters, 9 (1976) 595-604
52. Pellizzari, E. D., Bunch, J. E., Bursey, J. T., Berkley, R. E., Sawicki, E., and Krost, K., Estimation of N-Nitrosodimethylamine Levels in Ambient Air by Capillary Gas-Liquid Chromatography/Mass Spectrometry, Anal. Letters, 9 (1976) 579=594
53. Fine, D. H., Roundehler, D. P., Belcher, N. M., and Epstein, S. S., N-nitrosocompounds, detection in ambient air, Science, 192 (1976) 1328-1334
54. Anon, Nitrosamines Found in Herbicides: In Vivo Formation Documented, Toxic Materials News, 3 (1976) 148
55. Anon, Nitrosamines: EPA Refuses Suspension Request; NIOSH Promises Alert, Toxic Materials News, 3 (1976) 156
56. Anon, Nitrosamines Found in Commercial Pesticides, Chem. Eng. News, 54 (1976) 33
57. Anon, EPA Investigating Nitrosamine Impurities in 24 Herbicides, Chem. Eng. News, 54 (1976) 12
58. Anon, High Levels of Nitrosamines in Soils Lead to Government Task Force, Pesticide Chem. News, 4 (1976) 56
59. Dean-Raymond, D., and Alexander, M., Plant Uptake and Leaching of Dimethylnitrosamine, Nature, 262 (1976) 394-395
59a. Anon, N-nitrosamines found in toiletry products, Chem. Eng. News, 55 (1977) 7,8
60. Pasternak, L., Untersuchung Uber Die Mutagene Wirkung Verschiedner Nitrosamin Und Mitrosamid-Verbindungen, Arzneimittel-Forsch., 14 (1964) 802
61. Pasternak, L., Mutagene Wirkung Von Dimethyl Nitrosamin Bei Drosophila Melanogaster, Naturwissenschaften, 49 (1962) 381
62. Pasternak, L., The Mutagenic Effect of Nitrosamines and Nitroso Methyl Urea, Untersuchungen Uber Die Autagene Wirkung von Nitrosaminen Und Nitrosomethyl Harnstoff, Acta. Biol. Med. Ger., 10 (1963) 436-438
63. Fahmy, O. G., and Fahmy, M. J., Mutational Mosaicismin Relation to Dose With the Amine and Amide Derivatives of Nitroso Compounds in Drosophila, Mutation Res., 6 (1968) 139
64. Fahmy, O. G., Fahmy, M. J., Massasso, J., and Ondrej, M., Differential Mutagenicity of the Amine and Amide Derivatives of Nitroso Compounds in Drosophila Mutation Res., 3 (1966) 201
65. Veleminsky, J., and Gichner, T., The Mutagenic Activity of Nitrosamines in Arabidopsis Thaliana, Mutation Res., 5 (1968) 429
66. Geisler, E., Uber Die Wirkung Von Nitrosaminen Auf Mikroorganismen, Naturwissenschaften, 49 (1962) 380
67. Pogodina, O. N., O Mutagennoy Aktionosti Kancerogenoviz Gruppy Nitrosaminov, Cytologia, 8 (1966) 503
68. Marquardt, H., Zimmermann, F. K., and Schwaier, R., Die Wirkung Krebsauslösender Nitrosamine und Nitrosamide Auf Den Adenin-6-45-Rück Mutationssystem Von S. Cerevisiae, Z. Verebungslehre, 95 (1964) 82
69. Marquardt, H., Zimmermann, F. K., and Schwaier, R., Nitrosamide Als Mutagene Agentien, Naturwiss., 50 (1963) 625
70. Marquardt, H., Schwaier, R., and Zimmermann, F., Nicht-Mutagenität Von Nitrosaminen Bei Neurospora Crassa, Naturwiss., 50 (1963) 135
71. Malling, H. V., Mutagenicity of Two Potent Carcinogents, Dimethylnitrosamine and Diethylnitrosamine in Neurospora Crassa, Mutation Res., 3 (1966) 537

72. Yahagi, I., Nagao, M., Seino, Y., Matsushima, T., Sugimura, T., and Okada, M., Mutagenicities of N-nitrosamines on Salmonella, Mutation Res., 48 (1977) 121-130
73. Malling, H. V., Dimethyl Nitrosamine: Formation of Mutagenic Compounds by Interaction with Mouse Liver Microsomes, Mutation Res., 13 (1971) 425
74. Malling, H. V., and Frant, C. N., In Vitro Versus In Vivo Metabolic Activation of Mutagens, Env. Hlth. Persp., 6 (1973) 71
75. Czygan, P., Greim, H., Garro, A. J., Hutterer, F., Schaffner, F., Popper, H., Rosenthal, O., and Cooper, D. Y., Microsomal Metabolism of Dimethylnitrosamine and the Cytochrom P-450 Dependency of its Activation to a Mutagen, Cancer Res., 33 (1973) 2983
76. Bartsch, H., Malaveille, C., and Montesano, R., The Predictive Value of Tissue-Mediated Mutagenicity Assays to Assess the Carcinogenic Risk of Chemicals In "Screening Tests in Chemical Carcinogenesis" (eds) Montesano, R., Bartsch, H., and Tomatis, L., WHO/IARC Publ. No. 12, International Agency for Research on Cancer, Lyon (1976) pp. 467-491
77. Bartsch, H., Malaveille, C., and Montesano, R., In Vitro Metabolism and Microsome-mediated Mutagenicity of Dialkynitrosamines in Rat, Hamster, and Mouse Tissues, Cancer Res., 35 (1975) 644
78. Heath, D. F., The Decomposition and Toxicity of Dialkylnitrosamines in Rats, Biochem. J., 85 (1962) 72
79. Fahmy, O. G., Fahmy, M. J., and Wiessler, M., -Acetoxy-Dimethyl-Nitrosamin: Approximate Metabolite of the Carcinogenic Amine; Biochem. Pharmacol., 24 (1975) 1145
80. Rice, J. M., Joshi, S. R., Roller, P. P., and Wenk, M. L., Methyl (Acetoxymethyl) Nitrosamine: A New Carcinogen Highly Specific for Colon and Small Intestine (Abstract); 66th Meeting. Proc. Am. Assoc. Cancer Res., 16 (1975) 32
81. Wiessler, M., and Schmahl, D., Zur Carcinogenen Wirkling Von N-Nitroso-Verbindungen. V Acetoxymethyl-Methyl-Nitrosamin; Z. Krebsforsch. (1975) in press
82. Fahmy, O. G., and Fahmy, M. J., Mutagenic Selectivity of Carcinogenic Nitroso Compounds: III. N-Acetoxymethyl-N-Methyl Nitrosamine, Chem. Biol. Int., 14 (1976) 21
83. Montesano, R., and Bartsch, H., Mutagenic And Carcinogenic N-Nitroso Compounds Possible Environmental Hazards, Mutation Res. 32 (1976) 179
84. Neale, S., Mutagenicity of Nitrosamides and Nitrosamidines in Microorganisms and Plants, Mutation Res., 32 (1976) 229

CHAPTER 18

AROMATIC AMINES AND AZO DYES

A. Aromatic Amines

Benzidine, its salts, and analogs (e.g., substituted diaminobiphenyls, homologous toluidines and related derivatives) are of considerable importance as organic intermediates for the manufacture of a wide variety of organic chemicals and intermediates and dyes which are used in textile, leather and paper products.

1. Benzidine (4,4'-diaminobiphenyl, 4,4'-biphenyldiamine; 4,4'-diaminodiphenyl; $H_2N-\phi-\phi-NH_2$) is prepared primarily by the following sequential procedures as shown below: (1) reduction of nitrobenzene in the presence of NaOH and methanol to produce azoxybenzene; (2) reduction of azoxybenzene to hydrazobenzene in the presence of NaOH and zinc dust; (3) rearrangement of hydrazobenzene into benzidine hydrochloride in the presence of $HCl^{1,2}$. Impurities in the final product that can arise from the production of benzidine from hydrazobenzene include o,o- and o,p-diaminobiphenyls as well as semidines[3].

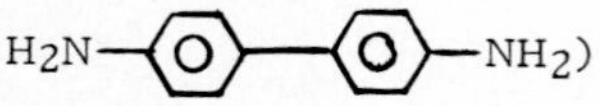

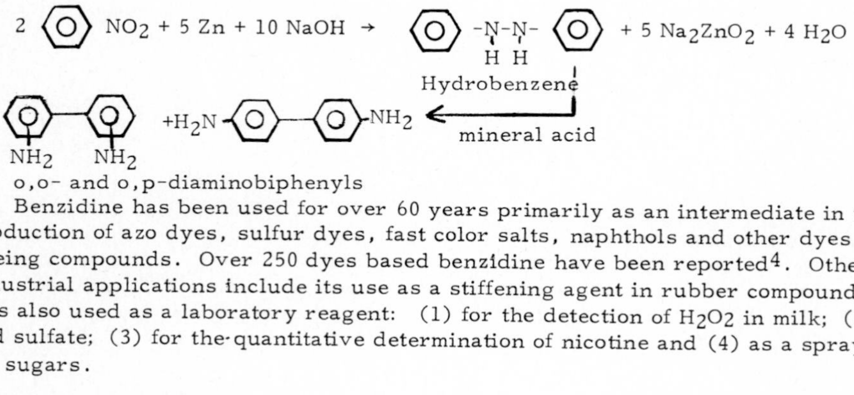

$$2\,\phi NO_2 + 5\,Zn + 10\,NaOH \rightarrow \phi\text{-N-N-}\phi + 5\,Na_2ZnO_2 + 4\,H_2O$$
$$\quad\quad\quad\quad\quad\quad\quad\quad\quad\quad\quad\quad\quad\quad\quad\text{H H}$$
$$\quad\quad\quad\quad\quad\quad\quad\quad\quad\quad\quad\quad\quad\quad\quad\text{Hydrobenzene}$$

$$\phi\text{-}\phi + H_2N\text{-}\phi\text{-}\phi\text{-}NH_2 \xleftarrow{\text{mineral acid}}$$
$$NH_2\ NH_2$$

o,o- and o,p-diaminobiphenyls

Benzidine has been used for over 60 years primarily as an intermediate in the production of azo dyes, sulfur dyes, fast color salts, naphthols and other dyes and dyeing compounds. Over 250 dyes based benzidine have been reported[4]. Other industrial applications include its use as a stiffening agent in rubber compounding. It is also used as a laboratory reagent: (1) for the detection of H_2O_2 in milk; (2) HCN and sulfate; (3) for the quantitative determination of nicotine and (4) as a spray reagent for sugars.

The estimates of the annual production of benzidine differ widely. For example, the U.S. production of benzidine in the U.S. was estimated to range from 1.5[5] to 10.4[6] million pounds. The number of manufacturers of benzidine in the U.S. and worldwide is not known definitively, but eight possible manufacturers have been identified. Three manufacturers estimate that they produce 45 million pounds of azo dyes from benzidine per year. The dyes are used by about 300 major manufacturers of textiles, paper and leather. The most important dye produced from benzidine is Congo Red (Direct Red 28, CI 22120)[7].

Estimates of the number of people exposed to benzidine in its production or use applications as well as levels of exposure is not definitively known. It has been suggested that 62 people in the U.S. are exposed to benzidine in its production per se[5,8].

OSHA in 1973 called for the regulation of 14 chemical compounds including the aromatic amine derivatives: benzidine, 3,3'-dichlorobenzidine, 4-aminodiphenyl, alpha- and beta naphthylamines, 2-acetylaminofluorene, and 4,4'-methylene-bis(2-chloroaniline), known to cause or suspected of causing human cancer from occupational exposure and hence there is no assigned TLV for these compounds in the U.S.[9].

Benzidine may enter the environment from benzidine production facilities from downstream chemical processing (primarily dye manufacture and application) and from use of products containing benzidine or benzidine derivatives. While industrial measurements in the past indicated that discharges of benzidine in waste water from production facilities usually did not exceed one pound per day, there have been significant episodes of accidental releases[7]. Benzidine is also believed present in sludge from industrial waste pretreatment plants with land disposal being the final method of disposal of this sludge in the past[7].

While the levels of benzidine emissions from downstream industrial processes are not well defined it is noted that free benzidine is present in the benzidine-derived azo dyes. Quality control specifications have required that the level not exceed 20 ppm and in practice the level of benzidine in azo dyes is usually below 10 ppm. Products dyes with the azo dyes could potentially release benzidine during their use[7].

It has been estimated that at the 300 uses facility sites a total of 450 pounds per year or about 1.5 pounds/year per facility of benzidine is discharged, assuming all of the free benzidine is discharged in the liquid effluent[7]. It has also been estimated that based on a 1972 production figure of 10.4 million pounds, a release rate of 0.02 million pounds of benzidine per year can be calculated. This figure probably includes atmospheric processes and fugitive emissions calculated from vapor pressure or other physical data[6].

Benzidine has been reported in the River Rhine. There are no reports of benzidine having been found in drinking water[7]. There is a paucity of data concerning the concentration of benzidine in ambient air or occupational atmospheres. In a 1951 study, benzidine was found at levels of 0.024 mg/m^3 in workroom atmospheres in a chemical plant[11].

Benzidine neither bioaccumulated nor was transferred through food chains to higher levels when tested in laboratory model ecosystems. It was degraded largely by N-acetylation and N-methylation. Benzidine levels in the test organisms were not affected appreciably by the presence of mixed function oxidase inhibitors[12].

Benzidine may enter the body by percutaneous absorption, by ingestion or by inhalation[2,13]. Epidemiological studies in several countries have shown that occupational exposure alone was strongly associated with bladder cancer[1,3,8,9,13-23]. In some studies, exposure to naphthylamine alone was similarly associated with bladder cancer.

Case et al[17,23] reported a high incidence of bladder tumors in workmen exposed to benzidine, or benzidine and aniline in British chemical factories, with a mean latent period of approximately 16 years (time was not influenced by severity or duration of exposure, but is characteristic of the causal agent). The susceptibility was found to be greater in older men; with exposure of even one year to benzidine found to be hazardous. Overall, there were 10 death certificates where only 0.72 would have been expected from the whole male population of England and Wales ($P < 0.001$). The morbidity was 34 cases[3,17]. Scott[14] found bladder tumors in 123 of 667 men exposed to benzidine for over 6 months. The latent period varied from one to 40 years, averaging 18 years. Moreover, in men exposed for 30 years or more, the incidence was 71%.

In a cohort study of a factory population it was reported that benzidine had an "attack rate" of 237 per 100,000 which was lower than that of 2-naphthylamine[20]. A retrospective survey by Goldwater et al[18] in 1965 of a single factory showed that 17 out of 76 workmen exposed to benzidine alone developed bladder tumors. von Ubelin and Pletscher[19] reported 20 cases of bladdertumors in Basel dyestuff workers exposed to benzidine and 2-naphthylamine. The average latent time for tumor development was 18.6 years. Benzidine was also said to be the cause of tumors of the stomach, rectum, lung and prostrate in this study[19].

Barsotti and Vigliani[21] reported that the maximum period of time from ending of benzidine exposure and tumor development in a group of Italian workers was 16 years with acute exposure not as important as chronic exposure. Processing of benzidine base was believed to most likely result in tumor formation.

A 13-year survey of workers in a benzidine plant in the U.S. was reported by Zavon et al[1] in 1973. Of the exposed group of 25 workers, 13 developed bladder tumors, with the average induction time being 16.6 years. Renal pelvic metastasis was observed in 3 cases.

Billiard-Duchesne[22] reported 17 cases of benzidine tumors and 15 cases of benzidine azo dye bladder tumor in French dyestuff workers.

Benzidine is carcinogenic in the mouse, rat and hamster, and possibly in the dog[3,8]. Cholangiomas and liver-cell tumors were found in rats fed 0.017% benzidine in the diet for 424 days[24], while benzidine and benzidine dihydrochloride administered to hamsters at levels of 0.1% in the diet throughout their life-span induced cholangiomas hepatomas and liver-cell carcinomas[25,26]. Bladder carcinoma has been induced after 7 to 9 years in 3 of 7 dogs given a total dose of 325g in 5 years (200 and then 300 mg/day, 6 days/week)[27,28].

Earliest mutagenic studies of benzidine have shown it to be converted to a bacterial mutagen by a liver enzyme preparation in the Salmonella/microsome assay, although the mutagenic metabolite(s) were not identified[20]. Benzidine is mutagenic in S. typhimurium TA 1538[30,31,33] and TA 98[32,33] only in the presence of liver post-mitochondrial S-9 fractions from induced pretreated rats[30,31,33] and uninduced rats[32]. In a comparative in vitro metabolic activation study of benzidine and 3,3'-dichlorobenzidine by rodent and human liver preparations benzidine was totally inactive, as a mutagen without activation in S. typhimurium TA 98 and TA 1538[34]. 3,3'-Dichlorobenzidine possessed much greater activity per n Mole than did benzidine and was mutagenic as well without activation, but that activity was greatly enhanced by metabolic activation. Human liver preparations were much less efficient in activating benzidine and 3,3'-dichlorobenzidine than were rat liver preparations. In experiments to detect any synergistic effects between benzidine and 3,3'-dichlorobenzidine, combinations of the two compounds exhibited greatly reduced mutagenic activity compared to the same concentrations of each compound when plated separately[24].

Benzidine in the absence of exogenous microsomal activation was found active in a differential toxicity test utilizing DNA repair proficient VA-13 human lung fibroblasts and repair dificient skin fibroblasts from xeroderma pigmentosum patient as a means of identifying chemicals which cause u.v. mimetic damage to DNA. This implies that the cells contain levels of enzymes which can activate aromatic amines (e.g., benzidine and 2-aminofluorene) to metabolites capable of DNA damage[34].

When benzidine was compared with that of two hair dye components (m-toluenediamine and 4-nitro-o-phenylenediamine) in <u>Drosophila melanogaster</u>, the mutagenicities and selectivities of the test compounds for rDNA gradually decreased in the order: benzidine > m-toluenediamine > 4-nitro-o-phenylenediamine, which correlated with the evidence, thus far, about their carcinogenicities[35]. In this study, the compounds were injected at equimolar dose ranges (5-20 mM) around the testes of adult males and their mutagenicities were measured separately in the various stages of spermatogenesis. All compounds exerted decisive mutagenicity both on the X-chromosome (induction of lethal and visible recessives) and the RNA genes, although their activities on the different genic sites varied between compounds and as a function of cell stage, but not in response to changes in dose within the investigated molarity range[35].

Benzidine has also been found mutagenic in the micronucleus test in rats inducing high incidences of micronucleated erythrocytes following both dermal application and subcutaneous injection[36].

Although benzidine is a recognized bladder carcinogen in exposed workers[1,3,8,9,13-23], and is a carcinogen in several species of laboratory animals[3,8,24-28] as discussed earlier, the nature of the <u>precise</u> mechanisms responsible for the induction of neoplasia following exposure to diverse aromatic amines is not known[37].

Clayson and Garner[37] recently reviewed the mechanism of activation of aromatic amines. Evidence exists that the metabolism of these compounds occurs via ring hydroxylation, N-hydroxylation of the monoacetyl derivative, and conjugation with sulfate and glucuronic acid[37-42]. On present evidence, N-hydroxylation is a prerequisite for carcinogenicity for this group of compounds this may be followed by esterification of the hydroxyl group in some but probably not all cases[36]. The final intermediate is unstable and breaks down to an electrophile which intereacts with the

tissue, viz., $ArNHR \rightarrow ArN\begin{subarray}{l}R\\OH\end{subarray}$ Ar = aryl; R= H, acyl or alkyl

$ArN\begin{subarray}{l}\\+\end{subarray} \xleftarrow{-R} ArN\begin{subarray}{l}R\\OX\end{subarray}$ X = ester group

In recent studies of hepatic microsomal N-glucuronidation and nucleic acid binding of N-hydroxy arylamines in relation to urinary bladder carcinogens it was found by Kadlubar et al[43] that arylamine bladder carcinogens are N-oxidized and N-glucuronidated in the liver and that the N-glucuronides are transported to the urinary bladder.

The hydrolysis of the glucuronides to N-hydroxy arylamines and the conversion of the latter derivatives to highly reactive electrophilic arylnitrenium ions in the normally acidic urine of dogs and humans may be critical reactions for tumor induction in the urinary bladder (Figure 1).

An additional summary of the metabolic pathways by which aromatic amines may modify nucleic acids and proteins is shown in Figure 2 using 4-aminobiphenyl as an illustrative example[44].

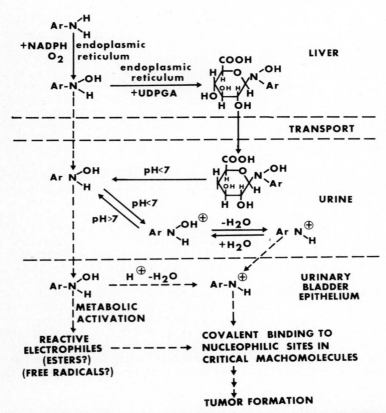

Chart 1 FORMATION AND TRANSPORT OF POSSIBLE PROXIMATE AND ULTIMATE CARCINOGENIC METABOLITES OF ARYLAMINES FOR THE INDUCTION OF URINARY BLADDER CANCER. Ar, ARYL SUBSTITUENT

FIGURE 2

PATHWAYS BY WHICH 4-AMINOBIPHENYL MAY BE ACTIVATED METABOLICALLY*

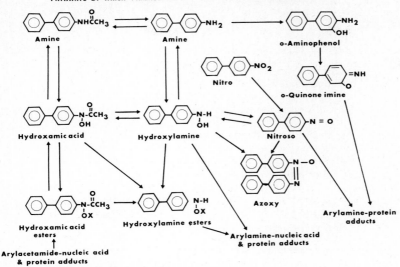

* Baetcke, K., Aromatic Amines Program, Mechanistic Approaches to Carcinogenics;
  NCTR Report, Nov. 6 (1976)

Tables 1, 2 and 3 summarize the biotransformation of benzidine and the physiologic changes induced by benzidine and congeners in various species respectively.

2. **3,3'-Dichlorobenzidine** (O,O'-dichlorobenzidine, 4,4'-diamino-3,3'-dichlorobiphenyl; 3,3'-dichloro-4,4'-biphenyldiamine; DCB; H_2N-⌬-⌬-NH_2 with Cl, Cl substituents) has been produced commercially for approximately 40 years. It can be produced by the following procedures[45]: (1) reduction of ortho-nitrochlorobenzene to the hydrazo-compound with subsequent rearrangement to DCB using mineral acid, and (2) via chlorination of diacetylbenzidine with hypochlorite followed by hydrolysis of the diamide to the diamine using HCl.

Production of DCB base and salts amounted to 4.612 million pounds in 1972, the latest year in which a production figure was reported[46]. Production figures were withheld for proprietary reasons in the U.S. since only two companies reported commercial production of DCB base and salts in recent years. Imports of DCB base and salts into the U.S. in 1972 and 1973 amounted to 664,085 and 291,687 pounds respectively and imports of this substance were not reported since[46]. Other countries that have produced DCB and its salts in recent years are Federal Republic of Germany, Italy, Japan[45] and the United Kingdom[46].

TABLE 1

Summary of Benzidine Biotransformation in Various Species[3]

Species	Metabolites
Mouse	Monoacetylated 3-OH ethereal sulfate Monoacetylated 3-OH glucuronide N-Hydrogen sulfate and/or glucuronide 3-OH-Benzidine glucuronide
Rat	3,3'-Dihydroxybenzidine (?) 4'-Acetamido-4-amino-3-diphenylyl hydrogen sulfate 4'-Amino-4-diphenylyl sulfamic acid 4'-Acetamido-4-diphenylyl sulfamic acid
Guinea pig	4'-Acetamido-4-aminodiphenyl N-glucuronide 4'-Acetamido-4-amino-3-diphenylyl hydrogen sulfate
Rabbit	3'-OH-Benzidine sulfate and glucuronide 4'-Acetamido-4-amino-3-diphenylyl hydrogen sulfate 4'-Amino-4-diphenylyl sulfamic acid 4'-Acetamido-4-diphenylyl sulfamic acid N-Glucuronides 4'-Acetamido-4-aminodiphenyl 3-OH-Benzidine
Dog	3-OH-Benzidine 3-OH-Benzidine hydrogen sulfate 4-Amino-4-hydroxybiphenyl Mono- and diacetylbenzidine 4,4'-Diamino-3-diphenyl sulfate and glucuronide
Monkey	Monoacetylbenzidine
Man	3,3'-Dihydroxybenzidine (?) Mono- and diacetylbenzidine 3-OH-Benzidine N-Hydroxy acetylaminobenzidine

TABLE 2

Types of Physiologic Changes in Various Species[3,8]

Species	Carcinogen	Physiologic change
Mouse	Benzidine	Hepatoma, lymphoma, bile duct proliferation
	3,3'-Dihydroxybenzidine	Hepatoma, lymphoma, bile duct proliferation, benign bladder papilloma
Rat	Benzidine and its sulfate	Cirrhosis of liver, hepatomas, carcinoma of Zymbal's gland, adenocarcinoma, degeneration of bile ducts, sarcoma, mammary gland carcinoma
	3,3'-Dichlorobenzidine	Extensive cancer
	3,3'-Dimethyoxybenzidine	Intestinal, skin, Zymbal, gland carcinoma, bladder papilloma
	3,3'-Dihydroxybenzidine	Hepatoma, adenocarcinoma of colon, carcinoma of fore stomach, Zymbal's gland carcinoma, bladder carcinoma
	Dianisidine	Zymbal's gland carcinoma, ovarian tumor
	o-Ditoluidine	Papilloma of stomach, Zymbal's gland carcinoma, mammary tumor leucoses
	3,3'-Benziniedioxyacetic	Papilloma of bladder, hepatic sarcoma
	N,N'-Diacetylbenzidine	Chronic glomerulonephritis
Hamster	Benzidine	Hepatoma, liver carcinoma, cholangiomas
	3,3'-Dichlorobenzidine	Transitional cell carcinomas of the bladder, liver cell tumors
	3,3'-Dimethoxybenzidine	Fore stomach papilloma, urinary bladder tumors
	o-Ditoluidine	Bladder cancer
Rabbit	Benzidine	Proteinuria hematuria, liver cirrhosis, myocardial atrophy, bladder tumor, gall bladder tumor
Dog	Benzidine	Recurrent cystitis, bladder tumor, convulsions, liver cirrhosis, hematuria
Monkey	Benzidine	No pathology
Man	Benzidine	Bladder tumor, papilloma, chronic cystitis, hematuria

TABLE 3. CARCINOGENICITY OF BENZIDINE AND ITS DERIVATIVES

Compound	Species	Route[a]	Adequacy[b]	Local	Bladder	Kidney	Liver	Intestine	Ear Duct	Breast	Other
Benzidine	dog	O	A	-	?+	-	-	-	-	-	---
	hamster	O	S	-	-	-	+	-	-	-	---
	rat	S.C.	A	-	-	-	+	+	+	-	---
	mouse	S.C.	S	-	-	-	+	-	-	-	---
Diacetylbenzidine	rat	O	S	-	-	-	-	-	-	?	---
o,o'-Tolidine	rat	S.C.	A	+	-	-	-	-	+	+	lymphoma, skin
o,o'-Dianisidine	hamster	O	S	-	?	-	-	-	-	-	forestomach
	rat	O	S	-	-	-	-	-	?	-	---
3,3'-Dichlorobenzidine	rat	O/S.C.	A	-	+	-	-	+	-	+	skin, lymphoma
	hamster	O	S	-	+	-	+	-	-	-	bone
3,3'-Dihydroxybenzidine	rat	O/S.C./Top	S	-	-	-	-	/,	-	-	non significant
3,3'-Benzidinedicarboxylic acid	rat	S.C.	S	-	-	-	?	-	-	?	?
	mouse			-	-	-	-	-	-	-	hemopoietic tissue
2-Methyldiacetylbenzidine	rat	O	S	-	-	-	-	+	+	+	---

[a] O, oral; Top, topical; S.C., subcutaneous injection
[b] A, tested in more than one institute; S, evidence less convincing
[c] +, tumors reported; -, tumors not reported; ?, evidence equivocal.

DCB, as well as the hydrochloride and sulfate salts, are used principally as intermediates in the dyestuffs industry. DCB can be used as an intermediate in the production of 13 dyes or pigments[47,48]. In the U.S., eight pigments were produced commercially in substantial quantities. The 1975 U.S. production quantities of four of these pigments were: C. I. Pigment Yellow 12, 6.028 million pounds; Pigment Yellow 14, 1.840 million pounds; Pigment Yellow 17, 415 thousand pounds; Pigment Yellow 13, 240 thousand pounds. These and Pigment Yellow 55 are used widely in printing inks, plastics, elastomers and in textile printing but are of secondary importance in the pain industry[46].

Of the remaining three pigments, C. I. Pigment orange, 209 thousand pounds produced in 1975, is used mainly to color rubber products, printing inks, paper and all types of plastic while C.I. Pigment Orange 34 (99 thousand pounds) is used in decorative finishes, transparent metal coatings, tin printing, foil lacquers, vinyl, polyethylene, and polypropylene plastics and textile printing inks. C.I. Pigment Red 38 is used primarily in rubber, plastics, inks and paints[46].

The demand for DCB-based pigments in the U.S. is expected to increase, especially if possible OSHA legislation should eliminate the use of lead chromate pigments (approx. 80 million pounds per year). The yellow pigments derived from DCB and its salts could be used as substitutes for the lead chromate pigments[46].

More recently, DCB has come to be used as a curing agent for isocyanate-containing polymers (e.g., polyurethane elastomers). In this application, DCB has been used alone or in blends with 4,4'-methylenebis(2-chloroaniline) (MOCA). U.S. consumption of DCB as a curing agent is believed to be no more than a few hundred thousand kg per year[45,46].

It has been suggested that about 2500 individuals are exposed to DCB in its production or use applications[8]. DCB is subject to an Emergency Temporary Standard on certain carcinogens under an order of OSHA on 26 April, 1973. It is also listed as a controlled substance in the U.K. Carcinogenic Substances Regulations 1967-Statutory Instrument (1967) No. 879[45].

While no definitive information is present concerning DCB in the general environment it may be present in the waste streams from plants where it is produced or used in the productions of dyes and pigments[45]. It may also be emitted into waste gases and water from its use in the production of cured polyurethane elastomers since curing agents are often melted before mixing into elastomer formulations and less than stoichiometric amounts of reactants are employed[45].

It is uncertain whether DCB should be accepted as a human carcinogen. While no epidemiological data are available, and since the production of DCB is frequently associated with the production of benzidine, the possibility cannot be excluded that DCB has contriubted to the incidence of bladder cancer attributed to benzidine[45]. Hence, it is highly probable that DCB has significant human carcinogenic potential[23]. However, it should also be noted that studies from one plant manufacturing this compound have thus far indicated no carcinogenic hazard to exposed workers who have worked with the compound for up to 30 years[49].

DCB is carcinogenic in the rat producing a high incidence of adenomas and carcinomas of the Zymbal gland and other organs (e.g., bladder) in animals fed 10-20 mg DCB in the diet 6 times/week for 12 months (total dose 4.5 g/rat)[50]. When male and female CLR-CF rats were fed 1,000 ppm DCB in a standard diet (23% protein) malignant mammary, skin, and acoustic duct tumors developed in both sexes and an excess of haemopoietic tumors arose in males compared to their incidence in control animals[51].

In lifetime studies in which Golden Syrian hamsters (30 male and 30 female) were fed 0.3% DCB in the diet, 4 transitional cell carcinomas of the bladder and some liver cell tumors were found which were not observed in control animals. In earlier studies where the dietary levels of 0.1% DCB in the diet were fed to a similar number of animals no tumors were found[25,26].

Subcutaneous administration of 15-60 mg DCB/rat in sunflower seed oil or glycerol and water at unspecificied intervals for 10 to 13 months has resulted in skin, sebaceous and mammary gland as well as (less frequently) intestinal, urinary bladder and bone tumors[52].

DCB is mutagenic without metabolic activation in S. typhimurium TA 1538[30,32] and TA 98[32]. However, in TA 1538 the mutagenic activity was increased over 50-fold by addition of a phenobarbitone pretreated rat liver mixed function oxidase preparation. In the pesence of the liver preparation DCB was approximately 10 times more active than benzidine, while, 3,3',5,5'-tetrafluorobenzidine was of approximately equipotency[30]. DCB and its dihydrochloride salt were found to be the most active mutagens compared to benzidine, 3,3'-dimethylbenzidine, 3,3'-dimethoxybenzidine, and 4-aminobiphenyl (as well as their respective salts) when tested at 50 and 100 µg/plate in S. typhimurium TA 98 with S-9 fractions from uninduced adult male BALB/C mice[32]. In this study, 3,3'-dimethoxybenzidine was the least mutagenic.

The metabolism of benzidine and DCB was compared in rats, dogs and monkeys after i.v. administration. Benzidine was excreted much more rapidly than was the dichloro derivative. The rat probably excretes both compounds via the bile. The dog excretes very little dichlorobenzidine via the urine and utilizes the same excretory mechanism of the rat[53].

A possible relationship between carcinogenicity of benzidine and its congener and their ability to reduce cytochrome C was recently reported by Hirai and Yasuhira[54]. Whereas benzidine, diorthotoluidine and dianisidine were moderate reductants, 3,3'-dichlorobenzidine could not reduce cytochrome C.

3. **3,3'-Dimethylbenzidine** (o-tolidine; 3,3'-tolidine; 4,4'-diamino-3,3'-dimethyl-biphenyl; 3,3'-dimethyldiphenyl-4,4'-diamine; H_2N-⟨○⟩(CH$_3$)-⟨○⟩(CH$_3$)-NH_2) and its salts have been widely used in the manufacture of dyestuffs and pigments based on the coupling of the tetrazotized base with phenols and amines. (More than 95 dyes are derived from o-tolidine). It has also been used as a laboratory reagent for the detection of blood and for the colorimetric determination of chlorine in air and water[55].

There are no production data available for o-tolidine in recent years. In 1962, the production in the U.S. was reported as 243 thousand pounds. Imports of o-tolidine into the U.S. have ranged from 2.3 thousand pounds in 1959 to 134.6 thousand pounds in 1962. In 1969 and 1970 the imports of o-tolidine into the U.S. were 80.6 and 97.8 thousand pounds respectively[55].

There are no data available as to the levels of o-tolidine in the environment, TLV's and MAC's for any country or any epidemiological studies.

o-Tolidine may enter the body by percutaneous absorption, ingestion or inhalation[13].

Purified o-tolidine has been found to be a systemic carcinogen in the rat (Sherman strain) when administered subcutaneously at a dose level of 60 mg/rat/week (total dose 5.5g). In contrast to rats given benzidine, no cirrhosis or hepatomas were observed among those receiving o-tolidine; Zymbal gland carcinomas were predominant[56]. Commercial o-tolidine did not induce tumors at a dietary level of 0.1% (3.0 g/animal per year) fed to male and female hamsters throughout their lifespans[25,26].

o-Tolidine with metabolic activation is weakly mutagenic in S. typhimurium TA 1538[31,32] and TA 98[32].

There is a paucity of data concerning the biotransformation of o-tolidine. Diacetyl-o-tolidine and a hydroxyamino metabolite (probably 5-hydroxy-o-tolidine) were identified in the urine of workers manufacturing o-tolidine[57].

Following i.p. administration of 70-100 mg/kg bw of o-tolidine to dogs, free o-tolidine (4%) and 40% of a metabolite (probably 5-ethereal sulfate of o-tolidine) were recovered in the urine within 3 days[58]. (These results confirm the fact that the dog is unable to acetylate aromatic amines)[55].

4. 3,3'-Dimethoxybenzidine (o-dianisidine; 4,4'-diamino-3,3'-dimethoxybiphenyl; di-p-amino-di-m-methoxydiphenyl; $H_2N\text{-}\phi(OCH_3)\text{-}\phi(OCH_3)\text{-}NH_2$) is prepared commercially via the reduction of the methyl ether of ortho-nitrophenol (ortho-nitroanisole) to hydrazoanisole thence rearrangement by acid to o-dianisidine[59]. o-Dianisidine has been produced commercially for at least 50 years. The last data on production in the U.S. of o-dianisidine was in 1967 when the total production of 5 companies amounted to 167 thousand kg[59,60]. Only two U.S. companies were producing o-dianisidine by 1971 and 124 thousand kg of the chemical was imported in that year[59,61].

There are no available data on the quantity of o-dianisidine produced in countries other than the U.S. Other countries known to have produced (or are still producing) o-dianisidine are: Federal Republic of Germany, one producer in 1967; Italy, one producer in 1969; the United Kingdom, two producers in 1970, and Japan, three producers in 1972[59].

The principal use for o-dianisidine and its dihydrochloride salt is in the production of dyes. o-Dianisidine can be used as an intermediate in the production of 89 dyes[62]. Eight of these dyes were produced in the following amounts (thousands of kg) in 1971[59,63,64]: Direct Blue 218, 479; Pigment Orange 16, 153; Direct Blue 1, 136; Direct Blue 15, 94; Direct Blue 8, 64; Direct Blue 76, 53; Direct Blue 98, 39 and Pigment Blue 25, 87. The o-dianisidine based dyes and pigments are widely used for dyeing leather, paper, plastics, rubber and textiles[48,59].

Another major application of o-dianisidine is as an intermediate in the production of o-dianisidine diisocyanate (3,3'-dimethoxy-4,4'-diphenylene diisocyanate and

3,3'-dimethyoxybenzidine-4,4'-diisocyanate). It is estimated that less than 500 thousand kg of o-dianisidine diisocyanate is produced per year by the only U.S. producer, and that it is used as a component of polyurethane elastomers and in isocyanate-based adhesive systems[59].

Other reported used of o-dianisidine include: the dyeing of acetate rayon; and the detection of a number of metals, thiocyanates and nitrites[65].

The number of individuals employed in the manufacture of o-dianisodine or its use applications are not known nor is information available as to its possible dissipation environmental levels. o-Dianisidine is listed as a controlled substance in the U.K. Carcinogenic Substances Regulations of 1967[59].

No conclusive epidemiological studies have been reported conerning the carcinogenicity of o-dianisidine in man[59]. Hueper[66] stated that the total of bladder tumors in dye workers had risen to 100 in 1938 and among the chemicals involved were benzidine[3], o-ditoluidine and dianisidine.

o-Dianisidine was carcinogenic in rats administered doses of 30 mg in sunflower seed oil by stomach tube 3 times/week for 13 months producing tumors of the Zymbal gland in addition to an ovarian tumor and a fibroadenoma of the mammary gland[67,68]. o-Dianisidine was also found carcinogenic in male and female Fischer rats in doses of 1, 3, 10 and 30 mg per animal in a "steroid" suspending vehicle[11,69].

o-Dianisidine has slight mutagenic activity in the presence of a liver mixed function oxidase preparation, but none in its absence when tested in S. typhimurium TA 1538[30,31] and TA 98[32].

Information relative to the biotransformation of o-dianisidine is extremely limited. It has been found in the urine of workers exposed to this compound[13,70,71]. Following administration of a dose of 1g of o-dianisidine to dogs, 0.4% of the free diamine and about 5% of a metabolite with properties similar to those of 3,3'-dihydroxybenzidine were isolated in the urine[58].

5. <u>1-Naphthylamine</u> (alpha-naphthylamine; 1-aminonaphthalane) has been produced commercially for approximately 50 years. In recent years it is believed to have been made exclusively via the catalytic hydrogenation of 1-nitronaphthalene using a nickel catalyst. Earlier procedures involved the reduction of 1-nitronaphthalene with iron and HCl[72]. It is not easily produced in pure form but usually contains a small proportion of the isomeric 2-naphthylamine as an impurity. Within the U.S., it is believed that 1-naphthylamine produced after 1953 did not contain more than 0.5% 2-naphthylamine as an impurity[23]. Certain grades of 1-naphthylamine produced in Europe probably contained up to 10% until its manufacture was finally abandoned[23].

Total U.S. production of 1-naphthylamine in 1948 (the last reported year) was 2.6 million kg[73]. It was estimated that U.S. production in 1963 was more than 500 thousand kg[74] and by 1972 only one U.S. company was still producing 1-naphthylamine[72]. Imports of 1-naphthylamine to the U.S. have grown from 2 thousand kg in 1967[75] to 27 thousand kg in 1971[72,76].

Although no production data are available for 1-naphthylamine produced elsewhere, other countries that have been involved with its production include: The Federal Republic of Germany (2 producers in 1967); the United Kingdom (2 producers in 1970); Italy (3 producers in 1969); and Japan (3 producers in 1972)[72].

The major areas of utility of 1-naphthylamine is in the production of dyes, herbicides and antioxidants. 1-Naphthylamine can be used in the production of a substantial number of dyes (estimated to range from 50[77] to 150[78]).

It is used as an intermediate in the manufacture of 1-naphthylamine-4-sulphonic acid (naphthionic acid) which is further used to make 1-naphthol-4-sulphonic acid, a coupling agent for azo dyes[72]. The total U.S. production of naphthionic acid in 1969 was 81 thousand kg[72,79].

The major herbicide produced from 1-naphthylamine is N-1-naphthylaminephthalamic acid, an estimated 450 thousand kg of which was used on all crops in 1966[72,80].

The rubber oxidants produced from 1-naphthylamine are N-phenyl-1-naphthylamine and the aldol-1-naphthylamine condensate. Other areas of utility of 1-naphthylamine are in the production of 1-naphthol and in the production of the rodenticide 1-naphthylthiourea (ANTU) (no longer produced in the U.S.)[72].

The number of workers that are employed in the production of 1-naphthylamine as well as its use applications are not known.

1-Naphthylamine is subject to an Emergency Temporary Standard on certain carcinogens under an order by OSHA on 26 April, 1973. 1-Naphthylamine containing less than 1% of 2-naphthylamine as a byproduct of a chemical reaction is a controlled substance under the U.K. Carcinogenic Substances Regulations 1967-Statutory Instrument No. 879[72].

Meagre data exists as to the environmental occurrence of 1-naphthylamine. It may be present in the waste streams from plants where it is produced or used. It has been found in coal-tar[81] as well as in cigarette smoke (e.g., 0.03 µg of 1-naphthylamine per cigarette)[82].

Occupational exposure to commercial 1-naphthylamine containing 4-10% 2-naphthylamine is strongly associated with bladder cancer[72]. Scott[83] recorded a total of 56 reported cases of bladder cancer attributed to 1-naphthylamine exposure and concluded that present evidence tends to incriminate 1-naphthylamine as a carcinogen. Case et al[17] reported that 5 years exposure to commercial 1-naphthylamine was necessary for the induction of bladder tumors. Six deaths were certificated as due to bladder cancer among 1-naphthylamine workers who had not also been engaged in the production of 2-naphthylamine or benzidine, where as only 0.70 would have been expected ($P < 0.005$). Other cases of bladder tumors attributed to 1-naphthylamine have been reported by Wignall[84], Evans[85], Gehrman et al[86], DiMaio[87], Barsotti[88], Billiard-Duchesne[89], and Goldblatt[90]. Nevertheless, the possibility that its carcinogenic effect is attributable to its 2-naphthylamine impurity cannot yet be discounted[23].

1-Naphthylamine has thus far induced only one bladder tumor in the dog experiments carried out by Bonser et al[91] after a prolonged exposure of 9 years. (Doses of 0.5 gm of 1-naphthylamine were given to 2 mongrel dogs 3 times/week for life). The experiments in dogs suggest that 1-naphthylamine, if carcinogenic at all, is less so to the bladder in this species than is the 2-isomer[72].

No carcinogenic effect was noted when 1-naphthylamine was administered in the diet at concentrations of 0.1% for life and 1% for 70 weeks to Syrian golden hamsters[25,26].

The carcinogenicity of N-(1-naphthyl)-hydroxylamine a metabolite of 1-naphthylamine in rats by i.p. administration[92] and in mice by bladder implantations[93] has been reported. Although 1-naphthylamine can be considered non-carcinogenic, or only weakly carcinogenic in animals, its N-hydroxy derivative is considerably more potent than 2-naphthylamine[92].

1-Naphthylamine is mutagenic in the Salmonella/microsome assay[94], and inactive in the transformation of Golden Syrian hamster embryo cells test[95]. N-hydroxy-1-naphthylamine although toxic was not mutagenic to intracellular T-4 phage[96]. Earlier studies indicated that N-hydroxy-1-naphthylamine caused mutations in bacteria[97,98], however, the significance of the mutagenesis data reported in these studies[97,98] was suggested to be marginal (e.g., less than a 10 fold increase in the frequency of revertants in back-mutation experiments).

The activation of N-hydroxyarylamines under acidic conditions may be a critical reaction in urinary bladder carcinogenesis by arylamines. Kadlubar et al[43,99] reported that N-glucuronides of N-hydroxy-1-naphthylamine were rapidly hydrolyzed to the N-hydroxy-1-naphthylamine and thenconverted to reactive derivative(s) capable of binding to nucleic acids. [G-^{3}H]N-hydroxy-1-naphthylamine was reacted with nucleic acids at pH 5 to form bound derivatives (3-10 naphthyl residues/10^3 nucleotides). The degree of binding was in the order: DNA > poly G > rRNA > sRNA > polA. Direct evidence for H$^+$-dependent electrophilic arylnitrenium ion formation (Figure 1) from N-hydroxy-1-naphthylamine and its selectivity toward macromolecules was further elaborated. It was suggested that this conversion to arylnitrenium ions in the normally acidic urine of dogs and humans may be critical reactions for tumor induction in the urinary bladder by a number of aromatic amines.

6. <u>2-Naphthylamine</u> (beta-naphthylamine; 2-aminonaphthalene, BNA, Fast Scarlet Base B; [structure with NH_2]) has been produced commercially for at least 50 years by what is believed to be the most common procedure, the reaction of 2-naphthol with ammonia and ammonium sulphite[72]. The commercial product can contain a number of polyaromatic heterocyclic compounds including: 3,4,5,6-dibenzophenazine (which is formed from 2-naphthylamine in the presence of air), and 2-amino-1,4-naphthoquinone-N_4-2-naphthylamine has also been reported as a contaminant[92].

Although commercial production of 2-naphthylamine has continued until early 1972[100], the production in recent years is believed to have been quite small[72]. The last available data for U.S. production was for the year 1955 when a total of 581 thousand kg were produced by 4 manufacturers[101]. The last reported U.S. imports of 2-naphthylamine of significance was in 1967 when they amounted to 17.4 thousand kg[72].

It is estimated that about 100 million kg of MDA is manufactured annually in the U.S. Approximately 90% of the MDA produced is consumed in crude form at its production site for the preparation of isocycnates and polyisocycnates[146]. The annual U.S. production of isolated MDA has been estimated to have been about 1 million kg in recent years[145]. In 1970, U.S. production of polymethylene polyphenylisocyanate is estimated to have been 52 million kg, while U.S. production of the refined 4,4'-methylenediphenylisocyanate was about 5.5 million kg[145]. NIOSH estimates that 2,500 workers are exposed to MDA in the U.S.[146].

While no data are available on the quantity of MDA produced in countries other than the U.S., the Federal Republic of Germany was reported to have one producer in 1967; the U.K. 3 producers in 1970, and Japan 2 producers in 1972[145].

Information regarding the carcinogenicity and mutagenicity of MDA is scant. An increase in the incidence of liver tumors in Wistar rats injected s.c. with doses of 30-50 mg/kg MDA at 1 to 3 week intervals over a period of 700 days (average total dose, 1.4 g/kg) was reported in one study[147]. In a later study by the same investigators, repeated s.c. injected over 410 days of 200-400 mg/kg (total dose, 7.3 g/kg) did not increase the incidence of benign and malignant tumors in treated rats compared with the incidence in controls. No tumors were found before 2 years in rats given MDA by gastric intubation in arachis oil 5 days/week for 121 days (total dose, 3.3 g/kg)[148]. Severe hepatoxic effects were noted among 84 persons who consumed bread baked from accidently contaminated flour with MDA[149].

9. **4,4'-Methylene bis(2-chloroaniline)** (4,4'-diamino-3,3'-dichlorodiphenylmethane; p,p'-methylenebis(ortho-chloroaniline); MOCA; DACPM; MBOCA;

$H_2N-\langle O \rangle-CH_2-\langle O \rangle-NH_2$) is prepared commercially by the reaction of formaldehyde with ortho-chloroaniline[145]. The commercial product can contain approximately 90% 4,4'-methylene bis(2-chloroaniline), about 10 percent polyamines

$(H_2N-\langle O \rangle-CH_2-\langle O \rangle-CH_2-\langle O \rangle-NH_2)$ and approximately 0.9% orthochloroaniline[150].

It is believed that virtually all the DACPM (perhaps 99%) produced in the U.S. is used as a curing agent for isocyanate-containing polymers. It is perhaps the most widely used agent for curing liquid-castable polyurethane elastomers suitable for molded mechanical articles and for potting and encapsulating purposes[151]. It is also frequently formulated with 3,3'-dichlorobenzidine or 4,4'-methylenedianiline to prepare curing agents. Perhaps less than 1% of the total consumed DACPM is believed to be used as a curing agent for epoxy and epoxy-urethane resin blends[145].

DACPM has been produced in quantity for approximately 15 years. It is estimated that the U.S. production for 1970 and 1972 was in the order of 1.5-2.5 million and approximately 3.5 million kg respectively[145]. There are indications that it is produced in the U.S. by two companies. Although no production data are available for DACPM produced in other countries other than the U.S., the U.K. was reported in 1970 to have one producer, while Japan was reported to have 3 producers in 1972[145].

The total number of workmen involved in the production of DACPM as well as its use application is not known. In one plant in the U.S., it was reported that over a 17 year period, 209 employees have had potential contact with this agent for varying periods of time, and seventy of exposure ranging from the casual experience to routine assignment in the finishing and packaging area[152].

DACPM has been subject to an Emergency Temporary Standard for certain carcinogens under an order of OSHA on 26 April, 1973[9], and a permanent standard in 1974[9]. However, the DACPM standard has been remanded for procedural reasons in December, 1974. Although at present OSHA has no standard or threshold limit value for DACPM, it can cite employers for excessive worker exposure under the "general duty clause" of the job safety and health act[153]. DACPM is listed as a controlled substance in the U.K. Carcinogenic Substances Regulations 1967-Statutory Instrument No. 879[145].

There are no data available regarding levels of DACPM in work atmospheres or in the general environment.

In the only epidemiological study reported involving 31 men whose exposure to DACPM ranged from 6 months to 16 years, no cytological evidence of a bladder cancer hazard was found[152]. Additionally, no cases of bladder cases were found among 178 other workers with unstated duration of exposure to DACPM who were also examined cytologically more than 10 years after exposure had ceased[145,152].

DACPM is carcinogenic in the dog[150], mouse[154] and rat[155-157] after oral administration. Of six female beagle dogs given a daily oral dose of 100 mg of DACPM (about 90% 4,4'-methylene bis(2-chloroaniline) about 10% polyamines and approximately 0.9% o-chloroaniline) for up to nine years, four developed bladder cancer, one developed cancer of the urethea, and three also developed "pre-malignant liver changes"[150]. While a _definitive_ comparison of the carcinogenic potency in dogs of DACPM with other aromatic amines cannot be made due to factors such as cystoscopic examination, daily dose, total dose, length of dosing period, and length of the observation period after dosing were not constant. Stula et al[150] estimated the following order of increasing potency: (1) benzidine, (2) DACPM, (3) 3,3'-dichlorobenzidine; (4) 2-acetylamino-fluorene, (5) 2-naphthylamine and (6) 4-aminobiphenyl. In making this estimate, the early appearance of tumors, using a low dose, as well as metastases to other organs, was considered to be an indicator of high carcinogenic potency. Production of tumors in more than one body system is also considered to be an indicator of greater carcinogenic potential.

The neoplastic effect of DACPM on the lung, liver, and the mammary gland noted in rats fed a daily diet containing 1000 ppm of DACPM for up to two years[157], was not seen in dogs[150]. This may represent a species difference in the metabolism and excretion of DACPM.

DACPM was found by Russfield et al[154] to be carcinogenic in CLR-CD-1 rats and mice. Mice fed DACPM at levels of 0.2% and 0.1% in the diet developed tumors in the liver but not in the lungs, which were the most sensitive organ in rats. Mice were found to tolerate a higher dose of DACPM than rats (e.g., 2000 ppm vs 1000 ppm in the diet)[154].

DACPM has been found mutagenic in the Salmonella/microsome test[94].

10. **4,4'-Methylene bis(2-methylaniline)** (4,4'-diamino-3,3'-dimethyldiphenylmethane; 3,3'-dimethyldiphenylmethane-4,4'-diamine; methylenebis(2-toluidine); 4,4'-methylenedi-ortho-toluidine; MBOMA; $H_2N-\underset{CH_3}{\bigcirc}-CH_2-\underset{CH_3}{\bigcirc}-NH_2$) has been produced commercially by the reaction of formaldehyde with ortho-toluidine. It has been used (as an unisolated intermediate) in the manufacture of the corresponding diisocyanate, 4,4'-methylene bis(ortho-tolylisocyanate) for use in the production of polyurethanes[145]. Commercial production of this diisocyanate was carried out in the U.S. by a total of three companies at various times prior to 1964, but no production has been reported since[158].

One Japanese company was reported to be producing MBOMA, but no data is available concerning the quantity produced; there is no evidence that MBOMA is produced in other countries[145].

MBOMA can be used in the synthesis of the dye C.I. Basic Violet 2 but information is lacking on the production of the dye in U.S. or elsewhere[145]. Although MBOMA has been investigated as a curing agent for epoxy resins, there is no indication that it was ever used commercially for this purpose[145].

There are no data available as to the numbers of workers who have been employed in the production or use applications of MBOMA or is there information as to its levels in the environment.

MBOMA is carcinogenic in the rat after oral administration (5 times/week for 10 months; total dose 10.2 g/kg) producing malignant liver tumors[159]. Rats administered 200 ppm MBOMA in a standard diet (23% protein) for about one year developed tumors of the lung, liver and skin[156,157]. The incidence of liver tumors, frequency of malignant tumors and the number of metastases were higher in females as compared with male rats, but this could have been due to an age difference in survival[145,156,157]. No data is available concerning the mutagenicity of MBOMA or its biotransformation.

11. **4-Aminobiphenyl** (p-aminobiphenyl; p-biphenylamine; p-phenylaniline; xenylamine; BPA; $\bigcirc-\bigcirc-NH_2$) which can be prepared by the reduction of 4-nitrobiphenyl was produced only in the U.S. from 1935 to 1955[23], and used primarily as a rubber antioxidant[160]. There is no current evidence that it is being commercially produced anywhere[160].

Bladder cancer has been strongly associated with occupational exposure to 4-aminobiphenyl. Melick et al[161] found 19 cases (11.1%) of bladder tumors in a population of 171 male workers who were exposed to this aromatic amine from periods ranging from 1.5 to 19 years. In a more recent study, Melick et al found 53 men with bladder tumors in a population of 315 male workers (the exact exposure times could not be established)[162].

Melamed[163] and Koss et al[164,165] found in their studies that in an exposed population of 503, histologically proved carcinoma of the bladder developed in 35 men. Included among the 503 workers studied by Koss et al[164,165] are the 315 workers previously described in the study of Melick et al[162]. Existing experimental epidemiological evidence seems to indicate now that the carcinogenic potency of 4-aminobiphenyl is at least equal to, and is possibly even greater than, that of 2-naphthylamine[23].

4-Aminobiphenyl is carcinogenic in the mouse[166], rat[167], rabbit[168] and dog[169-171]. Bladder and liver tumors in mice and bladder papillomas and carcinomas in rabbits and dogs were produced following its oral administration. In the rat, the number of mammary gland and intestinal tumors were significantly raised following its daily s.c. administration. (It should also be noted that 4-nitrobiphenyl which is used in the production of 4-aminobiphenyl induces carcinomas of the bladder when given orally to dogs)[172].

A number of metabolites of 4-aminobiphenyl have demonstrated carcinogenicity. These include N-hydroxy-4-acetamidobiphenyl[160]; 4-amino-4'-hydroxybiphenyl; 4-hydroxylaminobiphenyl; 4-amino-3-hydroxybiphenyl[173] and 3-hydroxy-4-aminobiphenyl sulfate[93,174].

4-Aminobiphenyl as well as its hydrochloride salt (50 and 100 µg/plate) are mutagenic with activation in S. typhimurium TA-98 and TA-100[32,94] causing both base pair substitution and frameshift mutations with activity more pronounced in TA-100 tester strain. The isomeric 2-aminobiphenyl is also mutagenic in the Salmonella/microsome assay[94] and is active in the in vitro transformation test of Golden Syrian hamster embryo cells[95].

The postulated biotransformation of 4-aminobiphenyl is illustrated in Figure 2. In dogs, 4-aminobiphenyl is converted to 4-amino-3-biphenylylhydrogen sulfate and 4-amino-3-biphenylylglucuronic acid[175,176].

4-Aminobiphenyl is a potent bladder carcinogen in dogs, but since this species does not acetylate aromatic amines, N-hydroxy-4-aminobiphenyl probably does not lie on the pathway leading to metabolic activation in this species[37]. The active metabolite 4-aminobiphenyl in dogs is likely to be the glucuronide of N-hydroxy-4-biphenyl amine[37]. The glucuronide of N-hydroxy-4-aminobiphenyl has been detected in the urine of dogs administered 4-aminobiphenyl[177,178].

Kadlubar et al[43] reported that the relative reactivities of a number of N-hydroxy amines and their glucuronides at acidic pH's (N-HO-1-naphthylamine > N-HO-4-aminobiphenyl > N-HO-2-naphthylamine) are consistent with the relative carcinogenicities of these N-hydroxyarylamines at sites of s.c. and i.p. injection. The hepatic microsomal N-glucuronidation and nucleic acid binding of N-hydroxy 4-aminobiphenyl in relation to urinary bladder carcinogenesis was additionally described by Kadlubar et al[43].

12. **2,4-Diaminotoluene** (2,4-toluenediamine; m-toluylenediamine; 4-methyl-1,3-benzenediamine; 2,4-TDA; MTD) is generally prepared by nitrating toluene to produce a dinitrotoluene mixture (containing approximately 80% 2,4-isomer and 20% of the 2,6-isomer with small amounts of 2,3- and 3,4-dinitrotoluene) followed by catalytic reduction to a mixture of the corresponding diamines[179].

2,4-TDA is used principally as an intermediate to make toluenediisocyanate (TDI) which is used in the production of polyurethane. Most of the 2,4-diaminotoluene produced in the U.S. is used as a part of an unisolated mixture of 80% 2,4- and 20% 2,6-diaminotoluene for the manufacture of TDI. Some 2,4-diaminotoluene is isolated for conversion to pure toluene-2,4-diisocyanate. A mixture of 65% of 2,4- and 35% 2,6-diamino toluene is also used for conversion to TDI[179]. In 1975, seven companies in the U.S. produced 217.4 million kg of an 80/20 TDI mixture[180]. In 1974, 233.6 million kg of an 80/20 TDI mixture produced in the U.S. was used for the production of flexible and rigid polyurethane foams (142 million kg TDI consumed); polyurethane coatings (5.5-7.3 million kg); cast elastomers, including fabric coatings (3.6-4.1 million kg); and polyurethane and other adhesives (0.9 million kg)[179].

Other major uses of 2,4-diaminotoluene include: as an intermediate for the production of dyes used for textiles, leather and furs and in hair-dye formualtions.

Although about 60 dyes can be produced from 2,4-diaminotoluene[179], 28 are currently believed to be of commercial world significance[181]. A number of these dyes (C. I. Basic Brown 4, Basic Orange 1, Direct Brown 154, Direct Black 4, Direct Black 9, Leuco Sulphur Brown 10, Leuco Sulphur Brown 26 and Sulphur Black 2)[180] are used to color silk, wool, paper, leather, cotton, bast fibers, and cellulosic fibers, in spirit varnishes and wood stains, as indicators, as biological stains and in the manufacture of pigments 2,4-Diaminotoluene was used as an oxidative hair dye component in the U.S. until 1971[182] in which application it was at a level of 60 mg per dyeing (1 mg/kg)[182].

2,4-Diaminotoluene has been produced commercially in the U.S. for over 50 years. In 1975, six companies in the U.S. produced 86.7 million kg[180] however, data for the portion of 2,4-diaminotoluene produced as an unisolated intermediate are not available[179]. Approximately 20 thousand kg of the diamine was imported into the U.S. in 1975.

Almost all of the 180-200 million kg of 2,4-diaminotoluene produced annually in Western Europe is used as an unisolated intermediate for the production of TDI[179]. Less than 100 thousand kg of 2,4-diaminotoluene are produced annually in the U.K. and less than 25 thousand kg are imported. In Japan production of 2,4-diaminotoluene during 1971-1975 amounted to approximately 120 thousand kg annually where it is used principally as a dye intermediate[183].

No information is available on the number of people who have been involved in the production of 2,4-diaminotoluene or in its use applications.

2,4-Diaminotoluene administered to male Wistar rats in diets at levels of 0.06% or 0.1% for 30-36 weeks produced hepatocellular carcinomas[184] and after its s.c. injection induced local sarcomas in rats[185]. 2,4-Diamonitoluene used topically along or mixed with a hair dye complex (2,5-toluenediamine, p-phenylenediamine and resorcinol) in a 2 year skin painting study with Swiss-Webster mice did not produce any abnormal proliferation and maturation of the squamous epithelium of the skin. 2,4-Diaminotoluene was found to be non-toxic and non-carcinogenic to the skin of mice, nor did it increase the incidence of lung neoplasms[186].

In an earlier study involving skin painting of Swiss-Webster mice with a hair-dye formulation containing among its constituents 0.2% 2,4-diaminotoluene, 3% 2,5-diaminotoluenesulphate and 1.5% para-phenylenediamine and 6% hydrogen peroxide, the incidence of lung tumors did not differ significantly from that in untreated control mice[187].

2,4-Diaminotoluene induced reverse mutations in S. typhimurium TA 1538 and TA 98 in the presence of a rat liver post-mitochondrial supernatant from animals pretreated with Aroclor 1254[182,188,189]. Strain TA 100, an indicator of mutation through base-pair substitution showed no mutagenicity when treated with 2,4-diaminotoluene with or without added S9 mix[189].

2,4-Diaminotoluene was shown to be weakly mutagenic in Drosophila melanogaster, inducing sex-linked recessive lethals when fed at a concentration of 15.2mM[190]. In a comparative study of the mutagenicity of hair dye components (2,4-diaminotoluene and 4-nitro-o-phenylenediamine) relative to the carcinogen benzidine in Drosophila melanogaster, all compounds were found to exert decisive mutagenicity both on the X-chromosome and the RNA genes, although their activities on the different genic sites varied between compounds and as a function of cell stage. The mutagenicities of the test compounds for rDNA gradually decreased in the order of benzidine > 2,4-diaminotoluene > 4-nitro-o-phenylenediamine, which correlated with the evidence (thus far) about their carcinogenicities[35].

2,4-Diaminotoluene induced morphological transformation in Syrian Golden hamster embryo cells[189].

The major urinary metabolite of 2,4-diaminotoluene in several rodent species is 2,4-diamino-5-hydroxytoluene. In addition N-acetyl and glucuronide conjugates were also found[191]. In rats, the major unconjugated urinary metabolites are 4-acetylamino-2-aminotoluene; 2,4-diacetylaminotoluene and 2,4-diacetylaminobenzoic acid[192]. Administration of [7-^{14}C]-2,4-diaminotoluene to rats resulted in the labelling (presumed to be due to covalent binding) of liver nuclear DNA, RNA and microsomal and soluble proteins[193].

13. **2,4-Diaminoanisole** (meta-diaminoanisole; 1,3-diamino-4-methoxybenzene; 4-methoxy-meta-phenylenediamine; 4-methoxy-1,3-benzenediamine;) can be prepared commercially via the reduction of 2,4-dinitroanisole with iron and acetic acid, or by the methylation of 2,4-dinitro-1-chlorobenzene followed by reduction with iron[179]. 2,4-Diaminoanisole is used as an intermediate for the production of dyes and pharmaceuticals and in hair-dye formulations[194,195]. The principal dye produced from 2,4-diaminoanisole is C.I. Basic Brown 2 which is used mainly to dye acrylic fibers, as well as cotton, viscase, wool, nylon polyester and leather[196].

In its use as a color modifier in certain permanent hair-dye compositions (it has been estimated that about 600 mg of 2,4-diaminoanisole are used in each dye application[182].

Although commercial production of 2,4-diaminoanisole in the U.S. dates back to 1933 and that of its sulfate salt to 1967, the production of both the free base and its salt has not been reported in the U.S. since 1940 and 1971 respectively[179]. However, it is known that 16 thousand kg and 5 thousand kg were imported into the U.S. in 1974 and 1975 respectively[179]. It is believed that currently about 12,000 kg a year are imported into the U.S.[197]. 2,4-Diaminoanisole has been produced commercially in Japan since 1974 and in 1975 one company produced approximately 7 thousand kg. Additionally, 2,4-Diaminoanisole is believed to have been made in the Federal Republic of Germany[179].

No data are available on the numbers of individuals involved in the production of 2,4-diaminoanisole and its salts as well as in their use applications.

NIOSH recommended in January 1978 that 2,4-diaminoanisole be handled in the workplace as if it and its salts were human carcinogens[197-199]. This recommendation was based primarily upon a preliminary analysis of NCI data indicating a statistically significant excess of site-speicific malignant tumors in rats and mice which were fed 2,4-diaminoanisole sulfate[200]. Approximately three out of four current oxidation or "permanent" hair dye formulations contain 2,4-diaminoanisole or its sulfate salt, accounting for about $200 million in annual retail sales[197]. NIOSH estimates that about 400,000 workers have potential occupational exposure, primarily hairdressers and cosmetologists. A relatively small number of fur dyers are also exposed to 2,4-diaminoanisole or its sulfate salt[197]. In addition, NIOSH's two epidemiologic studies suggested an elevated incidence of cancer among cosmetologists. Whereas one study indicated an excess of genital cancer among hairdressers and cosmetologists, the other which has not yet been completed, suggests excess cancer of a number of organ systems among this occupational group[197,201]. It is important to stress that these occupational groups are also exposed to a wide variety of other chemicals and that the epidemiologic studies do not clearly demonstrate a casual connection between hairdyes and cancer.

The NCI bioassay of technical grade 2,4-diaminoanisole sulfate was conducted using Fischer 344 rats and B6C3F1 mice of both sexes. The time-weighted average dietary concentrations were 0.2% for the low dosed rats and 0.5% for the high dosed rats. The dietary concentrations used for low and high dosed mice were 0.12 and 0.24% respectively. After a 78-week period of chemical administration, observations continued for an additional 29 weeks for rats and 19 weeks for mice.

Dietary administration of 2,4-diaminoanisole sulfate induced increased incidences of malignant tumors of the skin and its associated glands and malignant thyroid tumors in male and female rats while in mice the agent induced thyroid tumors in each sex[200].

2,4-Diaminoanisole induces reverse mutations in S. typhimurium TA 1538 when metabolically activated with S-9 fraction from rats pretreated with Aroclor 1254[182]. 2,4-Diaminoanisole was weakly mutagenic in Drosophila melanogaster, inducing sex-linked recessive lethals when fed at the highest concentration tested, 15.1 mM[190].

2,4-Diaminoanisole gave a questionable mutagenic response in studies on thymidine kinase heterozygous locus of L5178Y mouse lymphoma cells, 2,5-diaminoanisole, tested similarly in vitro without metabolic activation during a 24 hr chemical exposure period gave a negative response[202]. 2,4-Diaminoanisole without metabolic activation was a potent inducer of mitotic cross-over when tested in strain D-3 of S. cerevisiae[203]. (2,5-Diaminoanisole, 2-nitro-p-phenylenediamine and m-phenylenediamine were negative in similar experiments). In other short-term experiments (4 hr with liver enzyme preparations) in which either 2,4- or 2,5-diaminoanisole were mixed with dimethylnitrosoamine, it was found that the presence of the hair dye compound interfered with the enzymatic activation of the nitrosamine to a mutagen[203].

No dominant lethal effects were induced in Charles River CD rats treated i.p. 3 times/week for 8 weeks with 20 mg/kg 2,4-diaminoanisole (or 2,5-diaminoanisole) before mating[204]. 2,4-Diaminoanisole did not induce micronucleated cells in bone marrow when administered orally at 1000 mg/kg to two groups of 5 male and 5 female rats in two doses separated by a 24 hour interval[205].

Information on the biotransformation of 2,4-diaminoanisole is limited. In one study in which rats were given a single i.p. injection of 50 mg/kg of [^{14}C]-2,4-diaminoanisole, 94% of the dose was excreted within 48 hours (85% in the urine and 9% in the feces). The metabolites found in 24 hour urine samples included: 4-acetyl amino-2-aminoanisole; 2,4-diacetylaminoanisole and a small amount of diacetylaminophenol.

In addition, 2,4-diacetylaminophenol, hydroxylated derivatives of 2,4-diacetylaminoanisole and a small amount of 4-acetylamino-2-amino anisole were detected as glucuronides[206].

14. para-Phenylenediamine (4-aminoaniline; 1,4-benzenediamine; 1,4-diaminobenzene, para-diaminobenzene; 1,4-phenylenediamine; $H_2N-\langle O \rangle-NH_2$) is produced commercially via the reduction of 1-amino-4-nitrobenzene with iron and hydrochloric acid, iron and ferrous chloride or iron, ammonium polysulfide and hydrogen[207]. It is used primarily in dyestuff manufacture, in hair-dye formulations, in a variety of antioxidants and in photographic developers[207]. It is used as an intermediate in the production of 5 dyes which are believed to have commercial world significance. There are C.I. Direct Orange 27, Disperse Yellow 9, Solvent Orange 53, Sulphur Brown 23, and Leuco Sulphur Brown 23[208]. (None of these dyes are produced in the U.S.) It is also employed as an unisolated intermediate in the production of 10 dyes derived from 4-aminoacetanilide, 4-aminoformanilide, 4-aminooxanilic acid or 4-nitroaniline[208].

para-Phenylenediamine is used in a variety of hair-dye formulations at levels ranging from 0.20% to 3.75%[195]. It is also used with terephthaloyl chloride to produce polyparaphenylene terephthalamide (Fiber B) which is used primarily as tire cord. The annual use of Fiber B is estimated to be about 11 million kg in 1977[207]. A number of derivatives of para-phenylenediamine are important antioxidants in synthetic and natural rubbers, petroleum products, cellulose ethers and alfalfa meal[209].

para-Phenylenediamine has been produced commercially in the U.S. for over 50 years. Undisclosed amounts were reported in 1975 by two companies and of the hydrochloride by one company[210]. However, the amounts of para-phenylenediamine imported into the U.S. in 1972-1974 are known and amounted to (thousand kg) 75.5, 12.0 and 29.2 yearly respectively[207].

It is believed that at least four companies in Western Europe produce the diamine. About 10-100 thousand kg are produced annually in the U.K., with imports amounting to less than 100 thousand kg. About 10-100 thousand kg/year are imported into Switzerland where it is used as a dyestuff and industrial chemical intermediate[207]. The production of para-phenylenediamine in Japan during the period 1971-1975 (in thousand kg) was 80, 60, 114, 113 and 55 thousand kg/year respectively, where approximately 60% is used as a component of hair dyes and 40% as a dye intermediate[183].

The number of individuals involved in the production and use applications of para-phenylenediamine are not known.

The OSHA standard states that an employee's exposure to para-phenylenediamine should not exceed 0.1 mg/m^3 in the working atmosphere in any 8 hour work shift for a 40 hour workweek[211]. The MAC in the Federal Republic of Germany and Japan is also 0.1 mg/m^3 [207].

Definitive data on the carcinogenicity of para-phenylenediamine are meagre. According to IARC it has been inadequately tested[207] in albino mice by skin application[212] and in rats by oral[213] and s.c. administration[213]. Studies in mice in which para-phenylenediamine as a constituent of hair dye preparations wastested by skin application cannot be evaluated[207]. In one study the hair-dye formulation consisted of 1.5% para-phenylenediamine and 3% 2,5-diaminotoluene sulfate and either 0.2% 2,4-diaminotoluene, 0.38% 2,4-diaminoanisole sulphate or 0.17 meta-phenylenediamine and 6% hydrogen peroxide[187]. In another study the hair-dye formulation contained 1.5% para-phenylenediamine, 3% 2,5-diaminotoluene sulphate and either 0.2% or 0.6% 2,4-diaminotoluene[186].

para-Phenylenediamine induced reverse mutations in S. typhimurium TA 1538 in the presence of rat liver post mitochondrial supernatant fraction from animals pretreated with phenobarbital[214]. It did not induce micronucleated cells in bone marrow when 300 mg/kg were administered orally to rats in two doses separated by an interval of 24 hours[205]. No dominant lethal effects were induced in Charles River CD rats treated i.p. 3 times/week for 8 weeks with 20 mg/kg of para-phenylenediamine (or meta- or ortho-phenylenediamine)[204].

Information on the biotransformation of para-phenylenediamine is scant. N,N'-Diacetyl-para-phenylenediamine was identified as a urinary metabolite of para-phenylenediamine in dogs[215]. para-Phenylenediamine is excreted in the urine of dogs[216], and rats[217,218] following oral or s.c. administration of para-dimethylaminoazobenzene.

15. meta-Phenylenediamine (3-aminoaniline; 1,3-benzenediamine; 1,3-diaminobenzene, meta-diaminobenzene; 1,3-phenylenediamine; C$_6$H$_4$(NH$_2$)-NH$_2$) is produced commercially by the reduction of 1,3-dinitrobenzene with iron and hydrochloric acid or with iron, ammonium polysulfide and water gas[207,209]. The major areas of utility of meta-phenylenediamine include: the production of dyes, as a component of hair-dye formulations[195], as a curing agent for epoxy resins, and for the production of heat-resistant fibers.

It can be used for the production of over 140 dyes, 37 of which are believed to have commercial world significance[219]. Twelve of these are produced commercially in the U.S.: C. I. Direct Black 38 (984.3 thousand kg in 1975); C.I. Direct Black 22 (149.4 thousand kg); Basic Orange 2 (120.3 thousand kg); Basic Brown 1 (9.1 thousand kg); Basic Brown 2, Direct Brown 1, Direct Brown 44, Direct Black 9, Direct Black 19, Mordant Brown 1, Mordent Brown 12 and Solvent Orange 3[210]. These dyes are used to color various textile fibers and other materials.

meta-Phenylenediamine is added to isophthaloyl chloride to produce poly-meta-phenylene isophthalamide resin, a polyamide fibre (Nomex®) used in many high-temperature applications (e.g., electrical insulation, protective clothing). It is estimated that production of Nomex® in the U.S. will be approximately 4.5 million kg per year in 1977[207].

Additional areas of utility of meta-phenylenediamine include: in the manufacture of ion-exchange resins, rubber-curing agents, formaldehyde condensates, resinous polyamides, textile fibers, urethanes, rubber adidtives, corrosion inhibitors, petroleum additives[207] and as an analytical reagent for bromine, gold and as the hydrochloride salt for nitrite[220].

Although meta-phenylene diamine has been produced in the U.S. for over 50 years, no production quantities past or current are known. However, imports of the diamine into the U.S. in 1972, 1974 and 1975 (in thousands of kg) amounted to 146.7, 90.1 and 35.2 respectively[207]. meta-Phenylenediamine is believed to be produced in Western Europe by at least 2 companies. Between 10-100 thousand kg are produced annually in the U.K. and less than 50 thousand kg are imported between 100-1000 kg/year are imported into Switzerland[207]. Japan produced 296 thousand kg in 1974 and in 1975 3 companies produced 171 thousand kg[183,207] where 80% is used in the production of heat-resistant fibre and 20% in the production of dyes.

There are no data available concerning the number of persons involved in the production and rise applications of meta-phenylenediamine.

Chronic studies in mice[187] and rats[221] with meta-phenylenediamine per se, its hydrochloride salt or the diamine as a constituent in hair-dye formulations are considered by IARC to be inadequate to permit an evaluation of the carcinogenicity of this compound[20]

meta-Phenylenediamine is mutagenic in S. thyphimurium TA 1538 in the presence of a rat liver post-mitochondrial supernatant fraction from animals pretreated with either Aroclor 1254[182] or with phenobarbital[214]. It did not induce dominant lethals in Charles River rats injected i.p. 3 times/week for 8 weeks with 20 mg/kg before mating[204]. meta-Phenylenediamine was mutagenic at the thymidine kinase locus of L5178Y mouse lymphoma cells, exhibiting a dose response at 25-100 μg/ml[202].

Information on the biotransformation of meta-phenylenediamine is extremely scant. In a study in 1922, it was found to be excreted rapidly in man, unchanged and with little absorption[222].

16. 2-Nitro-para-phenylenediamine (1,4-diamino-2-nitrobenzene; 2-nitro-1,4-benzenediamine; 4-amino-2-nitroaniline; 2-nitro-1,4-phenylenediamine; 2-NPPD,

$H_2N-\langle O \rangle-NH_2$) can be prepared by the hydrolysis of 1,4-diamino-2-nitrobenzene
          NO_2
or via the acetylation of para-phenylenediamine followed by nitration and hydrolysis[223]. It is used primarily for dyeing furs and in hair-dye formulations.

No data on production quantities are available for 2-NPPD produced in the U.S. In 1973 only one U.S. company reported an undisclosed amount[223]. U.S. imports of 2-NPPD are reported to be about 1400 kg in 1974 and 200 kg in 1975[223]. In 1976 one Japanese manufacturer reported production of 650-750 kg, and in 1975 imports to Japan of 2-NPPD amounted to about 100-200 kg[183] where it is used as a dye intermediate in hair dyes.

No data are available on the number of persons engaged in the production of 2-NPPD or in its use applications. IARC has stated that no evaluation of the carcinogenicity of this compound can currently be made[223]. In the only reported study, a hair-dye formulation containing 2-NPPD mixed with 1,2-diamino-4-nitrobenzene (4-NPPD) was tested in DABb mice by skin application and did not increase the incidence of tumors in treated over control animals[224,225]. No carcinogenicity data on 1,4-diamino-2-nitrobenzene alone are currently available.

2-NPPD induces reverse mutations in S. typhimurium TA 1538 without metabolic activation[182,214,226]. The mutagenic activity was reduced upon addition of liver post-mitochondrial supernatant fraction from rats pretreated with Aroclor 1254 or phenobarbital[214,226]. 2-NPPD induced mutations (in a dose-dependent manner from 25-75 µg/ml, without metabolic activation) at the thymidine kinase locus in mouse lymphoma cell line L51178Y[202]. It induced chromatid breaks and chromosome aberrations in a hamster cell line A(T1)Cl-3[227], chromosome aberrations in CHMP/E cells[228], chromatid breaks and gaps in cultured human peripheral lymphocytes[226] and sister chromatid exchanges (SCE's) in Chinese hamster (CHO) cells[229]. It did not induce micronucleated cells in bone marrow following oral administration or 2000 mg/kg to female rats in two doses separated by an interval of 24 hours[205].

No data is currently available on the biotransformation of 2-NPPD.

17. <u>4-Nitro-ortho-phenylenediamine</u> (1,2-diamino-4-nitrobenzene; 4-nitro-1,2-diaminobenzene; 2-amino-4-nitroaniline; 4-nitro-1,2-phenylenediamine, 4-NOPD;

$O_2N-\langle O \rangle-NH_2$ with NH_2) can be produced by the reduction of 2,4-nitroaniline with alcoholic ammonium sulfide[223]. It is used primarily in semi-permanent and permanent hair coloring products[194,195]. Additionally, it has also been used as an analytical reagent for α-keto acids[220] and for the determination of ascorbic and dehydroascorbic acid in foods.

Although only one U.S. company reported an undisclosed amount of 4-NOPD, it is believed that 2 companies currently manufacture this diamine. U.S. imports of 4-NOPD in 1973 and 1975 amounted to 1900 kg and 1100 kg respectively[223]. No data on its production elsewhere is available nor is there information as to the numbers of individuals involved in its production or in its use applications.

IARC reported that no evaluation of the carcinogenicity of 4-NOPD could be currently made[223]. In the only reported study, a hair-dye formulation containing 4-NOPD mixed with 2-NPPD (1,4-diamino-2-nitrobenzene) was tested by skin application and no difference in the incidence of tumor formation was noted between treated and control mice[224,225].

4-NOPD induced reverse mutations in S. typhimurium TA 1538 without metabolic activation. This activity was reduced upon addition of liver post-mitochondrial supernatant fraction from rats pretreated with Aroclor 1254 or phenobarbital[182,214,216]. It induced sex-linked recessive lethals in <u>Drosophila melanogaster</u>[35,198] when fed at a concentration of 1.2 mM[190], and at 5-20 nM by microinjection technique[35], and induced mutations at the thymidine kinase locus (in a dose-dependent manner from 50-200 µg/ml) in mouse lymphoma cell line L51178Y[202]. 4-NOPD did not induce dominant lethals in Charles River rats treated i.p. 3 times/week for 8 weeks with 20 mg/kg before mating[204], nor did it induce micronucleated cells in bone marrow when 5000 mg/kg were administered orally to male and female rats in two doses separated by an interval of 24 hours[205]. Whereas 4-NOPD induced chromatid breaks in a hamster cell line A(T$_1$)Cl-3[227] chromosome aberrations in CHMP/E cells[228] and sister chromatid exchanges in Chinese hamster (CHO) cells[229], it did not induce chromosomal aberrations when incubated with cultured human peripheral lymphocytes[226].

No data is currently available on the biotransformation of 4-NOPD.

18. <u>ortho-Toluidine</u> (2-amino-1-methylbenzene; 2-methyl-benzeneamine; 2-aminotoluene; <u>ortho</u>-methylaniline; <u>ortho</u>-tolylamine; ⟨O⟩-NH$_2$) is produced commercially
                                                                                CH$_3$
by the reduction of 1-methyl-2-nitrobenzene either by catalytic hydrogenation or with iron filings[231]. The major use of <u>ortho</u>-toluidine (as well as its hydrochloride salts) are as intermediates in the production of more than 50 dyes of which 17 are believed to be produced in the U.S.[232]. (The most significant dye in the U.S. is C.I. Vat Red 1 of which over 38.6 thousand kg were sold in 1975)[231].

<u>ortho</u>-Toluidine and its hydrochloride salt are also used in significant quantities as intermediates in the production of 4-<u>ortho</u>-tolylazo-<u>ortho</u>-toluidine (ortho-aminoazotoluene) and 4-(4-amino-<u>meta</u>-tolylazo)meta-toluenesulphonic acid, which are used as raw material for the production of 11 dyes in the U.S.[231]. The most important include: C.I. Acid Red 115 (1974 U.S. sales, 12.7 thousand kg); C.I. Direct Red 72 (1975 U.S. production, 62.6 thousand kg) and C.I. Solvent Red 26 (1974 U.S. sales, 58.1 thousand kg)[231,232].

Other areas of utility of <u>ortho</u>-toluidine include: as an antioxidant in the manufacture of rubber; production of special dyes for color photography; printing of textiles; and as a analytical reagent to test blood samples for glucose[220].

Although <u>ortho</u>-toluidine has been produced commercially in the U.S. for over 50 years, annual production quantities are not available. Imports of <u>ortho</u>-toluidine and <u>ortho</u>-toluidine hydrochloride to the U.S. in 1974 were 11.8 thousand kg and 21.9 thousand kg respectively[231].

Total production by at least 9 companies in Wester Europe amounts to 2-5 million kg annually., 1-5 million kg in the Federal Republic of Germany and the U.K. (4 companies); 100 thousand-1 million kg in France, Italy and Switzerland; 10-100 thousand kg in Spain; less than 10 thousand kg are produced annually in Austria and Benelux. It is estimated that 1-5 million kg are used annually in western Europe in the manufacture of antioxidants, dyes, pigments, and as an agricultural and industrial chemical intermediate[231].

In Japan there are currently 3 commercial producers of <u>ortho</u>-toluidine and production during the period of 1971-1975 amounted to about 1 million kg/year[183] where about 50% is used in the manufacture of dyes, 20% for pigments, 20% in antioxidants and 10% for other applications[231].

Information on the numbers of individuals involved in the production and use applications of <u>ortho</u>-toluidine is not available.

The current OSHA time-weighted average concentration of <u>ortho</u>-toluidine for a normal 8-hr workday or 40 hr workweek to which workers in the U.S. may be exposed to cannot exceed 5 ppm (22 mg/m3)[233]. The work environment hygiene standards for ortho-toluidine in terms of an 8 hr time-weighted average (mg/m3) for a number of countries are: Federal Republic of Germany, 10 (1973); German Democratic Republic, 22 (1974); Czeckoslovakia 5 (1975)[234]. The MAC for ortho-toluidine in the USSR[234] and Romania[235] is 3 mg/m^3 and in Japan, the workplace tolerance concentration is 5 ppm (22 mg/m3)[231].

ortho-Toluidine has been found at levels of 0.5-28.6 mg/m^3 in the air of the working environment in a toluidine manufacturing plant in the USSR[236]. It has also been found as a component of the tar produced by low-temperature carbonization of coal[237], in the gasoline fraction of hydrocracked Arlon petroleum[238], as well as in the volatile aroma components of black tea[239] and in the steam volatiles from the distillation of leaves of Latakia tobacco (believed to be widely used in pipe tobaccos)[240].

Most epidemiological studies dealing with worker populations potentially exposed to ortho-toluidine have dealt with workers who also had mixed exposure to known carcinogenic amines (e.g., benzidine and 2-naphthylamine) in Germany[241-244], Switzerland[245] and the United Kingdom[17,246].

A fourth study reported bladder tumors in workers engaged in the production of ortho-toluidine and/or para-toluidine in the USSR during the 1960's[236]. While the finding of a small number of cases of workers with bladder tumors (papillomas) was suggestive of a carcinogenic effect, the information was considered to be incomplete. IARC considered these four studies did not allow any firm conclusion to be drawn concerning the carcinogenicity of ortho-toluidine to human[231].

Although tumors have been reported in mice[247], rats[247,248], rabbits[249,250] and guinea pigs[249,250] after administration of ortho-toluidine by various routes, according to IARC the data has been insufficiently described hence currently not allowing an evaluation of the carcinogenicity of this compound to be made[231].

ortho-Toluidine is not mutagenic in S. typhimurium TA 1535, TA 1537, TA 1538, TA 98 or TA 100 in the presence or absence of a rat liver preparation[182,214]. o-Toluidine, however, has been shown to be a very strong mutagen in S. typhimurium TA 98 in the presence of norharman and an S-9 mix from the liver of rats pretreated with polychlorinated biphenyl[251]. It is of interest to note that neither meta-, nor para-toluidine demonstrated mutagenicity even in the presence of norharman and S-9 mix in S. typhimurium TA 98[251]. Aniline, which is non-mutagenic by itself, was also found to be mutagenic in the presence of norharman and S-9 mix in TA 98. Since norharman itself is non-mutagenic in this system it could be termed a "comutagen" according to Nagao et al[251,252].

The biotransformation of ortho-toluidine has been studied to a limited extent. The following urinary metabolites have been recovered from male CDF rats treated s.c. with ortho-toluidine-[methyl^{14}C]hydrochloride: sulfate conjugates of 4-amino-m-cresol and N-acetyl-4-amino-m-cresol (40 and 10% respectively); the glucuronides of 4-amino-m-cresol and N-acetyl-4-amino-m-cresol (3 and 1% respectively) and the non-ionic metabolites, ortho-toluidine (2.8% of dose), N-acetyl-o-toluidine (0.2%) and o-aminobenzyl alcohol (1.9%)[253]. Evidence of N-hydroxylation or epoxide formation in the remaining metabolites as possible routes of the activation of o-toluidine is being currently sought[253].

It is useful in the above context to compare the status of the mutagenicity of para-toluidine and a number of its derivatives. Although p-toluidine, acetyl-p-toluidine, p-nitrotoluene, and p,p'-dimethylazobenzol were negative toward the induction of gene conversion in S. cerevisiae, N-hydroxy-p-toluidine, N-hydroxyacetyl-p-toluidine, N-acetoxyacetyl-p-toluidine and p-nitrosotoluidine were genetically active[254]. These studies suggest that a breakdown product of p-toluidine may be the proximately active mutagen rather than the parent compound. para-Toluidine has recently been found to induce mitotic crossing over Saccharomyces cerevisiae only in the presence of an Udenfriend hydroxylation medium[255]. It has also been found carcinogenic in rats[247], with evidence that supports a suggestion that an N-hydroxylation mechanism may play a role.

19. Miscellaneous Aromatic Amines

Although 18 major aromatic amines, representative of a number of structural sub-categories, have been examined in this chapter, it should be noted that this major category includes a far larger number of compounds which have been reported to be carcinogenic and/or mutagenic. In most cases the amounts of these materials as well as their use applications are smaller, and hence the potential population at risk are smaller.

There are an additional number of aromatic amine hair dyes that have been found to be mutagenic in S. typhimurium TA 1538[182]. These include: 2,5-diaminoanisole; 2-amino-5-nitrophenol; 2-amino-4-nitrophenol; and 2,5-diaminotoluene, which are mutagenic after oxidation by hydrogen peroxide. It has been estimated that about 30% of women in the U.S. dye their hair, often monthly and for many years and it was suggested by Ames et al[182] that the possible risk of cancer and of genetic abnormalities appears sufficiently large that large scale epidemiological studies should be done to determine any relationship between cancer incidence and years of hair dyeing or of industrial exposure in workers. There is evidence both from animal and human studies that constituents of semi-permanent hair colorants, at concentrations found in proprietary formulations, can be absorbed by the skin and appear in the urine in appreciable amounts[256]. Fragmentary epidemiologic data based on studies in Leeds (UK)[258], New York[258] and New Orleans[259] indicate an excess of cancer of the urinary bladder in hairdressers and beauticians. However, one must emphasize the myriad number of compounds and/or mixtures that these individuals are exposed to thus making it extremely difficult to unequivocally pin-point the causative agent(s).

B. Azo Dyes

Azo dyes form the largest and most versatile class of all dyes. They are a well defined group of compounds characterized by the presence of one or more azo groups (-N=N-). Chemically, the azo class is subdivided according to the number of azo groups present, into mono-, di-, tris-, tetrakis-, and higher azo derivatives.

Azo dyes have a multitude of uses depending on their chemical structures and method of application. Their areas of utility include: dyeing of wool, silk, leather, cotton, paper and the synthetic fibers (e.g., acetate, acrylics, polyamides, polyesters, viscose rayon); for the coloring of paints, plastics, varnishes, printing inks, rubber, cosmetics, food, drugs; for color photography; diazotypy and for staining polish as well as absorbing surfaces[260,261].

The toxicity and biological properties of a variety of azo dyes have been reviewed by IARC[261], Radomski[262], Truhaut[263], Miller and Miller[264], Terayama[265] and Clayson and Garner[37].

Several of the intermediates used in dye manufacture have been identified as human carcinogens (e.g., 2-naphthylamine, benzidine and 4-aminobiphenyl)[37]. While various potentially carcinogenic and/or mutagenic aromatic amines are manufactured for azo dye production and the processes are closely controlled, the use of the resultant dyes is not. It should be noted that since azo-reduction has been shown to occur in the body, catalyzed by both mammalian cell enzymes[266,267] and gut microorganisms[268,269] it is possible that the carcinogenic aromatic amines could be released by reduction of their precursor azo dyes. Indeed free benzidine has been detected in the urine of monkeys fed benzidine derived azo dyes[7,267] establishing a potential for reconversion of azo dyes to benzidine. The possibility arises that the metabolism of benzidine (as well as other carcinogenic aromatic amine)-derived azo dyes may be similar in humans[7].

1. Azobenzene (diphenyldiazene; ⟨O⟩—N=N—⟨O⟩) can be produced by the reduction of nitrobenzene under a variety of conditions. It has been used in the U.S. in the past as an unisolated intermediate in the production of benzidine and its salts. (In 1972 about 700,000 kg of benzidine were produced)[261]. It has also been widely used as an intermediate in the manufacture of insecticides, dyes, rubber accelerators and pyrazolone derivatives[261,270].

No production figures for azobenzene are available. It is believed to be available in the U.S. at present only in small research quantities[261]. The two major commercial producers of azobenzene in western Europe are in the Federal Republic of Germany. Azobenzene is no longer produced commercially in Japan, although it is believed that it had been produced formerly by about 8 companies[261].

Limited carcinogenicity studies of azobenzene in mice and rats have been reported with inconclusive results. In the oral study in mice it produced an excess of liver-cell tumors over the controls in males, but not in females in C56Bl/6XC3H/AnF)F1 strain. Hepatomas and tumors at other sites occurred in both sexes of (C56Bl/6XAKR)F1 mice to the same extent as in the controls[143]. Subcutaneous studies in the same strains of mice[143] and in female Sherman rats[27] were negative, but according to IARC[261], they cannot be evaluated because the adequacy of the dose used could not be assessed.

Azobenzene (of undefined purity) has been reported to be non-mutagenic when tested in Drosophila for the production of X-linked recessive lethals[271], and mutagenic in the Salmonella (TA 100)/microsome test[94].

Limited studies have been reported on the biotransformation of azobenzene. Following its i.p. administration to rats, aniline and a water-soluble compound that formed benzidine on acidification were found in the urine[272]. Following administration of 500 mg/kg of azobenzene to rabbits, 30% was found in the feces and of the absorbed azobenzene, 60% was excreted in the urine as the glucuronide, 20% as the sulfate and 23% unchanged[273].

2. para-Aminoazobenzene (4-(phenylazo)benzenamine; AAB; ⟨O⟩—N=N—⟨O⟩—NH$_2$) can be prepared by the diazotization of aniline and coupling the resulting diazoaminobenzene with a mixture of aniline and aniline hydrochloride; however, it is not known

whether this is the commercial procedure[261]. It is used as a dye for lacquers, varnishes, wax products, oil stains and styrene rubbers, and as an intermediate in the production of acid yellow, diazodyes and indulines[261].

No production quantities of para-aminoazobenzene are available but it is known that in 1972 one manufacturer reported production of this material. Although there may be as many as 7 producers of para-aminoazobenzene in western Europe, the current production is believed to be small. It is believed not to be currently produced in Japan although it has been imported on an irregular basis[261].

Liver tumors have been induced in male Wistar rats on a low protein diet fed the highest tolerated dose of para-aminoazobenzene (0.2-0.3% in the diet)[274]. Application of 1 ml of a 0.2% solution of AAB in acetone twice weekly on the dorsal skin of albino rats for life (mean length of treatment, 123 weeks) produced a variety of epidermal tumors[275].

para-Aminoazobenzene is mutagenic in S. typhimurium TA 1538[29,94], TA 100[276] and TA 98[276] only in the presence of S-9 rat liver microsomal systems. N-hydroxy-para-aminoazobenzene was also found to be mutagenic in S. typhimurium TA 100 and TA 98 strains, only in the presence of a S-9 rat liver microsomal system (induced by polychlorinated biphenyl, KC-500)[276]. This compound has not been shown to be carcinogenic[277].

para-Aminoazobenzene was negative when tested in the in vitro transformation of Golden Syrian hamster embryo cells[95], and was non-mutagenic when tested in Drosophila for the production of X-linked recessive lethals[271].

The biotransformation of para-aminoazobenzene involves N-hydroxylation, hydroxylation of the aromatic rings, N-acetylation and O-conjugation with sulphuric and glucornic acids[217,261,277]. N-hydroxy-N-acetyl-AAB has been found in conjugated form in the urine of rats, mice or hamster injected with para-aminoazobenzene[277]. Following administration of para-aminoazobenzene to rats by stomach tube, the following metabolites were found in the urine: 4'-hydroxy-AAB sulphate; 3-hydroxy-AAB sulphate; 3,4'-dihydroxy-AAB sulphate; N-acetyl-4'-hydroxy-AAB sulphate; para-acetamidoacetanilide and conjugated forms of ortho- and para-aminophenol[217].

3. ortho-Aminoazotoluene (2-methyl-4-[(2-methylphenyl)azo]-benzenamine; 4-(o-tolylazo)-o-toluidine; AAT; ⟨O⟩-CH$_3$-N=N-⟨O⟩-CH$_3$-NH$_2$) can be prepared by the diazotization of ortho-toluidine but it is not known whether this is the commercial procedure[261]. AAT is used primarily to color oils, fats and waxes. Production data are not available although it is believed that in 1972 two U.S. manufacturers produced this material and in western Europe there may be about 5 producers of AAT; Japan has neither produced nor imported AAT during the past 5 years[261].

AAT is the first carcinogenic azo compound to be discovered, with the report by Yoshida in 1932 that it induced hepatomas in rats[278]. AAT is carcinogenic in mice[279-281] rats[278,282-284], hamsters[285] and dogs[286] following its oral administration producing principally tumors of the liver, gall-bladder, lung and urinary bladder[37,261]. AAT is also carcinogenic in mice[287,288] and rats[289] when administered by other routes such as subcutaneous or intraperitoneally. It is carcinogenic in single s.c. doses in newborn mice[290] and there is some evidence that it produces papillomas of the bladder in rabbits[291] following its administration by direct bladder instillation and in mice after bladder implantation[292].

AAT is mutagenic in S. typhimurium TA 1538[29], TA 100[276] and TA 98[276] in the presence of a rat-liver microsomal system. It produces mutants resistant to T1 bacteriophage in E. coli B/r[293] and auxotrophic mutants in Neurospora[294,295], and was active in the in vitro transformation of Golden Syrian hamster embryo cells at a dose of 1.0 μg/ml[95].

AAT appears to be N-hydroxylated in vivo because 4,4'-bis(o-tolylazo)-2,2'-dimethylazobenzene has been isolated from the liver of mice treated with AAT[278]. This metabolic activation is confirmed indirectly by the demonstration that AAT interacts in vivo with DNA, RNA and protein[296-298].

The following metabolites have been identified in the bile of rats dosed with AAT: the N-glucuronide of AAT, the N-glucuronide of the corresponding 2-hydroxymethyl derivative (e.g., 4-amino-2-hydroxymethyl-3'-methylazobenzene), the sulfate and glucurondies of 4-amino-4'-hydroxyazotoluene and the N-glucornide-O-sulfate of 4'-hydroxyazotoluene[299]. In rabbits, para-toluenediamine is formed and is excreted mainly as the diacetyl derivative[300].

4. para-Dimethylaminoazobenzene (N,N-dimethyl-4-aminoazobenzene; N,N-dimethyl-4-(phenylazo)-benzenamine; N,N-dimethyl-4-phenylazoaniline; butter yellow; DAB: ⟨O⟩—N=N—⟨O⟩—N(CH$_3$)(CH$_3$)) can be synthesized via the reaction of aniline with dimethylaniline; followed by the addition of sodium nitrite in a sodium hydroxide solution but it is not known that this is the commercial procedure[261]. DAB is used for coloring polishes and other wax products, polystyrene, gasoline, soap and as an indicator for the determination of free hydrochloric acid in gastric juice.

Although three U.S. producers in 1972 reported production of DAB, separate production data were not reported and it was included in a miscellaneous category with at least 20 other colors with a total production of 465,500 kg[301]. Although currently there may be as many as 14 producers of DAB in western Europe, production is believed to be small. In Japan, 3 manufacturers reported production of 5000 kg in 1972[261].

DAB and its derivatives have been extensively employed in experimental studies on azo dye carcinogenesis[37,261,302]. DAB is carcinogenic in rats producing liver tumors by oral, s.c. and i.p. routes of administration[303-310]. Dose-response aspects of the production of liver tumors in rats following oral administration of DAB was highlighted in the classic studies of Druckrey[305-308]. When daily dosages of 1, 3, 10, 20 or 30 mg DAB/rat were administered for the lifespan, all doses produced liver tumors, the induction time being inversely proportional to the daily dose and ranging between 34 days in rats given 30 mg DAB/day to 700 days in animals given 1 mg/DAB/day[308]. DAB is carcinogenic in dogs[311] producing bladder tumors following its oral administration. In studies of short duration, where the adequacy of the dose levels used was not known, DAB was reported to be non-carcinogenic in hamsters[312], and guinea pigs[313], while the results of oral administration in mice were doubtful[314,315].

DAB is mutagenic in S. typhimurium TA 100 and TA 98 only when activated with an S-9 mix obtained from rat liver induced with the PCB, KC-500[276]. It is also mutagenic in the aerobic liquid test with S. typhimurium TA 100 when activated with S-9 mix from rat liver microsome prepration from animals induced with Aroclor 1254[316]. Although DAB did not induce reverse mutations in S. typhimurium TA 1538 in the presence of rat liver microsomal systems, they were produced in this strain when a urinary metabolite from rats fed DAB, following treatment with β-glucuronidase, was dissolved in DMSO and assayed in the presence of a rat-liver microsomal system[317]. DAB was non-mutagenic when tested in Drosophila for the production of X-linked recessive lethals[271]. It produced mutations at specific DNA regions (r-RNA and t-RNA genes) in Drosophila[318,319]. N-methyl-4-aminoazobenzene, a demethylated metabolite of DAB was only mutagenic in S. typhimurium TA 100 and TA 98 in the presence of S-9 mix from rat liver microsomes[276], but was inactive in producing forward mutations in B. subtilis transforming DNA[320]. N-acetoxy-N-methyl- and N-benzoyloxy-N-methyl-4-aminoazobenzene (and their 4'-methoxycarbonyl derivatives) were also mutagenic in S. typhimurium TA 100 and TA 98, but did not require metabolic activation by an S-9 mix[276]. The correlation of the mutagenicity of DAB and its derivatives with carcinogenicity can be tabulated as follows:

Compound	Mutagenicity TA100 S-9 mix (-)+	TA98 (-)+	Carcinogenicity
p-Dimethylaminoazobenzene (DAB)	- +	- +	+
3'-methyl-DAB	- +	- +	+
N-methyl-4-aminoazobenzene (MAB)	- +	- +	+
2-methyl-DAB	- +	- +	+
4-aminobenzene (AB)	- +	- +	+
o-aminoazotoluene (o-AT)	- +	- +	+
3-methoxy-AB	- +	- +	+
N-hydroxy-AB	- +	- +	-
N-acetoxy-MAB	+ +	+ +	
N-benzoyloxy-MAB	+ +	+ +	+
4'-methoxycarbonyl-N-acetoxy-MAB	+ +	+ +	
4'-methoxycarbonyl-N-benzoyloxy-MAB	+ +	+ +	
N-hydroxy-MAB	- +	+ +	
4-methoxycarbonyl-N-hydroxy-MAB	- +	+ +	
4'-methoxycarbonyl-MAB	- -	- -	

All the carcinogenic azo dyes and their derivatives tested were mutagenic[276]. These findings (as well as those reported by others[37,40,261]) suggest that azo dyes are metabolized to ultimate carcinogens which modify bases in DNA. This alteration in DNA bases probably results in mutation as well as carcinogenesis. According to Yahagi et al[276], N-acetoxy-MAB would appear to be a likely ultimate carcinogen.

The biotransformation of DAB has been extensively studied[37,261]. The major reactions appear to be: stepwise demethylation[321-324], acetylation, C-hydroxylation[325], and reductive splitting of the azo linkage[326-328] (Figure 3)[37].

FIGURE 3. Major Routes of Metabolism of
para-Dimethylaminoazobenzene (DAB)[37]

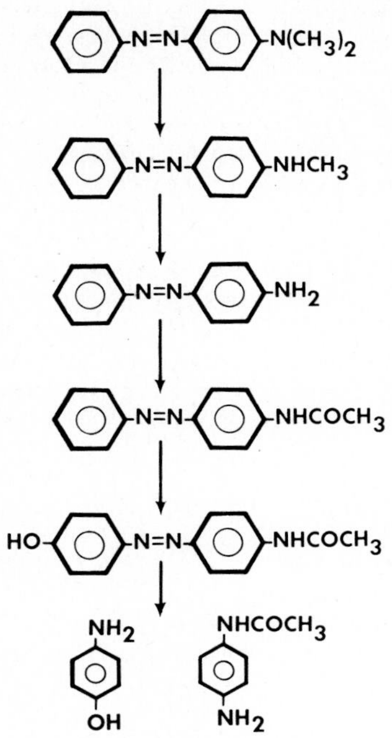

DAB and its metabolite, para-monomethylaminoazobenzene (MAB) are demethylated before reduction of the azo linkage[321]. The N-demethylase (as well as the azo reductase) is localized in the microsomal liver fractions and requires NADPH as electron donor[322,327].

While it has been reported that azo compounds that are demethylated slowly are weak carcinogens,[323] no clear-cut correlation between rate of demethylation and carcinogenic action has been found[324].

Ring hydroxylation by rat liver of DAB and its metabolites, MAB, aniline and para-phenylenediamine derivatives, has been demonstrated in vivo and in vitro[325].

The urine of salt fed DAB contained para-aminophenol, N-acetyl-para-aminophenol, para-phenylenediamine and N,N'-diacetyl-para-phenylenediamine, illustrating that the azo linkage is reduced in vivo[326]. Rat liver homogenates containing DAB formed N,N'-dimethyl-para-phenylenediamine and aniline by reductive cleavage[327]. The azo linkage of DAB is readily reduced by many strains of bacteria[328].

Although N-hydroxylation has been demonstrated in rat liver for several carcinogenic aromatic amines[40] there is no direct evidence for the N-hydroxylation of MAB. Attempts to prepare N-hydroxy-MAB were unsuccessful because of its instability. However, N-benzoyloxy-MAB is a potent carcinogen and reacts with nucleic acids and proteins to give the same products as obtained from rat liver nucleic acids and protein after treatment with DAB or MAB in vivo[329]. Data concerning N-hydroxylation and the type of interaction of N-benzoyloxy-MAB with proteins and nucleic acids have led to the suggestion that esterification of N-hydroxy-MAB may give rise to reactive metabolites[40]. In vitro experiments in which N-hydroxy MAB was converted by soluble sulphotransferase of rat liver to a reactive metabolite which reacted with methionine to form 3-methylmercapto-4-MAB has tended to further support this hypothesis[330].

The detailed mechanism by which DAB (or the azo dyes) induce cancer has not been fully elucidated[37]. The azo group in DAB is essential to the carcinogenic action of the chemical. At least one methyl groups appears to be essential for carcinogenicity to rat liver since N-methyl-aminoazobenzenes are in the main, equipotent to the corresponding dimethylaminoazobenzene whereas 4-aminoazobenzene itself is apparently noncarcinogenic. The potential relevance of protein and DNA binding by DAB to carcinogenesis has also been suggested[37,40,261].

5. <u>Chrysoidine</u> (4-(phenylazo)-1,3-benzenediamine, monohydrochloride, C.I. Basic Orange 2, (⟨O⟩-N=N-⟨O⟩-NH$_2$)HCl with NH$_2$) can be synthesized by coupling diazotized aniline with <u>meta</u>-phenylenediamine, although it is not known whether this is the method used for commercial production[261]. It is used as a colorant in textiles, paper, leather, inks and wood and biological stains[261].

In 1972, seven U.S. manufacturers reported a total production of about 203,000 kg. Although chrysoidine may be produced in western Europe by as many as 12 manufacturers it has been estimated that a few hundred thousand kg per year are produced currently. In Japan, one manufacturer produced 70,000 kg of chrysoidine in 1972 and 67,000 kg in 1973[261].

Chrysoidine is carcinogenic in C57Bl mice following its oral administration, producing liver-cell tumors, leukemia and reticulum-cell sarcomas[331]. It is mutagenic at 50 and 100 μg/plate in <u>S. typhimurium</u> TA 1538 only in the presence of a rat liver post-mitochondrial fraction[214]. One of the possible reduction products of chrysoidine, 1,2,4-triaminobenzene was also mutagenic in the same system while aniline or its p-hydroxylated metabolite, p-aminophenol were inactive[214].

6. <u>Sudan II</u> (1-[(2,4-dimethylphenyl)azo]-2-naphthalenol; 1-(2,4-xylylazo)-2-naphthol; C.I. Solvent Orange-7; H$_3$C⟨O⟩-N=N-⟨O⟩⟨O⟩ with CH$_3$ and OH) can be synthesized via coupling of diazotized 2,4-dimethylaniline with 2-naphthol, although it is not known whether this method is used for commercial production[261]. It is used for coloring of oils, waxes, hydrocarbon solvents for polishes, candles and polystyrene resins. In Japan it is used to color petroleum products, plastics, shoe polish, cosmetics and some drugs[261].

In 1971, three producers in the U.S. reported production of 34,500 kg and in western Europe, there may be as many as eight producers of Sudan II. One producer in the U.K. is believed to manufacture some several thousand kg of the dye per year while in Japan, three manufacturers produced 3300 kg in 1972[261].

Although Sudan II has been tested for carcinogenicity in mice[28] and rats[332-334] by oral subcutaneous routes, IARC[261] considers that the results of these studies cannot be evaluated because of the inadequacy either of the number of animals used, the duration of the experiment or the degree of reporting. Sudan II induced a high incidence of bladder carcinomas when tested in mice by bladder implantation [292].

Sudan II was mutagenic in S. typhimurium TA 1538 only in the presence of a rat liver post-mitochondrial fraction[214].

Information as to the biotransformation of Sudan II is limited. After its administration to rats in corn oil by gavage, 1-amino-2-naphthyl sulfate and 8 other metabolites were excreted in the urine[335]. When 8-^{14}C-labelled 1-xylylazo-2-naphthol was administered to rats, the urine contained 14% and the feces 86% of the total radioactivity. Eleven metabolites were identified including 1-amino-2-naphthylsulfate and 1-amino-2-naphthyl glucuronide[336].

6a. Ponceau MX (4-[(2,4-dimethylphenyl)azo]-3-hydroxy-2,7-naphthalene disulphonic acid, disodium salt, C.I. Food Red 5; H$_3$C-⌬(CH$_3$)-N=N-⌬(SO$_3$Na)(SO$_3$Na)) is produced commercially by coupling diazotized 2,4-dimethylaniline with 2-naphthol-3,6-disulphonic acid (R-acid). In the U.S. it is used principally as a textile and leather dye, as well as to color inks, paper, pigment and wood stains. It has been used to color foods in many countries throughout the world[261].

In 1972, three manufacturers in the US produced 17,000 kg. While total production data are not available for western Europe, it is believed that there may be as many as 15 producers of Ponceau MX; several thousand kg per year are believed manufactured in the U.K. In Japan, one manufacturer produced 3000 kg in 1972 and 4000 kg in 1973[261].

Ponceaux MX is carcinogenic in mice producing liver-cell tumors and possibly intestinal tumors following its oral administration[337]. In Wistar rats, fed 10,000 or 50,000 mg of the dye in the diet, it produced liver tumors[338]. In both mouse and rat studies, a dose-response effect was noted.

Ponceaux MX was inactive when tested in S. typhimurium TA 1538 with or without metabolic activation[214].

Data on the biotransformation of Ponceaux MX are limited. When rabbits were fed the dye, 2.5% unchanged material, 35% 2,4-dimethylaniline, 6% 3-methyl-4-acetamidobenzoid acid, 3-methyl-4-aminobenzoic acid and 2,4-dimethylphenylsulphamate were recovered in the urine. The dye is not hydroxylated by rabbits[339].

7. __Trypan Blue__ (3,3'-{[3,3'-dimethyl(1,1'-biphenyl)-4,4'-diyl]bis(azo)}bis(5-amino-4-hydroxy-2,7-naphthalenedisulphonic acid, tetrasodium salt; C.I. Direct Blue 14;

NaO$_3$S—[naphthalene]—SO$_3$Na N=N—[phenyl-CH$_3$]—[phenyl-CH$_3$]—N=N—NaO$_3$S—[naphthalene]—SO$_3$Na
     NH$_2$ OH   OH NH$_2$

can be produced by coupling tetrazotized ortho-toluidine with two equivalents of 8-amino-1-naphthol-3,6-disulphonic acid (H acid) in alkaline solution, but it is not known whether this is the commercial procedure. Samples of Trypan Blue have been found to contain varying amounts of its synthetic precursors including ortho-toluidine as well as components such as 8-amino-2-[4'-(3,3'-dimethylbiphenylazo)]-1-naphthol-3,6-disulphonic acid; 8-amino-2-[4'-(3,3'-dimethyl-4-hydroxybiphenylazo)]-1-naphthol-3,6-disulphonic acid; and 8-amino-2-[4'-(3,3'-dimethyl-4-aminobiphenylazo)]-1-naphthol-3,6-disulphonic acid[261]. Trypan Blue has been used for dyeing textiles, leather, and paper and as a stain in biological investigations. It is apparently not approved for food use anywhere.

Production data are unavailable for Trypan blue although it is known that four companies produced the material in 1972. It is believed that there may be as many as six producers of the dye in western Europe with an estimated total annual production of a few hundred thousand kg.

Trypan blue is carcinogenic in rats[340-343] following its i.p. or s.c. administration, producing reticulum-cell sarcomas, mainly of the liver, as well as fibrosarcomas at the site of injection.

Data on the mutagenicity of Trypan blue is extremely limited. It has been reported to produce deletions in __Aspergillus nidulans__[344,345]. No mutagenicity data are available involving mammalian metabolic systems.

Information is equally sparse concerning the biotransformation of Trypan blue. Following i.p. and s.c. injections into mice or rats, it is rapidly absorbed and widely distributed throughout the body. The dye appears bound to serum proteins, with an exponential decay of plasma levels due to rapid excretion in the urine and uptake by the reticulo-endothelial system[261,346,347]. Trypan blue is reduced __in vitro__ by a rat liver enzyme to __ortho__-toluidine and 2,8-diamino-1-naphthol-3,6-disulphonic acid[348].

8. __Amaranth__ (3-hydroxy-4-[(4-sulpho-1-naphthalenyl)azo]-2,7-naphthalenedisulphonic acid, trisodium salt, C.I. Food Red 9, FD and C Red No. 2,

NaO$_3$S—[naphthalene]—N=N—[naphthalene with OH, SO$_3$Na]—SO$_3$Na

can be produced commercially by the coupling of diazotized naphthionic acid (1-naphthylamine-4-sulphonic acid) with 2-naphthol-3,6-disulphonic acid[261]. It is used for dyeing textiles, paper, phenol-formaldehyde resins, wood and leather as well as for food use in many countries throughout the world[261].

The U.S. production of amaranth in 1972 by six manufacturers amounted to 440,000 kg. In western Europe there are an estimated 18 producers of the dye, with a combined total annual production of approximately 300,000 kg. In Japan in 1973, 5 manufacturers produced about 97,000kg of Amaranth[261].

The Joint FAO/WHO Expert Committee on Food Additives currently recommends an acceptable daily intake (ADI) of amaranth for man of 0-0.75 mg/kg body weight[261], 50% less than the previous ADI quoted by that committee[249].

Although amaranth has been tested in mice, rats and dogs by oral administration and in rats by s.c. administration, the shortcomings of the studies were considered by IARC to be of such a nature as to prevent an evaluation of the carcinogenicity of this compound[261]. For example, two oral studies in rats indicated a carcinogenic effect, however, in one study the compound used contained 25-35% of unspecified impurities[350] while in the other, the absence of spontaneous tumors in control animals after 33 months was considered to be very unusual[351]. Other oral studies in mice[252], rats[353,354] and dogs gave negative results but were inadequately reported[261].

In limited mutagenicity studies reported thus far, amaranth has been found non-mutagenic in S. typhimurium TA 1535, TA 100, TA 1537, TA 1538, and TA 98 with and without metabolic activation[316].

Amaranth is rapidly reduced by bacteria[355] rat liver homogenates, and intestinal contents[356]. The products of reductive cleavage of amaranth include: 1-amino-4-naphthalene sulphonic acid and 1-amino-2-hydroxy-3,6-naphthalene disulphonic acid (R-amino salt) were found in the urine of rats fed this agent[357].

9. <u>Miscellaneous Azo Dyes</u>. While no attempt has been made to examine more than a very small fraction of the azo dyes used in industry and in food technology it should be noted that an additional number have been reported to be mutagenic and/or carcinogenic. The structures of some addition azo dyes which are mutagenic or promutagenic for S. typhimurium are shown in Figure 4[316].

Acid Alizarin Yellow R is directly mutagenic in TA 1538 and TA 98 and mutagenic in TA 1537 with S-9 microsomal activation. Alizarian Yellow GG is mutagenic in TA 100, TA 1537, TA 1538 and TA 98 strains with or without metabolic activation. Acid Alizarian Red B is mutagenic in TA 1535, TA 100, TA 1538, and TA 98 with metabolic activation. Methyl Red is mutagenic in TA 1538 and TA 98 strains with metabolic activation. Acid Alizarin Violet N required prior chemical reduction and microsomal activation before being mutagenic in TA 1535, TA 100, TA 1538, and TA 98 strains while Sudan IV required the same pretreatment for its mutagenicity in TA 100, TA 1538 and TA 98 strains[316].

In addition to these compounds, four other azo dyes (e.g., Trypan blue, Fast Garnet GBC and Orange I) are considered likely to be mutagenic although the responses with these materials have been highly variable thus far.

Twelve azo dyes (e.g., Citrus Red No. 2, Orange B, Red No. 4, Red No. 40, Yellow No. 5, Yellow No. 6, Ponceau 3R, Red No. 2, Orange I, Ponceau S, Sudan I and Carmorsine) were all negative when tested with and without microsomal activation in TA 1535, TA 100, TA 1537, TA 1538 and TA 98 strains at concentrations of 50 to 500 µg/plate[316].

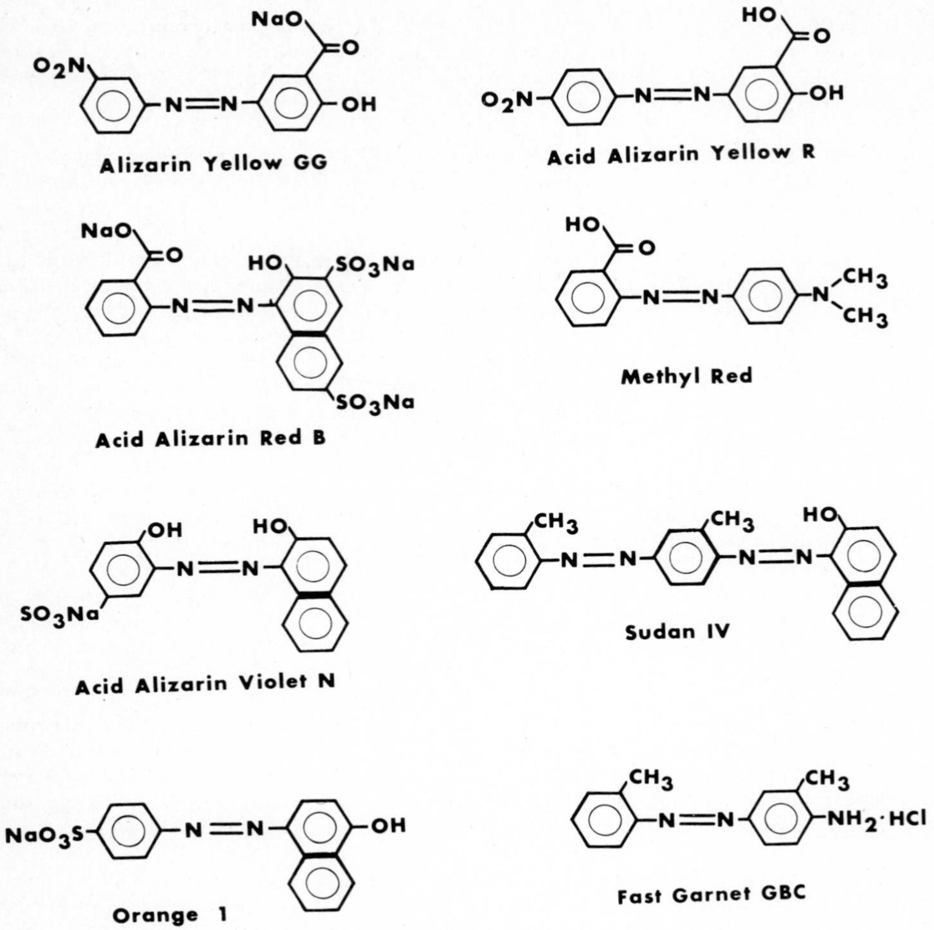

Fig 4 Structures of some azo dyes mutagenic or promutagenic for Salmonella typhimurium

Recently, the benzidine derived azo dyes Direct Black 38, Direct Brown 95 and Direct Blue 6 were reported as being hepatotoxic producing biliary hyperplasia and cholangiofibrosis in NCI 90-day feeding tests with Fisher 344 rats. All three produced focal cellular alteration and neoplastic nodules similar to those caused by many hepatic carcinogens[358]. The urine of the test animals did contain benzidine at the close of the study. These dyes are used in printing inks and perhaps in textiles. In 1975, U.S. production of Direct Black 38 was about 2,168,000 pounds, that of Direct Brown 95 was 346,000 pounds. In 1973, U.S. sales of Direct Blue 6 totalled 327,000 pounds[358].

References

1. Zavon, M. R., Hoegg, W., and Bingham, Benzidine exposure as a cause of bladder tumors, Arch. Environ. Hlth., 27 (1973) 1-11
2. Ehrlicher, H., von., Benzidin in arbeitsmedizinischer sicht, Ztb. Arbeitsmed. Arbeitsschutz., 8 (1958) 201
3. IARC, Monographs on the Evaluation of Carcinogenic Risk of Chemicals to Man, Vol. 1, International Agency for Research on Cancer, Lyon (1972) 80-86
4. U.S. Tariff Commission, Synthetic Organic Chemicals, U.S. Production and Sales, 1962, Government Printing Office, Washington, DC (1963)
5. Anon, Chem. Eng. News, Feb. 11 (1974) p. 12
6. Mitre Corporation, "Scoring of Organic Air Pollutants, Chemistry, Production and Toxicity of Selected Organic Chemicals", McLean, VA (1976)
7. Environmental Protection Agency, Status Assessment of Toxic Chemicals. 5. Benzidine, Industrial Pollution Control Division, Cincinatti, Ohio, Sept. 6 (1977)
8. Haley, T. J., Benzidine revisited: A review of the literature and problems associated with the use of benzidine and its congeners, Clin. Toxicol., 8 (1975) 13-42
9. U.S. Department of Labor, Occupational Safety and Health Standards, Part 1910 Carcinogens, Federal Register, 39 (20) (1974) 3756-3795
10. Shackelford, W. M., and Keith, L. H., Frequency of Organic Compounds Identified in Water, EPA Publication No. EPA/600/4-76/062, Environmental Protection Agency, Athens, GA (1976)
11. ACHIG, Documentation of the Threshold Limit Values, 3rd ed., American Conference of Governmental Industrial Hygienists, Cincinatti, Ohio (1971)
12. Lu, P. Y., Metcalf, R. L., Plummer, N., and Mandel, D., The environmental fate of three carcinogens: benzo-(α)-pyrene, benzidine, and vinyl chloride evaluated in laboratory model ecosystems, Arch. Environ. Contam. Toxicol. 6 (1977) 129-142
13. Meigs, J. W., Sciarini, L. J., and Van Sandt, W. A., Skin penetration by diamines of the benzidine group, Arch. Ind. Hyg., 9 (1954) 122
14. Scott, T. S., The incidence of bladder tumours in a dyestuff factory, Brit. J. Ind. Med., 9 (1952) 127
15. Scott, T. S., Occupational bladder cancer, Brit. Med. J., 2 (1964) 302
16. Hueper, W. C., Occupational and Environmental Cancers of the Urinary System, Yale University Press, New Haven and London (1969)
17. Case, R. A. M., Hosker, M. E., McDonald, D. B., and Pearson, J. T. (1954) Tumors of the urinary bladder in workmen engaged in the manufacture and use of certain dyestuff intermediates in the British Chemical Industry. I. The role of aniline, benzidine, α-naphthylamine and β-naphthylamine, Brit. J. Ind. Med., 11 (1954) 75
18. Goldwater, L. J., Rosso, A. J., and Kleinfeld, M., Bladder tumors in a coal tar dye plant, Arch. Env. Hlth., 11 (1965) 814
19. Ubelin, F., von., and Pletscher, A., Atiologie und prophylaxe gewerblicher tumoren in der farbstoff-industrie, Schweiz. Med. Wschr., 84 (1954) 917
20. Mancuso, T. F., and El-attar, A. A., Cohort study of workers exposed to beta-naphthylamine and benzidine, J. Occup. Med., 9 (1967) 277
21. Barsotti, M., and Vigliani, E. C., Bladder lesions from aromatic amines, Statistical considerations and prevention, Arch. Ind. Hyg. Occup. Med., 5 (1952) 234
22. Billiard-Duchesne, J. L., Les amino-tumuers de la vessie, J. Urol. Med. Chir., 65 (1959) 748

23. Parkes, H. G., The Epidemiology of the Aromatic Amine Cancers, In "Chemical Carcinogens", ed., Searle, C. E., ACS Monograph No. 173, American Chemical Society, Washington, DC (1976) 462-480
24. Boyland, E., Harris, J., and Horning, E. S., The induction of carcinoma of the bladder in rats with acetamidofluorene, Brit. J. Cancer, 8 (1954) 647
25. Saffiotti, U., Cefis, F., Montesano, R., and Sellakumar, A. R., Induction of bladder cancer in hamsters fed aromatic amines, In: "Bladder Cancer: A Symposium" eds, Deichmann, W., and Lampe, K. F., Aesculapius Publ. Co., Birmingham, AL (1967) p. 129
26. Sellakumar, A. R., Montesano, R., and Saffiotti, U., Aromatic amines carcinogenicity in hamsters, Proc. Am. Ass. Cancer Res., 10 (1969) 78
27. Spitz, S., Maguigan, W. H., and Dobriner, K., The carcinogenic action of benzidine, Cancer, 3 (1950) 789
28. Bonser, G. M., Clayson, D. B., and Jull, J. W., The induction of tumors of the subcutaneous tissues, liver and intestine in the mouse by certain dyestuffs and their intermediates, Brit. J. Cancer, 10 (1956) 653
29. Ames, B. N., Durston, W. E., Yamasaki, E., and Lee, F. D., Carcinogens are mutagens: A simple test system combining liver homogenates for activation and bacteria for detection, Proc. Natl. Acad. Sci. (US) 70 (1973) 2281-2285
30. Garner, R. C., Walpole, A. C., and Rose, F. L., Testing of some benzidine analogues for microsomal activation to bacterial mutagens, Cancer Letters, 1 (1975) 39-42
31. Ferretti, J. J., Lu, W., and Liu, H. B., Mutagenicity of benzidine and related compounds employed in the detection of hemoglobin, Am. J. Clin. Pathol., 67 (1977) 526-527
32. Lazear, E. J., and Louie, S. C., Mutagenicity of some congeners of benzidine in the Salmonella typhimurium assay systems, Cancer Letters, 4 (1977) 21-25
33. Haworth, S., Lawlor, T., Voytek, P., Carroll, J., and Guarraia, L., Abstracts of 9th Annual Meeting Environmental Mutagen Society, San Francisco, CA, March 9-13 (1978) 34
34. Burrell, A. D., Howard, M. A., and Andersen, J. J., Metabolic activation of UV mimetic carcinogens in a differential toxicity test using cultured human cells, Abstract of 9th Annual Meeting Environmental Mutagen Society, San Francisco, CA, March 9-13 (1978) 38
35. Fahmy, M. J., Fahmy, O. G., Mutagenicity of hair dye components relative to the carcinogen benzidine in Drosophila melanogaster, Mutation Res., 56 (1977) 31-38
36. Urwin, C., Richardson, J. C., and Palmer, A. K., An evaluation of the mutagenicity of the cutting oil preservative Grotan BK, Mutation Res., 40 (1976) 43
37. Clayson, D. B., and Garner, R. C., Carcinogenic Aromatic Amines and Related Compounds, In "Chemical Carcinogens" ed., Searle, C. E., ACS Monograph No. 173, American Chemical Society, Washington, DC (1976) 366-461
38. Miller, J. A., and Miller, E. C., The metabolic activation of carcinogenic aromatic amines and amides, Progr. Exptl. Tumor Res., 11 (1969) 273
39. Miller, J. A., and Miller, E. C., Chemical carcinogenesis: Mechanisms and approaches to its control, J. Natl. Cancer Inst., 47 (1971) V-XIV
40. Miller, J. A., Carcinogenesis by chemicals: An overview-G.H.A. Clowes Memorial Lecture, Cancer Res., 30 (1970) 559-576
41. Gutmann, H. R., Malejka-Giganti, D., Barry, E. J., and Rydell, R. E., On the correlation between the hepato carcinogenicity of the carcinogen N-2-fluorenylacetamide and its metabolite activation by the rat, Cancer Res., 32 (1972) 1554-1560

42. Arcos, J. C., and Argus, M. F., "Chemical Induction of Cancer", Vol. 11B, Academic Press, New York (1974) pp. 23-37
43. Kadlubar, F. F., Miller, J. A., and Miller, E. C., Hepatic microsomal N-glucuronidation and nucleic acid binding of N-hydroxy arylamines in relation to urinary bladder carcinogenesis, Cancer Research, 37 (1977) 805-814
44. Baetcke, K., Aromatic Amines Program: Mechanistic Approaches to Carcinogenesis, National Center for Toxicological Research, Nov. 6 (1976)
45. IARC, Monographs on the Evaluation of Carcinogenic Risk of Chemicals to Man, Vol. 4, International Agency for Research on Cancer, Lyon (1974) 49-55
46. Stanford Research Institute, A study of industrial data on candidate chemicals for testing, Prepared for U.S. Environmental Protection Agency, EPA-560/5-77-06, Menlo Park, CA, August (1977)
47. Anon, Final rules set for exposure to carcinogens, Chem. Eng. News, Feb. 11 (1974) p. 12
48. The Society of Dyes and Colourists, "Colour Index", 3rd ed., 4 (1971) 3272-3275; 3290-3291; 3294-3295; 3304-3305; 4142
49. MacIntyre, I., Experience of tumors in a British plant handling 3,3'-dichlorobenzidine, J. Occup. Med., 17 (1975) 23-26
50. Pliss, G. B., The blastomogenic action of dichlorobenzidine, Vop. Onkol., 5 (1959) 524
51. Stula, E. F., Sherman, H., and Zapp, J. A., Jr., Experimental neoplasia in CLR-CD rats with the oral administration of 3,3'-dichlorobenzidine, 4,4'-methylenebis-(2-chloroaniline) and 4,4'-methylenebis(2-methylaniline), Toxicol. Appl. Pharmacol. 19 (1971) 380
52. Pliss, G. B., On some regular relationships between carcinogenicity of aminodiphenyl derivatives and the structure of substances, Acta. Un. Int. Cancer, 19 (1963) 499
53. Kellner, H. M., Christ, O. E., and Lotesch, K., Animal studies on the kinetics of benzidine and 3,3'-dichlorobenzidine, Arch. Toxikol., 31 (1973) 61
54. Hirai, K., and Yasuhira, K., Mitochondrial oxidation of 3,3'-diaminobenzidine and related compounds and their possible relationship to carcinogenesis, Gann, 63 (1972) 665
55. IARC, Monographs on the Evaluation of Carcinogenic Risk of Chemicals to Man, Vol. 1, International Agency for Research on Cancer, Lyon (1972) 88-91
56. Pliss, G. B., and Zabezhinsky, M. A., Carcinogenic properties of ortho-tolidine (3,3'-dimethylbenzidine), J. Natl. Cancer Inst., 45 (1970) 283
57. Dieteren, H. M. L., The biotransformation of o-tolidine, Arch. Environ. Hlth. 12 (1966) 30
58. Sciarini, L. J., and Meigs, J. W., Biotransformation of the benzidines. Studies on diorthotolidine, dianisidine and chlorobenzidine: 3,3'-disubstituted congeners of benzidine (4,4'-diaminobiphenyl), Arch. Environ. Hlth., 2 (1962) 584
59. IARC, Monographs on the Evaluation of Carcinogenic Risk of Chemicals to Man, Vol. 4, International Agency for Research on Cancer, Lyon (1974) 41-47
60. U.S. Tariff Commission, Synthetic Organic Chemicals, United States Production and Sales, 1967, T.C. Publ. No. 295, Government Printing Office (1968)
61. U.S. Tariff Commission, Imports of Benzenoid Chemicals and Products, 1971, T.C. Publ. No. 466, Government Printing Office (1972)
62. The Society of Dyes and Colourists, "Colour Index", 3rd ed, 4 (1971) 2221, 2223, 2226, 2248, 2256, 2299, 3292, 3293, 3348, 3349, 4742
63. U.S. Tariff Commission, Synthetic Organic Chemicals, United States Production and Sales of Dyes, 1971, Preliminary, Government Printing Office, Washington, DC, October (1972)

64. U.S. Tariff Commission, Synthetic Organic Chemicals, United States Production and Sales of Organic Pigments, 1971, Preliminary, Government Printing Office, Washington, DC, August (1972)
65. Lurie, A. P., Benzidine and Related Diaminobiphenyls, In Kirk-Othmer's Encyclopedia of Chemical Technology, 2nd ed., Vol. 3, John Wiley & Sons, New York (1964) p. 417
66. Hueper, W. C., "Aniline tumors" of the bladder, Arch. Pathol., 25 (1938) 856
67. Pliss, G. B., On some regular relationships between carcinogenicity of aminodiphenyl derivatives and the structure of substance, Acta. Unio. Int. Cancer, 19 (1963) 499
68. Pliss, G. B., Concerning carcinogenic properties of o-tolidine and dianisidine, Gig. Tr. Prof. Zabol., 9 (1965) 18
69. Hadidian, Z., Fredrickson, T. N., Weisburger, E. K., Weisburger, J. H., Glass, R. M., and Mantel, N., Tests for chemical carcinogens. Report on the activity of derivatives of aromatic amines, nitrosamines, quinolines, nitroalkanes, amides, epoxides, aziridines and purine antimetabolites, J. Natl. Cancer Inst., 41 (1968) 985
70. Meigs, J. W., Brown, R. M., and Sciarini, L. J., A study of exposure to benzidine and substituted benzidines in a chemical plant, Arch. Industry. Hyg., 4 (1951) 533
71. Ghetti, G., Escrezione urinaria di alcune ammine aromatiche in lavoratori addetti alla produzione ed all'impiego di benzodina, benzidina sostituite e loro sali, Med. d. Lavoro, 51 (1960) 102
72. IARC, Monographs on the Evaluation of Carcinogenic Risk of Chemicals to Man, Vol. 4, International Agency for Research on Cancer, Lyon, (1974) 87-95
73. U.S. Tariff Commission, Synthetic Organic Chemicals, United States Production and Sales, 1948, Report No. 164, Second Series, Govt. Printing Office, Washington DC (1949)
74. Shreve, R. N., Amination by Reduction, In Kirk-Othmer's Encylcopedia of Chemical Technology, 2nd ed., Vol. 2, John Wiley & Sons, New York (1963) p. 82
75. U.S. Tariff Commission, Imports of benzenoid chemicals and products, 1967, TC Publication 264, Government Printing Office, Washington, DC Sept (1968)
76. U.S. Tariff Commission, Imports of benzenoid chemicals and products, 1971, TC Publication 466, Government Printing Office, Washington, DC July (1972)
77. Anon, Market Newsletters, Chem. Week, May 23 (1973) 31
78. The Society of Dyers and Colourists, Colour Index, 3rd ed., 4 (1971) 4806
79. U.S. Tariff Commission, Synthetic Organic Chemicals, United States Production and Sales, 1969, TC Publication 412, Government Printing Office, Washington, DC (1971)
80. U.S. Department of Agriculture, Economic Research Service, Quantities of Pesticides Used by Farmers in 1966, Agricultural Economic Report No. 179, April 1970
81. Treibl, H. G., Naphthalene derivatives, In Kirk & Othmer's Encylcopedia of Chemical Technology, 2nd ed., Vol. 13, John Wiley & Sons, New York (1967) p. 707
82. Masuda, Y., and Hoffmann, D., Quantitative determiantion of 1-naphthylamine and 2-naphthylamine in cigarette smoke, Analyt. Chem., 41 (1969) 650
83. Scott, T. C., "Carcinogenic and Chronic Toxic Hazards of Aromatic Amines", Elsevier, Amsterdam, and New York (1962)
84. Wignall, T. H., Incidence of disease of the bladder in workers in certain chemicals, Brit. Med. J., ii (1929) 258
85. Evans, E. E., Causative agents and protective measures in aniline tumour of the bladder, J. Urol., 38 (1937) 212

86. Gehrmann, G. H., Foulger, J. H., & Fleming, A. J., Occupational carcinoma of the bladder, In Proceedings of the 9th International Congress on Industrial Medicine, London, 1948, Bristol, Wright (1949) p. 472
87. DiMaio, G., Affections of the bladder due to aromatic amines, In Proceedings of the 9th International Congress on Industrial Medicine, London, 1948, Bristol, Wright, (1949) p. 476
88. Barsotti, M., and Vigliani, E. C., Lesioni vescicali da amine aromatiche, Med. d. Lavoro, 40 (1949) 129
89. Billiard-Duchesne, J. F., Les amino-tumeurs de la vessie en France. In Proceedings of the 9th International Congress on Industrial Medicine, London, 1948, Bristol, Wright, (1949) p. 507
90. Goldblatt, M. W., Acute haemorrhagic cystitis and versical tumours induced by chemical compounds in industry. In Proceedings of the 9th International Congress on Industrial Medicine, London, 1948, Bristol, Wright, (1949) p. 497
91. Bonser, G. M., Clayson, D. B., and Jull, J. W., Some aspects of the experimental induction of tumours of the bladder, Brit. Med. Bull., 14 (1958) 146
92. Radomski, J. L., Brill, E., Deichmann, W. B., and Glass, E. M., Carcinogenicity testing of N-hydroxy and other oxidation and decomposition products of 1- and 2-naphthylamine, Cancer Res., 31 (1971) 1461
93. Boyland, E., Busby, E. R., Dukes, C. E., Grover, P. L., and Manson, D., Further experiments on implantation of materials into the urinary bladder of mice, Brit. J. Cancer, 18 (1964) 575
94. McCann, J. E., Choi, E. S., Yamasaki, E., and Ames, B. N., Detection of carcinogens in the Salmonella/microsome test: Assay of 300 chemicals, Proc. Natl. Acad. Sci. (US), 72 (1975) 5135-5139
95. Dunkel, V. C., Wolff, J. S., and Pienta, R. J., In vitro transformation as a presumptive test for detecting chemical carcinogens, Cancer Bulletin, 29 (1978) 167-174
96. Corbett, T. H., Heidelberger, C., Dove, W. F., Determination of the mutagenic activity to bacteriophage of T4 of carcinogenic and non-carcinogenic compounds, Mol. Pharmacol., 6 (1970) 667-679
97. Perez, G., and Radomski, J. L., The mutagenicity of the N-hydroxy naphthylamines in relation to their carcinogenicity, Ind. Med. Surg., 34 (1965) 714-716
98. Bellman, S., Troll, W., Teebor, G., and Mukai, F., The carcinogenic and mutagenic properties of N-hydroxyaminonaphthalenes, Cancer Res., 28 (1968) 535-542
99. Kadlubar, F. F., Miller, J. A., and Miller, E. C., Reactivity of the carcinogen N-hydroxy-1-naphthylamine with nucleic acids, Proc. Am. Assoc. Cancer, 18 (1977) 300
100. Hyatt, J. C., US mulls rules for handling of chemicals that can lead to cancer in plant workers, Wall Street J. April 11 (1973) p. 36
101. U.S. Tariff Commission, Synthetic organic chemicals, US Production and Sales, 1955, Report No. 198, Second Series, Government Printing Office, Washington, DC (1956)
102. Kilner, E., and Samuel, D. M., "Applied Organic Chemistry", Interscience, New York (1960) pp. 262 and 266
103. Takemura, N., Akiyama, T., and Nakajima, C., A survey of the pollution of the Sumida river espeically on the aromatic amines in the water, Int. J. Air. Water Poll. 9 (1965) 665
104. Battye, R., Bladder carcinogens occurring during the production of "town" gas by coal carbonisation, In Transcripts of the XV International Congress on Occupational Health, Vienna, 1966 Vol. VI-2 (1966) p. 153

105. Hoffmann, D., Masuda, Y., and Wynder, E. L., α-naphthylamine and β-naphthylamine in cigarette smoke, Nature, 221 (1969) 254
106. Hueper, W. C., Wiley, F. H., and Wolfe, H. D., Experimental production of bladder tumors in dogs by administration of beta-naphtylamine, J. Industr. Hyg., 20 (1938) 46
107. Hueper, W. C., Occupational Tumours and Allied Diseases, Springfield, Ill. Thomas (1942)
108. Temkin, I. S., Industrial Bladder Carcinogenesis, Oxford, London, New York, Paris, Pergamon
109. Tsuji, I., Environmental and industrial cancer of the bladder in Japan, Acta Un. Int. Cancer, 18 (1963) 662
110. Veys, C. A., Two epidemiological inquiries into the incidence of bladder tumors in industrial workers, J. Natl. Cancer Inst., 43 (1969) 219
111. Vigliani, E. D., and Barsotti, M., Environmental tumors of the bladder in some Italian dye-stuff factories, Acta Un. Int. Cancer, 18 (1961) 669
112. Goldwater, L. J., Rosso, A. J., and Kleinfeld, M., Bladder tumors in a coal-tar dye plant, Arch. Environm. Hlth., 11 (1965) 814
113. Williams, M. H. C., "Cancer", ed. Raven, R. W., Vol. 3 Butterworth, London (1958) p. 377
114. Boyland, E., Busby, E. R., Dukes, C. E., Grover, P. L., and Manson, D., Further experiments on implantation of materials into the urinary bladder of mice, Brit. J. Cancer, 18 (1964) 575
115. Conzelman, G. M., Jr., and Moulton, J. E., Dose-response relationships of the bladder tumorigen 2-naphthylamine: A study in beagle dogs, J. Nat. Cancer Inst., 49 (1972) 193
116. Bonser, G. M., Clayson, D. B., Jull, J. W., and Pyrah, L. N., The carcinogenic activity of 2-naphthylamine, Brit. J. Cancer, 10 (1956) 533
117. Bonser, G. M., Clayson, D. B., Jull, J.W., and Pyrah, L. N., The carcinogenic properties of 2-amino-1-naphthol hydrochloride and its parent amine 2-naphthylamine Brit. J. Cancer, 6 (1952) 412
118. Radomski, J. L., and Brill, E., The role of N-oxidation products of aromatic amines in the induction of bladder cancer in the dog, Arch. Toxicol., 28 (1971) 159
119. Bonser, G. M., Boyland, E., Busby, E. R., Clayson, D. B., Grover, P. L., and Jull, J. W., A further study of bladder implantation in the mouse as a means of detecting carcinogenic activity: Use of crushed paraffin wax or stearic acid as the vehicle, Brit. J. Cancer, 17 (1963) 127
120. Boyland, E., Dukes, C. E., and Grover, P. L., Carcinogenicity of 2-naphthylhydroxylamine and 2-naphthylamine, Brit. J. Cancer, 17 (1963) 79
121. Deichmann, W. B., and Radomski, J. C., J. Natl. Cancer Inst., 43 (1969) 263
122. Boyland, E., The biochemistry of cancer of the bladder, Brit. Med. Bull., 14 (1958) 153
123. Boyland, E., and Manson, D., The metabolism of 2-naphthylamine and 2-naphthylhydroxylamine derivatives, Biochem. J., 101 (1966) 84
124. Kadlubar, F. F., The role of N-hydroxy arylamine N-glucuronides in arylamine-induced urinary bladder carcinogenesis, Symposium on Conjugation Reactions in Drug Biotransformation, Turku, Finland, July 23-27 (1978)
125. IARC, Monographs on the Evaluation of Carcinogenic Risk of Chemicals to Man, Vol. 16, International Agency for Research on Cancer, Lyon (1978)
126. Kehe, H. J., and Kouris, C. S., Diarylamines, In: Kirk-Othmer's Encyclopedia of Chemical Technology, 2nd ed., Vol. 7, John Wiley & sons, New York (1965) 40-49

127. The Society of Dyers and Colourists, "Colour Index", 3rd ed., Vol. 4, Yorkshire, U.K. (1971) pp. 4223, 4454, 4808
128. U.S. International Trade Commission, Synthetic Organic Chemicals, U.S. Production and Sales, 1973, ITC PUblication 728, Washington, DC., US Government Printing Office, (1975) pp. 136, 139
129. U.S. International Trade Commission, Synthetic Organic Chemicals, U.S. Production and Sales, 1974, USITC Publication 776, Washington, DC., US Government Printing Office, (1976) pp. 42, 136, 139
130. U.S. International Trade Commission, Imports of Benzoid Chemicals and Products, 1974, USITC Publication 762, Washington, DC, US Government Printing Office, (1976) p. 25
131. US International Trade Commission, Synthetic Organic Chemicals, US Production and Sales, 1975, USITC Publication 804, Washington, DC, US Government Printing Office (1977) pp. 131, 134
132. US International Trade Commission, Imports of Benzoid Chemicals and Products, 1975, USITC Publication 806, Washington, DC, US Government Printing Office, (1977) p. 25
133. NIOSH (National Institute for Occupational Safety and Health) Current Intelligence Bulletin: Metabolic Precursors of a Known Human Carcinogen, beta-Naphthylamine, Rockville, MD (1976) pp. 1-3
134. The Chemical Daily Co., Ltd., 6376-Chemicals, Tokyo (1976) p. 536
135. Vyes, C. A., A study on the incidence of bladder tumors in rubber workers, Thesis for Doctorate of Medicine, Faculty of Medicine, University of Liverpool, Liverpool, U.K. (1973)
136. Mazanov, G. N., and Malakhova, T. V., Atmospheric conditions in the weighing department of a technical rubber product plant, Sb. Nauchn. Tr. Sanit. Tekh. 5 (1973) 166-169
137. Tsai, L. M., Hygienic characteristics of the rubber industry, Zdravokhr. Kas., 30 (1971) 57-58
138. Kvoryaninova, N. K., and Khorobrykh, V. V., Working conditions in a shop for the formation of SKMS-30 rubber at the Omsk synthetic rubber plant, Nauch. Tr. Omsk. Med. Inst., 88 (1969) 62-67
139. Stasenkova, K. P., Experimental materials for evaluating the toxicity of 1,3-butadiene rubber, Toksikol. Novykh. Prom. Khim. Veshchestv., 10 (1968) 90-99
140. Moore, R. M., Jr., Woolf, B. S., Stein, H. P., Thomas, A. W., and Finklea, J. F., Metabolic precursors of a known human carcinogen, Science, 195 (1977) 344
141. Kummer, R., and Tordoir, W. F., Phenyl-beta naphthylamine (PBNA) Another carcinogenic agent?, Tijdschr. Soc. Geneesk., 53 (1975) 415-419
142. Fox, A. J., and Collier, P. F., A survey of occupational cancer in the rubber and cablemaking industries: analysis of deaths occuring in 1972-1974, Brit. J. Ind. Med., 33 (1976) 249-264
143. Innes, J. R. M., Ulland, B. M., Valerio, M. G., Petrucelli, L., Fishbein, L., Hart, E. R., Pallotta, A. J., Bates, R. R., Falk, H. L., Gart, J. J., Klein, M., Mitchell, I., and Peters, J., Bioassay of pesticides and industrial chemicals for tumorigenicity in mice: a preliminary note, J. Nat. Cancer Inst., 42 (1969) 1101-1114
144. NTIS (National Technical Information Service) Evaluation of Carcinogenic, Teratogenic and Mutagenic Activities of Selected Pesticides and Industrial Chemicals, Vol. 1, Carcinogenic Study, Washington, DC, US Department of Commerce (1968)
145. IARC, Monographs on the Evaluation of Carcinogenic Risk of Chemicals to Man, Vol. 4, International Agency for Research on Cancer, Lyon (1974) 65-71, 79-85, 73-77

146. Anon, NIOSH warns that DDM may cause hepatitis in workers, Toxic Materials News, 3 (4) (1976) 26
147. Steinhoff, D., and Grundmann, E., Zur cancerogenen wirkung von 4,4'-diaminodiphenylmethan und 2,4'-diaminodiphenylmethan, Naturwiss, 57 (1970) 247
148. Munn, A., Occupational bladder tumors and carcinogens: Recent developments in Britian, In: "Bladder Cancer, A Symposium", (eds) Diechmann, W., and Lampe, K. F., Aesculapius Press, Birmingham, Ala. (1967) 187
149. Kopelman, H., Robertson, M. H., Sanders, P. G., and Ash, I., The epping jaundice, Brit. Med. J., i (1966) 514
150. Stula, E. F., Barnes, J. R., Sherman, H., Reinhardt, C. F., and Zapp, J. A., Jr., Urinary bladder tumors in dogs from 4,4'-methylene-bis(2-chloroaniline) (MOCA), J. Env. Pathol. Toxicol., 1 (1977) 31-50
151. Gianatasio, P. A., Polyurethane polymers, Part I., Chemistry and chacteristics, Rubber Age, July (1969) p. 51
152. Linch, A. L., O'Connor, G. B., Barnes, J. R., Killian, A. S., Jr., and Neeld, W. E., Jr., Methylene-bis-ortho-chloroaniline (MOCA®): Evaluation of hazards and exposure control, J. Am. Ind. Hyg. Assoc., 32 (1971) 802
153. Anon, Health research groups calls for emergency standard for MOCA, Toxic Materials News, 4 (31) (1977) 193
154. Russfield, A. B., Homburger, F., Boger, E., Weisburger, E. K., and Weisbirger, J. H., The carcinogenic effect of 4,4'-methylene-bis-(2-chloroaniline) in mice and rats, Toxicol. Appl. Pharmacol., 31 (1975) 47-54
155. Grundmann, E., and Steinhoff, D., Leber und lungen tumoren nach 3,3'-dichlor-4,4'-diaminodiphenylmethan bei ratten, Z. Krebsforsch, 74 (1970) 28
156. Stula, E. F., Sherman, H., and Zapp, J. A., Jr., Experimental neoplasia in CLR-CD rats with the oral administration of 3,3'-dichlroobenzidine, 4,4'-methylenebis (2-chloroaniline) and 4,4'-methylene-bis(2-methyl-aniline), Toxicol. Appl. Pharmacol., 19 (1971) 380
157. Stula, E. F., Sherman, H., Zapp, J. A., Jr., and Clayton, J. W., Jr., Experimental neoplasia in rats after oral administration of 3,3'-dichlorobenzidine, 4,4'-methylene-bis(2-chloroaniline) and 4,4'-methylene-bis-(2-methylaniline), Toxicol. Appl. Pharmacol., 31 (1975) 159-176
158. U.S. Tariff Commission, Synthetic Organic Chemicals, United States Production and Sales, 1963, TC Publ. 143, Government Printing Office, Washington, DC (1964)
159. Munn, A., Occupational bladder tumors and carcinogens: Recent developments in Britain, In: "Bladder Cancer" A Symposium (eds) Deichmann, W., and Lampe, K. F., Atsculapius Press, Birmingham, Ala. (1967) 187
160. IARC, Monographs on the Evaluation of Carcinogenic Risk of Chemicals to Man, Vol. 1, International Agency for Research on Cancer, Lyon (1972) 74-79
161. Melick, W. F. Escue, H. M., Naryka, J. J., Mezera, R. A., Wheeler, E. R., The first reported cases of human bladder tumors due to a new carcinogenxenylamine, J. Urol. (Baltimore) 74 (1955) 760
162. Melick, W. F., Naryka, J. J., and Kelly, R. E., Bladder cancer due to exposure to para-aminobiphenyl: A 17-year follow up, J. Urol. (Baltimore) 106 (1971) 220
163. Melamed, M. R., Koss, L. G., Ricci, A., and Whitmore, W. F., Cytohistological observations and developing carcinoma of urinary bladder in man, Cancer (Phila.) 13 (196) 67
164. Koss, L. G., Myron, A., Melamed, M. R., and Kelley, R. E., Further cytologic and histologic studies of bladder lesions in workers exposed to paraminodiphenyl: Progress Report, J. Natl. Cancer Inst., 43 (1969) 233

165. Koss, L. G., Myron, R., Melamed, M. R., Ricci, A., Melick, W. F., and Kelley, R. E., Carcinogenesis in the human urinary bladder, Observations after exposure to para-aminodiphenyl, New Engl. J. Med., 272 (1965) 767
166. Clayson, D. B., Lawson, T. A., and Pringle, J. A. S., The carcinogenic action of 2-aminodiphenylene oxide and 4-aminobiphenyl on the bladder and liver of C57X1F mouse, Brit. J. Cancer, 21 (1967) 755
167. Walpole, A. L., Williams, M. H. C., and Roberts, D. C., The carcinogenic action of 4-aminodiphenyl and 3:2'-dimethyl-4-aminodiphenyl, Brit. J. Ind. Med., 9 (1952) 255
168. Bonser, G. M., Precancerous changes in the urinary bladder, In: "The Morphological Precursor of Cancer" (ed) Servera, L., Perugia (1962) p. 435
169. Walpole, A. L., Williams, M. H. C., and Roberts, D, C., Tumours of the urinary bladder in dogs after ingestion of 4-aminodiphenyl, Brit. J. Ind. Med., 11 (1954) 105
170. Deichmann, W. B., Radomski, J. C., Anderson, W. A. D., Coplan, M. M., and Woods, F. M., The carcinogenic action of p-aminobiphenyl in the dog, Ind. Med. Surg., 27 (1958) 25
171. Deichmann, W. B., Radomski, J. L., Glass, E., Anderson, W. A. D., Coplan, M., and Woods, F. M., Synergism among oral carcinogens, Simultaneous feeding of four bladder carcinogens to dogs, Ind. Med. Surg., 34 (1965) 640
172. Deichmann, W. B., MacDonald, W. M., Coplan, M. M., Woods, F. M., and Anderson, W. A. D., Para-nitrobiphenyl, A new bladder carcinogen in the dog, Ind. Med. Surg., 27 (1958) 634
173. Gorrod, J. W., Carter, R. L., and Roe, F. J. C., Induction of hepatomas by 4-aminobiphenyl and three of its hydroxylated derivatives administered to newborn mice, J. Natl. Cancer Inst., 41 (1968) 403
174. Bonser, G. M., Bradshaw, L., Clayson, D. B., and Jull, J. W., A further study of the carcinogenic properties of ortho-hydroxy-amines and related compounds by bladder implantation in the mouse, Brit. J. Cancer, 10 (1956) 539
175. Bradshaw, L., and Clayson, D. B., Metabolism of two aromatic amines in the dog, Nature, 176 (1955) 974
176. Gorrod, J. W., Species differences in the formation of 4-amino-3-biphenylyl-β-D-glucosiduronate in vitro, Biochem. J., 121 (1971) 29P
177. Radomski, J. L., Rey, A. A., and Brill, E., Evidence for a glucuronic acid conjugate of N-hydroxy-4-aminobiphenyl in the urine of dogs given 4-aminobiphenyl, Cancer Res., 33 (1973) 1284-1289
178. Miller, J. A., Wyatt, C. S., Miller, E. C., and Hartman, H. A., The N-hydroxylation of 4-acetylaminobiphenyl by the rat and dog and the strong carcinogenicity of N-hydroxy-4-acetylaminobiphenyl in the rat, Cancer Res., 21 (1961) 1465-1473
179. IARC, Monographs on the Evaluation of the Carcinogenic Risk of Chemicals to Man, Vol. 16, International Agency for Research on Cancer, Lyon (1978) 83-95
180. U.S. International Trade Commission, Synthetic Organic Chemicals, US Production and Sales, 1974, USITC Publ. 776, Government Printing Office, Washington, DC (1977) 22, 36, 42, 49, 60, 62, 65, 74
181. The Society of Dyers and Colourists, Colour Index, revised 3rd ed., Vol. 5, Yorkshire, UK (1975) pp. 5079, 5091, 5126, 5128, 5132, 5133, 5197, 5209, 5285, 5287, 5288, 5292, 5294, 5295, 5297
182. Ames, B. N., Kammen, H. O., Yamasaki, E., Hair dyes are mutagenic: Identification of a variety of mutagenic ingredients, Proc. Natl. Acad. Sci. (US) 72 (1975) 2423-2427
183. Japan Dyestuff Industry Association, Statistics of Dyestuffs (1971-1975) Tokyo (1976)

184. Ito, N., Hiasa, Y., Konishi, Y., and Marugami, M., The development of carcinoma in liver of rats treated with m-toluylenediamine and the synergistic and antagonistic effects with other chemicals, Cancer Res., 29 (1969) 1137-1145
185. Umeda, M., Production of rat sarcoma by injections of propylene glycol solution of m-toluylenediamine, Gann, 46 (1955) 597-603
186. Giles, A. L., Chung, C. W., and Kommineni, C., Dermal carcinogenicity study by mouse-skin painting with 2,4-toluenediamine alone or in representative hair dve formulations, J. Toxicol. Environ. Hlth., 1 (1976) 433-440
187. Burnett, C., Larrman, B., Giovancchini, R., Welcott, G., Scala, R., and Keplinger, M., Long-term toxicity studies on oxidative hair dyes, Food Cosmet. Toxicol., 13 (1975) 353-357
188. Shah, M. J., Pienta, R. J., Lebherz, W. B., III, and Andrews, A. W., Comparative studies of bacterial mutation and hamster cell transformation induced by 2,4-toluenediamine, In: Proceedings of the 68th Annual Meeting of the American Association of Cancer Research, Denver, Color. 1977, Vol. 18 (eds) Weinhouse, S., Foti, M., and Bergbauer, P. A., Williams ¢ Wilkins, Baltimore (1977) Abst. No. 89, p. 23
189. Pienta, R. J., Shah, M. J., Lebherz, W. B., III, and Andrews, A. W., Correlation of bacterial mutagneicity and hamster cell transformation with tumorigencity induced by 2,4-toluenediamine, Cancer Letters, 3 (1977) 45-52
190. Blijleven, W. G. H., Mutagenicity of four hair dyes in Drosophila melanogaster, Mutation Res., 48 (1977) 181-186
191. Waring, R. H., and Pheasant, A. E., Some phenolic metabolites of 2,4-diaminotoluene in the rabbit, rat and guinea pig, Xenobiotica, 6 (1976) 257-262
192. Grantham, P. H., Glinsukon, T., Mohan, L. C. Benjamin, T., Roller, P. P., Mitchell, F. E., and Weisburger, E. K., Metabolism of 2,4-toluenediamine in the rat, Toxicol. Appl. Pharmacol., 33 (1975) 179
193. Hiasay, Y., m-Toluylenediamine carcinogenesis in rat liver, J. Nara. Med. Assoc. 21 (1970) 1-19
194. Anon, Modern trends in hair colourants, Soap, Perfumery and Cosmętics, May (1976) pp. 189-194
195. Wall, F. E., Bleaches, hair colorings and dye removers, In "Cosmetics Science and Technology", 2nd ed., Vol. 2 (eds) Balsam, M. S., and Sagarin, E., Wiley-Interscience, New York (1972) pp. 308-310, 313, 317
196. "The Society of Dyers and Colourists", "Colour Index", 3rd ed., Vol. 3, Yorkshire, UK (1971) p. 3262
197. Baier, E. J., Statement before the Subcommittee on Oversight and Investigations House Committee on Interstate and Foreign Commerce, Washington, DC, Jan. 23 (1978)
198. NIOSH, Current Intelligence Bulletin on 2,4-Diaminoaniside, Rockville, MD, Jan 13 (1978)
199. Anon, Hair dye element and cleaning solvent should be treated as carcinogens, NIOSH says
200. National Cancer Institute, Bioassay of 2,4-Diaminoanisole sulfate for possible carcinogenicity, National Indstitutes of Health, Bethesda, MD, DHEW PUbl. No. (NIH) 78-1334, Feb. 2 (1978)
201. Anon, NIOSH studies show excess cancers in cosmetologists, hairdressers, Chem. Reg. Reptr., 1 (46) (1978) 1626-1627
202. Palmer, K. A., Denunzio, A., and Green, A., The mutagenic assay of some hair dye components, using a thymidine kinase locus of L5178Y mouse lymphoma cells, J. Environ. Pathol. Toxicol., 1 (1977) 87-91

203. Mayer, V. W., and Goin, C. J., Genetic effects of some hair dye chemicals in Saccharomyces cerevisiae, Abstracts of 9th Annual Meeting Environmental Mutagen Society, San Francisco, CA, March 9-13 (1978), 53
204. Burnett, C., Loeks, R., and Corbett, J., Dominant lethal mutagenicity study on hair dyes, J. Toxicol. Environ. Hlth., 2 (1977) 657-662
205. Hossack, D. J. N., and Richardson, J. C., Examination of the potential mutagenicity of hair dye constituents using the micronucleus test, Experientia, 33 (1977) 377-378
206. Grantham, P. H., Mohan, L. C., Benjamin, T. T., Roller, P. P., and Weisburger, E. K., The metabolism of the dyestuff intermediate, 2,4-diaminoanisole in the rat, Toxicol. Appl. Pharmacol. (1977) in press
207. IARC, Monographs on the Evaluation of Carcinogenic Risk of Chemicals to Man, Vol. 16, International Agency for Research on Cancer, Lyon (1978) 111-142
208. The Society of Dyers and Colourists "Colour Index", 3rd ed., Vol. 4, Yorkshire, UK (1971) pp. 4644, 4822
209. Thirtle, J. R., Phenelediamines and toluenediamines, In Kirk-Othmer's Encyclopedia of Chemical Technology, 2nd ed., Vol. 15, John Wiley and Sons, New York (1968) pp. 216-224
210. U.S. International Trade Commission, Synthetic Organic Chemicals, US Production and Sales, 1975, USITC Publ. No. 804, Government Printing Office, Washington, DC (1977) p. 40, 49, 51, 60, 62, 64, 65, 72, 73
211. US Occupational Safety and Health Administration, Air Contaminants, US Code of Federal Regulations, Title 29, Part 1910.1000, pp. 581-582, 585
212. Boutwell, R. K., and Bosch, D. K., The tumor-promoting action of phenol and related compounds for mouse skin, Cancer Res., 19 (1959) 413-427
213. Saruta, N., Yamaguchi, S., and Nakatomi, Y., Sarcoma produced by subdermal administration of paraphenylenediamine, Kyushu J. Med. Sci., 9 (1958) 94-101
214. Garner, R. C., and Nutman, C. A., Testing of some azo dyes and their reduction products for mutagenicity using S. typhimurium TA 1538, Mutation Res., 4 (1977) 9-19
215. Kiese, M., Rachor, M., and Rauscher, E., The absorption of some phenylenediamines through the skin of dogs, Toxicol. Appl. Pharmacol., 12 (1968) 495-507
216. Ishidate, M., and Hashimoto, Y., The metabolism of p-dimethylaminoazobenzene and related compounds, I. Metabolites of p-dimethylaminoazobenzene in dog urine, Chem. Pharm. Bull., 7 (1959) 108-113
217. Ishidate, M., and Hashimoto, Y., Metabolism of 4-dimethylaminoazobenzene and related compounds, II. Metabolites of 4-dimethylaminoazobenzene and 4-aminoazobenzene in rat urine, Chem. Pharm. Bull., 10 (1962) 125-133
218. Stevenson, E. S., Dobriner, K., and Rhoads, C. P., The metabolism of dimethylaminoazobenzene (butter yellow) in rats, Cancer Res., 2 (1942) 160-167
219. The Society of Dyers and Colourists, "Colour Index", 3rd ed., Vol. 4, Yorkshire UK (1971) pp. 4643, 4825
220. Windholz, M. (ed) "The Merck Index", 9th ed., Merck & co., Rahway, NJ (1976) pp. 947-948
221. Saruta, N., Yamaguichi, S., and Matsuoka, T., Sarcoma produced by subdermal administration of metaphenylenediamine and metaphenylene diamineohydrochloride, Kyushu J. Med. Sci., 13 (1962) 175-180
222. Hanzlik, P. J., The pharmacology of some phenylenediamine, J. Ind. Hyg., 4 (1922) 448-462
223. IARC, Monographs on the Evaluation of Carcinogenic Risk of Chemicals to Man, Vol. 16, International Agency for Research on Cancer, Lyon (1978), 63-71, 73-82

224. Searle, C. E., and Jones, E. L., Effects of repeated applications of two semi-permanent hair dyes to the skin of A and DBA mice, Brit. J. Cancer, 36 (1977) 467-478
225. Venitt, S., and Searle, C. E., Mutagenicity and possible carcinogenicity of hair colourants and constituents, In: Environmental Pollution and Carcinogenic Risks, (eds) Rosenfeld, C., and Davis, W., IARC Scientific Publications No. 13, Lyon (1976) pp. 263-271
226. Searle, C. E., Harnden, D. G., Venitt, S., and Gyde, O. H. B., Carcinogenicity and mutagenicity tests of some hair colourants and constituents, Nature, 255 (1975) 506-507
227. Benedict, W. F., Morphological transformation and chromosome aberrations produced by two hair dye components, Nature, 260 (1976) 368-369
228. Kirkland, D. J., and Venitt, S., Cytotoxicity of hair colourant constituents: chromosome damage induced by two nitrophenylenediamines in cultured Chinese hamster cells, Mutation Res., 40 (1976) 47-56
229. Perry, P. E., and Searle, C. E., Induction of sister chromatid exchanges in Chinese hamster cells by the hair dye constituents 2-nitro-p-phenylenediamine and 4-nitro-o-phenylenediamine, Mutation Res., 56 (1977) 207-240
230. Bourgeois, C. F., Czornomaz, A. M., George, P., Belliot, J. P., Mainguy, P. R., and Watier, B., Specific determination of vitamin C (ascrobic and de-hydroascorbic acids) in foods, Analysis, 3 (1975) 540-548
231. IARC, Monographs on the Evaluation of Carcinogenic Risk of Chemicals to Man, Vol. 16, International Agency for Research on Cancer, Lyon (1978) 349-366
232. US International Trade Commission, Synthetic Organic Chemicals, US Production and Sales, 1974, USITC PUbl. 776, Government Printing Office, Washington, DC pp. 45, 51, 54, 57, 58, 61, 65, 66, 69, 70, 78, 81
233. US Occupational Safety and Health Administration, Air Contaminants, US Code of Federal Regulations, (1976) Title 29, part 1910,1000, p. 585
234. Winell, M., An international comparison of hygienic standards for chemicals in the work environment, Ambio, 4 (1975) 34-36
235. Goldstein, I., Studies on MAC values of nitro- and amino-derivatives of aromatic hydrocarbons and Psychotropic Drugs (eds) Horvath, M., Vol. 1, Elsevier, New York (1973) 153-154
236. Khlebnikova, M. I., Gladkova, E. V., Kurenko, L. T., Pshenitsyan, A. V., and Shalin, B. M., Problems of industrial health and health status of workers engaged in the production of o-toluidine, Gig. Tr. Prof. Zabol., 14 (1970) 7-10
237. Pichler, H., and Hennenberger, P., Untersuchung flussiger und gasformiger produkte der steinkohlenschwelung. III. Zusammensetzung eines innenabsaugols, eines vakuumschwelteers und eines rohbenzols der hochtemper aturverkokung, Brennst.-Chem., 50 (1969) 341-346
238. Ben'kovskii, V. G., Baikova, A. Y., Bulatova, B. T., Lyubopytova, N. S., and Popov, Y. N., Nitrogenous bases in gasoline from the hydrocracking of Arlan petroleum, Neftekhimiya, 12 (1972) 454-459
239. Vitzthum, O. G., Werkhoff, P., and Hubert, P., New volatile constituents of black tea aroma, J. Agr. Food Chem., 23 (1975) 999-1003
240. Irvine, W. J., and Saxby, M. J., Steam volatile amines of Latakia tobacco leaf, Phytochemistry, 8 (1969) 473-476
241. Gropp, D., Zur atiologie des sogenannten anilin-blasenkrebses. (Inaugural-Dissertation Thesis, Johannes-Gutenberg Universitat, Mainz, FRG) (1958)

242. Oettel, H., Zur frage des berufskrebes durch chemikalien. In Verhandlungen Der Deutschen Gesellschaft fur Pathologie, 43 Tagung, Stuttgart, Gustav Fischer-Verlag, (1959) 313-320
243. Oettel, H., Bladder cancer in Germany, In Lampe, K. F., (ed) Bladder Cancer A Symposium, Aesculapius Publishing Co., Birmingham, Alabama (1967) pp. 196-199
244. Oettel, H., Thiess, A. M., and Uhl, C., Beitrag zur problematik berufsbedingter lungenkrebse, Zbl. Arbeitsmed. Arbeitsschutz., 18 (1968) 291-303
245. Uebelin, F., and Pletscher, A., Atiologie und prophylaxe gewerblicher tumoren in der farbstoffindustrie, Schweiz. Med. Wschr., 84 (1954) 917-928
246. Case, R. A. M., and Pearson, J. T., Tumours of the urinary bladder in workmen engaged in the manufacture and use of certain dyestuff intermediates in the British chemical industry, II. Further consideration of the role of aniline and of the manufacture of auramine and magenta (fuchsine) as possible causative agents. Brit. J. Industr. Med., 11 (1954) 213-216
247. Russfield, A. B., Homburger, F., Weisburger, E. K., and Weisburger, J. H., Further studies on carcinogenicity of environmental chemicals including simple aromatic amines (Abstract No. 20), Toxicol. Appl. Pharmacol., 25 (1973) 446-447
248. Russfield, A. B., Boger, E., Homburger, F., Weisburger, E. K., and Weisburger, J. H., Effects of structure of seven methyl anilines on toxicity and on incidence of subcutaneous and liver tumors in Charles River rats, (Abstract No. 3467), Fed. Proc., 32 (1973) 833
249. Morigami, S., and Nisimura, I., Experimental studies on aniline bladder tumors, Gann, 34 (1940) 146-147
250. Satani, Y., Tanimura, T., Nishimura, I., and Isokawa, Y., Klinische und experimentelle untersuchung des blasenpapilloma, Gann, 35 (1941) 275-276
251. Nagao, M., Yahagi, T., Honda, M., Seino, Y., Matsushima, T., and Sugimura, T., Demonstration of mutagenicity of aniline and o-toluidine by Norkarman, Proc. Japan Acad. 53 Ser. B. (1977) 34-37
252. Nagao, M., Yahagi, T., Seino, Y., Sugimura, T., and Ito, N., Mutagenicities of quinoline and its derivatives, Mutation Res., 42 (1977) 335-342
253. Son, O. S., Weiss, L., Fiala, E. S., and Weisburger, E. K., Metabolism of the carcinogen o-toluidine, Proc. Am. Assoc. Cancer, 18 (1977) 123
254. Marquardt, H., Zimmermann, E. K., Dannenberg, H., Neumann, H. G., Bodenberger, A., and Metzler, M., Die genetische wirkung von aromatischen aminen und ihren derivaten: induktion mitotischer konversionen bei der hefe Saccharomyces cerevisiae, Z. Krebsforsch., 74 (1970) 412-433
255. Mayer, V. W., Induction of mitotic crossing over in Saccharomyces by p-toluidine, Mol. Gen. Genet., 151 (1977) 1-4
256. Wernick, T., Lanman, B. M., and Fraux, J. L., Chronic toxicity, teratologic and reproduction studies with hair dyes, Toxicol. Appl. Pharmacol. (1977) in press
257. Anthony, H. M., and Thomas, G. M., Tumours of the urinary bladder: an analysis of the occupations of 1030 patients in Leeds, England, J. Natl. Cancer Inst., 45 (1970) 879-895
258. Wynder, E. L., Onderdonk, J., and Mantel, N., An epidemiological investigation of cancer of the bladder, Cancer, 16 (1963) 1388-1407
259. Dunham, L. J., Rabson, A. S., Stewart, H. L., Frank, A. S., and Young, J. L., Rates, interview and pathology study of cancer of the urinary bladder in New Orleans, J. Natl. Cancer Inst., 41 (1968) 683-709
260. Johnson, R. F., Zenhausern, A., and Zollinger, H., Kirk-Othmer "Encyclopedia of Chemical Technology" Vol. 2, 2nd ed., Interscience, New York (1963) pp. 868-910

261. IARC, Monographs on the Evaluation of Carcinogenic Risk of Chemicals to Man, Vol. 8, International Agency for Research on Cancer, Lyon (1975) 41-291
262. Radomski, J. L., Toxicology of food colors, Ann. Rev. Pharmacol., 14 (1974) 127-137
263. Truhaut, R., (ed) Considerations generales sur les risques de cancerisation pouvant resulter de l'emploi de certains agents chimiques en therapeutique, IN: UICC Monograph Series, Vol. 7, Potential Carcinogenic Hazard from Drugs, Heidelberg, Springer-Verlag, (1967) 16
264. Miller, J. A., and Miller, E. C., The carcinogenic aminoazo dyes, Adv. Cancer Res., 1 (1953) 339-396
265. Terayama, H., Aminoazo carcinogenesis-methods and biochemical problems, Methods Cancer Res., 1 (1951) 399-449
266. Hernandez, P. H., Gillette, J. R., and Mazel, P., Studies on the mecahnism of action of mammalian hepatic azoreductase. I. Azo-reductase activity of reduced nicotinomide adenine dinucleotide phosphate-cytochrome c reductase, Biochem. Pharmacol., 16 (1967) 1859-1875
267. Rhinde, E., and Troll, W., Metabolic reduction of azo dyes to benzidine in the Rhesus monkey, J. Natl. Cancer Inst., 55 (1975) 181-182
268. Allan, R. J., and Roxon, J. J., Metabolism by intestinal bacteria: the effect of bile salts on azo reduction, Xenobiotica, 4 (1974) 637-643
269. Dubin, P., and Kright, K. L., Reduction of azo food dyes in cultures of Proteus vulgaris, Xenobiotica, 5 (1975) 563-571
270. Rompp, H., Chemie-Lexicon, 6th ed., Frank'scheverlagbuchshandlung, Stuttgart (1966) pp. 290-201
271. Demerec, M., Wallace, B., Witkin, E. M., and Bertani, C., In: "Carnegie Institution of Washington Yearbook, 1948-1949" No. 48, Washington, DC (1949) 154-166
272. Elson, L. A., and Warren, F. L., The metabolism of azo compounds. I. Azobenzene, Biochem. J., 38 (1944) 217-220
273. Bray, H. G., Clowes, R. C., and Thorpe, W. V., The metabolism of azobenzene and p-hydroxyazobenzene in the rabbit, Biochem. J., 49 (1951) 115
274. Kirby, A. H. M., and Peacock, P. R., The induction of liver tumors by 4-aminoazobenzene and its N'-N-dimethyl derivative in rats on a restricted diet, J. Patho. Bact., 59 (1947) 1-7
275. Fare, G., Rat skin carcinogenesis by topical application of some azo dyes, Cancer Res., 26 (1966) 2406-2408
276. Yahagi, T., Degawa, M., Seino, Y., Matsushima, T., Nagao, M., Sugimura, T., and Hashimoto, Y., Mutagenicity of carcinogenic azo dyes and their derivatives, Cancer Letters, 1 (1975) 91-96
277. Sato, K., Poirier, L. A., Miller, J. A., and Miller, E. C., Studies on the N-hydroxylation and carcinogenicity of 4-aminoazobenzene and related compounds, Cancer Res., 26 (1966) 1678-1687
278. Yoshida, T., Uber die experimentelle erzeugung von hepatom durch die futterung mit o-amido-azotoluol, Proc. Imp. Acad. Japan, 8 (1932) 464-467
279. Nishiyama, Y., Experimentelle hepatombildung durch futterung mit o-amido-toluol bei der maus, Gann, 29 (1935) 285-294
280. Shelton, E., Production of liver tumors in mice with 2-amino-5-azotoluene, Proc. Amer. Ass. Cancer Res., 1 (1954) 44
281. Andervont, H. B., White, J. W., and Edwards, J. E., Effect of two azo compounds when added to the diet of mice, J. Natl. Cancer Inst., 4 (1944) 583-586
282. Yoshida, T., Uber die serienweise verfolgung der veranderungen der leber bei der experimentellen hepatomerzeugung durch o-aminoazotoluol, Tr. Jap. Patho. Soc., 23 (1933) 636-638

283. Sasaki, T., ortho-Aminoazotoluene, Gann, 29 (1935) 52-64
284. Sumita, S., Studien uber die umstimmung des gewebsstoffwechsels, Mitt. Med. Ges. Tokyo, 49 (1935) 875-891
285. Tomatis, L., Della Porta, G., and Shubik, P., Urinary bladder and liver cell tumors induced in hamsters witho-aminoazotoluene, Cancer Res., 21 (1961) 1513-1517
286. Nelson, A. A., and Woodard, G., Tumors of the urinary bladder, gall bladder, and liver in dogs fed o-aminoazotoluene or p-dimethylaminoazobenzene, J. Natl. Cancer Inst., 13 (1953) 1497-1509
287. Andervont, H. B., Grady, H. G., and Edwards, J. E., Induction of hepatic lesions hepatomas, pulmonary tumors and hemangio-endotheliomas in mice with o-aminoazotoluene, J. Natl. Cancer Inst., 3 (1942) 131-153
288. Andervont, H. B., Induction of hemangio-endotheliomas and sarcomas in mice with o-aminoazotoluene, J. Natl. Cancer Inst., 10 (1950) 927-941
289. Furukawa, R., Experimentelle entstehung des leberkrebes durch subkutane injektion von olivenöllösung des o-amidoazotoluols, Nagasaki Igaku Zasshi, 17 (1939) 2370-2387
290. Nishiyama, Y., Experimentelle hepatombildung durch fütterung mit o-amidoazotoluol bei der maus, Gann, 29 (1935) 285-294
291. Yamasaki, J., and Sato, S., Experimentelle erzeugung von blasengeschwülste durch anilin und o-amidoazotoluol, Jap. J. Dermatol. Urol., 42 (1937) 332-342
292. Clayson, D. B., Pringle, J. A. S., Bonser, G. M., and Wood, M., The technique of bladder implantation: further results and an assessment, Brit. J. Cancer, 22 (1968) 852-882
293. Scherr, G. H., Fishman, M., and Weaver, R. H., The mutagenicity of some carcinogenic compounds for E. coli, Genetics, 39 (1954) 141-149
294. Barratt, R. W., and Tatum, E. L., An evaluation of some carcinogens as mutagens, Cancer Res., 11 (1951) 234
295. Barratt, R. W., and Tatum, E. L., Carcinogenic mutagens, Ann. NY Acad. Sci., 71 (1958) 1072-1084
296. Lawson, T. A., The binding of o-aminoazotoluene to deoxyribonucleic acid, ribonucleic acid and protein in the C57 mouse, Biochem. J., 109 (1968) 917-920
297. Lawson, T. A., The covalent binding of o-aminoazotoluene-^{3}H to C-57 mouse liver deoxyribonucleic acid, ribonucleic acid and protein, Biochem. J., 107 (1968) 14P
298. Lawson, T. A., and Dzhioev, F. K., Binding of o-aminoazotoluene in proliferating tissues, Chem. Biol. Interact., 2 (1970) 165
299. Samejima, K., Tamura, Z., and Ishidate, M., Metabolism of 4-dimethylamino azobenzene and related compounds, IV. Metabolites of o-aminoazotoluene in rat bile, Chem. Pharm. Bull., 15 (1967) 964-975
300. Hashimoto, T., Uber den abbau vono-amido-azotoluol im tierkörper, Gann, 29 (1935) 306-309
301. U.S. Tariff Commission, Synthetic Organic Chemicals, US Production and Sales, 1972, TC Publication 681, Government Printing Office, Washington, DC (1974) 64
302. Arcos, J. C., and Argus, M. F., Chemical induction of cancer, Vol. IIB, Academic Press, New York and London (1974) pp. 142-175
303. Kinosita, R., Researches on the cancerogenesis of the various chemical substance, Gann, 30 (1936) 423-426
304. Kinosita, R., Special report, Studies on the cancerogenic chemical substances, Tr. Jap. Path. Soc., 27 (1937) 625-727

305. Druckrey, H., Quantitative grundlagen der krebserzeugung, Klin. Wschr., 22 (1943) 532-534
306. Druckrey, H., Experimentelle beitrage zum mechanismos der cancerogenen wirkung, Arznei.-Forsch., 1 (1951) 383-394
307. Druckrey, H., Quantitative asepcts in chemical carcinogenesis, In: Potential Carcinogenic Hazards from Drugs (ed) Truhaut, R., UICC Monograph Series 7 (1967) 60-78
308. Druckrey, H., Küpfmüller, K., Quantitative analyse der krebsentstehung, Z. Naturforsch., 3b(1948) 254-266
309. Mori, K., and Nakahara, W., Effect of liver feeding on the production of malignant tumors based on injections of carcinogenic substances, Gann, 34 (1940) 188-190
310. Maruya, H., Morphological studies on the development of liver cancer by butter yellow, Japan J. Med. Sci. Path., 5 (1940) 83-105
311. Nelson, A. A., Woodward, G., Tumors of the urinary bladder, gall bladder and liver in dogs fed o-aminoazotoluene or p-dimethylaminoazobenzene, J. Natl. Cancer Inst., 13 (1953) 1497-1509
312. Terracini, B., and Della Porta, G., Feeding with aminoazo dyes, thioacetamide and ethionine, Arch. Path., 71 (1961) 566-575
313. Orr, J. W., The histology of the rat's liver during the course of carcinogenesis by butter-yellow (p-dimethylaminoazobenzene) J. Path. Bact., 50 (1940) 393-408
314. Akamatsu, Y., and Ikegami, R., Induction of hepatoma and systemic amyloidosis in mice by 4-(dimethylamino)azobenzene beeding, Gann, 59 (1968) 201-206
315. Jaffe, W. G., The response of mice to the simultaneous application of two different carcinogenic agents, Cancer Res., 7 (1947) 529-530
316. Brown, J. P., Roehm, G. W., and Brown, R. J., Mutagenicity testing of certified food colors and related azo, xantheal, and triphenylmethane dyes with the Salmonella/microsome system, Mutation Res., 56 (1978) 249-271
317. Commoner, B., Vithayathill, A. J., and Henry, J. I., Detection of metabolic carcinogen intermediates in urine of carcinogen-fed rats by means of bacterial mutagenesis, Nature, 249 (1974) 850-852
318. Fahmy, O. G., and Fahmy, M. J., Mutagenic selectivity for the RNA-forming genes in relation to the carcinogenicity of alkylating agents and polycyclic aromatics, Cancer Res., 32 (1972) 550-557
319. Fahmy, O. G., and Fahmy, M. J., Genetic properties of substituted derivatives of N-methyl-4-aminoazobenzene in relation to azo-dye carcinogenesis, Int. J. Cancer, 10 (1972) 194-206
320. Maher, V. M., Miller, E. C., Miller, J. A., and Szybalski, W., Mutations and decreases in density of transforming DNA produced by derivatives of the carcinogens 2-acetylaminofluorene and N-methyl-4-aminoazobenzene, Mol. Pharmacol., 4 (1968) 411-426
321. Miller, J. A., Miller, E. C., and Baumann, C. A., On the methylation and demethylation of certain carcinogenic azo dyes in the rat, Cancer Res., 5 (1945) 162-168
322. Mueller, G. C., and Miller, J. A., The oxidative demethylation of N-methylamino-azo dyes by rat liver homogenates, Cancer Res., 11 (1951) 271
323. Matsumoto, M., and Terayama, H., Studies on the mechanism of liver carcinogenesis by certain aminoazo dyes. V. N-demethylation of various aminoazobenzene derivatives by rat liver homogenate, with respect to the carcinogenic potency, Gann, 52 (1961) 239-245

324. MacDonald, J. C., Plescia, A. M., Miller, E. C., and Miller, J. A., Studies on the metabolism of various N-methyl-C^{14}-substituted aminoazo dyes, Cancer Res. 12 (1952) 280
325. Terayama, H., Aminoazo carcinogenesis-Methods and biochemical problems, Methods Cancer Res., 1 (1967) 399-449
326. Stevenson, E. S., Dobriner, K., and Rhoads, C. P., The metabolism of dimethylaminoazobenzene (butter yellow) in rats, Cancer Res., 2 (1942) 160-167
327. Mueller, G. C., and Miller, J. A., The reductive cleavage of 4-dimethylaminoazobenzene by rat liver: the intracellular distribution of the enzyme system and its requirement for triphosphopyridine nucleotide, J. Biol. Chem., 180 (1949) 1125-1136
328. Walker, R., The metabolism of azo compounds: A review of the literature, Food Cosmet Toxicol., 8 (1970) 659-676
329. Poirier, L. A., Miller, J. A., Miller, E. C., and Sato, K., N-benzyloxy-N-methyl-Y-aminoazobenzene: Its carcinogenic activity in the rat and its reactions with proteins and nucleic acids and their constituents in vitro, Cancer Res., 27 (1967) 1600-1613
330. Miller, E. C., The metabolic activation of aromatic amines and amides, In: Programs for XIth International Cancer Congress, Forence, 1974, Vol. 1, UICC, Geneva (1974) p. 11
331. Albert, Z., Effect of prolonged feeding with chrysoidin on the formation of adenoma and cancer of the liver in mice, Arch. Immunol. Ter. Dosw,, 4 (1956) 189-242
332. Allmark, M. G., Grice, H. C., and Mannell, W. A., Chronic toxicity studies on food colors. II. Observations on the toxicity of FD & C Green No. 2 (light green SF yellowish), FD & C Orange No. 2 (Orange SS) and FD & C Red No. 32 (oil red XO) in rats, J. Pharm. Pharmacol., 8 (1956) 417-424
333. Fitzhugh, O. G., Nelson, A. A., and Bourke, A. R., Chronic toxicities of two food colors, FD & C Red No. 32, And FD & C Orange No. 2, Fed. Proc., 15 (1956) 422
334. Nelson, A. A., and Davidow, B., Injection site fibrosarcoma production in rats by food colors, Fed. Proc., 16 (1957) 367
335. Radomski, J. L., The absorption, fate and excretion of Citrus Red No. 2 (2,5-dimethoxyphenyl-azo-2-naphthol) and Ext. D & C Red. No. 14 (1-xylylazo-2-naphthol), J. Pharmacol. Exp. Therap., 134 (1961) 100-109
336. Radomski, J. L., 1-Amino-2-naphthyl glucornide, a metabolite of 2,5-dimethoxyphenylazo-2-naphthol and 2-xylyl-2-naphthol, J. Pharmacol. Exp. Therap., 136 (1962) 378-385
337. Ikeda, Y., Horiuchi, S., Kobayashi, K., Furuya, T., and Kohgo, K., Carcinogenicity of Ponceaux MX in the mouse, Food Cosmet. Toxicol., 6 (1968) 591-598
338. Ikeda, Y., Horuichi, S., Furuya, T., and Omori, Y., Chronic toxicity of Ponceaux MX in the rat, Food Cosmet. Toxicol., 4 (1966) 485-492
339. Daniel, J. W., The excretion metabolism of edible food colors, Toxicol. Appl. Pharmacol., 4 (1962) 572-594
340. Gilman, J., Gilman, T., Gilbert, C., and Spence, I., The pathogenesis of experimentally produced lymphomata in rats (including Hodgkin's-like sarcoma), Cancer, 5 (1952) 792-846
341. Gilman, T., Kinns, A. M., Hallowes, R. C., and Lloyd, J. B., Malignant lymphoreticular tumors induced by trypan blue and transplanted in inbred rats, J. Natl. Cancer Inst., 50 (1973) 1179-1193
342. Papacharalampous, N. X., Befunde an der ratte nach langfristigen versuchen mit intraperitonealen injektionen von trypanblau, Beitr. Path. Anat., 117 (1957) 85-89

343. Papacharalampous, N. X., Die ubertraugung von reticulosarkomen der ratte nach trypanblau injektionen, Frankfurt. Z. Path., 75 (1966) 74-77
344. Cooke, P., Roper, J. A., and Watmough, W., Trypan blue induced deletions in deplication strains of Aspergillus nidulans, Nature (1970) 276-277
345. Roper, J. A., Aspergillus, In: "Chemical Mutagens: Principle and Methods for their Detection" (ed) Hollander, A., Chapter 12, Plenum Press, New York (1971) pp. 343-363
346. Uchino, F., and Hosokawa, S., Pathological study on amyloidosis. A new induction method of amyloidosis by Trypan blue, Acta. Path. Jap., 22 (1972) 131-140
347. Thilander, H., Disappearance rate of Trypan blue in rat plasma after intraperitoneal injection, Acta. Path. Microbiol. Scand., 57 (1963) 57-59
348. Lloyd, J. B., Beck, F., Griffiths, A., and Parry, L. M., The mechanism of action of acid bisazo dyes, In: "The Interaction of Drugs and Subcellular Components in Animal Cells" (ed) Campbell, P. N., J. & A. Churchill, Ltd., London (1968) pp. 171-2
349. FAO/WHO, 16th Report of the Joint FAO/WHO Expert Committee on Food Additives, Evaluation of Mercury, Lead, Cadmium, and the Food Additives Amaranth, Diethyl-pyrocarbonate and octyl gallate, WHO Food Additive Series No. 4, Wld. Hlth. Org. Tech. Rep. Ser., No. 505 (1972) p. 65
350. Baigusheva, M. M., Carcinogenic properties of the amaranth paste, Vop. Pitan. 27 (1968) 46-50
351. Andrianova, M. M., Carcinogenous properties of red food pigments-Amaranth SX Purple and 4R Purple, Vop. Pitan., 29 (1970) 61-65
352. Cook, J. W., Henett, C. L., Kennaway, E. L., and Kennaway, N. M., Effects produced in the livers of mice by azonaphthalenes and related compounds, Amer. J. Cancer, 40 (1940) 62-77
353. Willheim, R., and Ivy, A. C., A preliminary study concerning the possibility of dietary carcinogenesis, Gastroenterology, 23 (1953) 1-19
354. Mannell, W. A., Grice, H. C., Lu, F. C., and Allmark, M. G., Chronic toxicity studies on food colors IV. Observations on the toxicity of tartrazine, amaranth and sunset yellow in rats, J. Pharm. Pharmacol., 10 (1958) 625-634
355. Roxon, J. J., Ryan, A. J., and Wright, S. E., Reduction of water-soluble azo dyes by intestinal bacteria, Food Cosmet. Toxicol., 5 (1967) 367-369
356. Ryan, A. J., Roxon, J. J., Sivayavirojana, A., Bacterial azo reduction: a metabolic reaction in mammals, Nature, 219 (1968) 854-855
357. Radomski, J. L., and Mellinger, T. J., The absorption, fate and excretion in rats of the water-soluble azo dyes, FD & C Red No. 2, FD & C Red No. 4, and FD & C Yellow No. 6, J. Pharmacol. Exp. Therap., 136 (1962) 259-266
358. Anon, Three printing ink dyes are probably carcinogenic, EPA Medical Adviser Says, Pesticide Toxic Chem. News, 6 (1) (1977) 29-30

CHAPTER 19

HETEROCYCLIC AMINES AND NITROFURANS

A. Amines
1. Quinoline [benzo(b)pyridine; 1-benzazine; leucoline] occurs in small amounts in coal, tar and petroleum, and is a volatile component in roasted cocoa[1]. It is produced by many synthetic procedures[2] including: a) the Skraup synthesis by heating aniline with glycerine and nitrobenzene in presence of sulfuric acid and b) via the interaction of aniline with acetaldehyde and a formaldehyde hemiacetal[3]. Quinoline is a weak tertiary heterocyclic base and it and its derivatives exhibit reactions which are familiar in the benzene and pyridine series[4]. For example, electrophilic substitution occurs almost exclusively in the benzene ring (partly because of the deactivation of the pyridine ring by the hetero atom), while nucleophilic substitution occurs in the pyridine ring.

Quinoline is used in the manufacture of dyes[4], deodorants (e.g., 8-hydroxyquinoline sulfate; aluminum salt); local anesthetics (dibucaine); anti-malarials (4- and 8-aminoquinoline derivatives; 4-quinolinemethanols)[4]. Quinoline derivatives are the parent substances of quinine and other plant alkaloids. Quinoline is also used as a solvent for resins and terpenes and as a preservative for anatomical specimens[5].

Patented areas of suggested utility of quinoline include: anti-knock additive for gasoline[6], catalyst for hardening of epoxy resins[7], dimerization of isoprene[8], catalyst for the dehydrohalogenation of trihaloalkanes to dihaloalkenes[9] and in petroleum recovery[10].

2. 8-Hydroxyquinoline (8-quinolinol; oxyquinoline; hydroxybenzopyridine (I)) is used as a fungistat and in the analysis of and separation of metallic ions. The citrate salt of 8-hydroxyquinoline is employed as a disinfectant while the copper derivative (copper 8-quinolinolate; cupric-8-hydroxyquinolate) is used as a mildew proofing agent, as a fungicide in the treatment of textiles, as an ingredient of paints, in wood paper and plastics[11] preservation, in agriculture and in miscellaneous other uses. In the U.S., an estimated 75% of approximately 132,000 lbs (domestic and imported copper 8-quinolinolate)[12,13] is used in the treatment of textiles (e.g., in fabric, rope, thread, webbing and cordage).

Copper 8-quinolinolate can be prepared by mixing solutions of copper salts with 8-hydroxyquinoline[14], and is available in many forms including: as a 5.0% liquid concentrate in combination with 17.6% pentachlorophenol and 2.4% tetrachlorophenol and in combination with zinc petroleum sulfonate[15]. The technical grade of copper 8-quinolinate can contain both free copper and free 8-hydroxyquinoline as impurities.

Quinoline was recently reported to be carcinogenic in Sprague-Dawley rats inducing hepatocellular carcinomas and hemangioendotheliamas in the livers of animals fed a basal diet containing 0.05, 0.10 or 0.25% quinoline for about 16 to 40 weeks[16]. 2-Chloroquinoline did not induce any nodular hyperplasia or other neoplastic changes when rats were treated analogously[16]. While there are no data on the possible formation of quinoline metabolites in the liver, it is belived that quinoline may be activated only in the liver, perhaps by the formation of the N-oxide. It is also possible that quinoline derivatives may be the proximal carcinogen(s) of quinoline. The lack of carcinogenicity of 2-chloroquinoline is suggested to possibly be related to its more difficult conversion to an N-oxide[16].

The carcinogenicity of 8-hydroxyquinoline appears to be conflicting. While it has been reported to induce tumors when implanted into mouse bladder as a pellet with cholesterol[17], or instilled into rat vagina[18], it has also been shown to be non-carcinogenic in mice or intraveginal administration[19,20], in hamsters by intratesticular injection[21] or when fed to mice or rats under various experimental conditions[22-25].

While no increase in the incidence of tumors was noted in two strains of mice following oral administration of copper 8-hydroxyquinoline[25,26], a significantly increased incidence of reticulum cell sarcomas was observed in males of one strain of mice following single subcutaneous injection of copper-8-hydroxyquinoline[25,26].

4-Nitroquinoline-1-oxide (4NQO) (III) and its related compounds have been shown to be carcinogenic in rats, mice, quinea pigs, hamsters and rabbits, producing such tumors as papillomas of the skin, lung carcinomas, and fiberosarcomas[27-32].

4-Nitroquinoline-1-oxide has been the most intensively studied of the quinolines, in regard to mutagenic activity. It is mutagenic for both prokaryotic organisms such as E. coli[33] and its phages[34,35], Salmonella typhimurium[36,37] and Streptomyces griseoflavus[38], and for eukaryotic microorganisms such as Aspergillus niger[39], Neurospora crassa[40], and Saccharomyces cerevisiae[41,42].

Base-pair substitutions arise at G·C base pairs which are the site of 4NQO attack. Hence, 4NQO induces G·C → A·T transitions. G·C → T·A transversions and possibly G·C → C·G transversions[43].

4NQO and its reduced metabolite 4-hydroxyaminoquinoline-1-oxide (4HAQO) (IV) bind covalently to cellular macromolecules such as nucleic acid and protein[44-48]. It has recently been reported that seryl-tRNA synthetase is an 4HAQO-activating enzyme[49] and that 4HAQO may be activated by both seryl- and prolyl tRNA synthetases which are capable of 4-HAQO-activation may possess a unique conformation enabling them to aminoacylate in vivo the N-hydroxyl group of the carcinogen[49] (e.g., carcinogenic aromatic amines and nitro compounds).

Quinoline is mutagenic in S. typhimurium TA 100[50-53] (and very weakly mutagenic in TA 98[50]) only in the presence of the metabolic activation S-9 mix; and reported to be non-mutagenic in TA 1535[51,52], TA 1537[51,52] and TA 98[51,52] tester strains with and without metabolic activation.

While isoquinoline was non-mutagenic with or without metabolic activation in TA 100[50] and TA 98[50], the isomeric methyl quinolines (e.g., 4,6,7, or 8 methyl derivatives) were all mutagenic to both tester strains, only when activated[50].

8-Hydroxyquinoline is mutagenic in S. typhimurium TA 100[50-53] and TA 98[50] in the presence of the metabolic activation system S-9 mix, it was reported to be non-mutagenic in TA 1535, TA 1537 and TA 98 tester strains with and without metabolic activation[51,52]. 8-Hydroxyquinoline was inactive in the induction of sex-linked recessive lethals in Drosophila[51,54], but induced chromatid breaks and achromatic lesions when tested at 5×10^{-6}M to 1×10^{-5}M, in human leukocytes[52].

The mutagenic activities of quinoline and 23 of its derivatives in the Ames bioassay with tester strains TA 98, TA 100, TA 1535 and TA 1537 with and without metabolic activation are compared in Table 1[50]. It is noted that quinoline and 5-hydroxy quinoline were mutagenic in TA 100 with a specific activity approaching that reported for benzidine[55]. Of the 16 quinoline derivatives listed in this table under industrial chemicals category, 4(H)-quinoline, quinoxaline, 2,4-quinolinediol, 5-hydroxyisoquinoline and 8-hydroxyquinoline-N-oxide were weakly positive in various tester strains.

Structure activity studies show that addition of hydroxy groups at C-5 or C-8 of the quinoline nucleus or the addition of methyl groups at C-4, C-6, C-7, or C-8[50] does not abolish or markedly reduce the TA 100 mutagenic activity. By contrast, methyl or hydroxyl substitution at C-2 or C-3 abolishes the mutagenicity of quinoline in TA 100, the only strain in which it is expressed. This suggests that C-2 and C-3 regions of quinoline are critical sites for the enzymatic conversion of quinoline to its principal mutagenic intermediate. However, it should be noted that some of the quinoline derivatives are weakly mutagenic in other tester strains which suggests alternate routes of activation, possibly independent of the C-2 and C-3 regions, may exist for quinoline and its derivatives. It was suggested by Hollstein et al[51], that quinoline, is coverted to quinoline epoxide(s) (analogous to the conversion of benzo(a)pyrene conversion to mutagenic epoxides by a P-448 AHH system)[56,57]. The intermediate(s) may then react with nucleic acids to cause mutations in the S. typhimurium test strains. Quinoline is mutagenic only in TA 100 (although reported in one study to be weakly mutagenic in TA 98[50]). The structure activity results obtained with TA 100 strongly suggests that unsubstituted C-2 and C-3 regions are critical for the expression of this mutagenicity[51]. Hence the principal mutagen arising from quinoline activation may be quinoline-2,3-epoxide. The correlation of mutagenicity with the formation of water-soluble metabolites and the observation that quinoline is metabolized to dihydrodiol[58] appear to be consistent with this conclusion.

The additional observation in the above studies of Hollstein et al[51] that human liver enzymes can convert quinoline to a mutagen (in the Ames test) suggests that quinoline and its mutagenic derivatives are potential hazards in the industrial environment.

B. Nitrofurans

Nitrofuran derivatives constitute a large category of important heterocyclic compounds many of which have been widely used as food additives, feed additives, human medicines and veterinary drugs[59-61]. Many of the nitrofuran derivatives which have been studied for their mutagenic and/or carcinogenic activities are chiefly classified into four groups[60]: (1) 5-thiazole derivatives, (2) and (3) consist of those compounds which have vinyl or acryl residues and azomethin residues, respectively, next to position 2 of the 5-nitrofuran (NO_2-⟨O⟩) and (4) other derivatives not

Table 1. Mutagenicity in the Ames Bioassay of Quinoline and 23 Derivatives

Compound (dose[a]/plate)	TA 98		TA 100		TA 1535		TA 1537	
	-S-9	+S-9	-S-9	+S-9	-S-9	+S-9	-S-9	+S-9
Industrial chemicals:								
Quinoline (100 μg)	18	59	157	960 (6,000)	5	23	0	0
2-Hydroxyquinoline (100 μg)	10	48	114	128	12	11	5	12
3-Hydroxyquinoline (100 μg)	26	18	82	70	10	10	3	9
4-Hydroxyquinoline (100 μg)	9	17	113	135	25	21	2	7
5-Hydroxyquinoline (100 μg)	8	49	111	439	8	14	8	17
6-Hydroxyquinoline (100 μg)	13	30	115	167	13	23	23	12
7-Hydroxyquinoline (100 μg)	18	32	130	135	10	6	10	9
Quinoline-N-oxide (1 mg)	27	47	80	171	13	12	7	32
8-Hydroxyquinoline-N-oxide (1 mg)	17	30	129	174	17	21	21	12
Isoquinoline (100 μg)	17	29	78	100	9	13	7	5
Isocarbostryil (1 mg)	22	9	102	94	7	15	2	26
5-Hydroxyisoquinoline (100 μg)	24	12	64	109	12	20	13	17
2,4-Quinolinediol (1 mg)	35	**101**	100	123	9	10	5	19
4(H)-Quinoline (1 mg)	17	54	122	**249**	14[d]	96[d]	6	8
8-Hydroxyquinaldine (100 μg)	16[d]	12[d]	75	100	13[d]	6[d]	1	14
Quinoxaline (1 mg)	17	**138**	85	87	8	17	9	10
3-Quinoline carbonitrile (1 mg)	11	27	135	203	7	15	5	
Endogenous compounds:								
Quinaldic acid (1 mg)	10	31	118	119	8	11	4	2
8-Hydroxyquinaldic acid (1 mg)	14	36	125	193	9	11	9	5
Xanthurenic acid (1 mg)	22	13	117	134	14	27	13	23
Food and drug products:								
Vioform (1 μg)	21	16	129	117	13	10	0	10
8-Quinolinesulfonic acid (1 mg)	20	18	166	163	7	12	7	6
8-Hydroxyquinoline (100 μg)	20	78	68	685[e]	0	0	0	0
Quinoline yellow (1 mg)	19	22	100	141	7	6	3	6
Dimethyl sulfoxide (solvent control)	24	35	110	121	12	24	8	12

[a] Dose at which the highest mutagenic activity was observed, or the highest nontoxic dose tested.
[b] Numbers underlines indicate positive results; numbers in boldface indicate marginal values.
[c] Number in parentheses: number of TA 100 revertants at a test dose of 1 mg.
[d] Number of revertants at a test dose of 10 μg; higher concentrations toxic to this strain.
[e] Number of revertants at a test dose of 40 μg; higher doses yielded a decrease in mutagenic activity.

classified in groups (1), (2) and (3). The potential carcinogenicity of nitrofurans has been known since 1966[62]. Some nitrofuran[63-65], nitrothiophene[66], nitroimidazole[67,68] and nitrothiazole[64] derivatives were found to be carcinogenic recently.

Nitrofurazone (NO_2-⟨furan⟩-CH=NNHCONH$_2$; 5-nitro-2-furaldehyde semicarbazone) which as been used extensively in human and veterinary medicine has been known as a mutagen in E. coli since 1964[69]. However, the mutagenicity of nitrofurans in microorganisms has received extensive attention only recently[60,70-75] and has been recently reviewed by Tazima et al[60].

Data compiled on structure and activity relationship strongly suggested that the nitro group was responsible for the carcinogenicity and mutagenicity of nitrofurans. Hydroxylaminofuran was proposed as the active intermediate. Nitroreduction of nitrofuran[77-79], nitrothiophene[79], and nitrothiazole[77,80] by animal tissues has been shown. Nitroreduction of nitrofuran was also observed in bacteria[81].

Nitrofurantoin (1-[(5-nitrofurylidene)amino]-hydantoin, furantoin) the nitrofuran utilized most in antimicrobial chemotherapy has recently been shown to be a mutagen in S. typhimurium TA 100 and TA-FR 1 strains following its metabolism by rat liver nitroreductase[82]. Recently, the presence of mutagenic activity was demonstrated (utilizing S. typhimurium TA 100) in the urine of patients given metronidazole [1-(2-hydroxyethyl)-2-methyl-5-nitroimadazole] a drug commonly used in treatment of amebiases. Trichomonas vaginitis and infections caused by anerobic microbes[83]. This activity was due to unmodified metronidazole and at least four of its metabolites (including a hydroxyamino derivative). Metronidazole has been shown previously to induce lung tumors and malignant lymphomas in mice[84].

The list of nitrofurans tested for carcinogenicity is shown in Table 2[60], while Table 3 lists the nitrofuran derivatives whose mutagenicity and/or DNA damaging capacity in bacteria have been studied[60]. Table 4 illustrates the correlation between prophage inducibility, mutagenicity and carcinogenicity of nitrofurans and related compounds in E. coli[60,70].

Ong[85] recently reported on the mutagenicity and mutagenic specificity of 12 nitrofurans in the adenine-3 (ad-3) test system of Neurospora crassa. With this test system recessive lethal mutations resulting from multilocus deletions or point mutations at ad-3A and/or ad-3B can be recovered. The studies indicated that the carcinogenic nitrofurans (e.g., nitrofurazon, AF-2, FANFT and SQ18506) are mutagenic in Neurospora crassa and that the NO_2 group at the C-5 position in the furan ring is necessary for the mutagenic activity of nitrofurans. No conclusion could be drawn from these studies as to the relationship between the side chain at the C-2 position and the mutagenicity of nitrofurans. However, the data seemed to indicate that compounds with a 5-membered heterocyclic ring bridged by a vinyl group (-C=C-) to the nitrofuran moiety (AF-2 and SQ18506) or with a 5-membered ring that binds directly to nitrofuran (FANFT) possess potent mutagenic activity and that a 6-membered ring bridged by a vinyl group to nitrofuran moiety has moderate mutagenic activity. Compounds with straight chain which bind to the C-2 position of nitrofuran via a vinyl group appear to have weak mutagenic activities.

The activity of a number of 5-nitroimidazoles and 5-nitrofurans with respect to the induction of sex-linked recessive lethals in <u>Drosophila</u> was investigated by Kramers[86]. While a number of nitrofurans such as nitrofurantoin and furazolidine showed mutagenic activity in <u>Drosophila</u>, in most cases the response was weak. There was no apparent relation between the mutagenic activity and the chemical group involved, neither was there in the case of positive results, a consistency with respect to the germ cell stage which was most sensitive, or with respect to the way of administration which was most effective.

TABLE 2
LIST OF NITROFURANS TESTED FOR CARCINOGENICITY[60]

Compound	Group	Test Animal	Carcinogeniticy	Principal tumour site	Reference
FNT	I	Rat	+	Mammary Kidney	22,32 33,71
		Mouse	+	Leukemia Stomach Lung Mammary	16
HNT	I	Rat	+	Mammary	32
		Mouse	+	Forestomach	19
DMNT	I	Rat	+	Mammary	18,19,20
FANFT	I	Rat	+	Bladder Renal Mammary	
		Hamster	+	Bladder	22
		Mouse	+	Bladder Leukemia	19,29
		Dog	+	Bladder Ureter Renal Gall bladder Mammary	27
NFTA	I	Rat	+	Mammary	28
		Mouse	+	Leukemia	17,19
		Hamster	+	Bladder	22
		Dog	+	Mammary Gall bladder	27
2,2,2-Trifluoro-N-[4-(5-nitro-2-furyl)-thiazolyl]-acetamide	I	Mouse	+	Forestomach	19
ANFT	I	Mouse	+	Forestomach	19
4-(5-Nitro-2-furyl)thiazole	I	Not described		Not described	110
2-Methyl-4-(5-nitro-2-furyl)thiazole	I	Not described	+	Not described	110
2-[(Dimethylamino)methylimino]-5-2-(5-nitro-2-furyl)vinyl-1,3,4-oxadiazole	II	Not described	+	Not described	110
2-(2-Furyl)-3-(5-nitro-2-furyl) acrylamide (AF$_2$)	II	Mouse	+	Forestomach	107
Nitrofurazone	III	Rat	+	Mammary	33,34,71
Nitrofurantoin	III	Rat	−		20,33,71
5-Nitro-2-furamidoxime	III	Rat	−		20

TABLE 2 (Continued)

Compound	Group	Test animal	Carcinogenicity	Principal tumour site	Reference
4-Methyl-1-[(5-nitrofurylfurylidine)-amino]-2-imidazolidinone	III	Rat		Mammary Breast	20
Furmethonol	III	Rat		Mammary Breast Ovary Kidney pelvis Lymphooma	20
1-(2-Hydroxyethyl)-3-[5-nitro-furfurylidine)-amino]-2-imidazolidine	III	Rat	+	Mammary Breast Kidney Uterus	20
Nifuradene	III	Rat	+	Mammary Braest Salivary gland Lymphoma Adenoma	20
5-Acetamido-3-(5-nitro-2-furyl)-6-H-1,2,4-oxadiazine	IV	Rat	+	Hemangioendo-thelial sarcoma	31
5-Nitro-2-furanmethandiol diacetate	IV	Rat	−		71
4,6-Diamino-2-(5-nitro-2-furyl)-s-triazine	IV	Rat	+	Mammary	20
5-Acetamido-3-(5-nitro-2-furyl)-6-H-1,2,4-oxadiazone	IV	Rat	+	Hemangioendo-thelial sarcoma	31
N-[5-(5-Nitro-2-furyl)-s-thiadiazol-2-yl]acetamide	IV	Mouse	+	Forestomach	19
N-N'-[6-(5-Nitro-2-furyl)-s-triazine-2,4-diyl]bisacetamide	IV	Rat	+	Mammary Breast Intestine Uterus	20
N-([3-(5-nitro-2-furyl)-1,2,4-oxadiazol-5-yl]methyl)acetamide	IV	Not described	+	Not described	110
Formic acid 2-[4-(2-furyl)-2-thiazolyl]acetamide	I	Rat	−		110
Formic acid 2-(4-methyl-2-thiazolyl)hydrazide	I	Rat	−		110
Furaldehyde semicarbazone	III	Rat	−		33

TABLE 3

LIST OF NITROFURAN DERIVATIVES WHOSE MUTAGENICITY AND/OR DNA DAMAGING CAPACITY IN BACTERIA HAVE BEEN STUDIED[60]

Compound	Group	Repair test[a]		S. typhimurium TA1978/ TA1538	B. subtilis rec$^+$/rec45	Mutation test[b] E. coli WP$_2$ try$^-$:uvr$^+$ and/or uvrB (her)
		E. coli rec$^+$/recA	rec$^+$/recB			
FNT	I	+++	+	+++	++	+
HNT	I					+
DMNT	I	++++	++	+++		++
FANFT	I	++++	+	+++	++	++
NFTA	I	++++	++	+++	++	++
2,2,2-Trifluoro-N-[4-(5-nitro-2-furyl)-2-thiazolyl]-acetamide	I	+++		+++		+
ANFT	I	+++	+	+++		+
N-[4-(5-nitro-2-furyl)-2-thiazolyl]phenylamine	I					+
2-Chloro-4-(5-nitro-2-furyl)thiazole	I					++
4-(5-Nitro-2-furyl)thiazole	I	++++	+	+++	++	++
2-Methyl-4-(5-nitro-2-furyl)thiazole	I	++++		+++		+
2-Formyl hydrazine-4-(2-furyl) 1,3-thiazole	I				−	
2-Amino-4-phenyl thiazole	I				−	
NFT	II					++
NF-416	II			+		+
5-Nitro-2-furyl acrylic acid	II	+++	+	++		++
3-(5-Nitro-2-furyl)acrylamide	II	++	++			+
2-Amino-4-[2-(5-nitro-2-furyl)vinyl]-1,3-thiazole	II	+++	−	+++		+
2-(2-Phenyl)-3-(5-nitro-2-furyl)acrylamide	II	++++	++	+		+
2-Formylamino-4-[2-(5-nitro-2-furyl)vinyl]-1,3-thiazole	II	++++	+	+++		+
2-Acetylamino-4-[2-(5-nitro-2-furyl)vinyl]1,3-thiazole	II	++++	−	++		++
2-[(Dimethylamino)methylimino]-5-[2-(5-nitro-2-furyl)vinyl]-1,3,4-oxadiazole	II	+++	+	++	+++	++
AF$_2$	II	++++	++	++	++	++
Nitrofurazone	III	+++	++	−		+
Furmethonol	III					+
Nitrofurantoin	III	+++	+	++		++
Furazolidone	III				++	++
Nifuroxime	III					++
1-(5-Nitro-2-furfurylidine)-3-N,N-diethyl-prophyl-aminourea HCl	III					+
Nifuratrone	III					++
2-(5-Nitro-2-furfurylidene) amino ethanol	III	−	−	−		++
2-(5-Nitro-2-furfurylidene) amino ethanol N-oxide	III					+
5-Nitro-2-furamidoxime	III	+++	++	++		+
4-Methyl-1-[(5-nitro-furfurylidene)amino-2-imidazolidinone	III	+++	+	+++		+
5-Acetamido-3-(5-nitro-2-furyl) 6-H-1,2,4-oxadiazine	IV	+++	++	++		++

TABLE 3 (continued)

Compound	Group	Repair test[a]		S. typhimurium TA1978/ TA1538	B. subtilis rec+/rec 45	Mutation test[b] E. coli WP₂ try⁻ :uvr⁺ and/or uvrB (her)
		E. coli				
		rec+/recA	rec+/recB			
1-(5-Nitro-2-furyl)-3-piperidino- propan-1-one-semicarbazone HCl	IV					++
3-(5-Nitro-2-furyl)-4-H-1,2,4- trizole	IV					+
5-Nitro-2-furanmethandiol diacetate	IV					++
5-Nitro-2-furoic acid	IV	−		−		−
4,6-Diamino-2-(5-nitro-2-furyl)-s- triazine	IV	+++	+	+++		++
5-Acetamido-3-(5-nitro-2-furyl)- 6-H-1,2,4-oxadiazine	IV	+++	++	+		++
N-[5-(5-Nitro-2-furyl)-1,3,4- thiadiazol-2-yl]acetamide	IV	+++	++	+++		+
N-([3-)5-nitro-2-furyl)-1,2,4- oxadiazol-5-yl]methyl)acetamide	IV	+++	++	+		++
5-Nitro-2-methyl furan	IV				++	

[a] Difference in diameters of inhibition zones between wild and mutant strains: ++++, ≥ 16 mm; +++, ≥ 10 mm; ++, ≥ 6 mm; -, ≥ 3 mm; −, < 3 mm.
[b] ++, Strongly mutagenic; +, moderately or weakly mutagenic; −, non mutagenic

TABLE 4

CORRELATION BETWEEN PROPHAGE INDUCIBILITY, MUTAGENICITY AND CARCINOGENICITY OF NITRO-FURANS AND RELATED COMPOUNDS IN E. coli

Compound	Group	Inducibility of prophage[a] ($\mu g/ml$)[b]	Mutagenicity ($try^- \rightarrow try^+$) per plate	Carcino-genicity
Nitrofuran derivatives				
NET	II	0.1	300	+
FANFT	I	0.1	201	+
HNT	I	0.1	83	+
FNT	I	0.25	89	+
2,2,2-Trifluoro-N-[4-(5-nitro-2-furyl)-2-thiazolyl]acetamide	I	0.5	44	+
5-Acetamido-3-(5-nitro-2-furyl)-6-H-1,2,4-oxadiazine	IV	0.5	27	+
DMNT	I	1.0	111	+
Nitrofurazone	III	1.0	12	+
Furan derivatives				
Formic acid-2-[4-(2-furyl)-2-thiazolyl]hydrazide	I	10.0	0	−
Formic acid-2(4-methyl-2-thiazolyl)hydrazide	I	10.0	0	−

[a] Strains used: E. coli T_{44}
[b] Minimum concentrations of compound required to induce mass lysis in E. coli.
[c] Strains used: E. coli WP_2 and its derivative uvr A^-.

References

1. Vitzhum, O. G., Werkhoff, P., and Hubert, P., Volatile components of roasted cocoa, Basic Fraction, J. Food Sci., 40 (1975) 911-916
2. Manske, R. H. F., The chemistry of quinolines, Chem. Revs., 30 (1942) 113
3. Cislak, F. E., and Wheeler, W. R., Quinoline, U. S. Patent 3,020, 281, Feb. 6, 1962, Chem. Abstr., 56 (1962) P14248e
4. Kulka, M., "Quinoline and Isoquinoline", in Kork-Othmer, Encyclopedia of Chemical Technology, 2nd eds., Vol. 16, Interscience Publishers, New York, pp. 865-886 (1968)
5. Merck & Co., Merck Index, 9th ed., Rahway, NJ (1976) p. 7682
6. Friberg, S. E., and Lundgren, L. E. G., Motor Fuel with High Octane Number, Ger. Offen. 2,440,521, March 6, 1975, Chem. Abstr., 83 (1975) 63261V
7. Wynstra, J., and Stevens, J. J., Jr., Hardenable Resin Composition, Ger. Offen. 2,065,701, May 22, 1975, Chem. Abstr., 83 (1975) 148382K
8. Mori, H., Imaizumi, F., and Hirayanagi, S., Isoprene Dimers, Japan Patent 7,511,884, May 7, 1975, Chem. Abstr., 83 (1975) 148079S
9. Tomio, A., Fukui, E., Yokoyama, I., Ioka, M., and Ohkoshi, Y., Dihaloalkenes, Japan Patent 7,433,166, Sept. 5, 1974, Chem. Abstr., 83 (1975) 6164C
10. Brown, A., Chichakli, M., and Fontaine, M. F., Improved Vertical Conformance via a Stream Drive, Canadian Patent 963,803, March, 4, 1975, Chem. Abstr. 83 (1975) 134786H
11. Turner, N. J., Industrial Fungicides, In Kirk-Othmer, Encyclopedia of Chemical Technology, 2nd ed., Vol. 16, Interscience Publishers, New York, (1966) pp. 231-236
12. U.S. International Trade Commission, Synthetic Organic Chemicals, U. S. Production and Sales, Pesticides and Related Products, 1975 Preliminary, U.S. Government Printing Office, Washington, D.C., (1976) p. 4
13. U.S. International Trade Commission, Imports of Benzenoid Chemicals and Products, 1974, USITC Publication No. 762, U.S. Government Printing Office, Washington, D.C., (1976) p. 26
14. Spencer, E. Y., Guide to the Chemicals Used in Crop Protection, 6th eds., University of Western Ontario, Research Branch, Agriculture Canada, Univ. of Western Ontario, London, Ontario, Publ. No. 1093 (1973)
15. Environmental Protection Agency, EPA Compendium of Registered Pesticides, Vol. II, Fungicides and Nematicides, U. S. Govt. Printing Office, Washington, D.C. (1973) pp. C-54-00.01-Z-07-00.01
16. Hirao, K., Shinohara, Y., Tsuda, H., Fukushima, S., Takahashi, M., and Ito, N., Carcinogenic activity of quinoline on rat liver, Cancer Research, 36 (1976) 329-335
17. Allan, M. J., Boyland, E., Dukes, C. E., Horning, E. S., and Watson, J. G., Cancer of the urinary bladder induced in mice with metabolites of aromatic amines and tryptophan, Brit. J. Cancer, 11 (1957) 212-218
18. Hoch-Ligeti, C., Effect of prolonged administration of spermicidal contraceptives on rats kept on low protein or on full diet, J. Natl. Cancer Inst., 18 (1957) 661-685
19. Boyland, E., Charles, R. T., and Gowing, N. F. C., The induction of tumors in mice by intraveginal application of chemical compounds, Brit. J. Cancer, 15 (1961) 252-256

20. Boyland, E., Roe, F. J. C., and Mitchley, B. C. V., Test of certain constituents of spermicides for carcinogenicity in genital tract of female mice, Brit. J. Cancer 20 (1966) 184-187
21. Umeda, M., Screening tests of various chemical substances as carcinogenic hazards (Report 1), Gann, 48 (1957) 57-64
22. Hadidian, Z., Fredrickson, T. N., Weisburger, E. K., Weisburger, J. H., Glass, R. M., and Mantel, N., Tests for chemical carcinogens. Report on the activity of aromatic amines, nitrosamines, quinolines, nitroalkanes, amides, epoxides, aziridines and purine antimetabolites, J. Natl. Cancer Inst., 41 (1968) 985-1036
23. Trihaut, R. C., Researchers sur les risques de nociuite a long terme de l'hydroxy-8-quinoleine au cours de son utilisation comme conservateur alimentaire, Ann. Pharm. Fr., 21 (1963) 266
24. Yamamoto, R. S., Williams, G. M., Frankel, H. H., and Weisburger, J. M., 8-Hydroxyquinoline: Chronic toxicity and inhibitory effect on the carcinogenicity of N-2-fluorenylacetamide, Toxicol. Appl. Pharmacol., 19 (1971) 687-698
25. Innes, J. R. M., Ulland, B. M., Valerio, M. G., Petrucelli, L., Fishbein, L., Hart, E. R., Pallotta, A. J., Bates, R. R., Falk, H. L., Gart, J. J., Klein, M., Mitchell, I., and Peters, J., Bioassay of pesticides and industrial chemicals for tumorigenicity in mice: A preliminary note, J. Natl. Cancer Inst., 42 (1969) 1101-1114
26. National Technical Information Service, Evaluation of Carcinogenic, Teratogenic and Mutagenic Activities of Selected Pesticides and Industrial Chemicals, Vol. I, Carcinogenic Study, U. S. Dept. of Commerce, Washington, D.C.
27. Mori, K., Induction of pulmonary and uterine cancers, and leukemia in mice by injection of 4-nitroquinoline, Gann, 56 (1965) 513-518
28. Mori, K., Kondo, M., Koibuchi, E., and Hashimoto, A., Induction of lung cancer in mice by injection of 4-hydroxyaminoquinoline-1-oxide, Gann, 58 (1961) 105-106
29. Takahashi, M., and Sato, H., Effect of 4-nitroquinoline-1-oxide with alkyl benzene sulfonate on gastric carcinogenesis in rats, Gann Monograph 8 (1969) 241-246
30. Magao, M., and Sugimura, T., Molecular biology of the carcinogen, 4-nitroquinoline-1-oxide, Adv. Cancer Res., 23 (1976) 131-169
31. Endo, H., Carcinogenicity, In "Results in Cancer Research: Chemistry and Biological Actions of 4-Nitroquinoline-1-oxide" (eds) Endo, H., Ono, T., and Sugimura, T., Vol. 34, Springer-Verlag, Berlin, Heidelberg, New York (1971) pp. 32-52
32. Takayama, S., Effect of 4-nitroquinoline-n-oxide painting on azo dye hepato carcinogenesis in rats, with note on induction of skin fibro sarcoma, Gann, 52 (1961) 165-171
33. Kondo, S., Ichikawa, H., Iwo, K., and Kato, T., Base-change mutagenesis and prophage induction in strains of E. coli with different DNA repair capacities, Genetics, 66 (1970) 187-217
34. Ishizawa, M., and Endo, H., Mutagenesis of bacterio phage T4 by a carcinogen 4-nitroquinoline-1-oxide Mutation Res., 12 (1971) 1-8
35. Ishizawa, M., and Endo, H., Mutagenic effect of a carcinogen, 4-nitroquinoline-1-oxide, in bacteriophage T4, Mutation Res., 9 (1970) 134-137
36. Ames, B. N., Lee, F. D., and Durston, W. E., An improved bacterial test system for the detection and classification of mutagens and carcinogens, Proc. Natl. Acad. Sci., USA 70 (1973) 782-786

37. Hartman, P. E., Levine, K., Hartman, Z., and Berger, H., Hycanthone: A frameshift mutagens, Science, 172 (1971) 1058-1060
38. Mashima, S., and Ikeda, Y., Selection of mutagenic agents by the Streptomyces reverse mutation test, Appl. Microbiol., 6 (1958) 45-49
39. Okabayashi, T., Studies on antifungal substances, VIII. Mutagenic action of 4-nitroquinoline-1-oxide, J. Ferment. Technol., 33 (1955) 513-516
40. Matter, B. E., Ong, T., and DeSerres, F., Mutagenic activity of 4-nitroquinoline-1-oxide and 4-hydroxyaminoquinoline-1-oxide in Neurospora crassa, Gann, 63 (1972) 265-267
41. Epstein, S. S., and St. Pierre, J. A., Mutagenicity in yeast of nitroquinolines and related compounds, Toxicol. Appl. Pharmacol., 15, (1969) 451-460
42. Nagai, S., Production of respiration-deficient mutants in yeast by a carcinogen, 4-nitroquinoline-1-oxide, Mutation Res., 7 (1969) 333-337
43. Prakash, L., Stewart, J. W., and Sherman, F., Specific induction of transition and transversions of G·C base pairs by 4-nitroquinoline-1-oxide in iso-1-cytochrome C mutants of yeast, J. Mol. Biol., 85 (1974) 51-65
44. Sugimura, T., Okabe, K., and Endo, K., The metabolism of 4-nitroquinoline-1-oxide. I. Conversion of 4-nitroquinoline-1-oxide to 4-aminoquinoline-1-oxide by rat liver enzymes, Gann, 56 (1965) 489-501
45. Tada, M., and Tada, M., Interaction of a carcinogen, 4-nitroquinoline-1-oxide, with nucleic acids: Chemical degradation of the adducts, Chemico-Biol. Interactions, 3 (1971) 225-229
46. Tada, M., Tada, M., and Takahashi, T., Interaction of a carcinogen, 4-hydroxy-aminoquinoline-1-oxide with nucleic acids, Biochem. Biophys. Res. Commun. 29, (1967) 469-477
47. Ikegami, S., Nemoto, N., Sato, W., and Sugimura, J., Binding of ^{14}C-labeled 4-nitroquinoline-1-oxide to DNA in vivo, Chemico-Biol. Interactions, 1 (1969/1970) 321-330
48. Andoh, T., Kato, K., Takaoka, T., and Katsuta, H., Carcinogenesis in tissue culture XIII Binding of 4-nitroquinoline-1-oxide-3H to nucleic acids and proteins of L.P3 and JTC-25.P3 cells, Int. J. Cancer, 7 (1971) 455-467
49. Tada, M., and Tada, M., Seryl-t-RNA synthetase and activation of the carcinogen 4-nitroquinoline-1-oxide, Nature, 255 (1977) 510-512
50. Nagao, M., Yahagi, T., Seino, Y., Sugimura, T., and Ito, N., Mutagenicities of quinoline and its derivatives, Mutation Res., 42 (1977) 335-342
51. Hollstein, M., Talcott, R., and Wei, E., Quinoline: Conversion to a mutagen by human and rodent liver, J. Natl. Cancer Inst., 60 (1978) 405-410
52. Epler, J. L., Winton, W., Ho, T., Larimer, F. W., Rao, T. K., and Hardigree, A. A., Comparative mutagenesis of quinolines, Mutation Res., 39 (1977) 285-296
53. Talcott, R., Hollstein, M., and Wei, E., Mutagenicity of 8-hydroxyquinoline and related compounds in the Salmonella typhimurium bioassay, Biochem. Pharmacol., 25 (1976) 1323-1328
54. Vogel, E., Lack of mutagenic effectiveness of 8-hydroxyquinoline sulfate in D. melanogaster, Drosophila information Service, 46 (1971) 109
55. McCann, J., Choi, Yamasaki, E., and Ames, B. N., Detection of carcinogens as mutagens in the Salmonella/microsome test: Assay of 300 chemicals, Proc. Natl. Acad. Sci. (US) 72 (1975) 5135-5139
56. Wood, A. W., Levin, W., and Lu, A. Y., Metabolism of benzo(a)pyrene and benzo(a)pyrene derivatives to mutagenic products by highly purified hepatic microsomal enzymes, J. Biol. Chem., 251 (1976) 4882-4890

57. Wood, A. W., Wislocki, P. G., Chang, R. L., et al., Mutagenicity and cytotoxicity of benzo(a)pyrene benzo-ring epoxides, Cancer Res., 36 (1976) 3358-3366
58. Posner, H. S., Mitoma, C., and Udenfriend, S., Enzymatic hydroxylation of aromatic compounds, Arch. Biochem. Biophys., 94 (1961) 280-290
59. Grunberg, E., and Titsworth, E. H., Chemotherapeutic Properties of Heterocyclic Compounds: Monocyclic Compounds With Five-Membered Rings, Ann. Rev. Microbiol., 27 (1973) 317-346
60. Tazima, Y., Kada, T., and Murakami, A., Mutagenicity of Nitrofuran Derivatives, Including Furyl Furamide, A Food Preservative, Mutation Res., 32 (1975) 55-80
61. Miura, K., and Reckendorf, H. K., The Nitrofurans, Prog. Med. Chem., 5 (1967) 320-381
62. Stein, R. J., Yost, D., Petroliunas, F., and Von Esch, A., Carcinogenic Activity of Nitrofurans. A Histologic Evaluation, Federation Proc., 25 (1966) 291
63. Cohen, S. M., and Bryan, G. T., Carcinogenesis Caused by Nitrofuran Derivatives, In: "Pharmacology and the Future of Man." Proceeding of the 5th Int. Congress on Pharmacology, Vol. 2., pp. 164-170, Basel, Switzerland: S. Karger, A. G., (1973)
64. Cohen, S. M., Ertürk, E., Von Esch, A. M., Crovetti, A. J., and Bryan, G. T., Carcinogenicity of 5-Nitrofurans and Related Compounds with Amino-Heterocyclic Substituents, J. Natl. Cancer Inst., 54 (1975) 841-850
65. Morris, J. E., Price, J. M., Lalich, J. J., and Stein, R. J., The Carcinogenic Activity of Some 5-Nitrofuran Derivatives in the Rat, Cancer Res., 29 (1969) 2145-2156
66. Cohen, S. M., and Bryan, G. T., Carcinogenicity of 5-Nitrothiophene Chemicals in Rats, Federation Proc., 32 (1973), 825
67. Cohen, S. M., Ertürk, E., Von Esch, A. M., Crovetti, A. J., and Bryan, G. T., Carcinogenicity of 5-Nitrofurans, 5-Nitroimidazoles, 4-Nitrobenzenes, and Related Compounds, J. Natl. Cancer Inst., 51 (1973) 403-417
68. Rustia, M., and Shubik, P., Induction of Lung Tumors and Malignant Lymphomas in Mice by Metronidazole, J. Natl. Cancer Inst., 48 (1972) 721-729
69. Zampieri, A., and Greenberg, J., Nitrofurazone as a Mutagen in E. Coli, Biochem. Biophys. Res. Commun., 14 (1964) 172-176
70. McCalla, D. R., and Voutsinos, D., On the Mutagenicity of Nitrofurans, Mutation Res., 26 (1974) 3-16
71. McCalla, D. R., Voutsinos, D., and Olive, P. L., Mutagen Screening with Bacteria: Niridazole and Nitrofurans, Mutation Res., 31 (1975) 31-37
72. Yahagi, T., Matsushima, T., Nagao, M., Seino, Y., Sugimura, T., and Bryan, G. T., Mutagenicity of Nitrofuran Deivatives on a Bacterial Tester Strain with an R Factor Plasmid, Mutation Res., 40 (1976) 9
73. Yahagi, T., Nagao, M., Hara, K., Matsushima, T., Sugimura, T., and Bryan, G. T., Relationships Between the Carcinogenic and Mutagenic or DNA-modifying Effects of Nitrofuran Derivatives, Including 2-(2-Furyl)-3-(5-nitro-2-furyl)-acrylamide, A Food Additive, Cancer Res., 34 (1974) 2266-2273
74. Wang, C. Y., Muraoka, K., and Bryan, G. T., Mutagenicity of Nitrofurans, Nitrothiophenes, Nitropyrroles, Nitroimidazole, Aminothiophenes, and Aminothiazoles in Salmonella Typhimurium, Cancer Res., 35 (1975) 3611
75. Rosenkranz, H. S., Jr., Speck, W. T., Stambaugh, J. E., Mutagenicity of Metronidazole: Structure Activity Relationships, Mutation Res., 38 (1976) 203
76. Kada, T., E. coli mutagenicity of furylfuramide, Jap. J. Genet., 48 (1973) 301
77. Morita, M., Feller, D. R., and Gillette, J. R., Reduction of Niridazole by Rat Liver Xanthine Oxidase, Biochem. Pharmacol., 20 (1971) 217-226

78. Wang, C. Y., Behrens, B. C., Ichikawa, M., and Bryan, G. T., Nitro-reduction of 5-Nitrofuran Derivatives by Rat Liver Xanthine Oxidase and Reduced Nicotinamide Adenine Dinucleotide Phosphate-Cytochrome c Reductase, Biochem. Pharmacol., 23 (1974) 3395-3904
79. Wang, C. Y., Chiu, C. W., and Bryan, G. T., Nitroreduction of Carcinogenic 5-Nitrothiophenes by Rat Tissues, Biochem. Pharmacol., 24 (1975) 1563-1568
80. Anon, Niridazole by Rat Liver Microsomes, Biochem. Pharmacol., 20 (1971) 203-215
81. Asnis, R. E., The Reduction of Furacin by Cell-free Extracts of Furacin-resistant and Parent-susceptible Strains of E. Coli, Arch. Biochem. Biophys., 66 (1957) 203-216
82. Rosenkranz, H. S., and Speck, W. T., Activation of Nitrofurantoin to a Mutagen by Rat Liver Nitroreductase, Biochem. Pharmacol., 25 (1976) 1555
83. Speck, W. T., Stein, A. B., and Rosenkranz, H. S., Mutagenicity of Metronidazole: Presence of Several Active Metabolites in Human Urine, J. Natl. Cancer Inst., 56 (1976) 283
84. Rustia, M., Shubik, P., Induction of Lung Tumors and Malignant Lymphomas in Mice by Metronidazole, J. Natl. Cancer Inst., 48 (1972) 721
85. Ong, T. M., Mutagenic activities of nitrofurans in Neurospora crassa, Mutation Res., 56 (1977) 13-20
86. Kramers, P. G., Mutagenicity of nitro compounds in Drosophila melanogaster, Mutation Res., 52 (1978) 213

CHAPTER 20

NITROAROMATICS AND NITROALKANES

A. <u>Nitroaromatics</u>

1. <u>2,4-Dinitrotoluene</u> (α-dinitrotoluene; DNT; O_2N-C$_6$H$_3$(NO$_2$)-CH$_3$) while not employed as a single explosive has utility as an ingredient in military and commercial explosives for its gelatinizing and water proofing action, as well as its explosive potential. 2,4-Dinitrotoluene is prepared from toluene by dinitration methods which yield an oil containing approximately 75% of the 2,4-isomer[1]. This oil (after removal of nitrating acid mixture) is used in commercial compositions or is purified by fractional freezing to obtain a solid product of desired purity. 2,4-DNT is an intermediate in the production of dyes.

2. <u>2,4,5-Trinitrotoluene</u> (α-trinitrotoluene; TNT; Triton, Tritol; O_2N-C$_6$H$_2$(NO$_2$)$_2$-CH$_3$) is the most important of modern high explosives and can be prepared from toluene by one-, two-, or three-stage nitration processes with the latter process the most widely employed[1]. The quality of the crude TNT can vary widely depending on the degree of control and efficiency of the third stage of nitration. For example, if this stage is not complete, a significant amount of dinitrotoluene can be present. The chief impurities in crude TNT are β or 2,3,4-trinitrotoluene (m.p. 112°C) small amounts are trinitrobenzoic acid, trinitrobenzene and tetranitromethane due to the oxidizing action of nitric acid[1]. Purification of crude TNT is an important step in production, since the crude material in the oily state is relatively insensitive to initiation.

Practical-grade 2,4-dinitrotoluene administered in the diet of male and female Fisher 344 rats at levels of 0.02 and 0.008% for 78 weeks induced benign tumors (e.g., fibroma of the skin and subcutaneous tissue in males and fibroadenoma of the mammary gland in females. No evidence was provided for the carcinogenicity of the compound in B6C3F1 mice of either sex when treated similarly[2].

The discharge of 2,4-dinitrotoluene and TNT into rivers and streams from munitions plants as well as exposure to those involved in their production also raises concern about the potential adverse health effects (e.g., mutagenesis and reproductive performance) in exposed humans.

2,4-Dinitrotoluene has been found capable of producing base-pair mutations in <u>S. typhimurium</u> TA 1535 strain[3]. Female rats mated with males fed a diet containing 0.2% 2,4-DNT for 13 weeks, showed a significant increase in the number of dead implants/total implants over control animals. While no increase in the frequency of translocation or chromatid breaks were observed in either lymphocyte or kidney cultures derived from these animals, significant increases in the frequency of chromatid breaks were observed, however, in kidney cultures after 5 weeks and in lymphocytes at 19 weeks. This would suggest that 2,4-DNT has a potential for inducing damage in somatic cells. <u>In vitro</u> studies with 2,4-DNT using the CHO-KI test system was negative[3].

When tested at levels of 10 or 40 mg/kg/day p.o. for 10 consecutive days in a dominant lethal study in Sprague-Dawley rats, 2,4-DNT was considered not a potent mutagen[4].

The biological activity of 2,4-DNT may be in part due to its conversion to the diamine with the rate of its enzymatic conversion being a limiting factor.

2,4,5-Trinitrotoluene was detected as a frameshift mutagen when tested in S. typhimurium TA-98 strain. In contrast, its major microbial metabolites (e.g., 2,6-dinitro-4-hydroxyaminotoluene; 2,6-dinitro-4-aminotoluene; 2-nitro-4,6-diaminotoluene; 2,4-dinitro-6-hydroxyaminotoluene; 2,4-dinitro-6-aminotoluene; 2,2',4,4'-tetranitro-6,6-azoxytoluene and 2,2',6,6'-tetranitro4,4'-azoxytoluene) appeared to be non-toxic and non-mutagenic[5].

A number of compounds present in wastewater from munitions plants were examined by Simmon et al[6] before and after bacteriocidal treatments (ozonation or chlorination) to determine whether they were mutagenic before treatment and/or whether the mutagenic activity, if any, was altered by the treatment. Several photolytic or metabolic products of TNT were also examined for mutagenic activity. The test compounds included TNT, TNT condensate water and components of condensate water (e.g., 1,3-dinitrobenzene; 2,4-dinitrotoluene; 3,5-dinitrotoluene), RDX, HMX, photolyzed TNT, PETN and trinitroresorcinol. The in vitro mutagenic assays were histidine reverse mutation in 5 strains of S. typhimurium (TA 1535, TA 1537, TA 1538, TA 98, and TA 100) and mitotic recombination in the yeast S. cerevisiae. A mammalian metabolic activation system using the liver of rats pretreated with Aroclor 1254 was used in each assay. The bacteriocidal treatments in most cases did not effect mutagenicity while some of the by-products of TNT production were mutagenic.

B. Nitroalkanes

1. 2-Nitropropane

Nitroalkanes (nitroparaffins) are derivatives of the alkanes in which the nitro group may be represented as a resonance hybrid, e.g., $R-N\overset{O^-}{\underset{O^-}{\rightleftharpoons}} \quad R-N\overset{O}{\underset{O^-}{\nearrow}}$

Primary and secondary mononitroalkanes are acidic substances which exist in tautomeric equilibria with their nitronic acids. These nitroalkanes undergo aldol-type condensations with aldehydes and ketones to yield nitroalcohols, reactions of nitroalkanes and primary or secondary amines yield Mannich bases[7].

2-Nitropropane (isonitropane; dimethylnitromethane, nitroisopropane, 2-NP, CH_3CHCH_3) is prepared by reacting nitric acid with an excess of propane[7]; the process
$\quad\;\; NO_2$
also yields nitromethane, nitroethane and 1-nitropropane. 2-Nitropropane (in concentrations ranging from 5 to 25%) is used as an industrial solvent in coatings (e.g., vinyl, epoxy, nitrocellulose, chlorinated rubber), printing inks and adhesives[8,9]. Other areas of utility include: the production of derivatives such as nitro alcohols, alkanol amines and polynitro compounds as a vehicle for other miscellaneous resins for printing on plasticized polyvinyl chloride films[7], as a stabilizer for chlorohydrocarbons[10,11] and as a corrosion inhibitor[12].

Occupational exposure to these products may occur in various industries including: industrial construction and maintenance, printing (rotogravure and flexographic inks) highway maintenance (traffic markings), ship building and maintenance (marine coatings), furniture, food packaging, and plastic products. NIOSH estimates that 100,000 workers are potentially exposed to 2-nitropropane in these and other industries[6]. Of the estimated 30 million pounds of 2-nitropropane produced annually in the U.S., 12 million pounds/year are sold domestically, the remainder is used internally by the sole manufacturer, IMC, or exported[8].

2-Nitropropane is not known to occur naturally but has been detected in tobacco smoke with other nitroalkanes, the levels were found to correlate with tobacco nitrate contents[12]. The smoke content of a filterless 85 mm U.S. blend cigarette was found to contain (μg): 2-nitropropane, 1.1; 1-nitropropane, 0.13; 1-nitrobutane, 0.71; nitroethane, 1.1; and nitromethane, 0.53.

NIOSH has recently reported that all Sprague-Dawley male rats exposed by inhalation to commercial grade 2-nitropropane (207 ppm) over a 6 month period developed hepatocellular carcinoma or hepatic adenoma[8,13]. No tumors were observed in New Zealand white male rabbits exposed to 27 and 207 ppm 2-nitropropane over the same period. Admitted shortcomings of the study were: (1) the 50 rats exposed to 207 ppm 2-nitropropane were weanling rats (younger and smaller than the other exposed rats to 27 ppm, and the control group) and (2) throughout the entire study, exposure to 2-nitropropane was conducted while food and water were present, hence introducing the potential for exposure by the oral route[8]. NIOSH believes that "it would be prudent to handle 2-nitropropane in the workplace as if it were a human carcinogen". The current OSHA standard for occupational exposure to 2-nitropropane is 25 ppm[8].

Information of the mutagenicity of 2-nitropropane is lacking.

References

1. Rinkenbach, W. H., Explosives, In Kirk-Othmer Encyclopedia of Chemical Technology, Vol. 8, 2nd ed., Wiley & Sons, New York (1965) pp. 581-658
2. National Cancer Institute, Bioassay of 2,4-dinitrotoluene for possible carcinogenicity, DHEW Publ. No. (NIH) 78-1304; Dec. 30 (1977)
3. Hodgson, J. R., Kowalski, M. A., Glennon, J. P., Dacre, J. C., and Lee, C. C., Mutagenicity studies on 2,4-dinitrotoluene, Mutation Res., 38 (1976) 387
4. Simon, G. S., Tardiff, R. G., and Borzelleca, J. F., Possible mutagenic effects of 2,4-dinitrotoluene: A dominant lethal study in the rat, Abstracts of 16th Annual Meeting of Society of Toxicology, Toronto, March 27-30 (1977)
5. Won, W. D., DiSalvo, L. H., and Ng, J., Toxicity and mutagenicity of 2,4,6-trinitrotoluene and its microbial metabolites, Appl. Env. Microbial., 31 (1976) 576-580
6. Simmon, V. F., Eckford, S. L., Griffin, A. F., Spanggord, R., and Newell, G. W., Munitions wastewater recycling: Do bacteriocidal treatments produce microbial mutagens? Abstracts of 16th Annual Meeting of Society of Toxicology, Toronto, March 27-30 (1977)
7. Martin, J C., and Baker, P. J., Jr., Nitroparaffins In Kirk-Othmer Encyclopedia of Chemical Technology, 2nd ed., Vol. 13, Interscience, New York, pp. 864-885
8. Finklea, J. F., Current NIOSH intelligence bulletin: 2-Nitropropane, Am. Ind. Hyg. Assoc., 38 (1977) A15-A19
9. Anon, 2-Nitropropane causes cancer in rats, Chem. Eng. News., May 2 (1977) p. 10
10. Hara, K., Stabilized methyl chloroform, Japan Kokai, 7455,606, 30 May (1974) Chem. Abstr., 82 (1975) 3756E
11. Sawabe, S., Genda, G., and Yamamoto, T., Stabilizing chlorohydrocarbons by addition of aliphatic alcohols, polyalkyl ethers and nitroalkanes, Japan Patent 7403,963, 29 Jan (1974), Chem. Abstr., 81 (1974) 151517X
12. Hoffman, D., and Rathkamp, G., Chemical studies on tobacco smoke, III. Primary and secondary nitroalkanes in cigarette smoke, Beitr. Tabakforsch., 4 (1968) 124-134
13. Busey, W. M., Ulrich, C. E., and Lewis, T. R., Subchronic inhalation toxicity of 2-Nitropropane in rats and rabbits, Abstracts 17th Annual Meeting of Society of Toxicology, San Francisco, CA, March 12-16 (1978) 135,137

CHAPTER 21

UNSATURATED NITRILES AND AZIDES

Nitriles, or cyanides, contain a non-bonded pair of electrons, and in contrast to amines, are exceedingly weak bases. Lacking hydrogen, nitriles cannot act as hydrogen donors and are such weak bases that they are not particularly effective as hydrogen bond acceptors[1]. Nitriles generally undergo a variety of reactions including: a) reduction to carboxylic acids and primary amines; b) hydrolysis to amides and c) as an electron-attracting group on reactive dienophiles in Diels-Alder reactions[2].

A. <u>Unsaturated Nitriles</u>

Acrylonitrile (vinyl cyanide; cyanoethylene, propenenitrile; $CH_2=CHCN$) can be prepared from propylene (SOHIO process) via the reaction of propylene, ammonia and air in the presence of catalyst[3,4], viz., $2CH_2=CHCH_3 + 2NH_3 + 3O_2 \rightarrow CH_2=CHCN + 6H_2O$. The catalyst can be a supportive molybdenum based agent such as 50 to 60% bismuth phosphomolybdate on silica (or bismuth-iron catalyst)[4]. Principal by products in the above synthesis of acrylonitrile are acetonitrile and hydrogen cyanide[3]. A variation of this process employs a mixture of propylene and nitric acid in stoichiometric amounts highly diluted with nitrogen and a catalyst of silver on silica[3]. All U.S. and Japanese acrylonitrile production is now based on the SOHIO process[4,5].

Acrylonitrile is also produced from (a) ethylene cyanohydrin and a catalyst such as alkali metal salts of organic acids, particularly formates[6] and (b) the reaction of acetylene and hydrogen cyanide in the presence of a catalyst (e.g., liquid phase catalysis with aqueous cuprous chloride and vapor phase catalysis with alkali and alkaline earth metal compounds)[6]. The principal by-products in the latter vapor phase catalysis reaction of acetylene and hydrogen cyanide are acetonitrile (CH_3CN) and propionitrile (CH_3CH_2CN).

Acrylonitrile is a very versatile chemical intermediate undergoing reactions of the cyano group alone, the activated double bond, or both groups. Acrylonitrile is readily hydrated as acrylamide ($CH_2=CHCONH_2$) a commercially important water soluble monomer. The carbonylation of acrylonitrile is the basis for the synthetic route to monosodium glutamate. The reactions of acrylonitrile with active hydrogen compounds introduces the cyanoethyl group into the reacting molecule. The active hydrogen molecules (AH) can be from many classes of compounds including alcohols, ammonia, amines, mercaptors, aldehydes, ketones, inorganic acids and their salts, some esters and aliphatic nitro compounds[4,6], viz., $AH + CH_2 = CHCN \rightarrow ACH_2CH_2CN$.

Although the manufacture of acrylonitrile in the United States began in 1940 and was primarily used for the production of nitrile rubber used in self-sealing liners of aircraft fuel tanks[6], a broad spectrum of uses has developed since, particularly in synthetic fibers, plastics, modified natural rubbers, hydrolyzed polymers as polyelectrolytes and as a chemical intermediate in the synthesis of antioxidants pharmaceuticals, dyes, surface coating agents and as a pesticide fumigant for stored grain.

Acrylic fibers (all acrylonitrile polymers or copolymers) have a substantive percentage of the total synthetic fiber market (approximately 20-25%)[6]. The utility of acrylonitrile plastics (e.g., styrene-acrylonitrile copolymers and acrylonitrile-butadiene-styrene (ABS) terpolymers) has expanded greatly in recent years[6,7]. These products are low-cost, all-around plastics which are moldable, tough and resistant to heat, light and solvents.

Another area of utility in the textile field is in cyanoethylation of cotton to make it more receptive to dyes and improve its properties of mildew, heat, acid, and abrasion resistance[3].

The total world production of acrylonitrile is estimated to have been about 2,400 million kg in 1976 with 4 U.S. companies producing 690 million kg; 6 Japanese companies produced 633 million kg and the total West European production amounts to an estimated 915 million kg[5]. The U.S. consumption pattern for acrylonitrile in 1976 is estimated to have been as follows (in %): for the manufacture of acrylic and modacrylic fibers (48%); acrylonitrile-butadiene-styrene (ABS), and styrene-acrolonitrile (SAN) resins (21%); adiponitrile (12%); and other applications (19%). Eighty-two percent of the acrylic and modacrylic fibers are used in the apparel and home furnishings markets and the remainder exported. The major markets for ABS resins are automotive and recreational vehicle components, large appliances and pipe fittings while SAN resins are used primarily in automobile instrument panels, and housware items (e.g., drinking tumblers)[5]. Adiponitrile is used almost exclusively as an intermediate for the production of Nylon 66 via hydrogenation to the intermediate hexamethylenediamine then reaction with adipic acid.

The percentage of total acrylonitrile consumed in the U.S. in other applications are (%): The manufacture of nitrile elastomers (4), acrylamide (4), barrier resins (3), and miscellaneous applications (e.g., polyether polymer poyols, fatty diamines (8).

Acrylamide is used primarily for the production of polyacrylamides used in waste and water treatment and as papermaking strengtheners and retention aids[5]. Nitrile barrier resins made from copolymers of acrylonitrile (70-75%) and other co-monomers such as styrene and methyl acrylate have been employed to make beverage containers (this usage was banned recently in the U.S.). Acrylonitrile is also used as a fumigant (in admixture with carbon tetrachloride) for stored tobacco.

The use pattern for acrylonitrile in Western Europe in 1977 is estimated to have been as follows (%): acrylic fibers (68), ABS/SAN resins (15), nitrile elastomers (5) and other applications (12). In Japan, the use pattern for acrylonitrile in 1976 is estimated to have been as follows (%): acrylic fibers (65), ABS/SAN resins and nitrile elastomers (17) and other applications (18)[5].

The use of acrylonitrile polymers and copolymers as components of a number of products when they are intended for use in contact with food is permitted by the U.S. Food and Drug Administration in the following applications: (1) vinyl resin coatings; (2) adhesives; (3) cellophane; (4) paper and paperboard components (limited to use as a size promoter and retention aid in aqueous and fatty foods only; (5) polyolefin films, (6) elastomers (for repeated use) and (7) rigid, semirigid, and modified acrylic and vinyl plastics (in conjunction with designated polymers and copolymers at maximum levels of 5 wgt % (acrylics and 50 wgt % (vinyl) of the toal polymer content). The amounts present may not exceed that which is reasonably required to produce the intended effect.

A styrene-acrylonitrile co-polymer until recently was test-marketed as a soft-drink and fruit-juice container and is already in use as an oleomargarine container in the U.S.[5,8,9].

FDA is proposing to lower the maximum amount of acrylonitrile permitted to migrate (leach) into the margarine tubs, vegetable oil product bottles, food wraps and other non-beverage packing made from acrylonitrile from a present maximum migration level of 300 parts-per-billion to 50 parts-per-billion[10]. An FDA ban on the use of acrylonitrile copolymers in the fabrication of beverage containers became effected on Sept. 19, 1977[11].

OSHA recently announced an emergency standard for acrylonitrile which limits employee exposure to an 8-hour time weighted average of 2 ppm of acrylonitrile in air. A ceiling level of 10 ppm was also set for any 15 minute period during an 8-hour shift[12]. (The previous agency standard permitted a 20 ppm, 45 mg/m^3, average concentration level).

NIOSH estimates that 125,000 people are potentially exposed to acrylonitrile in U.S. workplaces[13]. OSHA estimates that some 10,000 workers in industries including manufacture of various resins and plastics nitrile, elastomers and latexes are the most directly exposed to acrylonitrile[14].

The largest identified sources of acrylonitrile emissions are during production and transportation. With its relatively low boiling point (77.5-79°C) high losses are forseeable particularly during transport, especially if rail tank cars are used[4]. The total emissions of acrylonitrile to the U.S. workplace environment in 1974 was estimated to have been 14.1 million kg via sources such as production 6.4 million kg; end product manufacture 5.9 million kg and bulk storage 1.8 million kg[15]. Approximately 41 tons of acrylonitrile was estimated to be emitted to oceans in 1970 via losses during transportation (e.g., spills, ballasting and other transit losses)[4,16].

It should be noted that other potential toxicants are emitted during the production of acrylonitrile. These include (lb/ton): allyl chloride, 0.05; benzene, 0.60; toluene, 0.15; acetaldehyde, 0.06 and hydrogen cyanide, 1.32[4].

Acrylonitrile has been detected in workplace air in a number of USSR studies. For example, levels of 11.2-22.1 g hr has been reported in an acrylonitrile plant[17]; in a thermosetting plastics plant at a concentration of 1.4 mg/m^3[18]; at concentrations of 1-11 mg/m^3 in a rubber footwear plant[19] and in the workplace atmosphere of a large tonnage acrylonitrile plant[20]; a polyacrylonitrile plant[21] and in acrylic fiber plant at levels of 3-20 mg/m^3 in air[22].

In regard to the use of acrylonitrile in synthetic fibers, it should be noted that significant levels of acrylonitrile have been found in finished polyacrylonitrile, as well as "traces" of the monomer in the air surrounding wear-apparel products made of polyacrylonitrile in the USSR[23,24].

Concern is also manifest over acrylonitrile entering various waterways from product utilization. For example, acrylonitrile may be leached from fabrics during the laundering processes. In addition, polyacrylamide containing trace amounts of acrylonitrile used in water treatment may be a source of water pollution[4]. Acrylonitrile has been identified in various waterways[23]. Disposal by incineration or landfill of polymeric materials made from acrylonitrile may be a potential additional cause for concern.

Recent studies strongly suggests that acrylonitrile may be a human carcinogen[5,26-28]. O'Berg[26] reported to OSHA and NIOSH[13] in 1977 preliminary results of a study of the risk of cancer among 470 active and pensioned male wage employees initially exposed to acrylonitrile at a textile fibers plant in Camden, S. Carolina (at the plant's polymerization area where exposure potential is greatest) between 1950-1955. Seven deaths due to cancer were observed as compared with 3.4 expected on the basis of company wide mortality rates and 4.1 and 4.2 expected on the basis of national and regional rates respectively. (Four lung cancer deaths were observed whereas only 1.5 were expected using company wide rates.)[5] A subset of same cohort involving active male wage employees and studied 20 or more years following initial exposure revealed 14 cases of cancer compared to 4.9 and 5.7 expected on the basis of company wide cancer incidence rates and incidence rates of the Third National Cancer Survey respectively. A statistically significant increased incidence of both lung cancer and large intestine cancer were noted among acrylonitrile exposed workers.

These observations of an increase risk of cancer morbidity and mortality were most notable among wage employed as compared to salary employees. It should be noted that this excess of observed to expected cancers increased when consideration was given to latency. For example, with no consideration of latency the risk was least while with 15 years latency the risk was intermediate, and with 20 or more years, the risk was greatest[5].

A number of chronic oral and inhalation studies in rats are currently in progress[5]. Interim reports indicate that male and female Sprague-Dawley rats that ingested 100 or 300 ppm acrylonitrile drinking water for 12 months (half the study time) developed stomach papillomas (1/20 rats at 100 ppm, and 12/20 at 300 ppm); central nervous system tumors (2/20 at 35 ppm, 6/20 at 100 ppm, and 3/20 at 300 ppm) and Zymbal gland carcinoma (2/20 at 100 ppm, and 2/20 at 300 ppm), no such tumors were seen in control animals[5,29,30].

A mid-study report of a 2 year acrylonitrile inhalation study disclosed ear tumors and mammary region masses in female Sprague-Dawley rats at dose levels of 80 ppm and an "apparent increase of subcutaneous masses at 20 ppm"[27,31].

Acrylonitrile, in both aqueous and gas phases induced reverse mutations in S. typhimurium TA 1530, TA 1535, TA 1538 and TA 1978 in the presence of a 9000 g supernatant of liver from mice and rats[8,32].

Mutagenic effects were noted when the bacteria were exposed to an atmosphere containing acrylonitrile at levels as low as 57 ppm, less than 3 times the current TLV of 20 ppm (45 mg/m^3)[8].

It has been frequently speculated that an early stage in the metabolic conversion of a double-bonded chemical to an epoxide is important for mutagenic and/or carcinogenic activity. The projected epoxide from acrylonitrile was suggested by Milvy and

Wolff[8] to be glycidonitrile ($H_2C-C-CN$; oxirane carbonitrile) which is structurally similar to glycidaldehyde ($H_2C-C-CHO$) which is mutagenic in the Salmonella/microsome test system[33] and is also carcinogenic[34].

Acrylonitrile in solution produced reverse mutations in plate incorporation assays using E. coli WP2, WP2 UVRA and WP2 UVRApolA without metabolic activation[35]. Although these effects were weak in plate incorporation tests, assays using a simplified fluctuation test in E. coli WP2 and WP2 uvrApolA showed acrylonitrile to be significantly mutagenic at doses 20-40 times lower than the level used in the plate assay. The use of the different DNA-repair strains indicated that acrylonitrile causes DNA damage of the type exemplified by methyl methane sulfonate, e.g., non-excisable mis-repair damage[35].

Acrylonitrile has been previously found to inhibit growth and/or divison of S. cerevisiae[36] but was inactive in mutagenic tests using root tip meristems of Vicia faba[37]. Acrylonitrile was also reported to be weakly mutagenic (3 times the activity of controls) in Drosophila, with the toxicity of acrylonitrile to Drosophila limiting exposure to low values[38].

No chromatid or chromosomal aberrations were noted in rat bone marrow cells in animals receiving 35, 210, or 500 ppm of acrolonitrile in the drinking water for 90 days[39].

It has also been recently shown that administration of acrylonitrile by gavage to rats on days 6-15 of gestation produced teratological effects on the fetus at doses of 25 mg and 65 mg/kg/day[40].

Acrylonitrile is metabolized by mammals (rat, hamster, mouse and guinea pig) to cyanide, which is then transformed to thiocyanate, thence eliminated in the urine[24]. There is marked disagreement as to what percentage is thus metabolized with values ranging from 3.8% to 19.4% having been reported[42-44]. Gut et al[45] reported the biotransformation of acrylonitrile in female Wistar rats, conventional albino rats and Chinese hamsters given a single dose of acrylonitrile (0.5 to 0.85 mM/kg) to be influenced by the route of administration. The elimination of thiocyanate in the urine, indicated a decreasing proportion of biotransformation after oral (over 20%), intraperitoneal or subcutaneous (2 to 5%), and intravenous (1%) administration in rats. Oral administration of acrylonitrile in hamsters and mice was also followed by higher biotransformation than intraperitoneal administration. Acrylonitrile was found to be strongly bound in blood with this binding and cyanoethylation apparently being responsible for the unusually high influence of the different routes of administration on the metabolic fate of acrylonitrile. Acrylonitrile was more effectively metabolized to thiocyanate in mice than in rats after oral intraperitoneal and intravenous administration[45].

B. Azides

Azides (both inorganic and organic) are highly reactive nucleophilic agents that have been widely employed in the preparation of a variety of intermediates. The action of hydrazoic acid (HN_3) on a carboxylic acid, the action of sodium azide (NaN_3) on an acid chloride, or the action of hydrazine on an ester followed by treatment of the resulting hydrazide (RCONHNH$_2$) with nitrous acid, all produce the acyl or aryl azide (RCON$_3$ or ArCON$_3$). Acyl and aryl azides rearrange by thermal or photochemical processes to yield isocyanates via nitrene intermediates. Isocyanates are of importance in pharmaceutical, pesticide and polymer synthesis.

Areas of utility of sodium azide include: in fungicidal[46] and nematocidal compounds[47], as a gas-generating agent for inflating protective bags[48-51], as a preservative in diluents used with automatic blood cell counters and as a common reagent in hospitals and chemical laboratories.

It should be noted that the use of inflatable air bags as a "passive-restraint" will be mandatory for all U.S. automobiles produced in 5 years. There is concern that in this area of utility, the potential exists for faulty construction of the hermetically sealed container which carries the agent, Sodium azide, could as a consequence, leak out and be ingested or inhaled[51]. Another area of concern is the fate of sodium azide after autos (containing the air bags) are scrapped or junked.

Sodium azide effectively reverts Salmonella typhimurium strain TA 1530 indicating that it is a base-substitution mutagen[52]. Sodium azide is ineffective on strains TA 1531, TA 1532, and TA 1534 which are frameshift mutants[52]. Azide has been found highly effective in inducing his$^+$ revertants in uvr^B derivatives of S. typhimurium his G46 and in inducing high frequencies of 5-fluorouracil resistant mutants in uvr^A derivatives of E. coli B/r WP2[53]. In uvr$^+$ derivatives of these strains, azide was not effective or at best, only a marginal mutagen, hence demonstrating that the bacterial excision-repair system could repair azide-induced damage. Azide mutagenesis is suggested to be unique and different from ultra violet induced mutagenesis as shown by the inability of azide to revert ochre try$^-$ locus in E. coli WP2S. These results would suggest that the initial DNA damage induced by azide is different from UV induced DNA damage.

Sodium azide has also been reported to be a powerful and efficient mutagen when used on barley seeds[52] and has been suggested as a very useful mutagen for practical plant breeding applications[52]. The mutagenic action of sodium azide was not associated with chromosome aberrations[52,54]. Azide treatment has been shown to slightly increase the frequency of penicillin- and streptomycin- resistant mutants in Staphylococcus aureus (Micrococcus pyogenes var. aureus)[55]. However, in Drosophila, azide treatment alone induced no sex-linked recessive lethals[56,57] although in combination with carbon monoxide a slight increase in lethal mutations occured[58].

A heat and acid stable, highly water soluble mutagenic metabolite has been detected in barley embryos after sodium azide treatment. This metabolite was found to be mutagenic in S. typhimurium TA 1530 but whether or not it is the main agent through which azide mutagenesis occurs remains to be determined[59].

References

1. Gutsche, C. D., and Pasto, D. J., "Fundamentals of Organic Chemistry", Prentice-Hall, Englewood Cliffs, NJ (1975) pp. 426-427
2. Roberts, J. D., and Caserio, M. C., "Modern Organic Chemistry", W. A. Benjamin, Inc., New York (1967) 207-208
3. Lowenheim, F. A., and Moran, M. K., "Faith, Keyes and Clark's Industrial Chemicals", 4th ed., Wiley and Sons, New York (1975) pp. 46-49
4. EPA, Status Assessment of Toxic Chemicals. 1. Acrylonitrile, Environmental Protection Agency, Industrial Environmental Research Laboratory, Cincinatti, Ohio, Sept. 6 (1977)
5. IARC, Monographs on the Evaluation of the Carcinogenic Risk of Chemicals to Man, Vol. 18, Lyon (1978) in press
6. Fugate, W. O., Acrylonitrile, In Kirk-Othmer Encyclopedia of Chemical Technology, 2nd ed., Vol. 1, Interscience Publ., New York (1965) pp. 338-350
7. U.S. Code of Federal Regulations, Title 21, Food & Drugs, Parts 175.105, 175.300, 175.320, 176.170, 176.180, 177.1010, 177.1020, 177.1040, 177.1200, 177.2600, 178.3790, 180.22, Washington, DC (1977) pp. 445, 455, 465, 471-472, 482-487, 488, 496, 501-504, 555, 596, and 609-611
8. Milvey, P., and Wolff, M., Mutagenic studies with acrylonitrile, Mutation Res. 48 (1977) 271-278
9. Anon, FDA bans bottles made of acrylonitrile, Chem. Eng. News. March 14, (1977) p. 7
10. Anon, HEW News, March 7 (1977)
11. U.S. Federal Register, Acrylonitrile copolymers used to fabricate beverage containers; final decision. Vol. 42, No. 185, Sept. 23 (1977) pp. 48528-48544
12. U.S. Dept. of Labor, Occupational exposure to acrylonitrile, Federal Register, 42 (1977) pp. 33043-33044
13. NIOSH, Current intelligenic bulletin: Acrylonitrile, Rockville, MD, July 1 (1977)
14. Anon, Emergency standard set for acrylonitrile, Chem. Eng. News, Jan. 23 (1978) 4
15. Patterson, R. M., Bornstein, M. I., and Garşhick, E., Assessment of acrylonitrile as a potential air pollution problem, Vol. VI, U.S. Environmental Protection Agency Contract No. 68-02-1337, Task 8; NTIS, Springfield, VA, Publ. No. 258-358 (1976) pp. 20
16. National Academy of Sciences "Assessing Ocean Pollutants", Washington, DC (1975)
17. Musserakaya, A. N., and Boklag, E. P., Hygienic evaluation of ventilation in the modern production of acrylonitrile, Zdravookhr. Beloruss., 1 (1976) 51-53
18. Scupakas, D., Industrial hygiene conditions during the reworking of plastics and problems of improving them, Aktual. Vop. Gig. Tr. Prof. Patol. Mater. Konf. 1st (1968) 128-132
19. Volkova, Z. A., and Bagdinov, Z. M., Industrial hygiene problems in vulcanization processes of rubber production, Gig. Sanit., 34 (9) (1969) 33-40
20. Boklag, E. P., Hygienic characteristics of working conditions in the production of acrylonitrile, Zdravookhr. Beloruss, (8) (1975) 41-44
21. Stamova, N., Gincheva, N., Spasovskii, M., Bainova, A., Ivanova, S., et al., Labor hygiene during the production of bulana synthetic fibers, Khig. Zdraveopaz. 19 (1976) 134-140
22. Orusev, T., Bauer, S., and Popovski, P., Occupational exposure to acrylonitrile in a plant for the production of acrylic synthetic fibers, God. Zb. Med-Fak. Skopje 19 (1973) 445-449

23. Klescheva, M. S., Balandina, V. A., Usacheva, V. T., and Koroleva, L. B., Determination of the residual monomer content in some polystyrene plastics by gas chromatography, Vysokomal Soedin Ser. A, 11 (1967) 2595-2597
24. Rapaport, K. A., Ionkina, S. F., and Minteva, L. A., Hygienic evaluation of underwheat made of polyacrylonitrile fibers and the mixtures with natural fibers, Gig. Sanit. (1974) 85-86
25. Schackelford, W. M., and Keith, L. H., Frequency of organic compounds identified in water, EPA Rept. No. 600/4-76-062, U.S. Environmental Protection Agency, Athens, GA (1976)
26. O'Berg, M. T., Epidemiological study of workers exposed to acrylonitrile: Preliminary results, E.I. Dupont de Nemours & Co., Wilmington, Del. (1977)
27. Anon, Acrylonitrile looks like a human carcinogen: DuPont epidemiological study, Pesticide & Toxic Chem. News, 5 (16) (1977) 21-22
28. Anon, Carcinogenicity spectre raised by new data, Food Chemical News, 19 (1977) 26-27
29. Anon, Rats on 35 ppm acrylonitrile exhibit signs of toxicity, Food Chemical News, 19 (1977) 23-25
30. Quast, J. F., Humiston, C. G., Schwetz, B. A., Norris, J. M., and Gehring, P. J., Toxicological findings in rats maintained on water containing acrylonitrile (AN) for 24 months: Results of a 12 month interim sacrifice, Abstracts Int. Congr. Toxicology, Toronto, March 3-April 2 (1977) p. 19
31. Anon, Acrylonitrile data highlight "food additive" issue in hearing, Food Chemical News, 19 (1977) 40-41
32. deMeester, C., Poncelet, F., Roberfroid, M., and Mercier, M., Mutagenic activity of acrylonitrile, A preliminary study, Arch. Int. Physiol. Biochim. (1978) in press
33. McCann, J., Choi, E., Yamasaki, E., and Ames, B., Detection of carcinogens as mutagens in the Salmonella/microsome test: Assay of 300 chemicals, Proc. Natl. Acad. Sci. (US) 72 (1975) 5135-5139
34. Van Duuren, B. L., On the possible mechanisms of carcinogenic action of vinyl chloride, Ann. NY Acad. Sci., 246 (1975) 255-267
35. Venitt, S., and Bushell, Mutagenicity of acrylonitrile in bacteria, Mutation Res. 46 (1977) 241
36. Loveless, L. E., Spoerl, E., and Weisman, T. H., A survey of effects of chemicals on division and growth of yeast and E. coli, J. Bacteriol., 68 (1954) 637-644
37. Loveless, A., Qualitative aspects of the chemistry and biology of radiomimetic (mutagenic) substances, Nature, 167 (1951) 338-342
38. Benes, V., and Sram, R., Mutagenic activity of some pesticides in Drosophila melanogaster, Ind. Med. Surg., 38 (1969) 442-444
39. Cohen, M. M., and Hirschhorn, Cytogenic effects of acrylonitrile of rat bone marrow cells, In Chemical Mutagens, Principles and methods for their detection, ed. Hollaender, A., Plenum Press, New York (1971)
40. Murray, F. J., Mitschke, K. D., John, J. A., Smith, F. A., Quast, J. F., Blogg, C. D., and Schwetz, B. A., Teratologic evaluation of acrylonitrile monomer given to rats by gavage, Manufacturing Chemists Assoc., Washington, DC, cited in ref. 8
41. Brieger, H., Rieders, F., and Hodes, W. A., Acrylonitrile: Spectrophotometric determination, acute toxicity and mechanism of action, Arch. Ind. Hyg., 6 (1952) 128-140
42. Benes, V., and Cerna, V., Akrylonitril: Acute toxizität und wirkungs mechanisms, J. Hyg. Epidem. (Praha) 3 (1959) 106-116

43. Czajkowska, T., Acrylonitrile metabolites excretion following a single dose, Med. Pracy, 22 (1971) 381-385
44. Paulet, G., Desnos, J., and Battig, J., De la toxicite de l'acrylonitrile, Arch. Mal. Prof., 27 (1966) 849-856
45. Gut. I., Nerudova, J., Kopecky, J., and Holecek, V., Acrylonitrile biotransformation in rats, mice, and Chinese hamsters as influenced by the route of administration and by phenobarbital, SKF 525-A, cysteine, dimercaprol, or thiosulfate, Arch. Toxicol., 33 (1975) 151-161
46. McConnell, W. D., Control of Pythium SPP and Sclerotium SPP Using Azides, U. S. Patent 3,812,254, May 21 (1974), Chem. Abstr., 81 (1974) 73408W
47. Wilner, W. D., Nematocidal Mixtures, Ger. Offen., 2,359,226, June 6 (1974), Chem. Abstr., 81 (1974) P164782E
48. Klager, K., and Dekker, A. O., Non-toxic Gas Generation, U. S. Patent, 3,814,694, June 4 (1974) Chem. Abstr., 81 (1974) 172482X
49. Harada, I., Harada, T., Shiki, T., and Shiga, Y., Gas-generating Agent for Inflating Protective Bags for Passengers in Case of Automobile Accidents. Japan Kokai, 74 10, 887 (1972), Chem. Abstr., 81 (1974) 39546E
50. Anon, Explosion Warning Issued for Azides, Chem. Eng. News, 54 (1976) 6
51. Anon, Toxicity, disposal problem on sodium azide, clouding DOT airbag decision, Safety Letters, 7 (1977) 6
52. Nilan, R. A., Sideris, E. G., Kleinhofs, A., Sander, C., and Konzak, C. F., Azide-A Potent Mutagen, Mutation Res., 17 (1973) 142
53. Kleinhofs, A., Smith, J. A., and Nilan, R. A., Effect of excision repair on azide induced mutagenesis, Mutation Res., 38 (1976) 377
54. Sideris, E. G., and Argyrakis, The Effect of the Potent Mutagen Azide on Deoxyribonucleic Acid, Mutation Res., 29 (1975) 239
55. Berger, H., Haas, F. L., Wyss, O., and Stone, W. S., Effect of Sodium Azide on Radiation Damage and Photoreaction, J. Bact., 65 (1953) 538
56. Sobels, F. H., The Effect of Pretreatment with Cyanide and Azide on the Rate of X-ray Induced Mutations in Drosophila, Z. Vererbungslehre, 86 (1955) 399-404
57. Sobels, F. H., The Influence of Catalase Inhibitors on the Rate of X-ray Induced Mutations in Drosophila Melanogaster, Proc. 1st Intern. Photobiol. Congr., Amsterdam, (1954) 332-335
58. Clark, A. M., Genetic Effects of Carbon Monoxide, Cyanide and Azide on Drosophila, Nature, 181 (1958) 500-501
59. Owais, W. M., A mutagenic in vivo metabolite of sodium azide, 9th Annual Meeting Environmental Mutagen Society, San Francisco, CA, March 9-13 (1978) 35-36

CHAPTER 22

AROMATIC HYDROCARBONS

1. Benzene (benzole, coal naphtha, C_6H_6), the parent hydrocarbon of the aromatic group, is produced in enormous amounts principally from coal tar distillation, from petroleum by catalytic reforming of light naphthas from which it is isolated by distillation or solvent extraction, from pyrolysis gasoline produced at steam-cracking ethylene plants, and from hydrodealkylation of toluene. The petrochemical and petroleum refining industries together are responsible for 94% of the total U.S. production of benzene. The other 6% comes from coal, primarily as a by-product of the coking process in steel mills[1].

The broad spectrum of utility of benzene (commercially sometimes called "Benzol") includes the following[2]: extraction and rectification; intermediate for synthesis in the chemical and pharmaceutical industries; the preparation and use of inks in the graphic arts industries[3]; as a thinner for paints; as a degreasing and cleaning agent; as a solvent in the rubber industry; as an anti-knock fuel additive; and as a general solvent in various laboratories. Industrial processes involving the production of benzene and chemical synthesis usually are performed in sealed and protected systems. Figure 1 illustrates a flow diagram of products derived from benzene.

Benzene is consumed by the chemical industry in the U.S. at the rate of 1.4 billion gallons (over 14 billion pounds in 1976) annually[1] with the outlook in 1978 for benzene expected to increase by 5 to 7%[4]. Approximately 86% of the benzene produced in the U.S. is used in the production of major organic derivatives including styrene (50%), cyclohexane (15%), cumene (15%) and phenol used as intermediates to manufacture plastics and rubbers, nylons, resins, disinfectants and pharmaceuticals. The remaining 14% is used primarily in the manufacture of detergents, pesticides, solvents and paint removers. Benzene is also present as a component of motor fuels, averaging less than 2% in gasoline in the U.S. and up to 16% in some European countries[5].

Consumers may be expsoed unknowingly in the home through the use of commercial products that may contain benzene in concentrations of 10 to 100% in products such as rubber cement, brush cleaners, paint strippers and bicycle tire patching compounds[6]. Additional benzene containing consumer products are carburetor cleaners and art and craft supplies. The Consumer Product Safety Commission in the U.S. estimates that from 288,000 to 576,000 adults are exposed to paint remover containing benzene while 3,000 to 5,000 adults and children are exposed to benzene fumes from rubber cement[7].

NIOSH considers "that benzene is leukemogenic" and recommends that "for regulatory purposes it be considered carcinogenic in man"[8] and hence recommends that the Occupational Safety and Health Administration (OSHA) cut benzene exposure in plants from 10 ppm to 1 ppm, based on an 8 hr average, and reduce the permissible ceiling level from 25 ppm to 5 ppm[1,9]. According to government estimates, benzene is present in more than 24,000 workplaces, and about 2 million U.S. workers are potentially exposed to it[2]. However, only about 200,000 workers actually would be covered by the standard (Table 1).

446

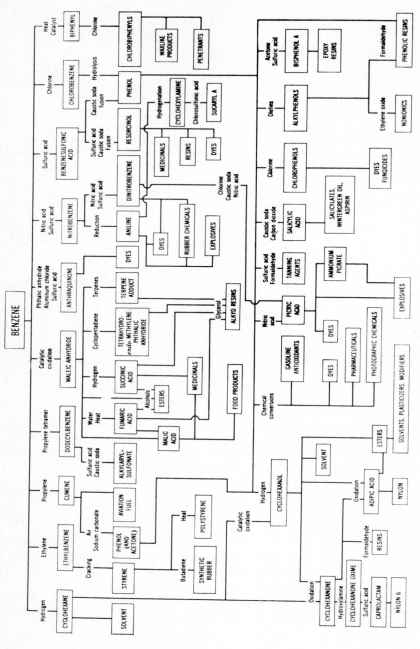

FIG. 1 PRODUCTS DERIVED FROM BENZENE*

*Ayers, G. W., Benzene, In Kirk-Othmer Encyclopedia of Chemical Technology, Vol. 3, 2nd ed., Wiley & Sons, New York (1964) pp. 367-401

Table 1. Benzene Exposure Rules to Affect 24,000 Facilities[1]

Industry	Facilities	Employees Exposed
Benzene Production		
Petroleum derivatives	48	2,160
Coke oven derivatives	65	2,470
Coke oven batteries	65	17,000
Benzene Transportation		
Barge	240	240
Tank car	100	100
Tank truck	200	200
Chemical Processing		
Chlorobenzenes	13	390
Cumene/Phenol	15	450
Cyclohexene	11	330
Dodecybenzene	5	150
Ethylbenzene/Styrene	18	540
Maleic Anhydride	9	270
Nitrobenzene	8	240
Other	13	390
Laboratories	12,296	25,000
Petroleum Refining		
Refineries	275	98,000
Terminal storage of gasoline	1934	11,604
Bulk storage	8594	12,891
Rubber Industry		
Tire Manufacture	206	11,400
Other	197	13,050
TOTAL	24,312	196,875

While this action will markedly reduce occupational exposure it should be noted that it will not effect environmental exposure of humans to benzene. No definitive assessment of the number of people currently at risk from exposure to benzene emissions is possible until definitive studies are made concerning true ambient levels and the actual risks involved with such exposure[10]. It is informative to consider a number of aspects of environmental sources and concentrations of benzene[10,11]. It should be noted that based on its vapor pressure and chemical stability, benzene is expected to be quite mobile and persistent. Reforming operations in the petroleum refining industry has contributed 17.5 gallons of benzene to the environment in 1975[12]. Major benzene contamination sources include: synthetic chemical plants, oil spills and emissions from coke ovens and automobiles. Approximately 122 million pounds of benzene were released in the vicinity of coke ovens in the U.S. during 1974[11]. Routine tanker discharges, accidents in port and during oceanic exploration and production, discarded lubricants, pipeline breakage, incompletely burned fuel, and untreated industrial and domestic sewage contributes 11,000-12,000 million pounds per year of oil to the ocean[12a]. Since crude oil contains an estimated 0.2% benzene, the above releases add 22-24 million pounds of benzene to the seaways[13]. Table 2 lists the losses of benzene during chemical synthesis of by-products[11]. EPA estimated benzene emissions to

Table 2. Losses of Benzene During Chemical Synthesis of By-Products

By-Product	Benzene Consumption 10^6	Amount of Benzene Lost - 10^6 lb
Ethyl benzene	3709	111
Phenol	1610	290
Cyclohexane	1311	0
Maleic Anhydride	325	140
Detergent Alkylate	323	65
Aniline	297	21
Dichlorobenzene	94	14
DDT	43	17
Other Fuel Uses	676	
TOTAL	8388	658

include: 830 million pounds per year from the production, transportation, storage and use of benzene; 1 billion pounds per year from the refueling and operation of motor vehicles, and 22-24 million pounds per year from oil spills as well as emissions from power plants and the use of insecticides and miticides containing benzene[10]. In limited studies, average levels of benzene detected in air were found to be in the low parts-per-billion range[2,10]. In an EPA survey of organic compounds in the drinking water of 10 cities, benzene was detected in water from 4 cities at concentrations ranging from 0.1-0.3 µg/liter[2,14].

A number of recent reviews on the toxicity of benzene have appeared[2,11,15-25]. The most severe long-term hazard is to the haemopoietic system. While continued exposure often results in an irreversible progression into chronic pancytopenia, cessation of contact with benzene can lead to recovery of an apparently normal blood picture. A more acute form of pancytopenia develops in some cases and, occasionally, acute or chronic leukemia has been reported.

An association between long-term exposure to high concentrations of benzene and leukemia has been extensively described[2,11,15,16,20,24,25] and includes cases of sub-acute myeloblastic[26,27] and also, with less convincing evidence, lymphatic leukemia[29] and Hodgkin's disease[30]. Symptoms of leukemia are frequently preceded by aplastic anaemia, and it has been suggested by Sanita[28] that many benzene haemopathies diagnosed on the basis of peripheral blood picture as pancytopenia might, in fact, have been aleukaemic leukemia.

The extent of the problem is difficult to assess by most accounts[2,15,16,20,24,25] It is not possible, from the available literature, to estimate the incidence of leukemia in the industrial population, or even the proportion of the industrial population chronically exposed to benzene[2,15,16], although, as noted previously, NIOSH estimates that 2 million workers in the U.S. have a potential exposure to benzene[2]. A review in 1974[24] documented 250 cases in which leukemia was diagnosed in subjects who had a record of chronic exposure to benzene[2]. Dean[14] has noted that only rarely can the extent and duration of benzene exposure to calculated from published reports on leukemia and in almost all recorded cases the affected individuals were exposed to high concentrations of other chemical agents in addition to benzene[2].

Despite this complicating factor, Goldstein[20] suggests that it appears to be more than reasonably demonstrated that benzene does produce hematotoxicity in humans. The evidence includes the high degree of association of hematopoietic disorders with benzene exposures, the apparent lack of such association with other known volatile agents present in the same occupational setting, outbreaks of hematotoxicity temporally related to the introduction of benzene to an industry which respond to replacement of benzene with other solvents, and the ability to produce bone marrow toxicity in various animal species exposed solely to benzene. It is generally agreed that elucidation of the benzene-leukemia question has been severely hindered by the lack of a suitable animal model. While it is possible to reproduce several of the haematological disorders found in man[31], attempts to induce leukemia by exposure of laboratory animals to benzene have been in the main unsuccessful[32] but with one reported exception. Preliminary reports of studies of Maltoni in Bologna indicate that of 60 rats administered 50 mg/kg of benzene in an olive oil vehicle for 76 weeks, two had zymbal gland tumors and one developed adrenal tumor; none of the more than 300 rats in the control group developed tumors[33-34].

It has been suggested that benzene induced leukemia occurs only in genetically sensitive individuals or only when benzene exposure (at rather high concentrations) occurs in conjunction with some other chemical physical or vial agent[2]. Dose-response relationships in chronic exposure of humans to benzene and details of the extent of exposures are generally considered to be either lacking or inadequate[2].

Conflicting epidemiological surveys relating to a correlation between leukemia and benzene exposure should be cited[2,15,30,35-40]. For example, in the first major epidemiological survey[35], a study of 28,500 shoe workers showed an annual incidence of leukemia of 13/100,000 compared to 6/100,000 in the general population. However, an epidemiologic study[38] on 38,000 petroleum workers who had potential exposures to benzene failed to indicate an increase of leukemia.

A recent controversial report of EPA's Cancer Assessment Group (CAG) estimates that between 30 and 80 cases of acute and myelogenous and myonocytic leukemia a year in the general population may be due to atmospheric benzene[41,42]. This translates to a risk of between one and 2.5 cases per year per 100,000 people exposed to benzene over a lifetime. The upper limit of 80 yearly cases amounted to 1.2% of the total leukemia cases reported in the U.S. in 1971. About 85% of the cases are directly related to the use of gasoline; from 14 to 35 cases of leukemia are caused by urban exposure to all types to atmospheric benzene, principally from automobile tailpipe emissions and evaporation of benzene from auto gasoline tanks[41,42]. Other sources of leukemia attributable to benzene exposures (in descending order to the number of predicted cases per year) were: chemical manufacturing plants (2.8 to 7.1 yearly cases); coke ovens (1.2 to 3.0 cases); petroleum refineries (0.46 to 1.2 cases); self-service gasoline stations and solvent operations (both less than 1 case per year)[42]. For estimating the health effects, the Group used two epidemiological studies; one conducted by Infante of NIOSH of pliofilm workers at 2 plants and the other by Ott of Dow Chemical Co. In a retrospective study of mortality among a cohort of 748 white male workers in the two Ohio plants manufacturing a natural rubber film product, Infante found a statistically significant higher rate of leukemia than both of two control groups. The individuals in the cohort all worked in the factories before 1950 and were followed until 1975. From among the 748 workers, 160 of whom were dead at the time of the study, seven leukemia deaths were found (4 acute myelogenous, one chronic myelogenous and two nonocytic). This leukemia mortality rate was 5.06 times the general U.S. white male population standardized for age and time period of the cohort exposure and 4.74 times higher than the cohort of 1474 white male employed at an Ohio fibrous construction products factory[41,42]. To produce the higher risk the CAG assumed that workers in Infante's cohort were exposed to 23.7 ppm benzene for 25 years, while the lower risk estimates were produced by assuming the workers were exposed to 82 ppm for 10 years and then 23.7 ppm for 25 years (averaging 40.36 ppm for 35 years)[42]. In the Infante study the lifetime risk to an individual continuously exposed to benzene at a level of 1 ppm was estimated by CAG to be between 10^{-5} and 2.5×10^{-5}, depending on which exposure.

The study by Ott reported no excess leukemia risk (for 102 employees who died among a cohort of 8,000 employees), but according to CAG was "consistent with the Infante study on statistical grounds given their relative exposures". The Ott study suggested that the level of calculated risk in the general population is not likely to be higher than 10^{-5} to 2.5×10^{-5} but it could be lower[41].

A risk assessment report of EPA states that the annual average exposure of an individual in the U.S. to ambient benzene from all sources is 1.03 parts-per-billion[43]. High industrial exposure, equivalent to 4.4 ppm is expected to cause 2,177 lifetime cases of leukemia in the U.S. and 31.1 cases/year while low industrial exposure, equivalent to 1.8 ppm is expected to cause 5,436 lifetime leukemia cases and 77.7 yearly cases[41-43].

The CAG listed the disadvantages of relying on the Infante study for determining general population risks. These included: (1) no estimates of worker exposure to say that the levels were less than the prevailing recommended occupational limits at the time various monitoring surveys were made; (2) the cohort studied actually worked at two separate plants but air monitoring information in the Akron plant is almost non-existent and hence the exposure to half the members of the cohort is almost completely unknown; and (3) other studies submitted by industry at the OSHA hearings claimed that over 400 workers known to be exposed to low benzene levels were deliberately excluded from the cohort[41].

The role of benzene-induced chromosome aberrations is not currently definitive[2,16]. Interpretation of studies carried out on workers exposed to benzene is complicated by the lack of quantitative data on the concentrations of benzene involved. Levels of 125 to 532 ppm are referred to by Forni et al[44] and all cases suggest extremely high concentrations, probably in the hundreds of ppm. Dean[16] in his review of human chromosome studies, states that even a detailed examination of the published data does not allow a close correlation between occupational exposure to benzene and the persistence of chromosomal aberrations[2], except where clinical syndromes were present, when an association between benzene-induced hemopathy and chromosome aberrations is clearly demonstrated.

A number of studies have been carried out on the chromosomes of peripheral blood leucocytes or bone-marrow cells from workers with known, and often extensive exposure to benzene. Such studies fall into two categories, viz., (1) chromosome investigations of patients with blood dyscrasias associated with benzene exposure (benzene hemopathies) and (2) the more wide-ranging studies of exposed working populations without clear evidence of clinical effects[2,15,16]. Chromosomal aberrations of both the stable and unstable type have been noted[44-46]. In general, the chromosomal aberrations were higher in peripheral blood lymphocytes of workers exposed to benzene than in the controls even in the absence of avert signs of bone-marrow damage. The stable type of chromosomal aberrations persisted several years after recovery from benzene hemopathy.

Numerous studies involving benzene-induced lymphocyte chromsome damage and hemopathies have been cited[16,47-54]. It should be re-stressed that no quantitative data on total benzene exposure were available on any of the above studies on chromosome aberration in humans, with all indications suggesting very high levels (e.g., several hundred ppm) of benzene[2,15,16]. In general, no correlation was found between the persistence of chromosomal changes and the degree of benzene poisoning[2,15,16,54].

Dean recently reviewed chromosome studies in laboratory animals exposed to benzene[16]. Rabbits injected subcutaneously with a dose of 0.2 ml/kg/day of undiluted benzene until sequential blood counts showed tha the peripheral blood leucocytes were reduced to below $1000/mm^3$. A high incidence of chromatid- and chromosome-type aberrations was found in all the rabbits analyzed, persisting up to 60 days after the end of dosing with benzene. Chromatid deletions were demonstrated in metaphase chromosomes from bone-marrow cells of rats analyzed 12 and 24 hrs after subcutaneous injections of 2 ml/kg undiluted benzene[58].

When rats were dosed with benzene or toluene or a mixture of both it was shown that the degree of chromosome damage induced by 0.2 gm/kg/day of benzene was similar to that induced by 0.8 gm/kg/day of toluene and that the effect as additive when both compounds were given[58].

During 4 months daily 4 hr inhalation of 300 mg benzene, 610 mg toluene or 300 mg benzene plus 610 mg toluene/m^3 air, the percentage of metaphases with damaged chromosomes in rat bone marrow gradually increased to 27.42, 21.56 and 41.21% respectively from 4.02% in controls. In addition benzene caused leukocytopenia whereas toluene and the mixture caused lukosis. One month after termination of the inhalation, the frequency of chromosome damage was still elevated, whereas the morphological composition of blood had almost completely returned to normal[59].

No induced dominant lethality was detected after intraperitoneal dosing of male rats with 0.5 ml benzene/kg body weight. In the bone-marrow study, no consistent numerical chromosome changes were observed but both chromatid and chromosome aberrations were significantly increased over the control values up to 8 days after dosing[60]. A micronucleus assay in rats given i.p. doses of 0.025, 0.05 and 0.25 ml/kg benzene on each of 2 successive days showed that animals in the 0.05 and 0.25 ml/kg groups had significantly higher micronuclei counts than had corresponding controls 6 hr after the final dose[60].

Exposure of cultured human leukocytes and Hela cells to 2.2×10^{-3}M benzene has resulted in a decrease in DNA synthesis. Cultured human leukocytes exposed to dose levels of 1.1×10^{-3}M and 2.2×10^{-3}M exhibited chromosome aberrations consisting of breaks and gaps[61].

Benzene was reported by Lyon[61,62] to be non-mutagenic when tested both directly and in the present of S9 microsomal fractions from rat liver in S. typhimurium strains TA 98 (reverted by frameshift mutagens) and TA 100 (reverted by base-substitution mutagens). Benzene was also reported by Shakin[63] to be non-mutagenic when tested in S. typhimurium strains TA 98, TA 100, TA 1535, TA 1537 and TA 1538.

Benzene was also non-mutagenic in a host-mediated assay (with S. typhimurium TA 1950) in mice given two subcutaneous doses of 0.1 ml benzene[60,61].

In the most comprehensive mutagenicity study of benzene to date employing the Salmonella/microsome assay, the mouse lymphoma forward mutation assay in L5178Y (TK+/-) cells, in vitro and in vivo cytogenetic and an in vitro sister chromatid exchange assay in mammalian cells, the tests for gene mutation using bacteria and L5178Y mouse lymphoma cells were negative under all test conditions[64], but clastogenic activity was observed. The data clearly showed genetic activity for benzene but it appeared to be specific for chromosome effects and little or no gene mutations were induced. The authors concluded that benzene is in a class of chemicals which may show false negative results in genetic screening tests where only gene mutation assay systems are employed hence demonstrating the need for measuring both chromosome and specific locus gene mutations in short-term test batteries[64].

Lutz and Schlatter[65] recently reported an irreversible covalent interaction of a benzene metabolite with DNA in vivo. Proposed likely intermediates were benzene oxide or a phenol epoxide. The alkylating potency of benzene to liver DNA was found to be about 3000 times lower than that of dimethyl nitrosamine (DMN). For example, a benzene dose of 60 mg/kg (30 times higher than the DMN dose) yielded only 1.5μ mole alkylated nucleotides per mole DNA phosphate.

Benzene has been shown to reduce DNA synthesis and it has been suggested by Morimoto that it might inhibit DNA repair in cultured human leucocytes[66].

The role of benzene metabolism in its toxicity as well as the significance of benzene-induced chromosome aberrations appears to be undefined[2,16,25]. Urinary excretion products following benzene exposure include phenol, hydroquinone, catechol, hydroxyhydroquinone, trans-trans muconic acid, and L-phenylmercapturic acid[25]. The major route of metabolism in all species tested was conjugation which included both ethereal sulfate and glucuronide conjugates. Figure 2 illustrates the probable routes of the biotransformation of benzene in mammals[16].

The rate of benzene metabolism depends on the dose administered as well as the presence of compounds which either stimulate or inhibit benzene metabolism[2,25].

Although the mechanism of benzene hydroxylation has not been definitively determined, it has been suggested that the reactions occur via an arene oxide intermediate[25]. While benzene oxide has not been found in liver microsomes (probably due to its extreme lability) it should be noted that incubation of benzene oxide with microsomes yields the metabolic products of benzene and that naphthalene oxide has been isolated from the incubation of naphthalene with microsomes[25].

In summary, it has been established that exposure to commercial benzene or benzene-containing mixtures may result in damage to the haematopoietic system[2,14,16,25], although the mechanism by which benzene acts is not known. The major problem area remains the association between benzene exposures, myelotoxicity, chromosome damage and leukemia[16].

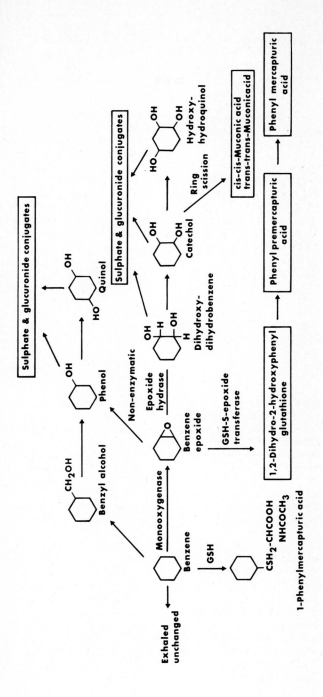

Fig. 2 Probable routes of the biotransformation of benzene in mammals[16]

In advanced stages the result can be pancytopenia due to bone marrow aplasia. DNA synthesis is reduced in bone marrow of benzene-treated animals either because of inhibition of enzymes involved in DNA synthesis or because a lesions revealed as reduced incorporation of tritiated thymidine in DNA occuring at some point in the cell cycle.

A relationship between exposure to benzene or benzene-containing mixtures and the development of leukemia is suggested by many case reports[2,15,16,25]. However, it would appear that more definitive data are required to enable a more accurate assessment of the myelotoxic, leukemogenic and chromosome-damaging effects of benzene[2,15,16,25].

2. <u>Toluene</u> (toluol, methylbenzene, phenyl methane, $C_6H_5CH_3$) is produced in enormous quantities (e.g., 1,562 million gallons in U.S. in 1976) mainly from coal or petroleum sources. Toluene from coal is principally a by-product of metallurgical coke manufacture. Toluene is produced from petroleum mainly by the hydroforming of selected petroleum naphthas. This involves catalytic dehydrogenation in the presence of hydrogen (which reduces coke formation) to yield a mixture of aromatic hydrocarbons chiefly toluene[67]. The principal uses of toluene include (%): the production of benzene (51), solvents (10), explosives (e.g., trinitrotoluene) (9), isocyanates (e.g., 2,4-toluene diisocyanate) (5) phenol (1) and gasoline pool and intermediates (24)[67]. The highest potential for toluene growth is in the production of toluene diisocyanate manufacture for polyurethane foams which is projected to show growth of about 12%/year[67]. Toluene is also used for the preparation of numerous derivatives and by-products such as benzoic acid, benzyl chloride, benzaldehyde, ring-chlorinated derivatives, and alkylated toluenes which are widely used in medicinals, plastics, dyes, germicides, perfumes and flavors[68].

Absorption of toluene over long periods may cause minor CNS effects and general malaise[69] but does not induce the severe haemopoietic injury associated with chronic benzene exposure[2,15,16,70]. In extensive animal studies, no evidence of bone-marrow damage in a variety of species after chronic inhalation exposure to toluene was reported by Fabre et al[71]. Toluene is currently employed in several industries as a "safe" replacement for benzene[44].

Conflicting results have been reported on the effects of toluene on mammalian chromosomes. No significant increase in chromosome aberrations in workers exposed to toluene for periods up to 15 years was reported by Forni et al[44]. Atmospheric concentrations of toluene during this period were generally close to the maximum allowable concentration (MAC) of 200 ppm although, on occasions, workers were exposed to much higher concentrations.

The persistence of chromosome aberrations after inhalation exposure of rats to 160 ± 42 mg/m^3 of toluene has been reported by Dobrokhotov and Enikoev[72]. Chromosome damage was detected in bone-marrow cells earlier and to a greater degree in peripheral blood leucocytes. A month after cessation of exposure, chromosome aberrations were still present although blood counts had returned to normal. The observed chromosome damage was the sum of the effects of both compounds when benzene and toluene were given together[58,72]. It was estimated that 0.8 gm/kg/day of toluene induced the same frequency of chromosome damage in rats as 0.2 gm/kg/day of benzene[58].

Chromatid gaps and breaks were detected in 11.5% of cells analyzed (compared with 57.2% in benzene-treated rats and 3.9% in untreated animals following injection of rats with 1 gm of toluene/kg body weight each day for 12 days.

In toluene-treated Bacillus subtilis, protein synthesis was inhibited as well as the absence of initiation of DNA synthesis noted[73].

Toluene was reported to be an enhancing agent in 7,12-dimethylbenz-(a)anthracene induced skin carcinogenesis[74].

The major route of metabolism of toluene is via oxidation to benzoic acid and the excretion as hippurric acid and benzoyl glucuronic acid. Small quantities of o-, m-, and p-cresyl sulphates and glucuronides are also excreted as a result of ring hydroxylation[75-77] (Figure 3)[16]. About 50% of an absorbed dose of toluene is exhaled unchanged[16].

The current TLV limit in the U.S. is 100 ppm.

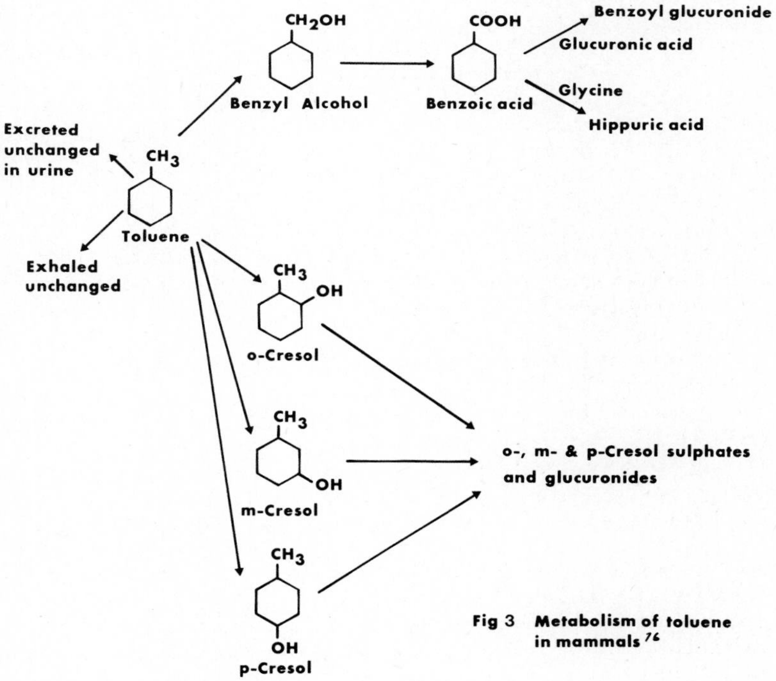

Fig 3 Metabolism of toluene in mammals[76]

References

1. Anon, OSHA Pushes Stiffer Benzene Exposure Rules, Chem. Eng. News, Aug. 1, (1977) 12-13
2. National Academy of Sciences, A Review of Health Effects of Benzene, National Academy of SCiences, Washington, DC, June (1976)
3. Kay, K., Toxicologic and cancerogenic evaluation of chemicals used in the graphic arts industries, Clin. Toxicol., 9 (1976) 359-390
4. Anon, Key Chemicals: Benzene, Chem. Eng. News, Nov. 7 (1977) p. 18
5. Runion, H. E., Benzene in gasoline, Am. Ind. Hyg. Assoc. J., 36 (1975) 338-350
6. Young, R. J., Rinsky, R. A., Infante, P. F., and Wagoner, J. K., Benzene in consumer products, Science, 199 (1978) 248
7. Anon, CPSC instructs staff to develop ban on use of benzene in products, Chem. Reg. Reptr., 1 (1977) 987-988
8. Anon, NIOSH links benzene to leukemia, Chem. Ecology (Oct., 1976), p. 5
9. Concklin, B. M., OSHA issues guidelines for the control of benzene, Employment Safety and Health Guide, 10 (1977) 735
10. EPA, Status assessment of toxic chemicals: 4. Benzene, U.S. Environmental Protection Agency, Industrial Pollution Control Division, Cincinatti, Ohio, Sept. 6 (1977)
11. Haley, T. J., Evaluation of the health effects of benzene inhalation, Clin. Toxicol., 11 (1977) 531-548
12. Howard, P. H., and Durkin, P. R., Benzene environmental sources of contamination ambient levels and fate, Syracuse University, New York, Dec. (1974)
12a. Blumer, M., Sanders, H. C., Grassle, J. F., and Hampson, G. R., A small oil spill, Environment, 13 (1971) 2
13. Green, A. D., Morrell, C. E., "Petroleum Chemicals" In Kirk-Othmer Encyclopedia of Chemical Technology, Vol. 10, Wiley (Interscience), New York (1973) pp. 177-210
14. EPA, Preliminary Assessment of Suspected Carcinogens in Drinking Water, Report to Congress, U.S. Environmental Protection Agency, Washington, DC, Dec. 1975
15. IARC, Monographs on the Evaluation of Carcinogenic Risk of Chemicals to Man, Vol. 7, International Agency for Research on Cancer, Lyon (1974) 203-221
16. Dean, B. J., Genetic toxicology of benzene, toluene, xylenes and phenols, Mutation Res., 47 (1978) 75-97
17. Leong, B. K. J., Experimental benzene intoxication, J. Toxicol. Env. Hlth, Suppl. 2 (1977) 45-62
18. Freedman, M. L., The molecular site of benzene toxicity, J. Toxicol. Env. Hlth., Suppl. 2 (1977) 37-44
19. Rusch, G. M., Leong, B. K. J., and Laskin, S., Benzene metabolism, J. Toxicol. Env. Hlth., Suppl. 2 (1977) 23-36
20. Goldstein, B. D., Hematotoxicity in humans, J. Toxicol. Env. Hlth., Suppl. 2 (1977) 69-106
21. Berlin, M., Gage, J., and Johnson, E., Increased aromatics in motor fuels: A review of the environmental and health effects, Work Env. Hlth., 11 (1974) 1-20
22. Saita, G., Benzene induced hypoplastic anaemias and leukaemias, In Girdwood, R. H., ed. "Blood Disorders Due to Drugs and Other Agents" Amsterdam, Excepta Medica (1973) p. 127-146
23. U.S. Dept. of Health Education and Welfare, Public Health Service, National Institute for Occupational Safety and Health, Criteria for a Recommended Standard-Occupational Exposure to Benzene, HEW Publ. No. (NIOSH) 74-137, Washington, DC, U.S. Government Printing Office (1974)

24. Benzene in the Work Environment, Considerations bearing on the question of safe concentrations of benzene in the work environment (MAK-Wert). Communication of the Working Group "Establishment of MAK-Werte" of the Senate Commission for the Examination of Hazardous Industrial Materials, prepared in cooperation with Dr. Gertrud Buttner, Boppard, Germany, Harald Boldt Verlag (1974)
25. Snyder, R., and Kocsis, J. J., Current concepts of chronic benzene toxicity, CRC Critical Rev. Toxicol., 3 (1975) 265-288
26. Vigliani, E. C., and Saita, G., Benzene and leukemia, New Engl. J. Med., 271 (1964) 872-876
27. Forni, A., and Moreo, L., Cytogenetic studies in a case of benzene leukemia, Eur. J. Cancer, 3 (1967) 251-255
28. Sanita, G., Benzene induced hypoplastic anaemia and leukaemias, In "Blood Disorders Due to Drugs and Other Agents", (ed) Girdwood, R. H., Excerpta Medica, Amsterdam (1973)
29. Girard, R., Tolot, F., and Bourret, J., Malignant haemopathies and benzene poisoning, Med. Lav., 62 (1971) 71-76
30. Aksoy, M., Erdem, S., Dincol, T., Hepyüksel, T., and Dincol, G., Chronic exposure to benzene as a possible contributory etiologic factor in Hodgkin's disease, Blut, 28 (1974) 293-298
31. Kissling, M., and Speck, B., Chromosome aberrations in experimental benzene intoxication, Helv. Med. Acta, 36 (1971) 59-66
32. Ward, J. M., Heisburger, J. H., Yamamoto, R. S., Benjamin, T., Brown, C. A., and Weisburger, E. K., Long-term effect of benzene in C57Bl/6N mice, Arch. Environ. Hlth., 30 (1975) 22-25
33. Anon, EPA Advisory Group to consider next move on benzene, Environ. Health Letter, 17 (3) (1978) 5-6
34. Anon, Preliminary bioassay results show benzene to be carcinogenic, Chem. Reg. Reptr., 1 (1977) 1195
35. Aksoy, M., Erdem, S., and Dincol, G., Leukemia in shoe-workers exposed chronically to benzene, Blood, 44 (1974) 837-841
36. Aksoy, M., Dincol, K., Erdem, S., and Dincol, G., Acute leukemia due to chronic exposure to benzene, Am. J. Med., 52 (1972) 160-166
37. Viadana, E., and Bross, I. D. J., Leukemia and occupations, Prev. Med., 1 (1972) 513
38. Thorpe, J. J., Epidemiologic survey of leukemia in persons potentially exposed to benzene, J. Occup. Med., 16 (1974) 375-382
39. Vigliani, E. C., Leukemia associated with benzene exposure, Ann. NY Acad. Sci. 271 (1976) 143-151
40. Tabersham, I. R., and Lamm, S. H., Benzene and leukemia, Lancet II (1977) 867-868
41. Anon, 30-80 cases of leukemia a year attributed to atmospheric benzene, Environ. Hlth. Letter, 16 (22) (1977) 4-5
42. Anon, CAG benzene report predicts 30-80 leukemia cases yearly in general population, Pesticide & Toxic Chem. News, 5 Nov. 2 (1977) 7-9
43. Anon, EPA considering three reports on risks from benzene emissions, Chem. Reg. Reptr., 1 (36) (1977) 1297
44. Forni, A., Pacifico, E., and Limonta, A., Chromosome studies in workers exposed to benzene or toluene or both, Arch. Environ. Hlth., 22 (1971) 373-378
45. Vigliani, E. C., and Forni, A., Benzene, chromosome changes and leukemia, J. Occup. Med., 11 (1969) 148-149

46. Forni, A. M., Capellini, A., Pacifico, E., and Vigliani, E. C., Chromosome changes and their evolution in subjects with past exposure to benzene, Arch. Environ. Hlth., 23 (1974) 385-391
47. Forni, A., and Moreo, L., Cytogenetic studies in a case of benzene leukemia, Eur. J. Cancer, 3 (1967) 251-255
48. Forni, A., and Moreo, L., Chromosome studies in a case of benzene-induced erythroleukaemia, Eur. J. Cancer, 5 (1969) 459-463
49. Hartwich, G., Schwanitz, G., and Becker, J., Chromosome anomalies in a case of benzene leukaemia, Ger. Med. Monthly, 14 (1969) 449-450
50. Khan, H., and Khan, M. H., Cytogenetic studies following chronic exposure to benzene, Arch. Toxicol., 31 (1973) 39-49
51. Sellyei, M., and Kelemen, E., Chromosome study in a case of granulocytic leukaemia with "Pelgerisation" 7 years after benzene pancytopenia, Eur. J. Cancer, 7 (1971) 83-85
52. Tough, I. M., and Court Brown, W. M., Chromosome aberrations and exposure to ambient benzene, Lancet, 1 (1965) 684
53. Pollini, G., and Colombi, R., Lymphocyte chromosome damage in benzene blood dyscrasia, Med. Lav., 55 (1964) 641-654
54. Tough, I. M., Smith, P. G., Court Brown, W. M., and Harnden, D. G., Chromosome studies on workers exposed to atmospheric benzene, The possible influence of age, Eur. J. Cancer, 6 (1970) 49-55
55. Wolman, S. R., Cytologic and cytogenic effects of benzene, J. Toxicol. Env. Hlth. Suppl. 2 (1977) 63-68
56. Sellyei, M., and Keleman, E., Chromosome study in a case of graulocytic leukaemia with "Pelgerisation" 7 years after benzene pancytopenia, Eur. J. Cancer, 7 (1971) 83-85
57. Kissling, M., and Speck, B., Chromosome aberrations in experimental benzene intoxication, Helv. Med. Acata., 36 (1971) 59-66
58. Dobrokhotov, V. B., The mutagenic effect of benzol and toluol under experimental conditions, Gig. Sanit., 37 (1972) 36-39
59. Dobrokhotov, V. B., and Enikeev, M. I., Mutagenic effect of benzene, toluene and a mixture of three hydrocarbons in a chronic experiment, Gig. Sanit., 42 (1977) 32-34
60. Lyon, J. P., Mutagenicity studies with benzene, Ph D. Thesis, University of California, Berkeley (1975)
61. Koizumi, A., Dobashi, Y., Tachibana, Y., Tsuda, K., and Katsunuma, H., Cytokinetic and cytogenetic changes in cultured human leucocytes and HeLa cells induced by benzene, Ind. Health (Japan) 12 (1974) 23-29
62. Lyon, J. P., Mutagenicity studies with benzene, Dissertation Abstracts B 36 (1976) 5537
63. Shahin, M. M., Personal communication quoted in reference 16.
64. Lebowitz, H., Brusick, D., Matheson, D., Reed, M., Goode, S., and Roy, G., The genetic activity of benzene in various short-term in vitro and in vivo assays for mutagenicity, 9th Annual Meeting of Environmental Mutagen Society, San Francisco, CA, March 9-13 (1978) p. 81
65. Lutz, W. K., and Schlatter, C. H., Mechanism of the carcinogenic action of benzene: Irreversible binding to rat liver DNA, Chem. Biol. Interactions, 18 (1977) 241-245
66. Morimoto, K., Inhibition of repair of radiation-induced chromosome breaks, Japan J. Ind. Health, 17 (1975) 160-167
67. Lowenheim, F. A., and Moran, M. K., "Faith, Keyes, and Clark's Industrial Chemicals", 4th ed., Wiley-Interscience, New York (1975) pp. 821-830

68. Cier, H. E., Toluene, In "Kirk-Othmer Encyclopedia of Chemical Technology" 2nd ed., Vol. 20, Wiley Interscience, New York (1969) pp. 528-565
69. Gerarde, H. W., Toxicology and biochemistry of aromatic hydrocarbons, Elsevier Monograhs, Elsevier/North Holland Biomedical Press, Amsterdam (1960)
70. Matsushita, T., Experimental studies on the disturbance of haemopoietic organs due to benzene intoxication, Nagoya J. Med. Sci., 28 (1966) 204-234
71. Fabre, R., Truhaut, R., Caham, S., and Peron, M., Recherches toxicologiques surles solvents de remplacement du benzene, Arch. Maladies Profess. Med. Travail et Securite Sociale, 16 (1955) 197-215
72. Dobrokhotov, V. B., and Enikoev, M. I., The mutagenic action of benzol, toluol and a mixture of these hydrocarbons in a chronic test, Gig. Sanit., 41 (1976) 32-34
73. Winstor, S., and Matsushita, T., Permanent loss of chromosome inibration in toluene-treated Bacillus subtilis cells, J. Bacteriol., 123 (1975) 921-927
74. Frei, J. V., and Kingsley, W. F., Observations on chemically induced regressing tumours of mouse epidermis, J. Natl. Cancer Inst., 41 (1968) 1307-1313
75. Williams, R. F., "Detoxification Mechanism", Chapman and Hall, London (1959)
76. DeBruin, A., "Biochemical Toxicology of Environmental Agents", Elsevier/North Holland Biomedical Press, Amsterdam (1976)
77. Latham, S., Metabolism of Industrial Solvents, 1. The biotransformation of benzene and benzene substitutes, Occup. Health Rev., 21 (1970) 24-28

CHAPTER 23

ANTHRAQUINONES, QUINONES, AND POLYHYDRIC PHENOLS

A. 9,10-Anthraquinones

A large number of naturally occurring and synthetic anthraquinones have been widely employed as dyestuffs and coloring agents in textiles, foods, drugs, cosmetics, and hair dyes[1]. Table 1 lists the mono- to hexa- hydroxyanthraquinones and their respective common names.

Table 1. Hydroxyanthraquinones

Name	Position of OH group
Erythrohydroxyanthraquinone	1-
	2-
Alizarin	1,2-
Purpuroxanthin, xanthopurpurin	1,3-
Quinizarin	1,4-
Anthrarufin	1,5-
	1,6-
	1,7-
Chrysazin	1,8-
Hystazarin	2,3-
Anthraflavin	2,6-
Isoanthraflavin	2,7-
Anthracene Brown; Anthragallol	1,2,3-
Purpurin	1,2,4-
Alizarin Brilliant Bordeau R	1,2,5-
Flavopurpurin	1,2,6-
Anthrapurpurin	1,2,7-
Alizarin Cyanine R	1,2,4,5,8-
Anthracene Blue WR	1,2,4,5,6,8-

Alizarin (1,2-dihydroxy-) (I); quinizarin (1,4-dihydroxy-) (II); anthragallol (1,2,3-trihydroxy-) (III) and purpurin (1,2,4-trihydroxy-) (IV) anthraquinones are among the best known of the hydroxylated derivatives and have achieved importance as mordant dyestuffs and as intermediates for the manufacture of a number of important anthraquinone intermediates and as intermediates for the production of dyes for wool and synthetic fibers[1]. Because of their rather unique ability to form lakes with metallic ions, many hydroxyanthraquinones are also used for the detection and estimation of metals.

Other areas of utility of the 1,4- and 1,5-dihydroxyanthraquinones and 1,2,4-trihydroxyanthraquinones are in the production of acrylate-ethylene polymers for hot-melt adhesives and laminates[2] as light stabilizers for polystyrene[3] and in the case of the 1,4-dihydroxy derivative (quinizarin), as lubricants for pneumatic tools[4].

(I)   (II)   (III)   (IV)

Among the nitroanthraquinones that have utility in the preparation of amino-anthraquinones for dyestuffs are the 1-nitro-, 1,5-dinitro-, and 1,8-dinitro anthraquinones.

In general, anthraquinone per se is a relatively inert compound. In spite of its quinone structure, many reactions characteristic of quinone compounds, either do not occur, or if so, only with difficulty. However, it is the base material for the manufacture of a group of dyes.

The carcinogenic activity of the anthraquinones have been sparsely examined. 1-Amino anthraquinone has been reported to be carcinogenic in rats[5]. 1-Methyl amino anthraquinone fed intragastrically was carcinogenic in rats, while 2-amino anthraquinone induced cystic changes in the kidneys[6]. 2,6-Diamino anthraquinone was also tested in this study and found negative[6].

Brown and Brown[7] recently described the screening of ninety 9,10-anthraquinone derivatives and related anthracene derivatives for mutagenicity with 5 S. typhimurium tester strains, TA 1535, TA 100, TA 1537, TA 1538, and TA 98, with and without mammalian microsomal activation. Three patterns of mutagenesis were apparent in the approximately 35% of the compounds considered to be mutagenic. These are: (1) direct frameshift mutagenesis by certain derivatives bearing free hydroxyl groups. The most potent were anthragallol (1,2,3-trihydroxy-); purpurin (1,2,4-trihydroxy-) and anthrarufin (1,5-dihydroxy-) anthraquinones. While some hydroxy anthraquinones particularly at lower concentrations, exhibited activation by mammalian microsomal preparations, the majority of mutagenic hydroxy anthraquinones appeared to revert strain TA 1538 (his 3076) specifically. (2) Frameshift mutagenesis by certain derivatives with primary amino groups, and, in a few cases, with secondary amino groups. Frameshift mutagenesis was potentiated with mammalian microsomes, and activity with strain TA 100 (sensitive to base-pair substitution) was observed in a few cases, e.g., 1,2-diamino anthraquinone. (3) Anthraquinones with one or more nitro groups exhibited the least specificity with regard to tester strain reverted and to microsomal activation; all 7 nitro anthraquinones tested were mutagenic. In anthraquinones containing mixed "mutagenic" functional groups, the type of mutagenesis observed was usually $NO_2 > OH > NH_2$. Table 2 illustrates the screening of a number of anthraquinone derivatives and related compounds with S. typhimurium tester strains TA 1535, TA 100, TA 1537, TA 1538, and TA 98/mammalian microsomal test.

At present, it is not known whether hydroxy anthraquinones revert TA 1537 by simple intercolation or a more reactive process[7]. It was suggested that possible oxidative metabolites or chemical oxidation products involved in the latter process might include cyclic peroxides as precursors to cis-dihydrodiol-anthraquinones[8,9] or phenoxide free radicals[10,11].

TABLE II

SCREENING OF ANTHRAQUINONE DERIVATIVES AND RELATED COMPOUNDS FOR MUTAGENICITY WITH *SALMONELLA TYPHIMURIUM*/MAMMALIAN MICROSOMAL TEST

Test compound	µg [b]	S-9 [c]	Number of *His*⁺ revertants/plate [a]				
			TA1535	TA100	TA1537	TA1538	TA98
Control $\bar{X}$ ± SD (N) [d]	—	—	21 ± 15 (114)	98 ± 28 (114)	10 ± 5 (121)	13 ± 6 (122)	27 ± 17 (132)
	—	+	22 ± 24 (100)	92 ± 25 (106)	12 ± 4 (112)	36 ± 16 (113)	41 ± 23 (112)
(a) Hydroxylated anthraquinones and related compounds	10	—			++	—	—
		+			+	—	—
	20	—			+++	—	—
		+			++	—	—
1,8,9-Trihydroxyanthracene (Anthralin) (1) [e]	100	—	—	—	*	*	*
		+	—	—	±	±	±
	500	—	*	*	*	*	*
		+	—	—	*	*	*
	2000	—	*	*	*	*	*
		+	—	—	—	—	*
1,2-Dihydroxyanthraquinone (Alizarin) (0)	100	—	—	—	++	—	±
		+	—	—	++	—	—
	500	—	—	—	+	—	±
		+	—	—	+	—	±
	1000	—	—	—	+	—	—
		+	—	—	—	—	—
	2000	—	*	*	—	—	—
		+	*	*	±	—	—
1,4-Dihydroxyanthraquinone (Quinizarin) (1)	100	—	—	—	++++	—	±
		+	—	—	±	—	—
	500	—	—	—	++++	—	±
		+	—	—	+	—	—
	2000	—	—	—	++++	—	±
		+	—	—	+	—	—
1,5-dihydroxyanthraquinone (Anthrarufin) (1)	50	—	—	—	—	—	++
		+	—	—	++	±	—
	100	—	—	—	++	+	—
		+	—	—	++	±	—
	2000	—	—	±	++++	+++++	+++
		+	—	—	++	+++	+++
1,8-Dihydroxyanthraquinone (Chrysazin, Danthron) (2)	100	—	—	—	++++	—	±
		+.	—	—	++	—	—
	500	—	—	—	++++	±	—
		+	—	—	+	—	—
	1000	—	±	—	—	—	—
		+	—	—	+	—	—
	2000	—	*	*	*	*	*
		+	*	—	+	—	—

TABLE II (continued)

Test compound	μg [b]	S-9 [c]	Number of His+ revertants/plate [a]				
			TA1535	TA100	TA1537	TA1538	TA98
1,5-Dihydroxy-4,8 diamino-anthraquinone (1)	50	—	—	—	±	+	++
		+	—	—	+++	++	++
	100	—	—	—	+	+++	+++
		+	—	—	+++++	++++	++++
	500	—	—	±	+++	+++	+++
		+	—	+	++++++	++++	+++++
Leuco-1,4,5,8-tetrahydroxy-anthraquinone (2)	100	—	—	—	+++	±	—
		+	—	—	++	—	—
	500	—	—	—	++	—	—
		+	—	—	—	—	—
1,2,4-Trihydroxyanthraquinone (Purpurin) (1)	10	—	—	—	++++	—	±
		+	—	—	+	—	—
	50	—	—	±	++++	±	—
		+	—	—	++	±	—
	100	—	—	—	+++	+	+
		+	—	—	++	—	—
	500	—	*	*	*	*	*
		+	*	*	*	*	*
1-Hydroxy-4-amino-anthraquinone (Disperse Red 15) (2)	100	—	—	—	+++	±	±
		+	—	—	++	—	—
	500	—	—	—	+++	±	—
		+	—	—	++	—	—
	2000	—	—	—	+++	—	—
		+	—	—	+	—	—
1-N-Acetyl-4-hydroxyanthra-quinone (2)	50	—	—	—	++	—	—
		+	—	—	—	—	—
	100	—	—	—	+++	—	—
		+	—	—	—	—	—
	500	—	—	—	+++	—	—
		+	—	—	±	—	—

(b) Aminated anthraquinones

1,2-Diamoanthraquinone (3)	100	—	—	—	++++	+	++
		+	—	±	+++	++	++++
	500	—	—	—	+++	++	++
		+	—	±	++++	++	++++
	2000	—	—	—	+++	+	++
		+	—	++	++++	++++	++++++
1,4-Diaminoanthraquinone (1)	100	—	—	—	+++	—	—
		+	—	—	++	+++	++++
	500	—	—	—	++	—	±
		+	—	—	+++	++	++
	1000	—	—	—	+	—	—
		+	—	—	++	+++	+++
	2000	—	—	—	++	—	—
		+					+++

TABLE II (continued)

Test compound	μg [b]	S-9 [c]	Number of His+ revertants/plate [a]				
			TA1535	TA100	TA1537	TA1538	TA98
2,6-Diaminoanthraquinone (0)	100	−	−	−	−	−	−
		+	−	−	−	±	±
	500	−	−	−	−	−	−
		+	−	−	−	±	−
	2000	−	−	−	−	−	−
		+	−	−	+	++	++
1,4,5,8-Tetraaminoanthra- quinone (Disperse Blue 1) (0)	100	−	−	−	±	−	−
		+	−	−	+	−	−
	500	−	−	−	±	±	−
		+	−	−	±	−	−
	2000	−	−	−	+	−	−
		+	−	−	−	−	−
Anthraquinone-1-diazonium chloride (Fast Red A Salt) (0)	10	−	−	−	−	±	−
		+	−	−	−	−	−
	200	−	−	−	++	±	−
		+	−	+	+++	±	+
	400	−	*	*	*	*	*
		+	−	−	+++++	−	++
1-Benzamido-5-chloro- anthraquinone	50	−	−	−	−	±	±
		+	−	−	±	−	−
	100	−	−	−	−	−	±
		+	−	−	±	−	++
	500	−	−	−	±	±	+
		+	−	±	++	±	+++
1-N-Acetyl-4-O-acetyl- anthraquinone (4)	500	−	−	−	+++++	−	−
		+	−	−	++++	−	−
1,4-Diamino 2,3 dihydro- anthraquinone (4)	50	−	−	−	−	−	−
		+	−	−	−	−	−
	100	−	−	−	+	−	−
		+	−	−	−	−	±
	500	−	*	*	−	−	*
		+	*	*	+	−	*
1-Anilino-2-methyl- anthraquinone	500	−	−	±	−	−	−
		+	−	+	−	−	−
1,5-Diamino-anthraquinone (0)	100	−	−	−	−	−	−
		+	−	−	−	−	−
	500	−	−	−	−	−	−
		+	−	−	±	±	±
	2000	−	−	−	−	−	−
		+	−	−	−	+	+
(c) Nitrated anthraquinone derivatives							
1,8-Dihydroxy-4,5-dinitro- anthraquinone (0)	100	−	−	+	±	−	−
		+	−	+++	+	±	±
	500	−	−	++	++++++	++++	+++++
		+	−	++	+++++	++	+++
	2000	−	−	−	++++	++++	+++
		+	±	−	++++	+++	+++

TABLE II (continued)

Test compound	μg [b]	S-9 [c]	Number of His[+] revertants/plate [a]				
			TA1535	TA100	TA1537	TA1538	TA98
2,6-Dihydroxyanthraquinone (Anthraflavic Acid) (0)	100	−	−	−	−	−	−
		+	−	−	±	−	−
	500	−	−	−	−	−	−
		+	−	−	±	−	−
	1000	−	−	−	±	−	−
		+	−	−	±	−	−
1,8-Dihydroxy-3-methyl-anthraquinone (Chrysophanic Acid)	100	−	−	−	±	−	−
		+	−	−	+	−	−
	500	−	−	−	−	−	−
		+	−	−	+	−	−
	2000	−	−	−	−	−	−
		+	−	−	+++	−	−
Leuco-1,4-dihydroxyanthra-quinone (Leucoquinizarin) (2)	100	−	−	−	++	±	−
		+	−	−	++	−	−
	500	−	−	−	+++	−	−
		+	−	−	++	−	−
	2000	−	−	−	++	−	−
		+	−	−	++	−	−
1,2,3-Trihydroxyanthra-quinone (Anthragallol) (1)	20	−	−	−	−	++++	++
		+	−	−	±	++	+
	100	−	−	−	++	+++++	+++++
		+	−	−	+++	+++++	++++
	500	−	−	±	+++	+++++	+++++
		+	−	±	+++	+++	+++++
1,2-Dihydroxy-9-anthrone (Anthrarobin) (2)	100	−	−	−	++	−	±
		+	−	−	+	−	−
	500	−	−	−	+	−	−
		+	−	−	−	−	−
3-Methyl-1,8,9-trihydroxy-anthracene (Chrysarobin) (1)	50	−	−	−	±	−	−
		+	−	−	+	−	−
	100	−	−	−	++	−	−
		+	−	−	+	−	−
	500	−	−	−	±	−	−
		+	−	−	+++	−	−
1,2,5,8-Tetrahydroxyanthra-quinone (Quinalizarin) (0)	50	−	−	*	++	±	−
		+	−	−	*	+++	−	−
	100	−	−	*	+	±	±
		+	−	−	*	+++	−	−
	500	−	−	−	+++	−	−
		+	−	−	+++	−	−
1,3,8-Trihydroxy-6-methyl-anthraquinone (Emodin)	50	−	−	−	−	−	−
		+	−	−	++++	−	−
	250	−	−	−	−	−	−
		+	−	−	++	−	−
	2000	−	−	−	−	−	−
		+	−	−	−	−	−

TABLE II (continued)

Test compound	μg [b]	S-9 [c]	Number of His+ revertants/plate [a]				
			TA1535	TA100	TA1537	TA1538	TA98
1-Nitro-2-methyl-anthra-quinone (O)	100	−	−	+	±	−	−
		+	−	+++	+	±	±
	500	−	−	+	±	−	+
		+	−	+++	++	±	+
	1270	−	−	−	±	++	±
		+	±	+++	+	±	±
1-Nitro-5-sulfonato-AQ	500	−	−	−	−	++	+
		+	−	−	−	±	±
1-Nitro-2-ethyl-anthraquinone	1000	−	−	−	−	−	−
		+	−	−	+	±	±	±
1,4-Diamino-5-nitro-anthraquinone (O)	50	−	−	−	−	++	++	+
		+	*	*	*	+	++
	100	−	−	−	+++	+++	++
		+	−	−	*	*	±
	500	−	−	−	*	++	+++	−
		+	−	−	−	++	−
1-Amino-2-carboxylate-4-nitro-anthraquinone (O)	50	−	−	−	+	7+	8+	8+
		+	−	−	++	5+	6+	5+
	100	−	−	−	+	7+	8+	8+
		+	−	−	+++	7+	8+	8+
	500	−	−	−	−	±	+++++	+++
		+	+++	++	++	++++	+++
1-Nitro-6(7)sulfonato-anthraquinone (O)	50	−	−	−	−	+	++
		+	−	−	−	+	±
	100	−	−	−	±	++	++
		+	−	−	+	++	+
	500	−	−	−	±	++	++++	+++++
		+	−	−	+	++	++
(d) Miscellaneous including known mutagens							
1-Methoxy-anthraquinone	500	−	−	−	++	−	−
		+	−	−	++	−	−
1 Aminoanthracene	20	−	−	−			−
		+	−	++			+++
2-Aminoanthracene	1	−	−	−	−	−	−
		+	+++	6+	+++++	+++++	7+
1-Aminopyrene	10	−	−	−		−	+++
		+	−	++		++	+++++
3-Aminopyrene	10	−	−	−	++	±	+++
		+	−	±	+++	++	++++
Benzo(a)pyrene	5	−	−	−			
		+	−	++	+	−	+
Dibenzpyrene	10	−	−	−	−	−	−
		+	−	±	−	−	+
Benz(a)anthracene	20	−	−	−		−	−
		+	−	+		−	−
6-Aminochrysene	5	−	−	−	−	−	−
		+	−	−	+	−	++
9-Aminoacridine	12	−			8+		

B. Quinones

Although the quinones and polyhydric phenols are normally very readily interconvertible, the quinones are considered as conjugated cyclic diketones (dioxo derivatives) rather than aromatic compounds per se[12-14]. The most important and characteristic reaction of quinones is reaction to the corresponding dihydroxy aromatic compounds, viz.,

$$\text{O=}\underset{\text{O}}{\bigcirc}\text{=O} + 2H^+ \underset{2e}{\overset{2e}{\rightleftarrows}} \underset{\text{OH}}{\bigcirc}\text{-OH}$$

The 1,2- and 1,4-quinones are the most common of a variety of quinone-like compounds that have been prepared. Benzene derivatives which are most susceptible to oxidation to quinones are those that are disubstituted in the ortho or para positions with hydroxyl or amino groups.

As α, β-unsaturated ketones, quinones readily form 1,4-addition products (e.g., addition of hydrogen chloride and acid-catalyzed addition of acetic anhydride) analogously as their open chain analogs[12-14]. Quinones with one-double-bond that is not part of an aromatic ring also readily undergo Diels-Alder additions[12-14].

1. para-Quinone (1,4-benzoquinone; benzoquinone; 2,5-cyclohexadiene-1,4-dione; 1,4-cyclohexadiene dioxide; O=⟨⟩=O) can be prepared by the oxidation of benzene or a variety of benzene derivatives. A widely used process involves the oxidation of aniline with manganese dioxide and sulfuric acid[14], viz.,

$$2\ \text{C}_6\text{H}_5\text{NH}_2 + 4\text{MnO}_2 + 5\text{H}_2\text{SO}_4 \longrightarrow 2\ \text{C}_6\text{H}_4\text{O}_2 + 4\text{MnSO}_4 + (NH_4)_2SO_4 + 4H_2O$$

Other oxidizing agents that have been employed are sodium dichromate or lead dioxide and sulfuric acid and sodium chlorate in very dilute sulfuric acid with a trace of vanadium pentoxide.

The major applications of para-quinone include[14-15]: (a) as an intermediate in the production of hydroquinone; (b) the preparation of quinhydrone (a complex of paraquinone and hydroquinone used in an electrode for pH determinations); (c) preparation of 2,3-dichloro-5,6-dicyanobenzoquinone for use as a selective oxidizing agent in steroid synthesis; (d) the preparation of its dioxime and the dibenzoate of the dioxime for use as rubber accelerators; (e) as a polymerization inhibitor; (f) in photography as a hydroquinone-quinone oxidation-reduction system; (g) as a tanning agent; (h) in the manufacture of dyes; (i) as a chemical reagent; (j) degradation inhibitor for acrylonitrile-vinylidene chloride polymers[16];)k) stabilizer for acrylic acid[17]; (1) vulcanization agent for fluoro rubbers[18]; (m) polymerization catalyst for linseed oil[19]; (n) catalyst in the manufacture of polyalkenamenes[20]; (o) a constituent in thermographic copying material[21]; and (p) as a fixative for immunohistochemistry[22].

para-Quinone has been found as a pollutant in filtered surface and ground water at a water-works in the Federal Republic of Germany[23].

Data on the carcinogenicity of para-quinone are extremely limited. An increased incidence of skin tumors and lung tumors were found in mice treated with 0.1% para-quinone by skin application[24].

Injection site fibrosarcomas were found among rats of Saitama mixed strain receiving 32 subcutaneous injections propylene glycol solution of para-quinone (81 mg paraquinone and 16.5 ml propylene glycol)[25].

Two lung adenocarcinomas were observed in 25 mice exposed to 5 mg para-quinone for 1 hr 6 times/week compared to 1 adenoma in 25 untreated controls[26].

Mutagenicity data on para-quinone is similarly limited. para-Quinone was non-mutagenic in Neurospora at toxic concentrations[27] and did not induce dominant lethal mutations when administered i.p as a single dose of 6.25 mg/kg in male mice[28].

C. Polyhydric Phenols

Polyhydric phenols (benzene diols and triols) compose a group of compounds that have had an extensive spectrum of utility in the production of dyes, pharmaceuticals, plasticizers, textile and leather chemicals[29] for an extensive period of time. The structures of these agents, with their respective common names with which they are generally associated with are shown below:

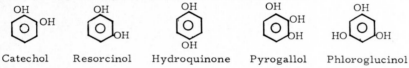

Catechol   Resorcinol   Hydroquinone   Pyrogallol   Phloroglucinol

These derivatives react typically as phenols hence undergoing ring electrophilic substitutions via halogenation, sulfonation, alkylation, nitration, Friedel-Crafts acylation, Kolbe and Reimer-Tiemann reactions or condensations with aldehydes, ketones or esters[29-30]. Polyhydric phenols with the hydroxyls in the ortho or para relationship are normally readily oxidized to quinones[29,30].

1. Catechol (ortho-hydroxypheonol; 1,2-benzenediol; ortho-hydroquinone; pyrocatechol) can be prepared by a variety of procedures including: (a) the alkali fusion of ortho-chlorophenol; (b) the oxidation of benzene with hydrogen peroxide; and (c) recovery from lignin-containing waste streams from wood pulping operations.

The major areas of utility of catechol include: (a) as an antioxidant and polymerization inhibitor; (b) photographic developer; (c) as an intermediate in the production of pharmaceuticals, pesticides, perfumes and resins; (d) as rubber compounding aids; (e) in the preparation of electron-sensitive copying papers; (f) as an additive in metal plating baths and (g) in the synthesis of tert.butyl-catechols which have been recommended as catalysts in polyurethan formulations[29].

Catechol has been widely found in drainage water from butaminous shale, coal, crude wood tar, cigarette smoke, and in the effluents resulting from the production of coal-tar chemicals in the U.S.S.R.[31].

2. Hydroquinone (para-hydroxyphenol; 1,4-dihydroxybenzene; p-hydroquinone; hydroquinol) can be produced by a variety of methods including: (a) the oxidation of aniline with MnO_2 and H_2SO_4 followed by reduction with iron dust and water; (b) the oxidation of benzene or numerous benzene derivatives followed by the reduction of the resulting para-quinone[29]; (c) the alkylation of benzene with propylene to produce a mixture of di-isopropyl benzene isomers then oxidation of the isolated para-isomer with oxygen to corresponding dihydroperoxide followed by treatment with acid to yield para-hydroquinone and acetone and (d) oxidation of phenol with H_2O_2 to produce a mixture of products from which both para-hydroquinone and pyrocatechol are isolated.

The major areas of utility of hydroquinone[29,32] are: (a) as a photographic developer; (b) as an antioxidant and polymerization inhibitor in materials such as fats, oils, paints, vitamins, unsaturated monomers and gasoline; (c) as a chemical intermediate in the preparation of hydroquinone ethers and diethers which are employed as stabilizing agents, antioxidants, plasticizers and in perfume and cosmetic preparations; (d) an intermediate for rubber-processing chemicals; (e) as a dye intermediate; (f) as a laboratory agent; and (g) as a component of dermatological agents used to bleach hyperpigmented skin blemishes.

Para-hydroquinone has been found in effluents resulting from the production of coal-tar chemicals in the U.S.S.R.[31] and in cigarette smoke[29].

The data on carcinogenicity and mutagenicity of the polyhydric phenols are very limited. Catechol in combination with benzo(a)pyrene enhanced the skin carcinogenic effects induced by benzo(a)pyrene alone in skin painting studies with female ICR/Ha Swiss mice[33].

Para-hydroquinone was inactive when tested as an initiator of skin carcinogenesis using albino male mice of the "S" strain[34]. However, in bladder implantation studies, para-hydroquinone in cholesterol pellets increased the incidence of bladder carcinomas in mice[35].

Resorcinol (meta-hydroquinol) showed no carcinogenic effect following skin application of 0.02 ml of a 5, 25 or 50% solution in acetone to female Swiss mice[36]. The hydroquinones have induced chromosome aberrations or kariotypic effects in Allium cepa[37,38], Vicia faba[39,40] and Allium sativium[40,41].

Parahydroquinine was found to be less toxic to E. coli W3110 (pol A^+) than to P3478 (pol A^-) which suggests that this agent can damage DNA.

Resorcinol at a dose of up to 1000 µg/plate was not mutagenic in S. typhimurium TA 1535, TA 1537, TA 98 or TA 100 in the presence or absence of rat liver microsomes[42].

References

1. Cofranesco, A., Anthraquinone derivatives, In Kirk-Othmer Encyclopedia of Chemical Technology, 2nd eds., Vol. 2, Interscience Publishers, New York (1966) pp. 465-477; 478-489; 501-533
2. Baumann, H., Bauer, P., and Glaser, R., Ger. Offen., 2,335,141, Feb. 28 (1974) Chem. Abstr., 81 (1974) 38559Z
3. Nakamura, K., and Honda, K., Photodegradable polystyrenes, Kobunshi Ronbunshu, 31 (1974) 373-376; Chem. Abstr., 81 (1974) 136580U
4. Hartmann, L. M., Pneumatic tool lubricant, U.S. Patent 3,801,503, April 2 (1974) Chem. Abstr., 81 (1974) 52113F
5. Laham, S., Grice, H. C., and Sinclair, J. W., Studies in chemical carcinogenesis III. β-Aminoanthraquinone, Toxicol. Appl. Pharmacol., 8 (1966) 346
6. Griswold, D. P., Casey, A. E., Weisburger, E. K., and Weisburger, J. H., The carcinogenicity of multiple intragastric doses of aromatic and heterocyclic nitro or amino derivatives in young female Sprague-Dawley rats, Cancer Res. 28 (1968) 924-933
7. Brown, J. P., and Brown, R. J., Mutagenesis by 9,10-anthraquinone derivatives and related compounds in Salmonella typhimurium, Mutation Res., 40 (1976) 203-224
8. Daly, J. W., Jernia, D. M., and Witkop, B., Arene oxides and the NIH shift: The metabolism toxicity and carcinogenicity of aromatic compounds, Experientia, 28 (1972) 1129-1149
9. Double, J. C., and Brown, J. R., The interaction of aminoalkylaminoanthraquinones with deoxyribonucleic acid, J. Pharm. Pharmacol., 27 (1975) 502-507
10. Nagata, C., Inomata, M., Kadoma, M., and Tagashira, Y., Electron spin response study on the interaction between the chemical carcinogens and tissue components. III. Determination of the structure of the free radical produced either by stirring 3,4 benzopyrene with albumin or incubating it with liver homogenates, Gann, 59 (1968) 289-298
11. Nagata, C., Tagashira, Y., Kodama, M., and Imamura, A., Free radical produced by interraction of aromatic hydrocarbons with tissue components, Gann, 57 (1966) 437-444
12. Roberts, J. D., and Caserio, M. C., "Modern Organic Chemistry", B. A. Benjamin, Inc., New York (1967) pp. 618-621
13. Gutsche, C. D., and Pasto, D. J., "Fundamentals of Organic Chemistry", Prentice-Hall, Englewood Cliffs, NJ (1975) pp. 357-358; 978-979
14. Thirtle, J.R., Quinones, In Kirk-Othmer Encyclopedia of Chemical Technology 2nd ed., Vol. 16, Wiley & Sons, New York (1968) 899-913
15. Merck & Co., The Merck Index, Ninth Edition, Merck & Co., Rahway, NJ (1976) p. 1051
16. Baker, A. W., Stabilizing vinylidene chloride polymers against metal induced degradation, U. S. Patent 3,882,081, 6 May, 1975, Chem. Abstr., 83 (1975) P115965Z
17. Otsuki, S., Hori, M., and Miyanoharo, I., Stabilization of acrylic acid in distillation, Japan Patent 7506,449; 14 March 1975; Chem. Abstr., 83 (1975) P58130X
18. Schmiegel, W. W., Vulcanizable fluoroelastomer composition, U. S. Patent 3,872,065 18 March, 1975, Chem. Abstr., 83 (1975) P11889M
19. Nagakura, M., Takada, A., Kai, Y., and Ogawa, Y., Polymerized oils. 1. Effects of various additives on thermal polymerization of linseed oil, Shikizai Kyokaishi, 48 (1975) 217-222; Chem. Abstr., 83 (1975) 133827K

20. Babitskii, B. D., Denisova, T. T., Kormer, V. A., Lapuk, I. M., et al., Method for preparing polyalkenamers, U. S. Patent 3,933,777, 20 Jan. 1976; Chem. Abstr. 84 (1976) P151452J
21. Endo, I., Matsuno, H., Kokado, H., Inoue, E., Nuhide, K., and Kinjo, K., Thermographic copying material. Ger. Offen 2,328,900, 19 December 1974, Chem. Abstr., 84 (1976) 67823Y
22. Pearse, A. G. E., and Polak, J. M., Bifunctional reagents as vapor and liquid phase fixations for immuno-histochemistry, Histochem. J., 7 (1975) 179-186
23. Thielemann, H., Thin-layer chromatographic results for identification of organic pollution components of shore-filtered surface and groundwater (Halle-Beesen Waterworks), Z. Chem., 9 (5) (1969) 189-190
24. Takizawa, N., Uber die experimentelle erzeugung der haut-und lungenkrebse bei der maus durch bepinselung mit chinone, Gann, 34 (1940) 158-160
25. Umeda, M., Production of rat sarcoma by injection of propylene glycol solution of p-quinone, Gann, 48 (1957) 139-144
26. Kishizawa, F., Carcinogenic action of para-quinone on the lung of mice by the experimental inhalation (Report 3), Gann, 47 (1956) 601-603
27. Reissig, J. L., Induction of forward mutants in the pyr-3 region of Neurospora, J. Gen. Microbiol., 30 (1963) 317-325
28. Röhrborn, G., and Vogel, F., Mutationen durch chemische einwirkung bei säuger und mensch., Dtsch. Med. Wschr., 50 (1967) 2315-2321
29. Raff, R., and Ettling, B. V., Hydroquinone, resorcinol and pyrocatechol, IN: Kirk-Othmer Encyclopedia of Chemical Technology, Vol. 11, 2nd ed., Wiley, New York (1966) pp. 462-492
30. Roberts, J. D., and Caserio, M. C., Modern Organic Chemistry, W. A. Benjamin, Inc. New York (1967) pp. 617-620
31. Umpelev, V. L., Kogan, L. A., and Gagarinova, L. M., J. Anal. Chem. USSR (Transl. of Zl. Anal. Khim.) 29 (1) (1974) 152-153
32. Anon, Cheaper route to hydroquinone, Chem. Week, Dec. 11 (1974) 51
33. Van Duuren, B. L., Katz, C., and Goldschmidt, B. M., Cocarcinogenic agents in tobacco carcinogenesis, J. Natl. Cancer Inst., 51 (1973) 703-705
34. Roe, F. J. C., and Salaman, M. H., Further studies on incomplete carcinogenesis: Triethylene melamine (TEM), 1,2-benzanthracene and β-propiolactone as initiators of skin tumour formation in the mouse, Brit. J. Cancer, 3 (1955) 177-203
35. Boyland, E., Busby, E. R., Dukes, C. E., Grover, P. L., and Manson, D., Further experiments on implantation of materials into the urinary bladder of mice, Brit. J. Cancer, 18 (1964) 575-581
36. Stenback, F., and Shubick, P., Lack of toxicity and carcinogenicity of some commonly use cutaneous agents, Toxicol. Appl. Pharmacol., 30 (1974) 7-13
37. Levan, A., and Tjio, J. H., Chromosome fragmentation induced by phenols, Hereditas, 34 (1948) 250-252
38. Levan, A., and Tjio, J. H., Induction of chromosome fragmentation by phenols Hereditas, 34 (1948) 453-484
39. Valadaud-Barrieu, M., and Izard, C., Modifications du cycle cellulaire sous l'influence de l'hydroquinone dans les meristemes radiculaires de Vicia faba, C.R. Acad. Sci. Paris, 276 (1973) 33-35
40. Sharma, A. K., and Chatterjee, T., Effect of oxygen on chromosomal aberrations induced by hydroquinone, Nucleus, 7 (1964) 113-124
41. Alarcon, M. I., and Moya, N. (1969) Variciones cariotipicas en Vicia faba inducidas por accion de algunes sustancies quimicas de importancia terapeutica, Bull. Soc. Biol. De Concepcion, 42 (1969) 287-306

42. McCann, J., Choi, E., Yamasaki, E., and Ames, B. N., Detection of Carcinogens as mutagens in Salmonella/microsome test: Assay of 300 chemicals, Proc. Natl. Acad. Sci., 72 (1975) 5135-5139

CHAPTER 24

PHTHALIC AND ADIPIC ACID ESTERS

The continual increase in the use of plastics in construction, food packaging and coating particularly in the U.S., Europe and Japan is well documented and is demonstrated by the production of about 4 to 5 x 10^9 pounds of vinyl chloride and its polymers (PVC) in 1972[1]. The annual production of PVC materials in 1972 was 20% greater than it was in 1971, and over the 19-year period 1962-1972, an average annual production increase of about 10.6% occurred[1]. It was estimated that approximately 10 billion kg of plastics were produced in 1973 in the U.S. alone, 50% more than in 1970. The annual rate of plastics production is expected to increase tenfold in less than 30 years[3].

The production of useful articles from PVC as well as other polymers requires the incorporation of plasticizers. These are high-molecular-weight monomeric liquids of high boiling point, added to materials such as vinyl and cellulosic plastics to improve their processability, and particularly their flexibility and softness[4,5]. They are generally used in concentrations of 20-50% or more of the total plastic composition.

A. Phthalic Acid Esters (PAES)

The largest group of plasticizers used in the U.S. are derivatives of phthalic acid. Phthalate esters are produced from phthalic anhydride and the appropriate alcohol. Phthalic anhydride is manufactured from either naphthalene or o-xylene by oxidation and the alcohols by either the OXO or the ald-OX process. The purity of the esters is considered to be 99.7 to 99.97%, the main impurities being isophthalic acid, terephthalic acid and maleic anhydride which can be converted to the corresponding ester[6]. Table 1 lists the main uses of plastics containing phthalates[6]. While the total weight of about 20 different PAES produced in the U.S. during 1973 was about 990 million pounds, about 50 million pounds of these PAES were used in non-plasticizer areas such as pesticide vehicles, insect repellants, in dyes, cosmetics and fragrances, in munitions and in lubrications[6,7].

Di-2-ethylhexyl phthalate (DEHP) is one of the most widely employed plasticizers for plastics and synthetic rubber (approximately 158.8 million kg in 1970 in the US).

In Japan, the production of PAES has increased sharply in recent years with more than 350,000 tons produced annually[9]. It is likely that the world production of phthalate esters is 3 to 4 times the U.S. production[10]. The major phthalic acid esters manufactured in decreasing order are di(2-ethylhexyl) (DEHP); diisooctyl-, diisodecyl-, dibutyl-, diethyl-, butyloctyl- and dimethyl phthalate esters[10]. (Figure 1)

The broad spectrum of plastic uses today ranges from home contruction, appliances, furnishings, automobiles and apparels to food containers and wrappings. It is important to note that the solid waste from such uses amounts to 4.31 billion kg/year in the U.S. representing a 2.5 fold increase over 1966[11]. It is estimated that by 1979, the annual rate should reach 12.7 billion kg[12].

Table 1. Uses of Phthalic Esters in the United States[6]

A. As plasticizers   (Millions of pounds)

Building and construction
    Wire and cable   185
    Flooring   150
    Swimming pool liners   20
    Miscellaneous   32
        Subtotal   387

Home furnishings
    Furniture upholstery   90
    Wall coverings   38
    Houseware   30
    Miscellaneous   45
        Subtotal   203

Cars (upholstery, tops, etc.)   114
Wearing apparel   72
Food wrapping and closures   25
Medical tubing and intravenous bags   21

TOTAL AS PLASTICIZERS   922

B. As nonplasticizers

Pesticide carriers   ---
Oils   ---
Insect repellant   ---
Miscellaneous   50
    Total as nonplasticizers   50

GRAND TOTAL   972

Table 2 shows that larger proportions of PAES were used in materials with longer useful lives so that PAES will be added to the environment at higher rates as more of the plastics become obsolete and come into direct contact with soil and water[8]. The PAE plasticizers, while they are in plastics, (e.g., in PVC in a resin ratio of usually 1:2) occur in monomeric forms only loosely linked to the polymers and hence are able to be eventually extruded. The PAES used are moderately volatile, lipid soluble and rather stable in the environment.

    There is a growing concern that PAES may represent a potential ecological threat as well as a health hazard because of their high rate of production, their wide spread use, their physical and chemical properties and in some instances, their demonstrated toxicity to aquatic organisms[8-21].

    Phthalate esters have been found in many water bodies[8,9,16] (including rivers, lakes, inland waters) soil, sediment, air and biota far removed from industrial-rural areas. DHEP and BGBP have been identified in concentrations increasing to hundreds of parts-per-billion in Escambia Bay and the Mississippi River Delta in the Gulf of Mexico[22]. An empirical calculation by Corcoran[22] indicated that the total DEHP in the Mississippi River effluent may be already as high as the 1973 total production of the PAES (158.8 million kg).

FIG. 1

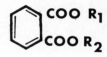

Phthalic acid diester (PAE)

Phthalate esters *Alcohol radicles*

Dimethyl (DMP)=R_1=R_2= .CH_3
Di-*n*-butyl(DnBP)=R_1=R_2= .$CH_2.CH_2.CH_2.CH_3$
Di-*iso*-butyl(DiBP)=R_1=R_2= .$CH_2.CH{:}(CH_3)_2$
Benzyl butyl (BBP)=R_1= .$CH_2.CH_2.CH_2.CH_3$
R_2= .$CH_2.C_6H_5$
Butyl glycolyl butyl (BGBP)=R_1=.$CH_2.CH_2.CH_2CH_3$
R_2= .$CH_2.COOCH_2.CH_2.CH_2.CH_3$
Dimethoxyethyl (DMEP) =R_1R_2=.$CH_2.CH_2.OCH_3$
Dioctyl (DOP)=R_1=R_2= .$CH_2.(CH_2)_6.CH_3$
Di-2-ethylhexyl (DEHP)=R_1=R_2= .$CH_2.CH.(CH_2)_4$
C_2H_5
.$CH_2.CH_2$
Dicyclohexyl (DCHP)=R_1=R_2= .CH CH_2
$CH_2.CH_2$

Table 2. Phthalate Esters: Intervals Between Formulation and Disposal States[8]

Major Usage Area	Approx. % of Total PAES used in U.S.	Useful Life (in years)
Building and construction	44	10 to 25
Home furnishings	23	5 to 15
Transportation	13	5 to 10
Apparel	8	2 to 5
Food and medicine	5	0 to 1
Non plastic uses	6	0 to 1

Average values (in nanograms/liter) for all samples of surface waters taken in the Gulf of Mexico were 90 for DEHP and 80 for DBP[16]. For comparison, values found for PCBs averaged 2 nanograms/liter and for DDTs, 1 nanogram/liter. Average sediment values decreased from 69 mg/g in the Mississippi delta to 2 ng/g in the open gulf for DEHP. DEHP was also found in fish taken in the Gulf of Mexico at a mean concentration of 4.5 ng/g[16].

Although the Mississippi River probably carries more industrial effluents than any other river in the U.S., PAES have been found in other rivers and lakes as well. For example, levels of PAE of 0.88 ppb and 1.9 ppb in water at the mouth of the Charles River in Massachusetts and 11.27 km upstream respectively have been reported[23].

Accumulation of DEHP in aquatic organisms as well as concentration through a food chain have been noted[8,9,20].

PAES found in air can arise directly partly via their use in sprays as well as from their incomplete decomposition during the incineration of plastics[8]. Since waste plastics comprise approximately only 2 to 3% of the total urban waste they may be incinerated or used in landfills, rather than separated for recycling or for specific treatment[12,13]. Not all of the PAES decompose in incinerators. For example, levels of 300 ng/m^3 of DEHP and 700 ng/m^3 of DBP and BGBP have been reported in air samples taken near a municipal incinerator[24].

Degassing of PAES from vinyl upholstery can result in the air of a new car containing PAES at nanogram/liter levels[6].

DEHP is known to be widely distributed in the environment having been detected in various forms of marine life[18,19], soil samples[25] and in both animal[26] and human tissues[27].

Stalling et al[18] reported levels of DEHP ranging from 0.2 to 10 ppb in fish collected from various locations in the U.S., these levels appeared to reflect the extent of industrial pollution of the habitats. Cereal, gelatin, corn starch and casein components of commercial fish feeds were found to contain 20 to 30 ppb DNBP and 140 to 400 ppb DEHP[18].

A recent survey of PAES from 10 different foodstuffs conducted by FDA indicates that most American foods are not contaminated with PAES except in some samples of milk and cheese[28]. The content of DEHP was 0-35.3 ppb (mean value, 5.4 ppb) for the fat from cheese and 0-31.4 ppb (mean value, 7.9 ppb) for the fat from milk[28].

DEHP has been found in human blood samples following exposure of individuals to various biomedical devices such as transfusion equipment and hemodialysis tubing[27,29].

It should also be noted that PAES occur in nature due to the presence of o-phthalic acid in wood, oxidation products of lignin, microorganisms, humic compounds and coal as well as the presence of aliphatic alcohols in plants and the fact that phthalates are intermediates in biochemical pathways[8].

There are no indications that any phthalates are carcinogenic although malignant tumors (the majority being fibrosarcomas) caused by commercial PVC samples have been reported[30,31].

The mutagenic and antifertility effects of DEHP and dimethoxyethyl phthalate (DMEP) in ICR mice have been reported[32-34]. Both phthalates produced some degree of dose and time dependent antifertility and mutagenic effects following their intraperitoneal administration as a single dose (representing 1/3, 1/2 and 2/3 of the acute LD_{50}) to male mice prior to the initiation of a mating period. The high dose of both phthalates produced a distinct reduction in incidence of pregnancies, with the effect being more persistent with DEHP than with DMEP. There was a reduction in the number of implantations/pregnancy and of litter size, particularly in the first few weeks (post meiotic stage) with the high dose of the compounds. Mutational effects, as expressed by an increase in early fetal deaths and reduced numbers of total implants were noted at various weeks during the study but most notably during the first few weeks. The mutagenic effect of DEHP was significantly greater than for DMEP in contrast to the teratogenic effect of these compounds.

Dibutyl phthalate has been reported non-mutagenic in _S. cerevisiae_ strain SV185-14C with or without metabolic activation[35].

DEHP and butylglycolbutylphthalate (BGBP) caused significant growth inhibition in cultures of the human diploid cell strain WI-38[36]. The ID50 (dose which causes 50% growth inhibition in tussue culture) values for DEHP and BHBP were 70 μM and 12 μM respectively for WI-38 cells. Toxic effects were greater in a replicating cell population than in a non-replicating, confluent cell layer. Toxic levels for DEHP were within the range of concentrations found in blood which had been stored in PVC blood bags for up to 21 days at 4^oC[36].

Dimethoxyethyl phthalate was distinctly more teratogenic of the 6 PAES (dimethyl-, dimethoxy ethyl-, diethyl-, dibutyl-, diisobutyl, and butylcarbobutoxymethyl- phthalates) administered i.p. to pregnant rats in doses of one-tenth, one-fifth and one-third of the adult LD_{50} dose on the 5th, 10th and 15th day of gestation[33,37]. The most embryotoxic compounds, dimethoxyethyl phthalate was the most water-soluble of the esters studied and the compound showing the lowest intrinsic cellular toxicity in tissue culture[33]. Dioctyl- and di-2-ethylhexylphthalates tended to have the least overall adverse effect on embryo-fetal development and were the least soluble of the compounds studied[33].

The activity of succinic dehydrogenase and adenosine triphosphatase was significantly reduced and that of β-glucuronidase was increased in rat gonads after treatment with DEHP. Focal degeneration of semiferous tubular and edema of interstitium in testis but no detectable alterations in the ovary of treated animals was noted. Such alterations were suggested by Seth et al[38] to perhaps be responsible for the reported reproductive dysfunction in experimental animals after exposure to this and other phthalate plasticizers.

Orally administered DEHP has been found to be metabolized by the rat[39,40] and ferret[41] to derivatives of mono(2-ethylhexyl)phthalate (MEHP) which are excreted in the urine both unconjugated and as glucuronides[39-41].

B. __Adipic Acid Esters__ $(ROOC(CH_2)_4COOR)$

The adipates, diesters of adipic acid and various alcohol, are used as plasticizers for many polymeric materials including PVC and its co-polymers, natural and synthetic rubber, polystyrene and cellulose derivatives[42]. Thirty adipate esters have been listed as plasticizers[43]. Di-2-ethylhexyl adipate (DEHA) is the principal adipic acid ester used as a plasticizer. In 1971, 44.9 million pounds of DEHA was produced in the U.S.[44]. Other commercially important adipate esters are: di-n-octyl-; di-2-butoxyethyl-, di-isobutyl-, di-ethyleneglycol monobutyl ether-, di-n-decyl-, di-n-butyl-, dicyclohexyl-, and di(3,5,5-trimethyl)-adipates.

Adipates of numerous branched chain alcohols and monohydric ether alcohols have found use as synthetic lubricants. Other applications for adipates include: as insect repellants, in DDT emulsions and as emulsifying agents[42]. Diallyl-, divinyl-, and allyl vinyl adipates can be polymerized in the presence of a catalyst such as benzoyl peroxide to yield transparent resins.

An important industrial application for polyesters of adipic acid is in the manufacture of polyurethane and elastomers utilizing diisocyanates (e.g., toluene and 1,5-naphthalene diisocyanates). The polyesters are prepared from ethylene glycol, diethylene glycol, 1,4-butanediol, propylene glycol and blends of these and other glycols[42].

There are no data on the carcinogenicity of the adipates. Data are meagre on the mutagenicity. Dominant lethal mutations and antifertility effects of di-2-ethylhexyl-adipate (DEHA) and diethyl adipate (DEA) in male mice were recently reported by Singh et al[45].

Both adipates produces dose-dependent antifertility and mutagenic effects as indicated by reduced percentage of pregnancies and increased numbers of early fetal deaths. Mutational effects, expressed by an increased in early fetal deaths/pregnancy occurred mainly during the postmeiotic stage of spermatogenesis in mice for both adeipates. Adverse effects upon the premeiotic stage were also observed for DEHA.

It has also been reported that some adipates can elicit embryo-fetal toxicity and teratogenic effects when administered i.p. to pregnant female rats[46].

References

1. Anon, Chem. Eng. News, May 7 (1973) p. 9
2. Anon, Chem. Eng. News, 50 (51) (1972) p. 10
3. Anon, Plastics heading for a boon, Chem. Ind., (13) (1973) 597
4. Brydson, J. A., "Plastic Materials", Van Nostrand, Princeton (1966)
5. Anon, Modern Plastics, 50 (a) (1973) 53
6. Graham, P. R., Phthalate ester plasticizers: Why and how they are used, Environ. Hlth. Persp., 3 (1973) 3-12
7. Anon, Phthalate effect on health still not clear, Chem. Eng. News, 50 (38) (1972) 14-15
8. Mathur, S. P., Phthalate esters in the environment: Pollutants or natural products, J. Environ. Quality, 3 (1974) 189-197
9. Tomita, I., Nakamura, Y., and Yagi, Y., Phthalic acid esters in various foodstuffs and biological materials, Ectotoxicol. Environ. Safety., 1 (1977) 275-287
10. Peakall, D. B., Phthalate esters: Occurrence and biological effects, Residue Revs. 54 (1975) 1-41
11. Srinivasan, V. R., Biodegradation of waste plastics, Tech. Rev., 74 (1972) 45-47
12. Milgrom, J., Identifying the nuisance plastics, New Scientist, 57 (1973) 184-186
13. Carpenter, E. J., and Smith, K. L., Jr., Plastics on the Sargasso Sea surface, Science, 175, (1972) 1240-1241
14. Dillingham, E. O., and Autlan, J., Teratogenicity, mutagenicity, and cellular toxicity of phthalate esters, Env. Hlth. Persp., 3 (1973) 81-89
15. Autian, J., Toxicity and health threats of phthalate esters: Review of the literature, Envrion. Hlth. Persp., 4 (1973) 3-26
16. Giam, C. S., Chan, H. S., Neff, G. S., and Atlas, E. L., Phthalate ester plasticizers: A new class of marine pollutant, Science, 199 (1978) 419-421
17. Mayer, F. L., Jr., and Sanders, H. G., Toxicology of phthalic acid esters in aquatic organisms, Environ. Hlth. Persp., 3 (1973) 153-158
18. Stalling, D. L., Hogan, J. W., and Johnson, H. L., Phthalate ester residues-their metabolism and analysis in fish, Environ. Hlth. Persp., 3 (1973) 159-173
19. Mayer, F. L., Stalling, D. L., and Johnson, J. L., Phthalate esters as environmental contaminants, Nature, 238 (1970) 411-413
20. Metcalf, R. L., Booth, G. M., Schuth, C. K., Hansen, D. J., and Lu, P. Y., Uptake and fate of di-2-ethylhexyl phthalate in aquatic organisms and in a model ecosystem, Environ. Hlth. Persp., 4 (1973) 27-34
21. Carter, S. A., the potential health hazard of substances leached from plastic packaging, J. Environ. Hlth., 40 (2) (1973) 73-79
22. Corcoran, E. F., Gas chromatographic detection of phthalic acid esters, Environ. Hlth. Persp., 3 (1973) 13-15
23. Hites, R., Phthalates in the Charles and Merrimack Rivers, Environ. Hlth. Persp., 3 (1973) 17-21
24. Thomas, G. H., Quantitative determination and confirmation of identity of trace amounts of dialkyl phthalates in environmental samples, Environ. Hlth. Persp., 3 (1973) 23-28
25. Ogner, G., and Schnitzer, Humic substances-fulfic acid-dialkyl phthalate complexes and their role in pollution, Science, 170 (1970) 317
26. Nazir, D. J., Alcaraz, A. P., Bierl, B. A., Beroza, M., and Nair, P. P., Isolation, identification and specific localization of di-2-ethylhexyl phthalate in bovine heart muscle mitochondria, Biochemistry, 10 (1971) 4228-4232

27. Jaeger, R. J., and Rubin, R. J., Migration of a phthalate ester plasticizer from polyvinyl chloride blood bags into stored human blood and its localization in human tissues, New Engl. J. Med., 287 (1972) 1114-1118
28. Food and Drug Administration, Compliance Program, Evaluation: Phthalate Esters in Food Survey, FY 1974, Bureau of Foods, Washington, DC (1975)
29. Jacobson, M. S., Parkman, R., Button, L. N., Jaeger, R. J., and Kevy, S. V., The toxicity of human serum stored in flexible PVC containers on human fibroblasts cell cultures: An effect of di-2-ethylhexylphthalate, Res. Commun. Chem. Patho. Pharmacol., 9 (1974) 315-321
30. Oppenheimer, B. S., Oppenheimer, E. T., Danishefsky, I., Scott, A. P., and Eirich, F. R., Further studies of polymers as carcinogenic agents in animals, Cancer Res., 15 (1955) 333
31. Oppenheimer, B. S., Oppenheimer, E. T., Stout, A. P., Danishefsky, I., and Willhite, M., The mechanism of carcinogenesis by plastic films, Acta. Unio. Internat. Conta. Cancum., 15 (1959) 659
32. Singh, A. R., Lawrence, W. H., and Autian, J., Mutagenic and antifertility sensitivities of mice to di-2-ethylhexyl phthalate esters, Env. Hlth. Persp., 3 (1973) 81-89
33. Dillingham, E. O., and Autian, J., Teratogenicity, mutagenicity and cellular toxicity of phthalate esters, Env. Hlth. Persp., 3 (1973) 81-89
34. Singh, A. R., Lawrence, W. H., and Autian, J., Mutagenic and antifertility sensitivities of mice to di-2-ethylhexyl phthalate (DEHP) and dimethoxyethyl phthalate (DMEP), Toxicol. Appl. Pharmacol., 29 (1974) 35-46
35. Shahin, M. M., and VonBorstel, R. C., Mutagenic and lethal effects of α-benzene hexachloride dibutyl phthalate and trichloroethylene in Saccharomyces cerevisiae, Mutation Res., 48 (1977) 173-180
36. Jones, A. E., Kahn, R. H., Groves, J. T., and Napier, E. A., Jr., Phthalate ester toxicity in human cell cultures, Toxicol. Appl. Pharmacol., 31 (1975) 283-289
37. Singh, A. R., Lawrence, W. H., and Autian, J., Teratogenicity of phthalate esters in rats, J. Pharm. Sci., 61 (1972) 51
38. Seth, P. K., Srivastava, S. P., Agarwal, D. K., and Chndra, S. V., Effect of di-2-ethylhexyl phthalate (DEHP) on rat gonads, Environ. Res., 12 (1976) 131-138
39. Albro, P. W., Thomas, R., and Fishbein, L., Metabolism of diethylhexyl phthalates by rats. Isolation and characterization of the urinary metabolites, J. Chromatog., 76 (1973) 321-330
40. Daniel, J. W., and Bratt, H., The absorption, metabolism and tissue distribution of di(2-ethylhexyl)phthalate in rats, Toxicology, 2 (1974) 51-65
41. Lake, B. G., Brantom, P. G., Gangoli, S. D., Butterworth, R., and Grasso, P., Studies on the effects of orally administered di(2-ethylhexyl)phthalate in the ferret, Toxicology, 6 (1976) 341-356
42. Standish, W. L., and Abrams, S. R., Adipic acid, In: Kirk-Othmers Encyclopedia of Chemical Technology, 2nd ed., Vol. 1, Wiley-Interscience, New York (1963) pp. 405-421
43. Gross, S., (ed) "Modern Plastics Encyclopedia", McGraw-Hill New York (1973)
44. U.S. Tariff Commission, Synthetic Organic Chemicals, Washington, DC (1972)
45. Singh, A. R., Lawrence, W. H., and Autian, J., Dominant lethal mutations and antifertility effects of di-2-ethylhexyl adipate and diethyl adipate in male mice, Toxicol. Appl. Pharmacol., 32 (1975) 566-576
46. Singh, A. R., Lawrence, W. H., and Autian, J., Embryonic-fetal toxicity and teratogenic effects of adipic acid esters in rats, J. Pharm. Sci., 62 (1973) 1596-1600

CHAPTER 25

CYCLIC ETHERS

1. <u>1,4-Dioxane</u> (1,4-diethylenedioxide; diethylene ether; di(ethylene oxide); dioxyethylene ether; tetrahydro-1,4-dioxin; $O\begin{smallmatrix}CH_2CH_2\\CH_2CH_2\end{smallmatrix}O$) can be made principally by three routes[1,2]: (a) the dehydration of ethylene glycol; (b) the dimerization of ethylene glycol and (c) the treatment of bis(2-chloroethyl)ether with alkali.

The principal use for 1,4-dioxane in the U.S. is as stabilizer for chlorinated solvents. The future growth for this area of utility was estimated in 1973 to be 7-8% per year[1,3]. It has also been widely used since 1930 as a solvent for lacquers, varnishes, resins, oils, waxes, cellulose acetate, ethyl cellulose, paints, fats and some dyes, as a solvent for agricultural and biological intermediates and for adhesives, sealants, rubber coatings, surface coatings, cosmetics, pharmaceuticals and rubber chemicals[1,4]. Lesser amounts are also employed as a solvent in laboratory synthesis, and in the preparation of tissues for histology.

In Japan the major uses of 1,4-dioxane are as a solvent, a stabilizer for trichloroethylene and as a surface-treating agent for artificial leather[1].

The pharmacokinetic and metabolic fate of 1,4-dioxane has been shown to be dose dependent in rats[5,6] due to a limited capacity to metabolize dioxane to β-hydroxyethoxyacetic acid (HEAA) the major urinary metabolite[5,6]. In rats given single I.V. doses, the half-life value was dependent on the dosage level with the lower doses of 3 and 10 mg/kg having half-life values of 1.1 and 1.5 hr respectively[7].

Young et al[8] recently determined the fate of dioxane in humans after exposure to 50 ppm for 6 hr in order to provide a rationale for extrapolating extensive toxicological information obtained with rats under a variety of conditions and dosages. This would allow a better assessment of the hazard of dioxane exposures to humans in the workplace. The pharmacokinetics profile of dioxane in humans was essentially as predicted from previous studies conducted in rats[6,7]. Dioxane was eliminated primarily by metabolism to HEAA which was subsequently rapidly eliminated in the urine. The half-life for overall elimination of dioxane in 4 healthy male volunteers was approximately 1 hr. According to the authors[8] the pharmacokinetic profile in humans demonstrates that at exposure concentrations encountered in the workplace, 50 ppm or lower, that dioxane will not accumulate in the body after continuous or repeated exposures. Since this study shows that the detoxification of dioxane in humans exposed to 50 ppm is not saturated, it was concluded that exposure at this level in the workplace will not cause adverse effects.

It should be noted, in contrast, that Woo et al[9,10] reported p-dioxane-2-one (also known as 2-hydroxyethoxy acetic acid δ-lactone) (IV Figure 1) as the major urinary metabolite of rats treated with dioxane (100-400 mg/100g body weight administered i.p.)

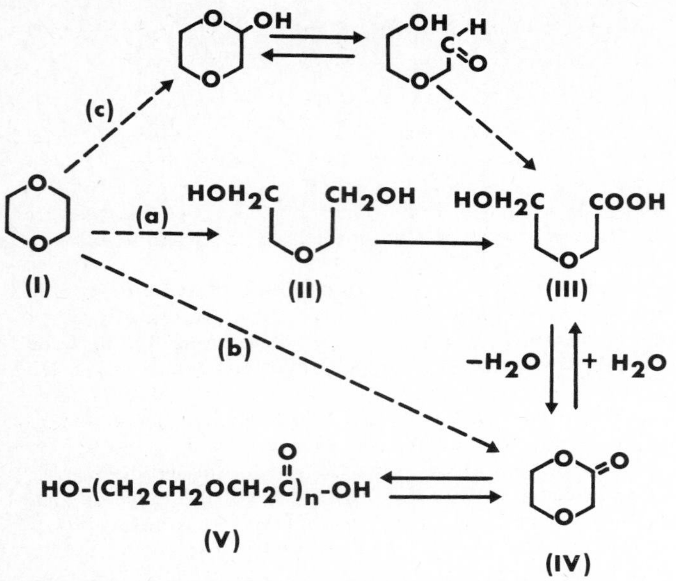

Fig. 1 Tentative metabolic pathway of dioxane[9]

Diethylene glycol administered to rats gave rise to the same metabolite. Figure 1 illustrates the tentative metabolic pathways of dioxane (I) as proposed by Woo et al[9]. These include: (a) hydrolysis to diethylene glycol (II) followed by oxidation of one of the hydroxyl groups, (b) direct conversion via a possible ketoperoxyl radical intermediate similar to the reaction scheme proposed by Lortentzen et al[11] for the oxidation of the carcinogen, benzo(a)pyrene to benzo(a)pyrene diones, and (c) through α-hydroxylation, followed by the oxidation of the hemiacetal or hydroxylaldehyde intermediate. The observation tht p-dioxane-2-one (IV) is excreted from rats given diethylene glycol (II) and at a rate faster than that of IV from dioxane (I) supports the pathway (a). However, the absence of diethyleneglycol (II) in the urine from rats given dioxane suggests either very rapid conversion of (II) to (IV) or that pathway (c) may be a more likely mechanism.

It has also been shown by Woo et al[12] that the metabolism of dioxane to p-dioxane-2-one can be significantly increased by pretreatment of rats with inducers of microsomal mixed function oxidases (MFO's) and decreased by inhibitors or repressors of MFO's suggesting the involvement of MFO's in the in vivo metabolism of dioxane. The metabolism of dioxane may be closely related to its toxicity and/or carcinogenicity since acute toxicity studies show that a number of agents that modify the LD_{50} of dioxane also modify its mtabolism in the same manner[9]. Para-dioxane-2-one is about 10 times more toxic than p-dioxane based on an LD_{50}[9,13]. The possibility that p-dioxane-2-one may be a proximate carcinogen is currently under study[9]. The carcinogenicity of a number of unsaturated and saturated lactones has long been known[14].

The principal toxic effects of dioxane have long been known to be centrilobular, hepatocellular and renal tubular, epithelial degeneration and necrosis[15-18].

More recent reports by Argus et al[19,20] and Hoch-Ligeti et al[21] described nasal and hepatic carcinomas in Sprague-Dawley and Wistar rats ingesting water containing large doses of dioxane (up to 1.8% of dioxane in the drinking water for over 13 months). For example, Argus et al[19] reported hepatomas in Wistar rats maintained in drinking water containing 1% dioxane for 63 weeks while Hoch-Ligeti et al[21] described the induction of nasal cavity carcinomas in Wistar rats maintained on drinking water containing from 0.75 to 1.8% dioxane for over 13 months.

Studies by Kociba et al[22] in 1974 indicated a dose response for the toxicity of dioxane in Sherman strain rats. Daily administration of 1% dioxane in drinking water to male and female rats (1015 and 1599 mg/kg/day respectively) for up to 2 years caused pronounced toxic effects including the occurrence of hepatic and nasal tumors. There was an induction of untoward effects and liver and kidney damage, but no tumor induction in male and female rats receiving 0.1% dioxane (equivalent to approximately 94 and 148 mg/kg/day respectively) in the drinking water, and female rats receiving 0.01% dioxane in the drinking water (equivalent to approximately 9.6 and 19.0 mg/kg/day, respectively) showed no evidence of tumor formation or other toxic effects considered to be related to treatment.

Carcinomas of the gall bladder were noted in male guinea-pigs which received drinking water containing 0.5-2% 1,4-dioxane over a 23 month period (total dose, 588-623g/animal)[23].

Gehring et al[24] postulated that the toxicity and carcinogenicity of 1,4-dioxane (as well as vinyl chloride) are expressed only when doses are sufficient to overwhelm the detoxification mechanisms. When such doses are given, there is a disproportionate retention of the compound and/or its metabolites in the body. Also observed are changes in the biochemical status of the animals consistent with accepted mechanisms for cancer induction[24].

Although dioxane is considered to be a weak to moderate hepatic carcinogen[19,20], the mechanism of its carcinogenic action has not been unequivocally elaborated[9,25]. Earlier suggestions were advanced that by virtue of the potent hydrogen bond breaking[26] and protein denaturing action[27] of dioxane, the molecular basis for carcinogenic action lies in the inactivation of key cellular macro-molecules involved in metabolic control. Although acute toxicity studies suggested involvement of microsomal mixed function oxidases, pre-treatment of rats with enzyme inducers had little or no effect on covelent binding[25]. No microsome-catalyzed dioxane binding to exogenous DNA was observed under conditions that allowed significant binding of benzo(a)pyrene. Incubation of isolated microsomes or nuclei also showed no enzyme-catalyzed binding of dioxane[25]. It had been earlier postulated by Hoch-Ligeti et al[21] that a reactive free radical or a carbonium ion may arise in the metabolism of dioxane and may represent a proximate carcinogen. Another possibility was that a peroxide of dioxane may account for its carcinogenicity[21]. The possibility that p-dioxane-2-one (a major metabolite of dioxane) may be a proximate carcinogen has been cited above[9].

A significant reduction in the existing United States dioxane limit of 100 ppm (360 mg/m^3) as a time-weighted average workplace was recently recommended by NIOSH[28]. NIOSH criteria document calls for an environmental limit of 1 ppm (3.6 mg/m^3) dioxane based on a 30 minute sampling period. This recommendation is the lowest concentration that can reliably be measured by current sampling and analytical methods. It was also noted that dioxane has been used since 1929 as a solvent in various industrial operations, as a stabilizer in 1,1,1-trichloroethane (methyl chloroform) where about 100,000 workers are potentially exposed. NIOSH estimates that an additional 2,500 people in the U.S. are exposed to dioxane in the workplace[28]. The current maximum allowable concentration (MAC) of dioxane in the USSR is 10 mg/m^3[29].

As has previously been cited, diethyleneglycol can be metabolized to p-dioxane-2-one the major metabolite of p-dioxane[9]. Diethyleneglycol is employed in extensive quantities (e.g., 472 million pounds in the U.S. in 1976), in areas including (%): polyurethane and unsaturated polyester resins (30); triethylene glycol (13); textile agents (12), petroleum solvent extraction (7), natural gas dehydration (7), plasticizers and surfactants (7), miscellaneous and export (8)[30].

Diethylene glycol is normally a co-product in the production of ethylene glycol or it can be produced from the reaction of ethylene glycol and ethylene oxide[31].

References

1. IARC, 1,4-Dioxane, In. Vol. 11, Cadmium, Nickel, Some Epoxides, Miscellaneous Industrial Chemicals and General Considerations on Volatile Anaesthetics, International Agency for Research on Cancer, Lyon (1976) pp. 247-256
2. Hawley, G. G., (ed) "Condensed Chemical Dictionary", 8th ed., Van Nostrand-Reinhold, New York (1971) p. 320
3. Anon, 1,4-Dioxane, Chem. Marketing Reporter, Feb. 5 (1973) pp. 11-12
4. Anon, Product Use Patterns, Union Carbide Corp., New York (1970)
5. Young, J. D., and Gehring, P. J., The dose-dependent fate of 1,4-dioxane in male rats, Toxicol. Appl. Pharmacol., 33 (1975) 183
6. Young, J, D., Braun, W. H., LeBeau, J. E., and Gehring, P. J., Saturated metabolism as the mechanism for the dose-dependent fate of 1,4-dioxane in rats, Toxicol. Appl. Pharmacol., 37 (1976) 138
7. Kociba, R. J., Torkelson, T. R., Young, J. D., and Gehring, P. J., 1,4-Dioxane, Correlation of the results of chronic ingestion and inhalation studies with its dose-dependent fate in rats, Proc. Sixth Conf. Environ. Toxicol., (1975) 345-354
8. Young, J. D., Braun, W.H., Rampy, L. W., Chenoweth, M. B., and Blau, G. E., Pharmacokinetics of 1,4-Dioxane in humans, J. Toxicol. Env. Hlth., 3 (1977) 507-520
9. Woo, Y. T., Arcos, J. C., Argus, M. F., Griffin, G. W., and Nishiyama, K., Structural identification of p-dioxane-2-one as the major urinary metabolite of p-dioxane, Naunyn-Schmiedeberg's Arch. Pharmacol., 299 (1977) 283-287
10. Woo, Y. T., Arcos, J. C., Argus, M. F., Griffin, G. W., and Nishiyama, K., Metabolism in vivo of dioxane: Identification of p-dioxane-2-one as the major urinary metabolite, Biochem. Pharmacol., 6 (1977) 1535-1538
11. Lorentzen, R. J., Caspary, W. J., Lesko, S. A., Ts'O, P. O. P., The autoxidation of 6-hydroxybenzo(a)pyrene and 6-oxobenzo(a)pyrene radical, reactive metabolites of benzo(a)pyrene, Biochemistry, 14 (1975) 3970-3977
12. Woo, Y. T., Argus, M. F., and Arcos, J. C., Effect of inducers and inhibitors of mixed-function oxidases on the in vivo metabolism of dioxane in rats, Proc. Am. Assoc. Cancer Res., 18 (1977) 165
13. Woo, Y. T., Personal communication to Office of Toxic Substances, Environmental Protection Agency, January 30 (1978)
14. Dickens, F., Carcinogenic lactones and related substances, Brit. Med. Bull., 20 (1964) 96-101
15. DeNavasquex, A., Experimental tabular necrosis of the kidney accompanied by liver changes due to dioxane poisoning, J. Hyg., 35 (1935) 540-548
16. Fairley, A., Linton, E. C., and Ford-Moore, A. H., The toxicity to animals of 1,4-dioxane, J. Hyg., 34 (1934) 486-501
17. Schrenk, H. H., and Yant, W. P., Toxicity of dioxane, J. Ind. Hyg. Toxicol. 18 (1936) 448-460
18. Kesten, H. D., Mulinos, M. G., and Pomerantz, L., Pathologic effects of certain glycols and related compounds, Arch. Pathol., 27 (1939) 447-465
19. Argus, M. F., Sohal, R. S., Bryant, G. M., Hoch-Ligeti, C., and Arcos, J. C., Dose-response and ultrastructural alterations in dioxane carcinogenesis, Eur. J. Cancer, 9 (1973) 237-243
20. Argus, M F., Arcos, J. C., and Hoch-Ligeti, C., Studies on the carcinogenic activity of protein-denaturing agents: Hepatocarcinogenicity of dioxane, J. Natl. Cancer Inst., 35 (1965) 949-958
21. Hoch-Ligeti, C., Argus, M. F., and Arcos, J. C., Induction of carcinomas in the nasal cavity of rats by dioxane, Brit. J. Cancer, 24 (1970) 164-170

22. Kociba, R. J., McColister, S. B., Park, C., Torkelson, T. R., and Gehring, P. J., 1,4-Dioxane, I. Results of a 2-year ingestion study in rats, Toxicol. Appl. Pharmacol., 30 (1974) 275-286
23. Hoch-Ligeti, C., and Argus, M. F., Effect of carcinogens on the lung of guinea pigs, In "Conference on the Morphology of Experimental Respiratory Carcinogenesis" Gatlinburg, TN (eds) Nettesheim, P., Hanna, M. G., Jr., and Deatherage, J. W., Jr., AEC Symposium Series No. 21, National Technical Information Service, Springfield, VA (1970) pp. 267-279
24. Gehring, P. J., Watanabe, P. G., and Young, J. D., The relevance of dose-dependent pharmacokinetics and biochemical alterations in the assessment of carcinogenic hazard of chemicals, Presented at Meeting of Origins of Human Cancer, Cold Spring Harbor, NY, Sept. 7-14 (1976)
25. Woo, Y. T., Argus, M. F., and Arcos, J. C., Dioxane carcinogenesis apparent lack of enzyme-catalyzed covalent binding to macromolecules, The Pharmacologist (1976) 158
26. Argus, M. F., Arcos, J. C., Alam, A., and Mathison, J. H., A viscometric study of hydrogen-bonding properties of carcinogenic nitrosamines and related compounds, J. Med. Pharm. Chem., 7 (1964) 460-465
27. Bemis, J. A., Argus, M. F., and Arcos, J. C., Studies on the denaturation of biological macromolecules by chemical carcinogens. III. Optical rotary dispersion and light scattering changes in oval bumin during denaturation and aggregation by water soluble carcinogens, Biochim. Biophys. Acta., 126 (1966) 275-285
28. Anon, NIOSH Criteria Document Recommends Reduction of Limits, Chem. Reg. Reptr., 1 (26) (1977) 893
29. Winell, M. A., An international comparison of hygienic standards for chemicals in the work environments, Ambio, 4 (1975) 34-36
30. Lowenheim, F. A., and Moran, M. K., "Farth, Keyes and Clark's Industrial Chemicals", 4th ed., Wiley-Interscience, New York (1975) pp. 397-402
31. Freifeld, M., and Hort, E. V., Glycols, In "Kirk-Othmer Encyclopedia of Chemical Technology", 2nd ed., Vol. 10, Wiley (Interscience), New York (1966) 638-676

CHAPTER 26

PHOSPHORAMIDES AND PHOSPHONIUM SALTS

A. Hexamethyl Phosphoramide

Hexamethyl phosphoramide (hexamethyl phosphoric-acid, triamide, tris-(dimethylamino)-phosphine oxide, HMPA) $((CH_3)_2N-P(=O)(N(CH_3)_2)(N(CH_3)_2))$ is used primarily in the following areas: (1) as a solvent for polymers, (2) polymerization catalyst, (3) stabilizer for polystyrene against thermal degradation, (4) as a selective solvent for gases, (5) as an additive for polyvinyl and polyolefin resins to protect against degradation by U.V. light and (6) as a solvent in organic and organometallic reactions in research laboratories[1,2]. In the U.S., hexamethyl phosphoramide is used by its major producer as a processing solvent for Aramid (aromatic polyamide fiber)[3]. The use of hexamethyl phosphoramide as a solvent in research laboratories has been reported to account for more than 90% of the estimated 5000 people who are occupationally exposed to this chemical in the U.S.[2]. Hexamethyl phosphoramide has also been evaluated as a chemosterilant for insects (e.g., houseflies, Musca domestica[4]) and to a lesser extent as an antistatic agent, flame retardant and as a de-icing additive for jet fuels[2].

Recent studies in 1975 have indicated that inhalation of HMPA vapor (400 or 4000 ppb by volume) in air 6 hrs/day for 5 days each week for 6 to 7 months produces dose-related squamous cell carcinomas of the nasal cavity in Charles River-CD rats[5]. In addition nasal tumors were first detected 9 months after inhalation of HMPA vapor at the 100 ppb level and 13 months at the 50 ppb level. The nasal tumors were predominantly squamous cell carcinoma and less frequently squamoadenocarcinoma or nasal papilloma. Some nasal tumors metasticized to the cervical lymph nodes and lungs or extended directly to the cerefrum. However, after 1 year of exposure the group of Charles River-CD rats at the 10 ppb level had not yet developed nasal tumors[6].

In an earlier study in 1973 involving a small number of 6 week old Sherman rats fed diets containing HMPA at concentrations of 6.25, 3.12, 1.56 and 0.78 mg/kg by weight for 2 years, a low incidence of tumors (mainly reticulum-cell or lymphosarcomas of the lungs) were noted in all groups)[7].

Although chromosome aberrations have been noted in insects treated with HMPA[8,9], no increase in chromosomal aberrations in Chinese hamster lung cells treated with $5 \times 10^{-3}M$ HMPA resulted in chromosomal aberrations in 11% of the cells, while only 6% of the control cells contained abnormal chromosomes[10]. These values are not significantly different at the 5% level of statistical significance[11]. HMPA induced a high frequency of recessive lethal mutations in the sperm of Bracon hebetor[12], testicular atrophy in rats[13] and a marked antispermatogenic effect in rats[13,14] and mice[14].

Ashby et al[15] recently compared HMPA with a number of its structurally related analogs such as the leukaemogen phosphoramide (I), the putative non-carcinogen phosphoric trianilide (II) and N,N,N'-trimethylphosphorothioic triamide (III) (a compound of unknown and unpredictable properties) in the S. typhimurium mutation assay of Ames[16] and the cell transformation assay of Styles[17]. In one experiment, HMPA gave a strong, dose-related, positive effect in S. typhimurium strains TA 1535 and TA 100 in the presence of S-9 mix, while phosphoramide (I) and the trianilide (II) were both negative. In the second experiment, HMPA together with compound (II) was negative. The unreliable nature of the Ames test response for HMPA was confirmed in subsequent experiments by Ashby et al[15] as well as in an independent contract laboratory. In contrast, the transformation assay found both HMPA and phosphoramide positive[15] while compounds (II) and (III) were negative by Jones and Jackson[18].

HMPA has been found to undergo in vivo and in vitro α-hydroxylation leading to formaldehyde formation. Using the S-9 liver fraction for metabolic activation, a dose-dependent relationship between the concentration of HMPA and formaldehyde formation was observed by Ashby et al[15]. These observations mimic those for metabolic α-hydroxylation of the carcinogen diethylnitrosamine as postulated by Druckrey[19].

$$\underset{(I)}{\underset{\underset{O}{\|}}{H_2N-\overset{NH_2}{P}-NH_2}} \qquad \underset{(II)}{\underset{\underset{O}{\|}}{\phi HN-\overset{NH\phi}{P}-NH-\phi}} \qquad \underset{(III)}{\underset{\underset{S}{\|}}{CH_3HN-\overset{NHCH_3}{P}-NHCH_3}}$$

B. <u>Phosphonium Salts</u>

1. <u>Tetrakis(hydroxymethyl)phosphonium Salts</u>

Flame retardants include many chemical classes such as organic compounds containing phosphorus, nitrogen, sulfur, halogens and inorganic compounds of antimony, boron and others[20-23]. It had been earlier estimated that the use of flame retardant plastics and textiles will grow 35 to 45% a year through 1975 from 800 million pounds in 1970 to 3.5-4.5 billion pounds by 1975[24].

Tetrakis(hydroxymethyl)phosphonium salts (I) are amongst the most widely used chemicals currently employed to make cotton flame resistant particularly in children's sleepwear[20,21-25,26].

$$\left[\begin{array}{c} CH_2OH \\ | \\ HOCH_2-P^+-CH_2OH \\ | \\ CH_2OH \end{array} \right] X^-$$

THPC: X=Cl
Pyroset TKP: X=$CH_3COO^-/H_2PO_4^-$
THPS: X=$SO_4^=$

$$X = \begin{array}{c} O \\ \diagdown \\ O \end{array} C-C \begin{array}{c} O^- \\ \diagup \\ O^- \end{array}$$

(I) Tetrakis(hydroxymethyl)phosphonium Salt Flame Retardants

Tetrakis(hydroxymethyl)phosphonium chloride (THPC) and the related compound Pyroset TKP are cross-linked either to the hydroxyl groups of cellulose or with nitrogen bases in order to bind them permanently to the cloth[25]. Commercial preparations of THPC and Pyroset TKP are generally in the form of aqueous solutions containing 80% and 70% solids[25,26]. The formaldehyde content of commercial THPC is pH dependent and has ranged from 3.79% at pH 0.4 to 14.10% at pH $\geq$ 4.5. Commercial THPC can contain 16.1% Cl$^-$ and possess a pH of 0.4[25]. Although the species required to produce bischloromethyl ether (BCME) can be present in large amounts, viz.,

$$\left[\begin{array}{c} CH_2OH \\ | \\ HOCH_2-P^+ \\ | \\ CH_2OH \end{array}\right] Cl^- \rightleftharpoons PH_3 + 4CH_2O + HCl \\ \downarrow \quad \uparrow \\ ClCH_2OCH_2Cl$$

BCME was not detected in concentrations above 1 ppm[25].

THPC exhibited a lower order of carcinogenicity when applied to mouse skin (female ICR/Ha Swiss mice) at a level of 2 mg THPC in 0.1 ml DMSO 3 times/week for 400 days[25]. THPC was inactive as an initiating agent in two stage mouse skin carcinogenesis with phorbol myristrate acetate as a promoter[25]. Both THPC and Pyroset TKP were active as tumor promoters using a single application of 7,12-dimethylbenz(a)anthracene (20 µg in 0.1 ml acetone) as initiator. With THPC as promoter (2 mg in 0.1 ml DMSO, thrice weekly), 3 of 20 mice bore papillomas which progressed to squamous carcinoma. With Pyroset TKP as promoter (7 mg in 0.1 ml DMSO) 7 of 20 mice bore papillomas of which two progressed to squamous carcinoma[25].

It should also be noted that some form of the THPC finish is being released from the fabric in aqueous medium[27]. It is also possible that common solutions such as sweat, urine, and saliva might extract some free THPC, and hence there is a possibility that some percentage of children could be at risk from long-term low level exposure to THPC from treated sleepwear[25].

It is of note that extracts of fabrics trested with tris-BP (tris-2,3-dibromopropyl)-phosphate, the major flame-retardant agent for children's sleepwear have been found mutagenic in S. typhimurium TA 1535 and TA 100 strains[28].

Information is lacking as to the mutagenicity of the phosphonium derivatives THPC and Pyroset TKP.

The reaction of guanosine with THPC to yield a 2-amino substituted product (II) (where the phosphonium salt has been reduced to a phosphine) has been recently reported by Loewengart and Van Duuren[29]. The reaction scheme below was suggested to be consistent with the known behavior of THPC and accounts for the formation of (II). If this reaction occurs with the guanine moiety of DNA in vivo, it was also suggested that the carcinogen bis(chloromethyl)ether (BCME) could potentially be formed from HCl and HCHO liberated during the formation of (II)[30].

It should be noted that guanosine is known to react with epoxides[31], lactones[32] and other alkylating agents[33,34]. While the nature and position of the reactions on guanosine have been suggested to be related to the mutagenicity and carcinogenicity of alkylating agents[35], no unequivocal relationship has been established[29].

$(HOCH_2)_4 P^+ Cl^- \xrightleftharpoons{pH\ 5.5} (HOCH_2)_3P: + HCl + CH_2O$

CH₂O + [guanine derivative with NH₂, R=β-D-ribofuronosyl] ⇌ [intermediate **2** with CH=N] $\xrightarrow{(HOCH_2)_3P:}$

[structure **3**: HOCH₂-P⁺(CH₂CH)₂-CH₂-NH-guanine-R] $+Cl^- +H_2O$ [structure **1**: (HOCH₂)₂P-CH₂-NH-guanine-R] $+HCl+CH_2O$

References

1. Robert, L. Properties and uses of hexamethyl phosphorotriamide, Chimie Indus.-Genie Chimique, 97 (3) (1967) 337-345
2. U.S. Dept. Health, Education, & Welfare, Background Information on Hexamethylphosphoric Triamide, National Institute for Occupational Safety & Health, Rockville, MD (1975) pp. 1-4
3. Anon, Hexamethyl phosphoramide causes cancer in laboratory animals, Chem. Eng. News, 53 (1975) 17
4. Borkovec, A. B., Insect chemosterilants Adv. Pest. Control Res., 7 (1966) 45
5. Zapp, J. A., Jr., HMPA: A possible carcinogen, Science, 190 (1975) 422
6. Lee, K. P., Trochimowicz, H. J., and Sarver, J. W., Induction of nasal tumors in rats exposed to hexamethyl phosphoramide (HMPA), Lab. Invest., 36 (3) (1977) 344-345
7. Kimbrough, R. D., and Gaines, T. B., The chronic toxicity of hexamethyl phosphoramide in rats, Bull. Env. Contam. Toxicol., 10 (1973) 225-226
8. Grover, K. K., Pillai, M. K. K., and Dass, C. M. S., Cytogenetic basis of chemically induced sterility in Culex Pipiensfatigans, Wiedemann III. Chemosterilant-induced damage in the somatic chromosomes, Cytologia, 38 (1973) 21-28
9. LeChance, L. E., and Leopold, R. A., Cytogenic effect of chemosterilants in housefly sperm: Incidence of polysptermy and expression of dominant lethal mutations in early cleavage divisions, Canad. J. Genet. Cytol., 11 (1969) 648-659
10. Sturelid, S., Chromosome-breaking capacity of TEPA and analogs in Vicia Faba and Chinese hamster cells, Hereditas, 68 (1971) 255-276
11. Chang, T. H., and Klassen, W., Comparative effects of tretamine, TEPA, Apholate, and their structural analogs on human chromosomes in vitro, Chromosoma, 24 (1968) 314-323
12. Palmquist, J., and LaChance, L. E., Comparative mutagenicity of two chemosterilants, tepa and hempa in sperm of Bracon Hebetor, Science, 154 (1966) 915-917
13. Kimbrough, R. M., and Gaines, T. B., Toxicity of hexamethyl phosphoramide in rats, Nature, 211 (1966) 146-147
14. Jackson, H., and Craig, A. W., Effects of alkylating agents on reproductive cells, Proc. 5th I.P.P.F. Conf. (1967) page 49
15. Ashby, J., Styles, J. A., and Anderson, D., Selection of an in vitro carcinogenicity test for derivatives of the carcinogen hexamethyl phosphoramide, Brit. J. Cancer, 36 (1977) 564-570
16. Ames, B. N., McCann, J., and Yamasaki, E., Methods for detecting carcinogens and mutagens with the Salmonella/mammalian-microsome mutagenicity test, Mutation Res., 31 (1975) 347
17. Styles, J. A., A method for detecting carcinogenic organic chemicals using mammalian cells in culture, Brit. J. Cancer, 36 (1977) 558-562
18. Jones, J. R., and Jackson, H., The metabolism of hexamethylphosphoramide and related compounds, Biochem. Pharmacol., 17 (1968) 2247
19. Druckrey, H., Chemical carcinogenesis of N-nitroso compounds, Gann. Monograph in Cancer Res., 17 (1975) 107
20. Pearle, E. M., and Liepins, R., Flame retardants, Env. Hlth. Persp., 11 (1975) 59-69
21. Jones, D. M., and Noone, T. M., Some approaches to the permanent flame proofing of cotton: Systems containing phosphorus, J. Appl. Chem. (London) 12 (1972) 397

22. Lyons, J. W., The chemistry and uses of fire retardants, Wiley-Interscience, New York (1970)
23. WHO, Health hazards from new environmental pollutants, Tech. Rept. Series No. 586, World Health Organization, Geneva (1976) pp. 83-96
24. Anon, Flame retardant growth flares up, Chem. Eng. News, Oct. 18 (1971) 16-19
25. Loewengart, G., and Van Duuren, B. L., Evaluation of chemical flame retardants for carcinogenic potential, J. Toxicol. Env. Hlth., 2 (1977) 539-546
26. Hooper, G., Nakajima, W. N., and Herbes, W. F., The use of various phosphonium salts for flame retardancy by the ammonia cure technique, In Proceedings of the 1974 Symposium on Textile Flammability, ed LeBlanc, R. B., LeBlanc Research Corp., East Greenwich, R.I. (1974) 30-46
27. Cutenmann, W. G., and Lisk, D. J., Flame retardant release from fabrics, During laundering and their toxicity to fish, Bull. Env. Contam. Toxicol., 14 (1975) 61-64
28. Prival, M. J., McCoy, E. C., Gutter, B., and Rosenkranz, H. S., Tris(2,3-dibromopropyl)phosphate: Mutagenicity of a widely used flame retardant, Science, 195 (1977) 76-78
29. Loewengart, G., and Van Duuren, B. L., The reaction of guanosine with tetrakis-(hydroxymethyl)phosphoniumchloride, Tetrahedron Letters, 39 (1976) 3473-3476
30. Van Duuren, B. L., Sivak, A., Goldschmidt, B. M., Katz, C., and Melchionne, Carcinogenicity of halo-ethers, J. Natl. Cancer Inst., 43 (1969) 481-486
31. Brookes, P., and Lawley, P. D., Reaction of mono- and difunctional alkylating agents with nucleic acids, Biochem. J., 80 (1961) 496-503
32. Roberts, J. J., and Warwick, G. P., Reaction of β-propiolactone with guanosine deoxyguanlyl acid, and RNA (ribonucleic acid), Biochem. Pharmacol., 12 (1963) 1441-1442
33. Van Duuren, B. L., Carcinogenic epoxides, lactones, chloroethers and their mode of action, Ann. NY Acad. Sci. (1969) 633-651
34. Goldschmidt, B. M., Blazej, T. P., and Van Duuren, B. L., Reaction of guanosine and deoxyguanosine with glycidaldehyde, Tetrahedron Letters, 13 (1968) 1583
35. Singer, B., The chemical effects of nucleic acid alkylation and their relation to mutagenesis and carcinogenesis, Prog. Nucleic Acid Mol. Biol., 15 (1975) 219-284

CHAPTER 27

MISCELLANEOUS POTENTIAL CARCINOGENS AND MUTAGENS

Although 150 industrial chemicals were examined individually in the previous chapters, it is recognized that in many instances, they are illustrative of categories of structurally related industrial agents that á-priori should warrant further scrutiny as potential carcinogens and mutagens.

1. Vinylic Derivatives

A category of particular concern are the olefinic compounds. These compounds, as well as aromatic compounds, can be transformed by mammalian microsomal monoxygenase(s) to alkene oxides (oxiranes, epoxides) from the epoxidation of an olefinic double bond, and arene oxides from the epoxidation of one of the double bonds of an aromatic nucleus[1-8]. This transformation to epoxides and arene oxides has been suggested to account for the observed carcinogenicity and/or mutagenicity of a number of environmental agents including chlorinated olefinic derivatives[5-13], e.g., vinyl chloride[5-9,12], vinylidene chloride[5,8,9,11,12], trichloroethylene[5,10,12-14], perchloroethylene[5], trans-1,4-dichlorobutene[10] and chloroprene (2-chloro-1,3-butadiene)[7,8].

Other olefinic compounds (vinyl analogs) that have been found carcinogenic and/or mutagenic include acrylonitrile (vinyl cyanide; $CH_2=CHCN$), acrolein (vinyl aldehyde $CH_2=CHCHO$) and styrene (vinyl benzene; ⌬-$CH=CH_2$) via its metabolite styrene oxide.

Hence, related industrial agents of broad utility (e.g., as monomers, intermediates, solvents) and used in relatively large amounts that are of potential concern would include: (a) vinyl alcohol (ethanol; $CH_2=CHOH$); (b) vinyl bromide ($CH_2=CHBr$); (c) vinyl esters such as vinyl acetate ($CH_2=CHOOCCH_3$); (d) methyl vinyl ether ($CH_2=CHOCH_3$); (e) vinyl acetonitrile (allyl cyanide; $CH_2=CHCH_2CN$); (f) vinyl formic acid (acrylic acid; 2-propenoic acid; $CH_2=CHCOOH$); (g) vinyl ethylene (divinyl; 1,3-butadiene; $CH_2=CHCH=CH_2$); (h) vinyl ether (divinyl ether; $CH_2=CHOCH=CH_2$) and (i) vinyl cyclohexene ⌬-$CH=CH_2$).

a. Vinyl bromide (bromoethane) is a reactive flame retardant used in small amounts as a comonomer with acrylonitrile and other vinyl monomers in modacrylic fibers. Modacrylic fibers containing vinyl bromide are used in fabrics and fabric blends with polyesters for children's sleepwear and other apparel, home furnishings, and industrial applications[15].

Vinyl bromide has been detected as an impurity in commercial vinyl chloride[16,17].

It was recently found inactive when tested as an initiator and as a complete carcinogen in a two-stage skin carcinogenesis study in female ICR/Ha Swiss mice at a dose of 15 mg in 0.1 ml acetone per application[18]. It was recently found inactive when tested in mice by subcutaneous administration with doses of 25 mg vinyl bromide in 0.5 ml trioctanoin/animal once weekly for 48 weeks and observed up to 420 days[18]. When polyvinyl bromide was tested in an analogous manner in ICR/Ha mice, no tumors were produced by application to the surface of the skin but subcutaneous administration produced liposarcomas at the injection site[18].

Exposure of S. typhimurium TA 1530[19] or TA 100[19,20] to gaseous mixtures of vinyl bromide in air caused mutagenic effects which were enhanced by the addition of liver supernatant fractions from phenobarbitone pretreated mice or from human liver biopsies[19]. In addition the epoxide formation from vinyl bromide (as well as vinyl chloride and chloroprene) was demonstrated by passing a gaseous mixture of the test compound and air through a mouse liver microsomal system and trapping volatile alkylating metabolites by reaction with excess 4-(4-nitrobenzyl)pyridine (NBP)[20,21].

b. <u>Vinyl acetate</u> (acetic acid vinyl ester; 1-acetoxyethylene) is produced commercially by two processes: in the ethylene process in which ethylene is reacted with acetic acid in the presence of oxygen; and in the acetylene process in which acetylene and acetic acid are reacted in the vapor phase over a catalyst bed[22].

In Western Europe, vinyl acetate produced by the gas phase ethylene process has the following typical specifications: vinyl acetate, 99.9% min.; ethyl acetate, 323 ppm; water, 240 ppm; methyl acetate, 175 ppm; acetaldehyde, 6 ppm; and acrolein, 1 ppm[22].

In 1976, six U.S. companies produced 673 million kg of vinyl acetate[23]. Annual production capacity for vinyl acetate in Western Europe is estimated to have been 285 million kg in 1976 and was composed by the following countries (capacity-millions of kg); the Federal Republic of Germany (150); France (60); United Kingdom (50) and Spain (25).

In Japan, 4 companies produced a total of 426 million kg of vinyl acetate in 1976. Annual production capacity in other countries in 1976 is estimated to have been as follows (capacity-millions of kg): Mexico (18), Australia (12) and Brazil (11).

The U.S. consumption pattern for vinyl acetate in 1976 is estimated to have been as follows (%): for the production of polyvinyl acetate homopolymer emulsions and resins (including copolymers with more than 60% vinyl acetate) (61); polyvinyl alcohol (22); polyvinyl butyral (6); vinyl chloride-vinyl acetate copolymers (5); ethylene-vinyl acetate resins and emulsions (5) and other uses (1). It is estimated that approximately 310 million kg of vinyl acetate were used in the U.S. in 1976 for the production of polyvinyl acetate emulsions and resins[22].

The Japanese consumption pattern for vinyl acetate in 1976 is estimated to have been as follows (5): polyvinyl alcohol (70); polyvinyl acetate resins and emulsions (20); ethylene-vinyl acetate resins (9); and vinyl chloride-vinyl acetate copolymers (1)[22].

Polyvinyl alcohol (PVA) is produced from vinyl acetate via polyvinylacetate by hydrolysis. In 1976, the U.S. consumption patterns for PVA is estimated to have been as follows (%): as a textile warp sizing and finishing agent (47); in adhesives (23); as a polymerization aid in polyvinyl acetate emulsions used in adhesives (14); as a paper sizing and coating (8); and for miscellaneous applications which include thickening and binding applications (8).

Data are not available concerning the numbers of workers involved in the production of vinyl acetate or in its use applications.

The American Conference of Governmental Industrial Hygienists (ACGIH) recommends that an employee's exposure to vinyl acetate should not exceed an 8 hr time weighted average of 10 ppm (30 mg/m^3) in the workplace air in any 8-hr work shift of a 40 hr week. An absolute ceiling concentrations limit of 20 ppm (60 mg/m^3) was recommended for any 15-min. period provided the 8-hr time weighted average limit is not exceeded[24].

Data are extremely limited concerning the chronic effects of vinyl acetate. A long-term inhalation (4 hr/day, 5 days/week, for 52 weeks) study of vinyl acetate using one dose level (2500 ppm) in Sprague-Dawley rats produced no evidence of carcinogenicity[22].

Vinyl acetate in air is non-mutagenic in S. typhimurium TA 100 in the presence or absence of a 9000 g supernatant liver fraction from phenobarbitone pretreated mice[19]. Vinyl acetate enhanced simian adenovirus SA7 transformation in Syrian hamster cells in vitro[25].

c. 1,3-Butadiene (divinyl; vinyl ethylene) is widely used as a polymer component in the manufacture of synthetic rubber and copolymeric plastics such as acrylonitrile-butadiene-styrene (ABS) used for the packaging of foodstuffs.

In the U.S., butadiene is produced largely from petroleum gases, e.g., by catalytic dehydrogenation of butene or butene-butane mixtures. It can be obtained directly by cracking naphtha and light oil[26].

U.S. butadiene demand is forecast to reach 4.5 billion pounds by 1981, representing a growth rate of 3.8%/year from the 1976 total of 3.77 billion pounds[27]. General purpose rubbers (styrene-butadiene-rubber, SBR, and polybutadiene) account for about 70% of butadiene demand. The U.S. consumption pattern for butadiene in 1977 was (%): SBR, (48); polybutadiene rubber, (20); neoprene (9); adiponitrile (7); ABS, (5); nitrile rubber, (4); other (7)[28].

Existing butadiene capacity in Europe is about 2 million tons per year; by 1980 it will be 2.25 million tons. Current production is 1.3 million tons, or 65% of capacity[29].

Data on the number of workers involved in the production of butadiene or in its use applications are not available.

The ACGIH recommends that an employees exposure to butadiene should not exceed an 8-hr time weighted average of 1,000 ppm (2,200 mg/m^3) in the workplace air in any 8-hr workshift of a 40 hr week[24].

The work environment hygienic standards in several countries for butadiene (mg/m^3)[30] are: the Federal Republic of Germany, 2200; the Democratic Republic of Germany, 500; Czechoslovakia, 500; and the MAC in the USSR is 100 mg/m^3.

No information is available on any carcinogenic studies having been performed with butadiene.

Incubation of S. typhimurium TA 1530 and TA 1535 in the presence of gaseous butadiene for 20 hrs increased the number of his+ revertant/plate[31]. This mutagenic effect occurred in the absence of fortified S-9 rat liver fractions. In its presence, the mutagenic effect seemed to be dependent on its composition. With butadiene monoxide, a reversion to histidine prototrophy was obtained without metabolic activation with strains TA 1530, TA 1535 and TA 100. The mutagenic effect was observed with TA 100 within a range of 1 μ mole/plate to beyond 250 μ moles/plate and with TA 1535 and TA 1530 within a range of 1 μ mole/plate to 100 μ moles of butadiene monoxide/plate. No significant mutagenic effect was found in tester strains TA 1537, TA 1538 and TA 98 strains when exposed to butadiene monoxide. The selective mutagenic effect of butadiene and butadiene monoxide in certain S. typhimurium strains such as TA 1530, TA 1535 and TA 100 (the latter with butadiene monoxide only) indicates that base-pair substitutions in the bacterial DNA are involved. Since the same strains TA 1530 and TA 1535 are sensitive to both butadiene and butadiene monoxide and because a similarity exists between the mutagenic properties of butadiene and vinyl chloride (existence of a mutagenic effect in the absence of a metabolic activating system) influence of the mix composition upon the reversion rate, it was speculated by de Meester et al[31] that butadiene monoxide is a probable primary metabolite of butadiene. Other possible mechanisms of activation, such as radical formation or oxidative metabolism by the bacteria themselves were not excluded[31].

2. Allylic Derivatives

Allyl derivatives are an important class of industrial compounds whose reactivity and possible metabolism to epoxide (oxirane) intermediates should also warrant additional consideration as potential carcinogens and mutagens. These would include in addition to allyl cyanide (vinyl acetonitrile) and acrylic acid (vinyl formic acid) cited above, the following important derivatives: (a) allyl alcohol (2-propen-1-ol; $CH_2=CHCH_2OH$); (b) acrylamide (propenamide; CH_2CHCNH_2); and (c) allyl ether
                                                                                                                     $\overset{\parallel}{O}$

(diallyl ether, 3,3'-oxybis(1-propene); $(CH_2=CHCH_2)_2O$. Allyl chloride has been shown to be mutagenic and its metabolism could involve epoxidation of the double bond to form epichlorohydrin, e.g., $CH_2=CHCH_2Cl \rightarrow CH_2-CHCH_2Cl$, a known carcinogen and mutagen. The subsequent oxidation products of epichlorohydrin are glycidol and glycidaldehyde and possibly acrylic acid[18].

a. Allyl alcohol (vinyl carbinol) can be prepared by heating glycerol with formic acid or via the hydrolysis of allyl chloride or by the selective reduction of acrolein. It is used in the production of resins and plastics; as a solvent; as a chemical intermediate; as a polymerization inhibitor; as a corrosion inhibitor for metals; as a catalyst for the polymerization of olefins and in soil sterilization.

Information is lacking on the carcinogenic potential of allyl alcohol. It has been found non-mutagenic when tested at 0.17 mg/plate in S. typhimurium TA 1535, TA 1538, TA 98 and TA 100 tester strains, with and without metabolic activation[32]. It was also inactive when tested at 85.4 mg/plate in Streptomyces coelicolor and at 17 mg/plate in Aspergillus nidulans for 8-azaguanine resistance[32].

b. Acrylamide (propenamide) is prepared from acrylonitrile by treatment with sulfuric acid or HCl, or via the catalytic hydration of acrylonitrile with copper[33].

Acrylamide has a wide range of industrial uses primarily in the production of polymers (polyacrylnitriles) which with acrylamide are used as paper-strengthening agents, flocculants, textile-finishing agents, adhesives, and as grouting agents in public works[33,34].

The importance of acrylamide to the polymer industry is evident from the fact that in North America alone, annual production is estimated to be 50 million pounds and to be increasing rapidly[35]. In Japan, the demand for acrylamide in 1974 was estimated to be approximately 57 million pounds[33].

The number of individuals employed in the production of acrylamide or in its use applications are not known.

At present in the U.S., it is recommended that factory workers should not be exposed to more than 0.5 mg/kg acrylamide monomer in an 8-hr shift of a 4-hr week, and that air levels should not exceed 0.3 mg/m^3[24,35].

Aside from presenting a potential hazard to those engaged in its manufacture and handling, an area of potential concern lies in the present practice of permitting some residual contamination of polymers by the acrylamide monomer. Up to 2% residual monomer is considered acceptable for some industrial applications of polyacrylamide, as in the flocculations of ore slimes[36]. The Food and Drug Administration in the U.S. has established a maximum level of 0.05% residual acrylamide monomer for polymers used in paper and paperboard in contact with foodstuffs[35]. Similar levels are considered satisfactory for the polymers used in the clarification of potable water and cane-sugar juice.

Acrylamide release to the environment usually ending up in surface and ground water, occurs at manufacturing sites, soil grouting sites, polymer application sites and in handling. While low-level exposure to acrylamide is considered likely to occur whenever polyacrylamides are utilized there are no data available on release rates into the environment or actual concentration levels.

Acrylamide has been shown to be a contaminant of ground-water used for drinking when it has been used as a grouting agent in public works[37].

Many examples of human acrylamide intoxication with peripheral neuropathy have been reported in individuals engaged both in the manufacture and in the use of acrylamide monomer[35,37-43]. A peripheral neuropathy can also be produced in a variety of experimental animals by repeated dosing with acrylamide[44,45].

No information is currently available on the potential carcinogenic or mutagenic effects of acrylamide.

There is a paucity of metabolic data on acrylamide as well. In limited studies, the molecule has not been found to be significantly metabolized[46,47].

3. <u>Diisocyanates</u>

A number of diisocyanates are used in extensive quantities primarily in the manufacture of polyurethanes. These products are formed by reacting active-hydrogen-containing polyfunctional molecules (e.g., polyether resins) with polyfunctional isocyanates. The product may be a foam, fiber, elastomer, coating or adhesive. The basic requirements for a commercial polyurethan product are: a resin (usually a polyether), isocyanate, catalyst and often a surfactant to facilitate homogenization[48].

a. Toluene-2,4-and 2,6-Diisocyanates (2,4-diisocyanatotoluene; 4-methyl-m-phenylene isocyanate, I) and 1,3-diisocyanate-2-methylbenzene (2,6-diisocyanato-toluene; 2-methyl-m-phenyleneisocyanate, II) particularly as a mixture of 80% 2,4-isomer

$$\text{(I)} \qquad \text{(II)}$$

and 20% 2,6-isomer is the major diisocyanate employed commercially. Toluene diiso-cyanate (80/20 mixture) is produced commercially by nitrating toluene with HNO_3-H_2SO_4 and catalytically reducing the resultant dinitrotoluene (a mixture of at least 76% 2,4-isomer; less than 24% 2,6-isomer and small amounts of 2,3- and 3,4-isomers) to a toluenediamine mixture. This mixture is dissolved in monochlorobenzene or ortho-dichlorobenzene and treated with phosgene to yield the diisocyanates[49].

In 1976, eight companies in the U.S produced 255 million kg of the 80/20 toluene diisocyanate mixture[50]. West European production of toluene diisocyanate amounted to an estimated 280 million kg in 1975 and was produced by the following countries (% of total W. European production): the Federal Republic of Germany (34), France, (25), Italy (12.5), United Kingdom (12.5), Belgium and the Netherlands combined (11) and Spain (5). In 1976, 6 Japanese companies produced a total of 64 million kg of toluene diisocyanate[49]. The U.S. consumption pattern for toluene diisocyanates in 1975 is estimated to have been as follows (%): for the production of polyurethane flexible foams (90-92); polyurethane coatings (4); polyurethane elastomers (3); and other applications (1). In 1975 Western Europe consumption pattern for toluene diisocyanate is estimated to have been as follows: for the production of polyurethane flexible foams (90-05%) and for polyurethane coatings (5-10%). The Japanese consumption pattern of toluene diisocyanate in 1976 is estimated to have been: for the production of polyurethane flexible foams (75%) and for polyurethane coatings and elastomers (25%)[49].

There are no available data on the numbers of workers engaged in the production of the toluene diisocyanates or in their use applications.

OSHA health standards for exposure to air contaminants requires that an employee's exposure to toluene-2,4-diisocyanate does not exceed a ceiling concentration limit of 0.02 ppm (0.14 mg/m^3) at any time[51].

Work environment hygiene standards in terms of 8-hr. time weighted averages for toluene-2,4-diisocyanate are as follows (mg/m^3)[30]: The Federal Republic of Germany (0.11); the German Democratic Repbulic (0.14); Czeckoslovakia (0.07); the MAC[30] of toluene 2,4-diisocyanate in Sweden is 0.07 mg/m^3 and in the USSR it is 0.5 mg/m^3.

Although no data are currently available on the carcinogenicity of 2,4- and 2,6-toluene diisocyanates, there is an on-going study to assess the carcinogenicity of a mixture of 80% 2,4-toluene diisocyante and 20% 2,6-toluene diisocyanate in mice and rats by oral administration[49,52]. Data on the mutagenicity of 2,4- and 2,6-diisocyanates as well as the 80/20 mixture of toluene diisocyanates are lacking.

It should be noted that toluene diisocyanates are highly reactive and react with water to form toluene diamine. 2,4-Toluene has been tested for carcinogenicity and mutagenicity.

2,4-Diaminotoluene induced hepatocellular carcinomas in Wistar rats after oral administration of 0.06 or 1% in the diet for 30-36 weeks[53]. In a separate study in rats fed various combinations of 0.1% 2,4-diaminotoluene and 0.25% ethionine in the diet, there was evidence of an increased incidence of hepatocellular carcinomas over that in rats treated with 2,4-diaminotoluene alone[53].

Local sarcomas were induced in rats injected subcutaneously with 0.5 ml of a 0.4% solution of 2,4-Diaminotoluene in propylene glycol at weekly intervals[54].

2,4-Diaminotoluene is mutagenic in S. typhimurium TA 1538 and TA 98 in the presence of a rat liver post-mitochondrial supernatant fraction from animals pretreated with Aroclor 254[55]. It is a weak mutagen in Drosophila melanogaster inducing sex-linked recessive lethals when fed at a concentration of 15.2 mM[56]. 2,4-Diaminotoluene induces morphological transformation in Syrian golden hamster embryo cells[57].

Polyurethanes of different chemical composition when tested for carcinogenicity in different physical forms (e.g., discs, strips, foams, powders, etc.) by subcutaneous and intraperitoneal implantation and inhalation of dust in rats, mice and hamsters caused a high incidence of fibrosarcoma at the site of implantation[49]. Intraperitoneal implantation also seemed to induce some epithelial tumors as well[49].

b. <u>1,5-Naphthalene diisocyanate</u> (1,5-diisocyanatonaphthalene; isocyanic acid; 1,5-naphthalene ester;) can be produced by the reaction of phosgene and 1,5-naphthalene dihydrochloride or via the nitration of naphthalene followed by reduction with iron to 1,5-naphthalene diamine which is then treated with phosgene[49]. 1,5-Naphthalene diisocyanate has never been produced commercially in the U.S. One company in the Federal Republic of Germany is estimated to produce approximately 1 thousand kg of 1,5-naphthalene diisocyanate per year. In Japan, one company produced about 70 thousand kg in 1976[49].

In Western Europe and Japan, 1,5-naphthalene diisocyanate is used for the production of polyurethane elastomers.

No data on the carcinogenicity and mutagenicity of 1,5-naphthalene diisocyanate <u>per se</u> or on its hydrolytic product 1,5-diaminonaphthalene are available.

c. <u>Miscellaneous Diisocyanates</u>

Other isocyanates which are widely used for urethan production include diphenylmethane-4,4-diisocyanate (MDI); hexamethylenediisocyante (HDI); dicyclohexylmethane-4,4-diisocyanate; and polymethylene polyphenylisocyanate.

In 1976, U.S. production of polymethylene polyphenylisocyanate prepared from 4,4'-methylenedianiline (MDA) is estimated to have been 52 million kg while U.S. production of the refined 4,4'-methylenediphenyl isocyanate from MDA was about 5.5 million kg[58]. The polymethylene polyphenyl isocyanate is used in rigid polyurethane foam, and this usage is continuing to grow at a significant rate. The refined 4,4'-methylenediphenylisocyanate is mostly used in the production of Spandex fibers where the growth rate is also considered high[58].

Dicyclohexylmethane-4,4-diisocyanate is made from the precursor 4,4'-methylenebis(cyclohexylamine) prepared from MDA. This saturated diisocyanate is used in light-stable, high performance polyurethane coatings.

There are no available data on the carcinogenicity or mutagenicity of the above diisocyanates per se. However, it is germane to consider their hydrolytic products (e.g., their precursors) the respective diamines.

4,4'-Methylenedianiline (MDA) was considered earlier in Chapter 18. The available experimental evidence in the rat, the only species tested, according to IARC, does not permit a definitive conclusion regarding the caricnogenicity of MDA in this species[58].

Data are also unavailable on the carcinogenicity and/or mutagenicity of hexamethylenediamine and 4,4'-methylenedicyclohexylamine.

4. Chlorophenols

Chlorinated phenols are widely employed as wood preservatives, antiseptics, disinfectants, fungicides, herbicides, insecticides, bactericides, molluscacides, slimicides as well as intermediates in the production of industrial chemicals and phenoxy acids such as 2,4-D and 2,4,5-T.

A. <u>Pentachlorophenol</u>

Pentachlorophenol (PCP) and sodium pentachlorophenate (PCP-Na) are the most widely employed of the chlorophenols with a broad spectrum of applications in agriculture and industry. Their utility as fungicides and bactericides includes the processing of cellulosic products, adhesives, starches, proteins, leather, oils, paints, rubber, textiles, and use in food processing and paper pulp plants to control mold and slime and in construction and lumber industries to control mold, termite infestation, and wood-boring insects.

PCP has been manufactured for over 40 years and is produced commercially by the chlorination of phenol and chlorophenols or by alkaline hydrolysis of hexachlorobenzene[59]. Approximately 50 million pounds of PCP were produced in the U.S. in 1972 and it is believed that production has increased in the interveing years[60]. There are about 800 companies nationwide that produce compounds that use PCP in some form[60]. PCP production is expected to increase as the use of alternatives such as creosote and toxic inorganic salts decreases.

No information is available on the numbers of workers involved in the production of pentachlorophenol or in its use applications.

PCP is used primarily (80% of total production) to protect structural wood and utility poles. Treated wood contains 0.3 to 0.5 pounds of PCP per cubic foot or approximately 8 pounds per utility pole. Losses of PCP to the environment (e.g., sludge waste entering water systems) have occurred in this process.

As a fungicide, PCP has been used in the processing of oils, leathers, paints, glues, and textiles, and has been incorporated into hair shampoos. PCP may also come directly into contact with food since it is a wood preservative in crates used for agricultural materials as a fungicide in beer vats, and in gaskets used to seal food containers.

Residues of pentachlorophenol have been found in municipal water supplies and wells with toxic symptoms described in 4 families after drinking and bathing in water obtained from a well containing 12.5 ppm PCP. It has not been determined whether or at what level trace impurities of PCP were also present in the above samples[61].

Investigations have shown that PCP is a threat to the aquatic environment[62]. Concentrations as great as 10 ppm have been measured in tributary waters of the Deleware River[63], and large fish kills have been observed after PCP was applied to rice fields in South America[64]. Many species of fish have been killed by exposure to PCP at concentrations of 0.2-0.6 ppm[65]. A summary of information on PCP demonstrates that it is toxic to a wide variety of aquatic biota[62,66].

Technical grade pentachlorophenol produced in the United States contains as impurities up to 13% of other chlorophenols, of which the isomeric tetrachlorophenols constitute the principle portion. PCP can be manufactured by alkaline hydrolysis of hexachlorobenzene at elevated temperatures and pressures and hence three factors, e.g., alkalinity, elevated temperature, and pressure, might be responsible for the formation of chlorinated dioxins from chlorinated phenols (Figure 1).

Plimmer et al[67] described the isolation and mass spectrometric identification of hepta- and octachlorinated dibenzo-p-dioxins and dibenzofurans in technical pentachlorophenol. In all, 17 ppm of the hexa-, 108 ppm of the hepta-, and 144 ppm of the octadioxins were found. Although octa- and hepta- chlorodibenzofurans were identified, their quantities were not determined.

While PCP samples examined in the above study were produced before 1960, it was pointed out that samples from a 1970 batch in another study contained dioxins (e.g., 0.5 to 37 ppm of the hexadioxin and 90 to 135 ppm of the heptadioxin).

Blaser et al[68] recently reported hexachloro- and octachlorodibenzo-p-dioxin at levels ranging from 9-18 ppm and 575-1980 ppm in 4 commercial samples of domestically manufactured pentachlorophenol.

Jensen and Renberg[69] reported the presence of chlorinated dioxins as well as predioxins (chlorinated hydroxydiphenyl ethers) in penta-chlorophenol formulations most often used in Sweden.

Basic structure of dibenzo-dioxin (dioxin). Chlorine may be attached, in various combinations, at the 1,2, 3,4,5,6,7,8 or 9 positions.

Basic structure of Dibenzofuran (furan)

Tetrachlorobenzene $\xrightarrow[\text{alkalinity}]{\text{high temperature high pressure}}$ Trichlorophenol $\diagup$ several additional steps $\rightarrow$ 2,4,5-T
$\diagdown$ reacts with itself $\rightarrow$ Tetra-dioxin

Hexachlorobenzene $\xrightarrow[\text{alkalinity}]{\text{high temperature high pressure}}$ pentachlorophenol $\diagup$ reacts with itself $\rightarrow$ Octa-dioxin
$\diagdown$ +hexachlorobenzene $\rightarrow$ Furans

Chlorine+Phenols $\xrightarrow{\text{high temperature}}$ PCP+dioxin contaminants

2,3,7,8-Tetrachlorodibenzo-p-dioxin (TCDD) can be formed as a by-product during the synthesis of 2,4,5-trichlorophenol if the reaction is allowed to heat to temperatures higher than the normal 180°C, ethylene glycol is distilled off or polymerized, and an exothermic reaction starts at about 230°C and proceeds rapidly and uncontrollably to about 410°C[93] (Figure 1). Since 2,4,5-trichlorophenol is the precursor of 2,4,5-T, another potential source of TCDD is during the manufacture of 2,4,5-T and its esters[91,93,94].

Hexachlorophene produced from 2,4,5-trichlorophenol has been reported to contain less than 15 µg/kg TCDD[93]. Although no TCDD has been found in commercial pesticidal samples of 2,4,5-trichlorophenol, hexa- and hepatchlorodibenzodioxins have been detected in these samples[95].

2,4,5-Trichlorophenol is tumorigenic to male mice when fed at a dose of 600 ppm in the diet[96].

2,4,5-Trichlorophenol has been identified as one of the urinary constituents of rats treated with alpha-hexachlorocyclohexane (alpha-HCH)[97], lindane[98], 4-(2,4,5-trichlorophenoxy)-butyric acid[99], 2,4,5-T[100] and hexachlorobenzene[101].

References

1. Oesch, F., Mammalian epoxide hydrases: Inducible enzymes catalyzing the inactivation of carcinogenic and cytotoxic metabolites derived from aromatic and olefinic compounds, Xenobiotica, 3 (1972) 305-340
2. Glatt, H. R., Oesch, F., Frigerio, A., and Garattini, S., Epoxides metabolically produced from some known carcinogens and from some clinically used drugs. I. Differences in mutagenicity, Int. J. Cancer, 16 (1975) 787-797
3. Daly, J. W., Jerina, D. M., and Witkop, Arene oxides and the NIH shift: The metabolism, toxicity, and carcinogenicity of aromatic compounds, Experientia 28 (1972) 1129-1264
4. Jerina, D. M., and Daly, Arene Oxides: A new aspect of drug merabolism, Science, 185 (1974) 573-582
5. Bouse, G., and Henschler, Chemical reactivity, biotransformation and toxicity of polychlorinated aliphatic compounds, CRC Crit. Revs. Toxicology, 4 (1976) 395-409
6. Bartsch, H., Malaveille, C., and Montesano, R., Human, rat and mouse liver-mediated mutagenicity of vinyl chloride in S. typhimurium strains, Int. J. Cancer, 15 (1975) 429-437
7. Bartsch, H., and Montesano, R., Mutagenic and carcinogenic effects of vinyl chloride, Mutation Res., 32 (1975) 93-114
8. Bartsch, H., Malaveille, C., Montesano, R., and Tomatis, L., Tissue-mediated mutagenicity of vinylidene chloride and 2-chlorobutadiene in S. typhimurium, Nature, 255 (1975) 641-643
9. Bartsch, H., Malaveille, C., Barbin, A., Planche, G., and Montesano, R., Alkylating and mutagenic metabolites of halogenated olefins produced by human and animal tissues, Proc. 67th Ann. Meeting Amer. Assoc. Cancer Res., Toronto, May 4-8, Proc. Am. Assoc. Cancer Res. (1976) p. 17
10. Van Duuren, B. L., Goldschmidt, B. M., and Siedman, I., Carcinogenic activity of di- and tri-functional α-chloroethers and of 1,4-dichlorobutene-2- in ICR/HA Swiss mice, Cancer Res., 35 (1975) 2553-2557
11. Hathway, D. E., Comparative mammalian metabolism of vinylidene and vinyl chloride in relation to outstanding oncogenic potential. NIEHS Conference on Comparative Metabolism and Toxicity of Vinyl Chloride Related Compounds, Bethesda, MD, May 2-4 (1977)
12. Greim, H., Bonse, G., Radwan, Z., Reichert, D., and Henschler, D., Mutagenic in vitro and potential carcinogenicity of chlorinated ethylenes as a function of metabolic oxirane formation, Biochem. Pharmacol., 24 (1975) 2013
13. Bonse, G., Urban, T., Reichert, D., and Henschler, D., Chemical reactivity metabolic oxirane formation and biological reactivity of chlorinated ethylenes in the isolated perfused rat liver preparation, Biochem. Pharmacol., 24 (1975) 1829-1834
14. Uehleke, H., Tabarelli-Poplawski, S., Bonse, G., and Henschler, D., Spectral evidence for 2,2,3-trichlorooxirane formation during microsomal trichloroethylene oxidation, Arch. Toxicol., 37 (1977) 95-105
15. LeBlanc, R. B., Flame resistant fibers, Fiber Producers April (1977) pp. 10,12,64
16. Sassu, G. M., Zilio-Grandi, F., and Conte, A., Gas-chromatographic determination of impurities in vinyl chloride, J. Chromatog., 34 (1968) 394-398
17. Kurosaki, M., Taima, S., Hatta, T., and Nakamura, A., Identification of high-boiling materials as by-products in vinyl chloride manufacture, Kogyo Kagaku Zasshi, 71 (1968) 488-491

18. Van Duuren, B. L., Chemical structure, reactivity and carcinogenicity of halohydrocarbons, Environ. Hlth. Persp., 21 (1977) 17-23
19. Bartsch, H., Malaveille, C., Barbin, A., Planche, G., and Montesano, R., Alkylating and mutagenic metabolites of halogenated olefins produced by human and animal tissues, Proc. Am. Assoc. Cancer Res., 17 (1976) 17
20. Barbin, A., Planche, G., Croisy, A., Malaveille, C., and Bartsch, H., Detection of electrophilic metabolites of halogenated olefins with 4-(4-nitrobenzyl)pyridine (NBP) or with Salmonella typhimurium, Mutation Res., 53 (1978) 150
21. Barbin, A., Bresil, H., Croisy, A., Jacquignon, P., Malaveille, C., Montesano, R., and Bartsch, H., Liver-microsome-mediated formation of alkylating agents from vinyl chloride and vinyl bromide, Biochem. Biophys. Res. Commun., 67 (1975) 596-603
22. IARC, Monographs on the Evaluation of Carcinogenic Risk of Chemicals to Man, Vol. 17, International Agency for Research on Cancer, Lyon, in press
23. U.S. International Trade Commission, Synthetic Organic Chemicals, U.S. Production and Sales, 1976, USITC Publ. 833, Government Printing Office, Washington, DC (1977) pp. 183, 187,301 and 327
24. ACGIH, Threshold Limit Values for Chemical Substances in Workroom Air Adopted by ACGIH, American Conference of Governmental Industrial Hygienists, Cincinatti, Ohio (1976) p. 30
25. Casto, B. C., Meyers, J., and DiPaolo, J. A., Assay of industrial chemicals in Syrian hamster cells for enhancement of viral transformation, Proc. Am. Assoc. Cancer Res., 18 (1977) 155
26. Bailey, R., "Butadiene", In: "Vinyl and Diene Monomers", part 2, (ed) Leonard, E. C., Interscience, New York (1971) pp. 757-995
27. Anon, United States styrene-butadiene rubber materials use, Chem. Age., 4 (1977) p. 11
28. Anon, U.S. Olefins end-uses, Chem. Mkt. Reptr. 3 (28) (1977) p. 30,32
29. Anon, Butadiene capacity in Europe, Chem. Eng. News, May 16 (1977) p. 18
30. Winell, M., An international comparison of hygienic standards for chemicals in the work environment, Ambio, 4 (1) (1975) 34-36
31. deMeester, C., Poncelet, F., Roberfroid, M., and Mercier, M., Mutagenicity of butadiene and butadiene monoxide, Biochem. Biophys. Res. Communs., 80 (1978) 298-305
32. Bignami, M., Cardamone, G., Carere, A., Comba, P., Dogliotti, E., Morpurgo, G., and Ortali, V. A., Mutagenicity of chemicals of industrial and agricultural relevance in Salmonella, Streptomyces and Aspergillus, Cancer Res. (1978) in press
33. Matsuda, F., Acrylamide production simplified, Chem. Tech., May (1977) 306-308
34. Schildknecht, C. E., "Vinyl and Related Polymers", Wiley, New York (1952) 314-322
35. Spencer, P. S., and Schaumburg, H. H., Nervous system degeneration produced by acrylamide monomer, Environ. Hlth. Persp., 11 (1975) 129-133
36. Bikales, N. M., Preparation of acrylamide polymers, In: Water-Soluble Polymers (Polymer Science and Technology, Vol. 2), Bikales, N. M. (ed), Plenum Press, New York (1973) p. 213
37. Igisu, K., Goto, I., Kawamura, Y., et al., Acrylamide encephaloneuropathy due to well water pollution, J. Neurol. Neurosurg. Psychiatry, 38 (1975) 581-584
38. Satoyoshi, E., Kinoshita, M., Yano, H., et al., Three cases of peripheral polyneuropathy due to acrylamide, Clin. Neurology (Tokyo), 11 (1971) 667-672

39. Morviller, P., An industrial poison not well known in France: acrylamide, Arch. Mal. Prof. Med. Trav. Secur. Soc., 30 (1969) 527
40. Graveleau, J., Loirat, P., and Nusinovici, V., Polynevrite par l'acrylamide, Rev. Neurol., 123 (1970) 62
41. Garland, T. O., and Patterson, M. W. H., Six cases of acrylamide poisoning, Brit. Med. J., 4 (1967) 134
42. Fujita, A., et al., Clinical observations on acrylamide poisoning, Nippon Iji Shimpo, 1869 (1960) 27
43. Takahashi, M., Ohara, T., and Hashimoto, K., Electrophysiological study of nerve injuries in workers handling acrylamide, Int. Arch. Arbeitsmed., 28 (1971) 1-11
44. Edwards, P. M., Neurotoxicity of acrylamide and its analogs and effects of these analogs and other agents on acrylamide neuropathy, Brit. J. Ind. Med., 32 (1975) 31-38
45. Hopkins, A., The effect of acrylamide on the peripheral nervous system of the baboon, J. Neurol. Neurosurg. Psych., 33 (1970) 805-816
46. Bikales, N. M., Acrylamide and related amides, In: Vinyl and Diene Monomers (High Polymer Series, Vol. 24, part 1) E. C. Leonard (ed), Interscience, New York (1970) p. 81
47. Hashimoto, K., and Aldridge, W. N., Biochemical studies on acrylamide: a neurotoxic agent, Biochem. Pharmacol., 19 (1970) 2591
48. Lowenheim, F. A., and Moran, M. K., Faith, Keyes and Clark's Industrial Chemicals, 4th ed., Wiley-Interscience, Wiley, New York (1975) 831-835
49. IARC, Monographs on the Evaluation of Carcinogenic Risk of Chemicals to Man, Vol. 17, International Agency for Research on Cancer, Lyon (1978) in press
50. U.S. International Trade Commission, Synthetic Organic Chemicals, U.S. Production and Sales, 1976, USITC Publ. 833, Government Printing Office, Washington, DC (1977) pp. 37 and 59
51. U.S. Occupational Safety and Health Administration, Occupational Safety and Health Standards Subpart Z-Toxic and Hazardous Substances, Code of Federal Regulations, Title 29, Chapter XVII, Section 1910.1000, Bureau of National Affairs, Washington, DC (1976) p. 31:8303
52. IARC, Toxicology Information Programme, 5 (1976) 17
53. Ito, N., Hiasa, Y., Konishi, Y., and Marugami, M., The development of carcinoma in liver of rats treated with m-toluylenediamine and the synergistic and antagonistic effects with other chemicals, Cancer Res., 29 (1969) 1137-1145
54. Umeda, M., Production of rat sarcoma by injections of propylene glycol solution of m-toluylenediamine, Gann, 46 (1955) 597-603
55. Ames, B. N., Kammen, H. O., Yamasaki, E., Hair dyes are mutagenic: Identification of a variety of mutagenic ingredients, Proc. Natl. Acad. Sci. (US) 72 (1975) 2423-2427
56. Blijleven, W. G. H., Mutagenicity of four hair dyes in Drosophila melanogaster Mutation Res., 48 (1977) 181-186
57. Pienta, R. J., Poiley, J. A., and Lebherz, W. B., III, Morphological transformation of early passage of golden Syrian hamster embryo cells derived from cryopreserved primary cultures as a reliable in vitro bioassay for identifying diverse carcinogens, Int. J. Cancer, 19 (1977) 642-655
58. IARC, Monographs on the Evaluation of Carcinogenic Risk of Chemicals to Man, Vol. 4, International Agency for Research on Cancer, Lyon (1974) 79-85
59. Bevenue, A., and Kawano, Y., Pesticides, pesticide residues, tolerances and the law (USA), Residue Revs., 35 (1971) 102-149

60. Hatfield, M., Letter to the Environmental Protection Agency concerning industrial wastes containing dioxins, Chem. Reg. Reptr., 1 (38) (1977) 1343-1345
61. Uede, K., Nagai, M., and Osafune, M., Contamination of drinking water with pentachlorophenol-analysis and removal, Osaka Shiritsu Eisei Kenkyli-sho Kenykyo Hokuku, 7 (1962) 19-23
62. Tagatz, M. E., Ivey, J. M., Moore, J. C., and Tobia, M., Effects of pentachlorophenol on the development of estuarine communities, J. Toxicol. Env. Hlth., 3 (1977) 501-506
63. Fountaine, J. E., Joshipura, P. B., and Keliher, P. N., Some observations regarding pentachlorophenol levels in Haverford Township, Pennsylvania, Water Res., 10 (1976) 185-188
64. Vermeer, K., Risebrough, R. W., Spaans, A. L., and Reynolds, L. M., Pesticide effects on fishes and birds in rice fields of Surinam, South America, Environ. Pollut., 7 (1974) 217-236
65. Goodnight, C. J., Toxicity of sodium pentachlorophenate and pentachlorophenol to fish, Ind. Eng. Chem., 34 (1942) 868-872
66. Becker, C. D., and Thatcher, T. O., Toxicity of power plant chemicals to aquatic life, Report. No. WASH-1249, U.S. Atomic Energy Commission, Battelle Pacific Northwest Laboratories, Richland Washington
67. Plimmer, J. R., Ruth, J. M., and Woolson, E. A., Mass spectrometric identification of the hepta- and octachlorinated dibenzo-p-dioxins and dibenzofurans in technical pentachlorophenol, J. Agr. Food Chem., 21 (1973) 90-96
68. Blaser, W. W., Bredeweg, R. A., Shadoet, L. A., and Stehl, R. H., Determination of chlorinated dibenzo-p-dioxins in pentachlorophenol by gas chromatography-mass spectrometry, Anal. Chem., 48 (1976) 984-986
69. Jensen, S., and Renberg, L., Contaminants in pentachlorophenol, chlorinated dioxins and predioxins, Ambio, 1 (1972) 62-65
70. Firestone, D., Ress, J., Brown, M. L., Barron, R. P., and D'Amico, J. N., Determination of polychlorodibenzo-p-dioxins and related compounds in commercial chlorophenols, J. Assoc. Off. Anal. Chem., 55 (1972) 85-92
71. Schwetz, B. A., Keeler, P. A., and Gehring, P. T., The effect of purified and commercial grade pentachlorophenol on rat embryonal and fetal development, Toxicol. Appl. Pharmacol., 28 (1974) 151-161
72. Villaneuva, E. C., Jennings, R. W., Burse, V. W., and Kimbrough, R. D., Evidence of chlorodibenzo-p-dioxin and chlorodibenzofuran in hexachlorobenzene, J. Agr. Food Chem., 22 (1974) 916-917
73. Norstrom, A., Anderson, K., and Rappe, C., Formation of chlorodibenzofurans by irradiation of chlorinated diphenylethers, Chemosphere, 1 (1976) 21-24
74. Rappe, C., and Nilsson, C. A., An artifact in the gas chromtographic determination of impurities in pentachlorophenol, J. Chromatog., 67 (1972) 247
75. Nilsson, C. A., and Renberg, L., Further studies on impurities in chlorophenols, J. Chromatog., 89 (1974) 325
76. Bevenue, A., and Beckman, H., Pentachlorophenol; A discussion of its properties and its occurrence as a residue in human and animal tissues, Residue Revs., 19 (1967) 83-134
77. Innes, J. R. M., Ulland, B. M., Valerio, M. G., Petrucelli, L., Fishbein, L., Hart, E. R., Pallotta, A. J., Bates, R. R., Falk, H. L., Gart, J. J., Klein, M., Mitchell, I., and Peters, J., Bioassay of pesticides and industrial chemicals for tumorigenicity in mice: A preliminary note, J. Natl. Cancer Inst., 42 (1969) 1101

78. Vogel, E., and Chandler, J. L. R., Mutagenicity testig of cyclamate and some pesticides in Drosophila melanogaster, Experientia, 30 (1974) 621
79. Fahrig, R., Comparative mutagenicity studies with pesticides, In: "Chemical Carcinogenesis Assays", Vol. 10, (eds) Montesano, R., and Tomatis, L., International Agency for Research on Cancer, Lyon (1974) 161-168
80. Buselmaier, W., Röhrborn, G., and Propping, P., Mutagenitatsuntersuchungen mit pestizidenim host-meidated assay und mit dem dominanten letaltest an der maus, Biol. Zentr., 91 (1972) 311-325
81. Jacobsson, I., and Yllner, S., Metabolism of (^{14}C)-pentachlorophenol in the mouse, Acta. Pharmacol., 29 (1971) 513
82. Ahlborg, U. G., Lindgren, J. E., and Mercier, M., Metabolism of pentachlorophenol, Arch. Toxicol., 32 (1974) 271
83. Arrhenius, E., Renberg, L., and Johansson, L., Subcellular distribution, a factor in risk evaluation of pentachlorophenol, Chem.-Biol. Interactions, 18 (1977) 23-34
84. Arrhenius, E., Renberg, L., Zetterquist, M. A., and Johansson, L., Disturbance of microsomal detoxication mechanisms in liver by chlorophenol pesticides, Chem.-Biol. Interact., 18 (1977) 35
85. Arrhenius, E., Effects of various in vitro conditions on hepatic microsomal N- and C-oxygenation of aromatic amines, Chem.-Biol. Interact., 1 (1969/70) 361
86. Beckett, A. H., and Belanger, P. M., The mechanism of metabolic N-oxydation of phentermin and chlorphentermine to the hydroxylamino- and nitroso compounds, J. Pharm. Pharmcol., 26 (1974) 1205
87. Arrhenius, E., Some aspects of microsomal N- and C-oxygenation of aromatic amines, Xenobiotica, 1 (1971) 478
88. Arrhenius, E., Comparative metabolism of aromatic amines, In: Montesano, R., and Tomatis, L., (eds) Chemical Carcinogenesis Essays, IARC Scientif. Publ. no. 10, Lyon (1974) pp. 25-37
89. Arrhenius, E., Effects in vitro of 2-aminofluorene or electrophilic agents on hepatic microsomal N- and C-oxygenation of aromatic amines, Chem.-Biol. Interact., 1 (1969/70) 381
90. Arrhenius, E., Effects on hepatic microsomal N- and C-oxygenation of aromatic amines by in vivo corticosteroid or aminofluorene treatment, diet or stress, Cancer Res., 28 (1968) 264
91. Milnes, M. H., Formation of 2,3,7,8-tetrachlorodibenzodioxin by thermal decomposition of sodium 2,4,5-trichlorophenate, Nature, 232 (1971) 395-396
92. U.S. Tariff Commission, Synthetic Organic Chemicals: Production and Sales, T.C. Publ. No. 479, Washington, DC (1969)
93. IARC, Monographs on the Evaluation of the Carcinogenic Risk of Chemicals to Man, Vol. 15, International Agency for Research on Cancer, Lyon (1977) 273-299
94. May, G., Chloracne from the accidental production of tetrachlorodibenzodioxin, Brit. J. Ind. Med., 30, (1973) 276-283
95. Woolson, E. A., Thomas, R. F., and Ensor, P. D. J., Survey of polychlorodibenzo-p-dioxin content in selected pesticides, J. Agr. Food Chem., 20 (1972) 351-358
96. Goto, M., Hattori, M., and Miyagawa, T., Toxicity of alpha-benzenehexachloride in mice, Chemosphere, 1 (1972) 153-154
97. Koransky, W., Munch, G., Noack, G., Portig, J., Sodomann, S., and Wirsching, M., Biodegradation of alpha-hexachlorocyclohexane I. Characterization of the major urinary metabolites, Nauny Schmiedebergs Arch. Pharmacol., 288 (1975) 65-78

98. Chadwick, R. W., and Freal, J. J., Comparative acceleration of lindane metabolism to chlorophenols by treatment of rats with lindane or with DDT and lindane, Food Cosmet. Toxicol., 10 (1972) 789-795
99. Boehme, C., and Grunow, W., Metabolism of 4-(2,4,5-trichlorophenoxy)-butyric acid in rats, Arch. Toxicol., 32 (1974) 227-231
100. Grunow, W., and Boehme, C., Metabolism of 2,4,5-T and 2,5-D in rats and mice, Arch. Toxicol., 32 (1974) 217-225
101. Renner, G., and Schuster, K. P., 2,4,5-Trichlorophenol, a new urinary metabolite of hexachlorobenzene, Toxicol. Appl. Pharmacol., 39 (1977) 355-356

SUMMARY

It is generally acknowledged that in the main, the spectrum of chemicals that has been introduced in the 20th century, has brought enormous benefits to man in terms of the enhancement of the quality and length of life achieved by the eradication, or control of diseases, providing a more abundant and nutritious food supply by the eradication or control of pests, and the introduction of a plethora of goods and services.

However, we must also acknowledge that living in an environment into which more and more chemicals have been introduced can, in some measure, be attendant with risks because of the possible hazards to health and safety of a number of these same chemicals.

The universe of potentially hazardous chemicals is quite broad and includes classes of agents drawn from use categories such as industrial chemicals, pesticides, food additives, and drugs per se, as well as their trace synthetic impurities and/or degradation products. Additionally, a broad spectrum of hazardous naturally occurring toxicants is known to exist. The increasing prevalence of toxic substances in the environment has become a matter of public interest as a complex potential public health problem. Many have cited the risk of cancer in the human population from the large and ever increasing multitude of chemicals entering the environment and the serious and perhaps virtually insurmountable problems caused thereby. It has been noted that because of the typical latency period of 15 to 40 years for cancer, it can be assumed that much of the cancer from recent industrial development is not yet observable. It has also been suggested that as this "lag period" for chemical carcinogenesis is almost over, a steep increase in the human cancer rate from suspect chemicals may soon occur.

Concomittant with the potential cancer risk of environmental agents, is the growing concern over the possibility that future generations may suffer from genetic damage by mutation-inducing chemical substances to which large segments of the population may unwittingly be exposed. Since mutations in reproductive cells may be carried in the recessive state through several generations, effects of mutagenic agents are likely to be obscured for decades. While specific agents may never be associated with specific genetic damage, it is widely held that increasing rates of mutation may have devastating consequences for the health of future generations.

Chemical carcinogens and mutagens represent a spectrum of agents varying in activity by a factor of at least 10^7 with strikingly different biological activities, ranging from highly reactive molecules that can alkylate macromolecules and cause mutations in many organisms to compounds that are hormonally active and have neither of these actions.

We do not know with precision what percentage of existing chemicals as well as those which enter the environment annually may be hazardous, primarily in terms of their potential carcinogenicity and mutagenicity. Although the etiology of human neoplasia, with rare exceptions, is unknown it is held by some that a large number of cancers can be directly or indirectly attributed to environmental factors. At present, the most influential single carcinogenic exposure appears still to be cigarette smoking. Modes of exposure to chemicals include diets, personal habits such as smoking, and external environmental air and water pollution.

It should also be noted that while the numbers of individuals directly involved in the preparation of these chemicals and their byproducts (e.g., plastics, polymers, etc.) are relatively small in number compared to many industrial segments and processes, the degree of exposure to potentially hazardous (carcinogenic and mutagenic) substances can be very substantial indeed. Substantially greater numbers of individuals may be indirectly exposed to these potential carcinogens and mutagens via (1) use applications which may contain entrained materials, (2) inhalation, ingestion or absorption of these agents via air, water and food sources resulting from escape into the atmosphere, leaching into water and food, etc.

It is estimated that of all the chemicals on the market, a relatively small proportion (approximately 6,000) have been tested to determine their cancer-causing potential.

Despite the converging tendency of chemicals to be both carcinogenic and mutagenic it cannot be known at present whether all carcinogens will be found to be mutagens and all mutagens, carcinogens, e.g., for classes of compounds such as base analogs which do not act via electrophilic intermediates and steroidal sex hormones which are carcinogenic in animals and not yet been shown to be mutagens, different cancer-inducing mechanisms may be implied.

In the preceding chapters, a total of 176 illustrative industrial organic chemicals have been considered and collated in terms of 21 major groupings and 38 structural sub-groupings. These include: 1) alkylating agents, 2) acylating agents, 3) peroxides, 4) halogenated unsaturated hydrocarbons, 5) halogenated saturated hydrocarbons, 6) halogenated aryl, 7) halogenated polyaromatics, 8) hydrazines, hydroxylamines, carbamtes, acetamides, thioacetamides and thioureas, 9) nitrosamines, 10) aromatic amines and azo dyes, 11) heterocyclic amines and nitrofurans, 12) nitroaromatics and nitroalkanes, 13) unsaturated nitriles and azides, 14) aromatic hydrocarbons, 15) anthraquinones, quinones and polyhydric phenols, 17) phthalic- and adipic acid esters, 18) cyclic ethers, 19) phosphoramides and phosphonium salts and 20) miscellaneous potential carcinogens and mutagens comprising of vinylic, allylic and chlorophenol derivatives.

The industrial chemicals considered in this report were limited to organic compounds and selected on factors including: their reported carcinogenicity and/or mutagenicity, their chemical structures and relationships to known chemical carcinogens and mutagens, their volume or use characteristics and suggested or estimated potential populations at risk.

Additionally germane aspects (where known) of their synthesis (primarily in terms of the nature of the possible hazardous trace impurities), production volumes and use patterns (U.S., Western Europe and Japan), chemical and biological reactivity and stability, environmental occurrence, national permissible worker exposure levels (TLV's and MAC's) have been included for cohesiveness of treatment and hopefully for enhanced utility for a broader spectrum of scientists and public health officials.

The carcinogenicity information presented focused on the dose, route of administration, strain and species and tumor type while the mutagenicity data centered on test systems primarily consisting of bacteria (Salmonella typhimurium and E. coli), yeast, Neurospora, Drosophila, mammalian and human cells in culture, dominant lethal and host-mediated assays.

In a number of cases there are no reports of a compound having been testing for carcinogenicity or mutagenicity or they are currently on test. In some cases, conflicting carcinogenicity and/or mutagenicity results for the same compound were reported in the literature and this information has been cited in this volume.

A major factor in the consideration of the validity of carcinogenicity and mutagenicity testing is the purity of the test substance which is regrettably infrequently reported. The possibility cannot be excluded that the effect observed in a number of instances could be caused by trace contaminants of the test chemical which were engendered during synthesis, formulation or storage. For example, trichloroethylene is known to contain amine compounds as stabilizers. The possibility exists that these stabilizers could contribute to the weak mutagenic activity of the compound observed in bacterial systems in the presence of mammalian metabolizing systems.

There are a number of factors that must be considered in a quantitative correlation of in vitro and in vivo tests by various chemicals. These factors include volatility, stability, capacities of compounds for differentially permeating cell membranes, hydrolysis and detoxification rates, detoxification by non-target metabolites in body fluids and cytoplasm and residence times of carcinogens in tissues caused by a combination of these factors.

There is a very considerable body of evidence that suggests that chemicals which are carcinogenic in laboratory animals are carcinogenic in human populations, if appropriate studies could be performed. In addition, there is tentative evidence that there may be a quantitative relationship between the amount of a chemical that is carcinogenic in man. We must, however, recognize the possible competing and confounding effects in chemical carcinogenesis and mutagenesis. Exposure of an individual to these chemical agents can be modified by genetic determinants such as enzyme levels, immune defense capability, and hormonal imbalance, cultural influences such as smoking, diet and other environmental factors and health status.

In animal bioassay, species and strain differences illustrate the major role of genetic background in determining susceptibility to carcinogenesis. One major source of variation is the known difference in activity of the drug-metabolizing enzymes among different species which can account for differences of carcinogenic potency, if the proximate carcinogen is metabolized to an active product, or if an active carcinogen has to reach a remote site of action before it is transformed to an inert metabolite.

There may in addition to species differences in the metabolism of compounds to active carcinogens, intrinsic differences to sensitivity of tissues to the carcinogenic process.

The influence of genetic constitution is revealed in marked strain differences within a single species. The pattern of genetic variation in susceptibility to cancer in general, susceptibility to cancer of particular tissues, and susceptibility to individual chemical carcinogens, may well also apply in the human.

In an evaluation of the effects of both carcinogenic and mutagenic chemicals, it is essential that four phases be distinguished, viz., (a) primary identification or detection of carcinogenic and/or mutagenic activity; (b) verification; (c) quantification and (d) extrapolation to man. It is argued that the extent and rigor of testing should be related to the extent to which the reproductive section of the human population is exposed to the agent in question.

A strong possibility exists that chemicals found mutagenic in non-mammalian test systems could have genetic consequences for humans since the genetic material and hereditary processes are similar in all living organisms.

Although mutation of somatic cells can have potentially serious health consequences, a more insidious and potentially more dangerous long-term problem is the effect of mutation in germ cells. Whether or not a mutagenic substance actually causes mutation in human germ cells depends on a number of circumstances including the body's usual manner of assimilating the substance (i.e., route of administration, blood and tissue barriers, site of metabolism, etc.); the degree of exposure to the mutagen both in terms of total amount delivered and amount delivered per unit time; the possibility of metabolic alterations of benign substances to intermediates with mutagenic properties; and the synergistic effects of combinations of substances.

Although variability in test systems, in experimental methods and in analysis may yield ambiguous information, even clearly positive results are not easily extrapolated to man.

There are a number of severe ambiguities in terms of statistics and biology. We need to know more about false positives and false negative rates. A major toxicological problem is to effect species conversion of data from animal to human, to effect the transition from high dose to low dose in the same species.

The validity of extrapolating results of tests made in rodents with high doses of chemicals to the environment and workplace exposure levels of humans is disputed and may never be completely resolved.

To obtain statistically significant data with a reasonable number of rodents, cancer incidences of 5 to 10% must be obtained, necessitating the administration of large doses. Hence, assessment of risk in man requires extrapolation of these data to a human population approximately 10^6 times larger where the exposure levels are likely to be in general approximately 10^4 times lower, and with a life span of about 30 times longer.

As has been repeatedly stressed while the animal population under test is both genetically and environmentally homogeneous, human populations differ in geography, age, diet, occupation, health, life-style and most importantly, in genetic constitution which may place certain sub-populations at a particularly high risk.

One of the major criticisms of the predictive value of experimental results be it short-term or chronic bioassays, stems from the possible overestimation of their significance or of their misuse. There must be recognition of what animal cancer tests and mutagenicity short-term assays are capable of doing and what they are not capable of doing in terms of assessing human hazard and the essential need for additional research to delineate mechanisms of carcinogenicity toward the better evaluation of human hazard. The existing national legislations on occupational carcinogens for example, indicate that there is no concensus on the criteria by which chemicals are considered sufficiently hazardous to man to be included in the legislation.

Most would hold to the concept that reason must prevail in the weighing of the benefits of chemicals versus the societal risk. A major goal has been to develop the data base which will better permit risk estimations for these are undoubtedly vital for rationale societal regulation of chemicals which both pose human benefits as well as human risks. Hence, in order to develop priorities for testing of the major potentially hazardous chemicals already in the environment and those which may be introduced it would appear useful to examine the collation and structural categorization of the agents discussed in the preceding chapters.

This assessment of potentially hazardous industrial chemicals and their derivatives cannot be complete, but it will attempt to focus on the possible correlative features (e.g., structural) of a number of significant industrial chemicals that have been reported to be carcinogenic and/or mutagenic and hence enable a more facile prediction of potential chemical hazards in the future. It is hoped that the enclosed tabular compilation will also provide a more rational basis for the prioritization of those compounds shown to be mutagenic in individual and/or tier systems and hence are potential candidates for long-term animal studies to more definitively ascertain their carcinogenicity.

The largest number of industrial agents that have been reported to be carcinogenic and/or mutagenic are alkylating and acylating agents. The major industrial class of demonstrated carcinogenicity and/or mutagenicity are the halogenated hydrocarbons comprising saturated and unsaturated derivatives including: alkanes, alkanols, ethers, vinyl, vinylidene and allylic analogs, and alkyl, aryl and polyaromatic derivatives.

While the observation of metabolic epoxidation of a chemical agent in the environment would prompt us to test these compounds for biological effects such as carcinogenicity and mutagenicity, we are cautioned to avoid indiscriminantly or unequivocally extrapolating the established adverse biological effects of the more thoroughly investigated epoxides to all epoxides.

Although the industrial organic potential chemical carcinogens and mutagens were considered as discrete entities, it is recognized that man is exposed to a broad galaxy of environmental agents and hence considerations relative to possible synergistic, potentiating, co-carcinogenic, co-mutagenic and/or antagonistic interactions of carcinogenic and mutagenic and non-carcinogenic and non-mutagenic chemicals are of vital importance.

It is also important to restress that the mutagenicity of a compound is important per se and suggestive to a degree of the compounds potential carcinogenicity. However, it is recognized that more definitive elaboration of a compound's carcinogenicity can only be obtained at present by long-term bioassay.

It should also be noted that although a relatively small number of industrial organic compounds were reviewed in this report many structurally related agents are currently in use. It would appear prudent to consider their potential toxicity as well in the event that they have been untested in regard to carcinogenicity and/or mutagenicity.

We cannot be absolutely certain of the significance to man of findings of neoplasia in test animals or positive mutagenic effects in a variety of test systems. However, it would also appear prudent to minimize the burden and risk of potentially carcinogenic and mutagenic agents in the environment, lest we bequeth to subsequent generations the risk of catastrophic exposures for which redress would be either difficult or impossible.

In this regard, we should especially require greater focus on potent carcinogens compared to weak carcinogens with requisite greater concern to a carcinogen to which millions of people are potentially exposed to rather than one to which a relatively small number may come into contact.

SUBJECT INDEX

AAT (see ortho-Aminoazotoluene)
Acetaldehyde 147, 149, 150, 159, 438
--,mutagenicity of 149, 150
--,occurrence of 149
--,preparation and utility of 149
Acetamide 316, 317
--,carcinogenicity of 317
--,preparation and utility of 316, 317
--,reactivity of 316
Acetylhydrazine 309
--,formation from hydrazine 309
Acetylene tetrachloride (see 1,1,2,2-Tetrachloroethane)
Acid Alizarin Red B 397
--,mutagenicity of 397
Acid Alizarin Violet N 397
--,mutagenicity of 397
Acid Alizarin Yellow R 397
--,mutagenicity of 397
Acrolein 99, 147, 148, 149
--,carcinogenicity of 148
--,metabolism of 149
--,mutagenicity of 148, 149
--,occurrence of 148
--,preparation and utility of 147
--,reactivity of 147
--,TLV's and MAC's of 149
Acrylamide 496, 497
--,metabolism of 497
--,occurrence of 497
--,production and utility of 496, 497
--,TLV's of 497
--,toxicity of 497
Acrylonitrile 190, 194, 436-440
--,carcinogenicity of 439
--,metabolism of 439, 440
--,mutagenicity of 440
--,occurrence of 438, 439
--,populations at risk to 438, 439
--,production and utility of 436-438
--,reactivity of 436
--,teratogenicity of 440
--,TLV's of 438

Acylating agents (see also Benzoyl chloride, Diethyl carbamoyl chloride, dimethyl carbamoyl chloride and Ketene)
--,reactivity of 155
Adipic acid esters 477-478 (see also Diethyl adipate and Di-2-ethylhexyl adipate)
--,antifertility effects of 478
--,mutagenicity of 478
--,production and utility of 477, 478
Aldehydes (see also Acetaldehyde, Acrolein, and Formaldehyde)
--,reactivity of 142, 147
Aliphatic sulfuric acid esters (see Diethyl sulfate and Dimethylsulfate)
Alizarin 460-463
--,mutagenicity of 461, 463
--,utility of 460, 461
Alizarin Yellow GG 397
--,mutagenicity of 397
Alkane halides (see Dibromochloropropane, Ethylene dibromide, Ethylene dichloride, Propylene dichloride)
Alkylating Agents (see also Aldehydes, Alkanesulfonic esters, Alkyl halides, Alkylsulfates, Aryl dialkyl triazenes, Aziridines, Diazoalkanes, Epoxides, Halogenated alkanols ethers, Lactones, Phosphoric acid esters
Alkylglycidyl ethers 103
--,preparation and utility of 103
Allyl alcohol 496
--,mutagenicity of 496
--,production and utility of 496
Allyl chloride 97, 194-196, 245, 247, 438, 496
--,carcinogenicity of 195
--,metabolism of 195, 196
--,mutagenicity of 496
--,populations at risk 195

--, production and utility of 194, 195
--reactivity of 194
Allylic derivatives 496, 497 (see also Acrylamide, Allyl alcohol and Allyl chloride)
Amaranth 396, 397
--, carcinogenicity of 397
--, metabolism of 397
--, mutagenicity of 397
--, production and utility of 397
3-Aminoaniline (see meta-Phenylenediamine)
4-Aminoaniline (see para-Phenylenediamine)
1-Aminoanthraquinone 461
--, carcinogenicity of 461
Aminoanthraquinones 461, 464-466
--, mutagenicity of 464-466
--, utility of 461
para-Aminoazobenzene 389, 390
--, carcinogenicity of 390
--, metabolism of 390
--, mutagenicity of 390
--, production and utility of 390
ortho-Aminoazotoluene 390, 391
--, carcinogenicity of 390, 391
--, metabolism of 391
--, mutagenicity of 391
4-Aminobiphenyl 45, 62, 357, 360, 376-378
--, carcinogenicity of 45, 376, 378
--, metabolism of 360, 378
--, mutagenicity of 378
--, populations at risk to 378
--, production and utility of 377
4-Aminoimidazole-5-carboxamide 126
2-Amino-1-methylbenzene (see ortho-Toluidine)
1-Aminonaphthalene, (see 1-Naphthylamine)
2-Aminonaphthalene (see 2-Naphthylamine)
2-Amino-4-nitrophenol 388
--, in hair-dye formulations 388
--, mutagenicity of 388
2-Amino-5-nitrophenol 388
--, in hair-dye formulations 388
--, mutagenicity of 388
Anthragallol 460, 461, 464

--, mutagenicity of 461, 464
--, utility of 460, 461
Anthraquinone 461
9,10-Anthraquinones 460-466 (see also Alizarin, Anthragallol, and Purpurin)
--, carcinogenicity of 461
--, mutagenicity of 456-466
--, utility of 460, 461
Aromatic amines 356-388 (see also 4-Aminobiphenyl, Benzidine, 2,4-Diaminoanisole, 2,4-Diaminotoluene, 3,3'-Dichlorobenzidine, 3,3'-Dimethoxybenzidine, 3,3'-Dimethylbenzidine, 4,4'-Methylenebis(2-chloroaniline), 4,4'-Methylenedianiline, 1-Naphthylamine, 2-Naphthylamine, 2-Nitro-para-phenylenediamine, 4-Nitro-ortho-phenylenediamine, N-Phenyl-2-naphthylamine, meta-Phenylenediamine, para-Phenylenediamine, ortho-toluidine
Aryldialkyltriazenes (see also 1-Phenyl-3,3-trimethyltriazene)
--, carcinogenicity of 126
--, metabolism of 127, 128
--, mutagenicity of 126
--, utility of 125
Arylnitrenium ions 360
Arsenic compounds 45
--, in carcinogenesis 45
Asbestos 45
Auramine 45
Azides 440, 441 (see also Sodium Azide)
Aziridine 118-120
--, carcinogenicity of 119
--, mutagenicity of 119
--, preparation and utility of 118, 119
--, TLV's and MAC's of 119
Aziridines (see Aziridine, 2-(1-Aziridinyl)ethanol, 2-Methylaziridine)
2-(1-Aziridinyl)ethanol 120
--, carcinogenicity of 120

--,mutagenicity of 120
--,preparation and utility of 120
Azobenzene 389
--,carcinogenicity of 389
--,metabolism of 389
--,mutagenicity of 389
--,production and utility of 389
Azo dyes 388-398 (see also Amaranth, para-Aminoazobenzene, ortho-Aminotoluene, Azobenzene, Chrysoidine, para-Dimethylaminoazobenzene, Ponceau MX, Sudan II and Trypan Blue

BCME 45, 60, 251-255, 489
--,carcinogenicity of 45, 60, 253, 255
--,formation of 252, 253
--,from THPC 489
--,mutagenicity of 255
--,populations at risk 253, 255
--,production and utility of 251
--,reactivity of 252, 255
Benzene 45, 438, 445-454
--,chromosomal aberrations from 450, 451
--,leukomogenicity of 448-450, 453, 454
--,metabolism of 452, 453
--,mutagenicity of 451, 452
--,occurrence of 447, 448
--,populations at risk to 445, 447-450
--,production and utility of 445, 446
1,3-Benzenediamine, (see meta-Phenylenediamine)
1,4-Benzenediamine (see para-Phenylenediamine)
1,2-Benzenediol (see Catechol)
Benzidine 45, 58, 356-361, 366-368, 376, 398
--,carcinogenicity of 45, 358, 363, 364
--,dyes from 356
--,metabolism of 359, 360-362
--,mutagenicity of 358
--,occurrence of 357
--,populations at risk to 357, 358
--,production and utility of 356
Benzidine derived azo dyes (see Direct Black 38, Direct Blue 6, and Direct Brown 95)
Benzol (see Benzene)
Benzo(b)pyridine (see Quinoline)
1,4-Benzoquinone (see para-Quinone)
Benzoyl chloride 155, 156
--,carcinogenicity of 155, 156
--,mutagenicity of 156
--,preparation and utility of 155, 156
Benzoyl peroxide 160, 161
--,populations at risk 161
--,preparation and utility 160, 161
--,reactivity 161
Benzoyl superoxide (see Benzoyl peroxide)
Benzyl chloride 278, 279, 286
--,carcinogenicity of 279
--,mutagenicity of 279
--,production and utility of 278, 279
--,reactivity of 278
--,TLV's and MAC's of 286
Benzyl halides 278, 279 (see also Benzyl chloride)
BGBP (see Butylglycolylbutylphthalate
BHT 36, 37
--,tumor inhibition by 36, 37
Bis(2-chloroethyl)ether 256, 257, 481
--,carcinogenicity of 256
--,metabolism of 257
--,mutagenicity of 256, 257
--,occurrence of 256
--,production of 256
--,reactivity of 256
Bis(2-chloroisopropyl)ether 256, 257
--,carcinogenicity of 256
--,mutagenicity of 257
--,occurrence of 256
--,production and utility of 256
--,reactivity of 256
Bis(chloromethyl)ether (see BCME)
BPL (see beta-Propiolactone)

Bromobenzene 277, 278
--, metabolism of 277, 278
--, mutagenicity of 277
--, occurrence of 277
--, production and utility of 277
Bromoethene (see Vinyl bromide)
Bromoform 227
Brominated benzenes (see Bromobenzene)
1,3-Butadiene 188, 189, 495, 496
--, metabolism of 496
--, mutagenicity of 496
--, production and utility of 495
--, TLV's and MAC's of 495
Butadiene diepoxide (see Diepoxybutane)
Butadiene dioxide (see Diepoxybutane)
Butadiene monoxide 496
--, mutagenicity of 496
1,4-Butane sultone 124
--, mutagenicity of 124
Butter yellow (see para-Dimethylaminoazobenzene)
Butylglycolylbutylphthalate 474, 477
--, occurrence of 474
--, toxicity of 477
tert.Butyl hydroperoxide 96, 158
--, mutagenicity of 158
--, utility of 158
Butylated hydroxytoluene (see BHT)
1,2-Butylene oxide 97
--, carcinogenicity of 97
--, mutagenicity of 97
--, preparation and utility of 97
1,3-Butylene sulfate 122, 123
--, mutagenicity of 123
--, reactivity of 122
--, utility of 122
beta-Butyrolactone 107
--, carcinogenicity of 107
--, mutagenicity of 107
--, preparation of 107
gamma-Butyrolactone 107 108
--, carcinogenicity of 108
--, occurrence of 108
--, preparation and utility of 107, 108

Cancer 1-3, 31, 32
--, definition of 31
--, origins of 31, 32
Carbamates 315 (see also Urethans)
--, utility of 315
Carbon tetrachloride 35, 187, 193, 212, 213, 217-220, 245, 270
--, carcinogenicity of 219
--, impurities in 217
--, in tumor promotion 35
--, metabolism of 220
--, mutagenicity of 219
--, occurrence of 218, 219
--, populations at risk 217-219
--, production and utility of 217, 218
--, reactivity of 220
--, TLV's and MAC's of 218
Carcinogenesis etiology 1, 2, 42-44, 47
Carcinogenic potency 3, 58, 59, 65, 512
Catechol 468-469
--, carcinogenicity of 469
--, occurrence of 468
--, production and utility of 468
Chemical carcinogens 9-20
--, metabolic activation of 9-20 (see also individual chapters)
Chloral 135, 186
Chlorinated benzenes 266-277
--, reactivity of 266
Chlorinated dibenzofurans 294 (see also PCB's)
Chlorinated naphthalenes (see PCN's)
Chloroacetaldehyde 172, 173, 174, 241, 248, 249
--, mutagenicity of 172
Chloroacetic acid 172, 173
--, mutagenicity of 172
Chloroallylene (see Allyl chloride)
Chlorobenzene 266
--, utility of 266
2-Chlorobutadiene (see Chloroprene)
2-Chloro-1,3-butadiene (see Chloroprene)

Chlorodibenzofurans 277
3-Chloro-1,2-dibromopropane (see DBCP)
Chloroethane (see Ethylchloride)
2-Chloroethanol 172, 241, 248, 249
--,carcinogenicity of 249
--,formation from fumigants 248
--,metabolism of 249
--,mutagenicity of 172, 249
--,production and utility of 248
--,TLV's and MAC's of 248
Chloroethene (see Vinyl chloride)
Chloroethylene oxide 172, 173
--,mutagenicity 172
Chloroform 212-217
--,carcinogenicity of 215, 216
--,impurities in 213
--,metabolism of 217
--,mutagenicity of 216
--,occurrence of 214, 215
--,populations at risk 214
--,production and utility of 213
--,reactivity of 214
--,TLV's and MAC's of 214
Chloromethylbenzene (see Benzyl chloride)
Chloromethyl methyl ether (see CMME)
Chlorophenols 500-505 (see also 2,4,5-Trichlorophenol and Pentachlorophenol)
Chloroprene 189-193
--,carcinogenicity of 190, 191
--,impurities in 189
--,mutagenicity of 192
--,occurrence of 190
--,populations at risk 190 191
--,production and utility of 189, 190
--,reactivity of 189
--,TLV's and MAC's
1-Chloro-2-propanol 249, 250
--,formation in food, 250
--,mutagenicity of 250
--,occurrence of 250
--,production and utility of 249, 250
3-Chloro-1-propene (see Allyl chloride)
3-Chloropropylene (see Allyl chloride)
2-Chloroquinoline 418
alpha-Chlorotoluene (see Benzyl chloride)
Chronic bioassays 3, 47, 48, 62
Chromium compounds 45
--,in carcinogenisis 45
Chrysoidine 394
--,carcinogenicity of 394
--,mutagenicity of 394
--,production and utility of 394
C.I. Basic Orange II (see Chrysoidine)
C.I. Direct Blue 14, (see Trypan Blue)
C.I. Food Red 5 (see Ponceau MX)
C.I. Food Red 9 (see Amaranth)
C.I. Solvent Orange-7 (see Sudan II)
CMME 45, 251-255
--,carcinogenicity of 45, 253
--,populations at risk 253, 255
--,production and utility of 251
--,reactivity of 252, 255
Coal naphtha (see Benzene)
Coal tar 60
Copper 8-quinolinolate 417
--,production and utility of 417
Cumene hydroperoxide 158
--,mutagenicity of 158
--,utility of 158
Cyclic aliphatic sulfuric acid esters (see also 1,3-Butylene sulfate, 1,2-Ethylene sulfate and 1,3-Propylene sulfate)
--,reactivity of 122
--,utility of 122
Cyclic ethers (see 1,4-Dioxane)

DAB (see para-Dimethylamino-azobenzene)
DACPM (see 4,4'-Methylene bis(2-chlrooaniline)
DBCP 245-248
--,antifertility effects of 247, 248
--,carcinogenicity of 246, 247
--,impurities of 247
--,mutagenicity of 247
--,occurrence of 246
--,populations at risk 246

--,production and utility of 245, 246
--,TLV's of 246
DCB (see 3,3'-Dichlorobenzidine)
DCE (see Vinylidene chloride)
DEHP (see Di-2-ethylhexyl-phthalate)
DES (see Diethylsulfate)
Diacetyl-o-tolidine 367
Diamine (see Hydrazine)
2,4-Diaminoanisole 380-382
--,carcinogenicity of 381
--,in hair-dye formulations 380, 381
--,metabolism of 382
--,mutagenicity of 381, 382
--,populations at risk to 381
--,production and utility of 381
2,5-Diaminoanisole 388
--,in hair-dye formulations 388
--,mutagenicity of 388
meta-Diaminoanisole (see 2,4-Diaminoanisole)
2,4-Diaminotoluene 378-380, 499
--,carcinogenicity of 379, 499
--,dyes from 379
--,formation from toluene diisocyanates 499
--,metabolism of 380
--,mutagenicity of 380, 499
--,production and utility of 378, 379
2,5-Diaminotoluene 388
--,in hair-dye formulations 388
--,mutagenicity of 388
1,4-Diamino-2-nitrobenzene (see 2-Nitro-para-phenylenediamine
1,2-Diamino-4-nitrobenzene (see 4-Nitro-ortho-phenylenediamine)
ortho-Dianisidine 366-368
--,carcinogenicity of 368
--,dyes from 367
--,metabolism of 368
--,mutagenicity of 368
--,production and utility of 367, 368
Diazoacetic ester 125
Diazoalkanes 124 (see also Diazomethane)
--,reactivity of 124
Diazomethane 124, 125

--,carcinogenicity of 125
--,mutagenicity of 125
--,preparation and utility of 124
--,reactivity of 124-126
--,TLV's of 125
Dibenzoyl peroxide (see Benzoyl peroxide)
para-Dibromobenzene 278
--,metabolism of 278
--,occurrence of 278
--,production and utility of 278
Dibromochloropropane (see DBCP)
1,3-Dibromo-3-chloropropane 136, 137
sym-Dibromoethane (see Ethylene dibromide)
1,2-Dibromoethane (see Ethylene dibromide)
1,2-Dibromo-3-chloropropane (see DBCP)
2,3-Dibromo-1-propanol 250, 251
--,metabolism of 251
--,mutagenicity of 251
--,occurrence of 250
--,production and utility of 251
Di-tert.butyl peroxide 158
--,mutagenicity of 158
--,utility of 158
ortho-Dichlorobenzene 266-268, 498
--,carcinogenicity of 267
--,occurrence of 267
--,metabolism of 268
--,production and utility of 266, 267
para-Dichlorobenzene 268, 269
--,carcinogenicity of 269
--,metabolism of 269
--,occurrence of 268, 269
--,production and utility of 268
--,TLV's and MAC's of 268
1,2-Dichlorobenzene (see ortho-Dichlorobenzene)
ortho-Dichlorobenzol (see ortho-Dichlorobenzene)
3,3'-Dichlorobenzidine 357, 361, 365, 366, 376
--,carcinogenicity of 365
--,dyes from 365
--,metabolism of 366
--,mutagenicity of 366

--, populations at risk to 365
--, production and utility of 361, 365
trans-1,4-Dichlorobutene 189, 193
--, carcinogenicity of 193
--, mutagenicity of 193
--, production and utility of 193
1,2-Dichloroethane (see Ethylene dichloride)
1,2-Dichloroethane 222
1,2-Dichloroethane (see Ethylene dichloride)
1,1-Dichloroethylene (see Vinylidene chloride)
cis-1,2-Dichloroethylene 188
trans-1,2-Dichloroethylene 188
Dichloroethylene (see Vinylidene chloride)
Dichlorofluoromethane 227-231 (see also Fluorocarbons)
1,2-Dichloropropane (see Propylene dichloride)
1,2-Dichloropropane 119 (see also Propylene dichloride)
Diepoxy butane 103, 104
--, carcinogenicity of 104
--, mutagenicity of 104
--, preparation and utility of 103
Diethyl adipate 478
--, antifertility effects of 478
--, mutagenicity of 478
--, toxicity of 478
Diethylcarbamoyl chloride 155
--, mutagenicity of 155
--, utility of 155
Di-2-ethylhexyladipate 477, 478
--, antifertility effects of 478
--, mutagenicity of 478
--, production and utility of 477, 478
--, toxicity of 478
Di-2-ethylhexylphthalate 473-477
--, metabolism of 477
--, mutagenicity of 477
--, occurrence of 474-476
--, in air, 476
--, in blood 476
--, in water 476
--, in food 476
--, production and utility of 473-475
--, toxicity of 477
Diethyl sulfate 121, 122
--, carcinogenicity of 122
--, mutagenicity of 122
--, preparation and utility of 121, 122
1,4-Diethylenedioxide (see 1,4-Dioxane)
Diethyleneglycol 482, 484
--, metabolism of 482
Diglycidyl resorcinol ether 102, 103
--, carcinogenicity of 103
--, preparation and utility of 102
1,2-Dihydroxy anthraquinone (see Alizarin)
Diisocyanates 497-500 (see also Dicyclohexylmethane-4'-diisocyanate, Diphenylmethane-4,4'-diisocyanate, 1,5-Naphthalenediisocyanate, Toluene-2,4-diisocyanate, and Toluene-2,6-diisocyanate)
DMS (see Dimethylsulfate)
DMN (see Dimethylnitrosamine)
3,3'-Dimethoxybenzidine (see ortho-Dianisidine)
Dimethoxyethylphthalate 477
--, mutagenicity of 477
para-Dimethylaminoazobenzene 391-394
--, carcinogenicity of 391, 393, 394
--, metabolism of 392-394
--, mutagenicity of 392
--, production and utility of 391
3,3'-Dimethylbenzidine (see ortho-Tolidine)
Dimethylcarbamoyl chloride 155
--, carcinogenicity of 155
--, mutagenicity of 155
--, occurrence of 155
--, production and utility of 155
1,1-Dimethylhydrazine 308
--, antifertility of 308
--, carcinogenicity of 308
--, occurrence of 308
--, production and utility of 308
1,2-Dimethylhydrazine 308
--, carcinogenicity of 308

Dimethyl nitrosamine 331–334
Dimethylsulfate 60, 120, 121
--, carcinogenicity of 121
--, mutagenicity of 121
--, preparation and utility of 120, 121
--, TLV's 121
2,4-Dinitrotoluene 432, 433
--, carcinogenicity of 432
--, metabolism of 433
--, mutagenicity of 432, 433
--, occurrence of 432, 433
--, production and utility of 432
1,4-Dioxane 60, 481–484
--, carcinogenicity of 483
--, metabolism of 481–484
--, production and utility of 481
--, TLV's and MAC's of 484
--, toxicity of 483
para-Dioxane-2-one 481–484
--, formation from 1,4-dioxane 481–484
--, toxicity of 482
Dioxybutane (see Diepoxybutane)
Dioxyethylene ether (see 1,4-Dioxane)
Diphenyldiazene (see Azobenzene)
Diphenylmethane-4,4'-diisocyanate 499, 500
--, utility of 499, 500
Direct Black 38 398
--, carcinogenicity of 398
Direct Blue 6 398
--, carcinogenicity of 398
Direct Brown 95 398
--, carcinogenicity of 398
Di-orthotoluidine 366
Divinyl (see 1,3-Butadiene)
DNT (see 2,4-Dinitrotoluene)
Dose-response modeling 48–56

EDC-tar 173
--, mutagenicity of 173
Environmental factors in carcinogenesis 1, 2, 32, 42, 43, 44, 66, 67
Epichlorohydrin 97–99, 186, 194, 196, 247, 496
--, carcinogenicity of 97–99
--, mutagenicity of 97, 98
--, preparation and utility of 97

--, reactivity of 97
--, TLV's and MAC's of 98
Epidemiology 2, 32, 42, 43, 62, 66, 67
Epoxides (see Butylene oxide, Diglycidyl resorcinol ether, Epichlorohydrin, Ethylene oxide, Glycidaldehyde, Glycidol, Propylene oxide, Styrene oxide)
Epoxidized esters 103
--, utility of 103
1,2-Epoxybutane (see 1,2-Butylene oxide)
1-Epoxyethyl-3,4-epoxycyclohexane 103
--, preparation and utility of 103
1,2-Epoxypropane (see Propylene oxide)
2,3-Epoxypropoxybenzene (see Diglycidyl resorcinol ether)
Equivalent chemical (rec) dose 59, 61, 62, 63
Enzyme inducers 35
Ethyl carbamate (see Urethan)
Ethyl chloride 225
--, production and utility of 225
--, TLV's and MAC's of 225
Ethyl sulfate (see Diethyl sulfate)
Ethylene bromide (see Ethylene dibromide)
Ethylene chloride (see Ethylene dichloride)
Ethylene chlorohydrin (see 2-Chloroethanol)
Ethylene dibromide 241–245
--, antifertility effects of 244
--, carcinogenicity of 243
--, metabolism of 244, 245
--, mutagenicity of 243, 244
--, occurrence of 242, 243
--, populations at risk 242, 243
--, production and utility of 241, 242
--, reactivity of 244
--, TLV's of 242
Ethylene dichloride 118, 165, 173, 181, 182, 220, 240, 241
--, carcinogenicity of 240, 241
--, mutagencity of 241

--, occurrence of 240
--, populations at risk 240
--, production and utility of 240
--, TLV's of 240
Ethyleneimine (see Aziridine)
Ethylene monochloride (see vinyl chloride)
Ethylene oxide 93-95, 120, 248, 256
--, carcinogenicity of 95
--, mutagenicity of 95
--, occurrence of 94
--, preparation and utility of 93, 94
--, reactivity of 93, 95
--, TLV's and MAC's of 94
1,2-Ethylene sulfate 122, 123
--, carcinogenicity of 122, 123
--, mutagenicity of 123
--, reactivity of 122
--, utility of 122
Ethylene thiourea 318, 319
--, carcinogenicity of 318
--, metabolism of 319
--, mutagenicity of 318, 319
--, occurrence in fungicides 318
--, preparation and utility of 319
--, toxicity of 319
ETU, (see Ethylene thiourea)

Fast Scarlet Base B (see 2-Naphthylamine)
FD and C Red No. 2 (see Amaranth)
Flame retardants (see THPC and Tris(2,3-dibromopropyl)phosphate)
Fluorocarbons 227-231
--, metabolism of 231
--, production and utility of 227
--, occurrence of 228
--, TLV's of 228
--, mutagenicity of 231
--, reactivity of 230
--, ozone reduction by 229, 230
Formaldehyde 104, 142-147, 251, 252, 488, 489
--, carcinogenicity of 146
--, formation from hexamethylphosphoramide 488

--, formation from THPC 489
Furantoin (see Nitrofurantoin)

Glycidaldehyde 99, 148, 196, 440
--, carcinogenicity of 99
--, mutagenicity of 99
--, preparation and utility of 99
--, reactivity of 99
Glycidol 99, 196
--, mutagenicity of 99
--, utility of 99
Glycidonitrile 440
Glycol sulfite 122, 123
Glycol sulfate (see 1,2-Ethylene sulfate)

Haloethers (see BCME, Bis(2-chloroethyl)ether, Bis(2-chloroisopropyl)ether and CMME)
Halogenated alkanols (see 2,3-Dibromo-1-propanol, 2-Chloroethanol, 1-Chloro-2-propanol)
Halogenated aryl derivatives (see Bromobenzene, Chlorobenzene, para-Dibromobenzene, ortho-Dichlorobenzene, para-Dichlorobenzene, Hexachlorobenzene, 1,2,4-Trichlorobenzene)
Halogenaetd polyaromatic derivatives (see PBBs, PCBs, PCNs, and PCTs)
Halogenated saturated hydrocarbons (see Carbon tetrachloride, Chloroform, 1,2-Dichloropropane, 1,2-Ethylene dibromide, 1,2-Ethylene dichloride, Hexachloroethane, Methyl chloride, Methyl iodide, Methylene chloride, 1,1,2,2-Tetrachloroethane, 1,1,1-Trichloroethane)
--, reactivity of 211
Halogenated unsaturated hydrocarbons (see Allyl chloride, Chloroprene, Hexachloro-

butadiene, Tetrachloro-
ethylene, Trichloroethylene,
Vinyl bromide, Vinyl
chloride, Vinylidene
chloride)
HCB (see Hexachlorobenzene)
HCBD (see Hexachlorobutadiene)
Hematite mining 46
Heterocyclic amines (see
Copper-8-quinolinolate,
8-Hydroxyquinoline, 4-Nitro-
quinoline-1-oxide and
Quinoline)
Hex wastes (see Hexachlorobutadiene)
Hexachlorobenzene 181, 270-277
--,carcinogenicity of 276
--,disposal of wastes of 272-275
--,formation of 270, 271, 273
--,metabolism of 277
--,mutagenicity of 276
--,occurrence of 272, 275, 276
--,production and utilit of 270, 272
--,toxicity of 276
--,trace contaminants in 277
Hexachlorobutadiene 193, 194
--,carcinogenicity 193, 194
--,formation from 193
--,mutagenicity of 194
--,occurrence of 193
--,utility of 193
Hexachloroethane 224
--,carcinogenicity of 224
--,production and utility of 224
--,reactivity of 224
--,TLV's of 224
Hexachlorophene 505
Hexamethylphosphoramide 60, 487, 488
--,carcinogenicity of 487, 488
--,metabolism of 488
--,mutagenicity of 488
--,production and utility of 488
Hexamethylenediamine 193
HMPA (see Hexamethylphos-
phoramide)
Hydrazine 307-310
--,carcinogenicity of 308, 311
--,metabolism of 309, 310
--,mutagenicity of 308, 310
--,production and utility of 307

--,reactivity of 307, 310
--,toxicity of 310
Hydrazine carboxamide (see
Semicarbazide)
Hydrazines (see 1,1-Dimethyl-
hydrazine, 1,2-Dimethyl-
hydrazine, Hydrazine,
Isonicotinylhydrazine,
Semicarbazide)
Hydrogen peroxide 159, 160
--,mutagenicity of 160
--,reactivity 159
--,utility 159, 160
Hydroquinone 469
--,carcinogenicity of 469
--,mutagenicity of 469
--,occurrence of 469
--,production and utility of 469
4-Hydroxyaminoquinoline-1-oxide 418
Hydroxybenzopyridine (see
8-Hydroxyquinoline)
Hydroxylamine 310-315
--,mutagenicity of 312, 314, 315
--,reactivity of 312, 313, 315
--,utility of 310, 311
Hydroxylamines 310-315 (see
Hydroxylamine, N-Methyl-
hydroxylamine, O-Methyl-
hydroxylamine)
ortho-Hydroxyphenol (see
Catechol)
-Hydroxypropionic acid lactone
(see -Propiolactone)
-Hydroxyquinoline 417-420
--,carcinogenicity of 418
--,mutagenicity of 419, 420
--,production and utility of 417
N-Hydroxy-p-toluidine 387
N-hydroxyurethan 316
--,mutagenicity of 316
--,reactivity of 316

Iatrogenic cancers 42
Idopathic cancers 42
Isocyanate polymers 375
Isoniazid (see Isonicotinyl
hydrazide)
Isonicotinyl hydrazide 309, 310
--,mutagenicity of 310
Isopropyloils 46

--, in carcinogenesis 46
Isoquinoline 418

Ketene 104, 156
--, mutagenicity of 156
--, preparation and utility of 156
--, reactivity of 156

Latency period 2, 43, 47, 512
Log-probit model 52, 53, 54

MDA (see 4,4'-Methylene-dianiline)
MDI (see Diphenylmethane-4,4'-diisocyanate)
ortho-Methylaniline (see ortho-Toluidine)
2-Methyl aziridine 119, 120
--, carcinogenicity of 120
--, mutagenicity of 120
--, preparation and utility of 119, 120
Methyl benzene (see Toluene)
Methyl chloride 211, 212
--, mutagenicity of 212
--, occurrence of 212
--, production and utility of 211, 212
--, TLV's and MAC's of 212
Methyl chloroform 220-222, 240
--, carcinogenicity of 222
--, metabolism of 222
--, mutagenicity of 223
--, occurrence of 221
--, populations at risk 221
--, production and utility of 220, 221
--, TLV's and MAC's of 221
Methyl chloromethyl ether (see CMME)
Methylethylenimine (see 2-Methylaziridine)
Methyl halides 211, 212, 224-227
--, carcinogenicity of 226
--, mutagenicity of 212, 227
--, production and utility of 211, 212, 226
--, TLV's and MAC's of 212, 227

N-Methylhydroxylamine 314
--, mutagenicity of 314
O-Methylhydroxylamine 314
--, mutagenicity of 314
Methyl iodide 225, 226
--, carcinogenicity of 226
--, mutagenicity of 226, 227
--, occurrence of 225
--, production and utility of 225
--, TLV's of 225
Methyl Red 397
--, mutagenicity of 397
Methyl sulfate (see Dimethyl sulfate
4,4'-Methylene bis(2-chloro-aniline) 365, 375, 376
--, carcinogenicity of 376
--, mutagenicity of 376
--, populations at risk to 376
--, production and utility of 375
4,4'-Methylenebis(cyclohexyl-amine) 374
4,4'-Methylenebis(2-methylani-line) 377
--, carcinogenicity of 377
--, production and utility of 377
Methylene chloride 212, 213
--, carcinogenicity of 213
--, metabolism of 213
--, mutagenicity of 213
--, production and utility of 212, 213
--, TLV's and MAC's of 213
4,4'-Methylenedianiline 374, 375, 500
--, carcinogenicity of 375, 500
--, populations at risk to 375
--, production and utility of 374, 375, 500
MOCA (see 4,4'-Methylenebis-(2-chloroaniline)
para-Monomethylaminoazobenzene 393, 394
Multi-hit model 51
Multi-stage model 51, 52
Multiple factors in carcino-genesis 2, 31-37
--, genetic effects in 32
--, nutritional effects in 2, 32, 513, 33
--, smoking effects in 2, 32
Mutagenic risk assessment 68-72

--, molecular dosimetry in 71
--, REC concept in 71
--, threshold levels in 72
Mutagenesis etiology 42

alpha-Naphthylamine (see
   1-Naphthylamine)
1-Naphthylamine 360, 368-370
--, carcinogenicity of 369, 370
--, metabolism of 360, 370
--, mutagenicity of 370
--, occurrence of 369
--, populations at risk to 369
--, production and utility of 368
beta-Naphthylamine (see
   2-Naphthylamine)
2-Naphthylamine 3, 60, 360,
   370-373, 376, 378
--, carcinogenicity of 371, 372
--, metabolism of 372, 373
--, mutagenicity of 373
--, occurrence of 371
--, populations at risk to 371
--, production and utility of
   370, 371
1,5-Naphthalene diisocyanate 499
--, production and utility of 499
Neoprene 189
Nickel 46
--, in carcinogenesis 46
Nitroalkanes (see 2-Nitro-
   propane)
Nitroanthraquinones 461, 465,
   466
--, mutagenicity of 465, 466
--, utility of 461
Nitroaromatics (see 2,4-Dinitro-
   toluene and 2,4,5-Trinitro-
   toluene)
5-Nitro-2-furaldehyde semicarbazone (see Nitrofurazone)
Nitrofurans 419-426 (see also
   Nitrofurantoin, Nitrofurazone)
Nitrofurantoin 421, 422, 424
--, carcinogenicity of 422
--, mutagenicity of 421, 422, 424
--, utility of 421
Nitrofurazone 421, 422
--, carcinogenicity of 422
--, mutagenicity of 421, 422, 424

--, utility of 421
4-Nitrobiphenyl 378
4-Nitro-ortho-phenylenediamine
   359, 385, 386
--, carcinogenicity of 385
--, mutagenicity of 385, 386
--, production and utility of 385
2-Nitro-para-phenylenediamine
   384, 385
--, carcinogenicity of 384, 385
--, in hair-dye formulations 384,
   385
--, mutagenicity of 385
--, production and utility of 384
2-Nitropropane 433, 434
--, carcinogenicity of 433
--, occurrence of 434
--, populations at risk to 434
--, production and utility of
   433, 434
--, reactivity of 433
4-Nitroquinoline-1-oxide 418
--, carcinogenicity of 418
--, mutagenicity of 418
--, reactivity of 418
Nitrosamines 331-351
--, carcinogenicity of 331, 334,
   335
--, exposure to 332
--, formation of 331
--, metabolism of 332
--, mutagenicity of 332, 336-351
--, occurrence of 332
--, reactivity of 331
--, utility of 331
2-Nitrosonaphthalene 372
para-Nitrosotoluidine 388
para-Nitrotoluene 388
4-NQO (see 4-Nitroquino-
   line-1-oxide)
Nutrition and carcinogenesis 2,
   32, 35, 36, 42, 44, 513

Octachlorodibenzo-p-dioxin 277
Organo chlorine pesticides 35
--, in tumor promotion 35
Oxirane (see Ethylene oxide)

PAES (see Phthalic acid esters)
PBBs 294-297

--,mutagenicity of 297
--,occurrence of 295, 296
--,production and utility of 295, 296
--,toxicity of 296
PCBs 35, 286-294, 297
--,carcinogenicity of 292, 293
--,in tumor promotion 35
--,in Yusho disease 292, 294
--,metabolism of 287-289, 293
--,mutagenicity of 293
--,occurrence of 290
--,produciton and utility of 286
--,reactivity of 286
--,toxicity of 289
PCN's 298, 299
--,metabolism of 299
--,production and utility of 298
--,toxicity of 298, 299
PCP (see Pentachlorophenol)
PCT's 297
--,in human samples 297
--,occurrence of 297
--,production of 294
Pentachlorophenol 500-504
--,carcinogenicity of 503
--,chlorodibenzofuran impurities in 503
--,chlorodioxin impurities in 501-503
--,metabolism of 503
--,mutagenicity of 503
--,occurrence of 501
--,production and utility of 500, 501
--,reactivity of 503, 504
Peracetic acid 97, 99, 159
--,mutagenicity of 159
--,production and utility of 159
Perchloroethane (see Hexachloroethane)
Perchlorobenzene (see Hexachlorobenzene
Perchloroethylene 214, 224, 270 (see also Tetrachloroethylene)
Peroxides 158-161 (see also Benzoyl peroxide, Cumene hydroperoxide, Di-tert.butyl peroxide, Hydrogen peroxide, Peracetic acid, Succinic acid peroxide and Tert.Butyl peroxide)
--,reactivity of 158
--,utility of 158
Peroxyacetic acid (see Peracetic acid)
1-Phenyl-3,3-dimethyltriazene 125, 126-128
--,mutagenicity of 127
--,utility of 125
Phenylglycidylether 104
--,preparation and utility of 104
N-Phenyl-1-naphthylamine 369, 371, 373, 374
--,carcinogenicity of 374
--,occurrence of 373, 374
--,populations at risk to 373, 374
--,production and utility of 373, 374
meta-Phenylenediamine 383, 384
--,carcinogenicity of 384
--,dyes from 383
--,metabolism of 384
--,mutagenicity of 384
--,production and utility of 383, 384
para-Phenylenediamine 382, 383
--,carcinogenicity of 383
--,in hair dye formulations 382
--,metabolism of 383
--,mutagenicity of 383
--,production and utility of 382
--,TLV's and MAC's 383
Phorbol diesters 33-34
Phosgene 188
Phosphoramides (see Hexamethylphosphoramide)
Phosphoric acid esters 135 (see also Triethylphosphate, Trimethylphosphate, and Tris-(2,3-dibromopropyl)phosphate
--,reactivity of 135
--,utility of 135
Phthalic acid esters 473-477 (see also Butyl glycolyl butyl phthalate, Di-2-ethyl hexylphthalate and Dimethoxyethyl phthalate)
--,antifertility effects of 477
--,disposal of 475
--,mutagenicity of 477

--, occurrence of 474-476
--, in air 476
--, in food 476
--, in water 474, 476
--, production and utility of 473-475
--, toxicity of 477
Polybrominated biphenyls (see PBBs)
Polychlorinated biphenyls (see PCBs)
Polychlorinated terphenyls (see PCTs)
Polychlorodiphenyl ethers 503
Polycyclic hydrocarbons 35
--, in tumor promotion 35
Polyhydric phenols 468, 469 (see also Catechol, Hydroquinone and Resorcinol)
Polyurethanes 497-500
--, carcinogenicity of 499
--, formation from diisocyanates 497-500
Polyvinyl alcohol 494
Ponceau MX 395
--, carcinogenicity of 395
--, metabolism of 395
--, mutagenicity of 395
--, production and utility of 395
Populations at risk 2 (see also individual chapters)
1,3-Propane sultone 123, 124
--, carcinogenicity of 124
--, mutagenicity of 124
--, preparation and utility of 123, 124
--, reactivity of 123, 124
Propenamide (see Acrylamide)
Propenenitrile (see Acrylonitrile)
beta-Propiolactone 104-107
--, carcinogenciity of 105, 106
--, mutagenicity of 106, 107
--, preparation and utility of 104, 105
--, reactivity of 105, 106
--, TLV's and MAC's of 105
Propyloxirane (see 1,2-Butylene oxide)
Propylene chlorohydrin (see 1-chloro-2-propanol)
Propylene dichloride 245

--, impurities in 245
--, mutagenicity of 245
--, production and utility of 245
--, TLV's and MAC's of 245
Propylenimine (see 2-Methyl aziridine)
Propylene oxide 96, 250, 256
--, carcinogenicity of 96
--, mutagenicity of 96
--, occurrence of 96
--, preparation and utility of 96
--, reactivity of 96
--, TLV's and MAC's of 96
1,3-Propylene sulfate 122, 123
--, mutagenicity of 123
--, reactivity of 122
--, utility of 122
Purpurin 460-462, 464
--, mutagenicity of 461, 464
--, utility of 460, 461
PVC 166-169, 171, 173

Quinazarin 460, 461, 463
--, mutagenicity of 461, 463
--, utility of 460, 461
Quinoline 417-420
--, carcinogenicity of 418
--, mutagenicity of 418-420
--, occurrence of 417
--, production and utility of 417
--, reactivity of 419
8-Quinoline (see 8-Hydroxyquinoline)
para-Quinone 467, 468
--, carcinogenicity of 468
--, mutagenicity of 468
--, occurrence of 467
--, preparation and utility of 467
Quinones 467, 468 (see also para-Quinone)
--, reactivity of 467
--, utility of 467

Regulated carcinogens 6
Resorcinol 469
--, carcinogenicity of 469
--, mutagenicity of 469
Retinoids 36
--, tumor inhibition by 36

Risk assessment 47-72, 515-517

Saran 178, 180 (see also Vinylidene chloride)
--,production of 178, 180
SDMH (see 1,2-Dimethylhydrazine)
Semicarbazide 308, 309
--,mutagenicity of 308
--,production and utility of 308, 309
--,reactivity of 309
Short-term bioassays 3-12
Smoking and carcinogenesis 2, 32, 513, 34, 35, 42, 43
Sodium azide 440, 441
--,mutagenicity of 441
--,reactivity of 440, 441
--,utility of 441
Styrene 100-102
--,carcinogenicity of 102
--,metabolism of 100
--,mutagenicity of 102
--,preparation and utility of 101
--,TLV's and MAC's of 101
Styrene oxide 99-100
--,formation from styrene 100
--,mutagenicity of 100
--,occurrence of 100
--,preparation and utility of 99
Succinic acid peroxide 159
--,mutagenicity of 159
--,utility of 159
Sudan II 394, 395
--,carcinogenicity of 395
--,metabolism of 395
--,mutagenicity of 395
--,production and utility of 395
Sudan IV 397
--,mutagenicity of 397
Sultones (see 1,3-Propane sultone and 1,4-Butane sultone)

1,1,2,2-Tetrachloroethane 223, 224
--,carcinogenicity of 223
--,mutagenicity of 224
--,production and utility of 223
--,reactivity of 223

--,TLV's and MAC's of 223
Tetrachloroethylene 187-189, 193, 245
--,carcinogenicity of 188
--,metabolism of 188, 189
--,mutagenicity of 188, 189
--,occurrence of 187, 188
--,populations at risk 187
--,production and utility of 187
--,TLV's and MAC's of 187
Tetrachlorohydroquinone 503
Tetrahydro-1,4-dioxin (see 1,4-Dioxane)
Tetrakis(hydroxymethyl)phosphonium chloride (see THPC)
THPC 488, 489
--,carcinogenicity of 489
--,production and utility of 488
--,reactivity of 488
"Threshold dose" 48, 49, 54-62
Thioacetamide 317
--,carcinogenicity of 317
--,metabolism of 317
--,preparation and utility of 317
Thiourea 317, 318
--,carcinogenicity of 318
--,mutagenicity of 318
--,preparation and utility of 318
Thioureas (see Ethylene thiourea, Thiourea)
Tier systems (see Short-term bioassays)
Time-to-tumor models 50, 65
TNT (see 2,4,5-Trinitrotoluene)
ortho-Tolidine 366, 367
--,carcinogenicity of 367
--,metabolism of 367
--,mutagenicity of 367
--,utility of 366
Toluol (see Toluene)
Toluene 438, 454, 455
--,chromosome aberrations from 454
--,metabolism of 455
--,mutagenicity of 454, 455
--,production and utility of 454
--,TLV's of 455
--,toxicity of 454
meta-Toluenediamine 359
2,4-Toluenediamine (see 2,4-

Diaminotoluene)
2,6-Toluenediisocyanate 498, 499
--,production and utility of 498
Toluene-2,4-diisocyanate 498, 499
--,carcinogenicity of 498
--,production and utility of 498
--,reactivity of 499
--,TLV's and MAC's of 498
ortho-Toluidine 386-388
--,carcinogenicity of 387
--,dyes from 386
--,metabolism of 387
--,mutagenicity of 387, 388
--,occurrence of 387
--,populations at risk from 387
--,production and utility of 386
--,TLV's and MAC's of 386, 387
para-Toluidine 387
Trichloroacetic acid 186-188
Trichloroacetyl chloride 188
1,2,4-Trichlorobenzene 269, 270
--,metabolism of 270
--,production and utility of 269
--,toxicity of 270
1,1,1-Trichloroethane (see Methyl chloroform)
1,1,2-Trichloroethane 178, 181, 222, 223
--,carcinogenicity of 223
--,mutagenicity of 223
--,production and utility of 222
--,TLV's and MAC's of 223
Trichloroethanol 186
Trichloroethylene 180, 182-187, 192, 193, 270
--,carcinogenicity of 185, 186
--,metabolism of 186, 187
--,mutagenicity of 186
--,occurrence of 183-185
--,populations at risk 184
--,production of 182, 183
--,TLV's and MAC's
Trichlorofluoromethane 227-231 (see also Fluorocarbons)
2,4,5-Trichlorophenol 504, 505
--,carcinogenicity of 505
--,chlorodioxin impurities in 505
--,production and utility of 504, 505
--,urinary metabolite of HCB,

lindane 505
Trichloropropane oxide 187
Triethylphosphate 135
--,mutagenicity of 135
--,utility of 135
Trihalomethanes 215, 227
1,2,3-Trihydroxyanthraquinone (see Anthragallol)
1,2,4-Trihydroxyanthraquinone (see Purpurin)
Trimethyl phosphate 135
--,mutagenicity of 135
--,utility of 135
2,4,5-Trinitrotoluene 432, 433
--,mutagenicity of 433
--,occurrence of 432, 433
--,production and utility of 432
Tris (see Tris(2,3-dibromopropyl)phosphate
Tris(2,3-dibromopropyl)phosphate 135-138, 489
--,carcinogenicity of 137, 138
--,impurities in 136
--,mutagenicity of 137, 138
--,populations at risk of 136
--,toxicity of 138
Trypan Blue 396
--,carcinogenicity of 396
--,metabolism of 396
--,mutagenicity of 396
--,production and utility of 396
Tumor inhibitors 36, 37
Tumor promoters 33-35

UDMH (see 1,1-Dimethylhydrazine)
Unsaturated nitriles (see Acrylonitrile)
Urethan 315-316
--,carcinogenicity of 315
--,metabolism of 315
--,mutagenicity of 315, 316
--,reactivity of 316
--,vinyl carbamate from 316
Urethanes 315-316 (see also N-hydroxyurethan, Urethan, Vinyl carbamate)

VCM (see Vinyl chloride)
Vinyl acetate 494, 495
--,carcinogenicity of 495